# INTRODUCTORY
# BIOSTATISTICS

# INTRODUCTORY BIOSTATISTICS

**CHAP T. LE**
Distinguished Professor of Biostatistics
and Director of Biostatistics
Comprehensive Cancer Center
University of Minnesota

WILEY-INTERSCIENCE

A JOHN WILEY & SONS PUBLICATION

Copyright © 2003 by John Wiley & Sons, Inc. All rights reserved.

Published by John Wiley & Sons, Inc., Hoboken, New Jersey.
Published simultaneously in Canada.

No part of this publication may be reproduced, stored in a retrieval system or transmitted in any form or by any means, electronic, mechanical, photocopying, recording, scanning or otherwise, except as permitted under Sections 107 or 108 of the 1976 United States Copyright Act, without either the prior written permission of the Publisher, or authorization through payment of the appropriate per-copy fee to the Copyright Clearance Center, 222 Rosewood Drive, Danvers, MA 01923, (978) 750-8400, fax (978) 750-4470. Requests to the Publisher for permission should be addressed to the Permissions Department, John Wiley & Sons, Inc., 111 River Street, Hoboken, NJ 07030, (201) 748-6011, fax (201) 748-6008, E-Mail: PERMREQ@WILEY.COM.

To order books or for customer service please, call 1(800)-CALL-WILEY (225-5945).

Limit of Liability/Disclaimer of Warranty: While the publisher and author have used their best efforts in preparing this book, they make no representations or warranties with respect to the accuracy or completeness of the contents of this book and specifically disclaim any implied warranties of merchantability or fitness for a particular purpose. No warranty may be created or extended by sales representatives or written sales materials. The advice and strategies contained herein may not be suitable for your situation. You should consult with a professional where appropriate. Neither the publisher nor author shall be liable for any loss of profit or any other commercial damages, including but not limited to special, incidental, consequential, or other damages.

For general information on our other products and services please contact our Customer Care Department within the U.S. at 877-762-2974, outside the U.S. at 317-572-3993 or fax 317-572-4002.

Wiley also publishes its books in a variety of electronic formats. Some content that appears in print, however, may not be available in electronic format.

*Library of Congress Cataloging-in-Publication Data Is Available*

ISBN 0-471-41816-1

Printed in the United States of America

10 9 8 7 6 5 4 3 2 1

*To my wife, Minhha, and my daughters, Mina and Jenna*
*with love*

# CONTENTS

# PREFACE

A course in introductory biostatistics is often required for professional students in public health, dentistry, nursing, and medicine, and for graduate students in nursing and other biomedical sciences, a requirement that is often considered a roadblock, causing anxiety in many quarters. These feelings are expressed in many ways and in many different settings, but all lead to the same conclusion: that students need help, in the form of a user-friendly and real data-based text, in order to provide enough motivation to learn a subject that is perceived to be difficult and dry. This introductory text is written for professionals and beginning graduate students in human health disciplines who need help to pass and benefit from the basic biostatistics requirement of a one-term course or a full-year sequence of two courses. Our main objective is to avoid the perception that statistics is just a series of formulas that students need to "get over with," but to present it as a way of thinking—thinking about ways to gather and analyze data so as to benefit from taking the required course. There is no better way to do that than to base a book on real data, so many real data sets in various fields are provided in the form of examples and exercises as aids to learning how to use statistical procedures, still the nuts and bolts of elementary applied statistics.

The first five chapters start slowly in a user-friendly style to nurture interest and motivate learning. Sections called "Brief Notes on the Fundamentals" are added here and there to gradually strengthen the background and the concepts. Then the pace is picked up in the remaining seven chapters to make sure that those who take a full-year sequence of two courses learn enough of the nuts and bolts of the subject. Our basic strategy is that most students would need only one course, which would end at about the middle of Chapter 8, after covering simple linear regression; instructors may add a few sections of Chapter 12. For students who take only one course, other chapters would serve as references to supplement class discussions as well as for their future needs. A subgroup of students with a stronger background in mathematics would go on to a second course, and with the help of the brief notes on the fundamentals would be able to handle the remaining chapters. A special feature of the book is the sections "Notes on Computations" at the end of most chapters. These notes cover uses of Microsoft's Excel, but samples of SAS computer programs are also included at the end of many examples, especially the advanced topics in the last several chapters.

The way of thinking called *statistics* has become important to all professionals: not only those in science or business, but also caring people who want to help to make the world a better place. But what is biostatistics, and what can it do? There are popular definitions and perceptions of statistics. We see "vital statistics" in the newspaper: announcements of life events such as births, marriages, and deaths. Motorists are warned to drive carefully, to avoid "becoming a statistic." Public use of the word is widely varied, most often indicating lists of numbers, or data. We have also heard people use the word *data* to describe a verbal report, a believable anecdote. For this book, especially in the first few chapters, we don't emphasize statistics as things, but instead, offer an active concept of "doing statistics." The doing of statistics is a way of thinking about numbers (collection, analysis, and presentation), with emphasis on relating their interpretation and meaning to the manner in which they are collected. Formulas are only a part of that thinking, simply tools of the trade; they are needed but not as the only things one needs to know.

To illustrate statistics as a way of thinking, let's begin with a familiar scenario: criminal court procedures. A crime has been discovered and a suspect has been identified. After a police investigation to collect evidence against the suspect, a presecutor presents summarized evidence to a jury. The jurors are given the rules regarding convicting beyond a reasonable doubt and about a unanimous decision, and then debate. After the debate, the jurors vote and a verdict is reached: guilty or not guilty. Why do we need to have this time-consuming, cost-consuming process of trial by jury? One reason is that the truth is often unknown, at least uncertain. Perhaps only the suspect knows but he or she does not talk. It is uncertain because of variability (every case is different) and because of possibly incomplete information. Trial by jury is the way our society deals with uncertainties; its goal is to minimize mistakes.

How does society deal with uncertainties? We go through a process called *trial by jury*, consisting of these steps: (1) we form an assumption or hypothesis (that every person is innocent until proved guilty), (2) we gather data (evidence against the suspect), and (3) we decide whether the hypothesis should be rejected (guilty) or should not be rejected (not guilty). With such a well-established procedure, sometime we do well, sometime we don't. Basically, a successful trial should consist of these elements: (1) a probable cause (with a crime and a suspect), (2) a thorough investigation by police, (3) an efficient presentation by a prosecutor, and (4) a fair and impartial jury.

In the context of a trial by jury, let us consider a few specific examples: (1) the *crime* is lung cancer and the *suspect* is cigarette smoking, or (2) the *crime* is leukemia and the *suspect* is pesticides, or (3) the *crime* is breast cancer and the *suspect* is a defective gene. The process is now called *research* and the tool to carry out that research is biostatistics. In a simple way, biostatistics serves as the biomedical version of the trial by jury process. It is the *science of dealing with uncertainties using incomplete information.* Yes, even science is uncertain; scientists arrive at different conclusions in many different areas at different times; many studies are inconclusive (hung jury). The reasons for uncertainties remain the same. Nature is complex and full of unexplained biological variability. But most important, we always have to deal with incomplete information. It is often not practical to study an entire population; we have to rely on information gained from a *sample.*

How does science deal with uncertainties? We learn how society deals with uncertainties; we go through a process called *biostatistics,* consisting of these steps: (1) we form an assumption or hypothesis (from the research question), (2) we gather data (from clinical trials, surveys, medical record abstractions), and (3) we make decision(s) (by doing statistical analysis/inference; a guilty verdict is referred to as *statistical significance*). Basically, a successful research should consist of these elements: (1) a good research question (with well-defined objectives and endpoints), (2) a thorough investigation (by experiments or surveys), (3) an efficient presentation of data (organizing data, summarizing, and presenting data: an area called *descriptive statistics*), and (4) proper statistical inference. This book is a problem-based introduction to the last three elements; together they form a field called *biostatistics.* The coverage is rather brief on data collection but very extensive on descriptive statistics (Chapters 1 and 2), especially on methods of statistical inference (Chapters 4 through 12). Chapter 3, on probability and probability models, serves as the link between the descriptive and inferential parts. Notes on computations and samples of SAS computer programs are incorporated throughout the book. About 60 percent of the material in the first eight chapters are overlapped with chapters from *Health and Numbers: A Problems-Based Introduction to Biostatistics* (another book by Wiley), but new topics have been added and others rewritten at a somewhat higher level. In general, compared to *Health and Numbers,* this book is aimed at a different audience—those who need a whole year of statistics and who are more mathematically prepared for advanced algebra and pre-calculus subjects.

I would like to express my sincere appreciation to colleagues, teaching assistants, and many generations of students for their help and feedback. I have learned very much from my former students, I hope that some of what they have taught me are reflected well in many sections of this book. Finally, my family bore patiently the pressures caused by my long-term commitment to the book; to my wife and daughters, I am always most grateful.

CHAP T. LE

*Edina, Minnesota*

# 1

# DESCRIPTIVE METHODS FOR CATEGORICAL DATA

Most introductory textbooks in statistics and biostatistics start with methods for summarizing and presenting continuous data. We have decided, however, to adopt a different starting point because our focused areas are in biomedical sciences, and health decisions are frequently based on proportions, ratios, or rates. In this first chapter we will see how these concepts appeal to common sense, and learn their meaning and uses.

## 1.1 PROPORTIONS

Many outcomes can be classified as belonging to one of two possible categories: presence and absence, nonwhite and white, male and female, improved and non-improved. Of course, one of these two categories is usually identified as of primary interest: for example, presence in the presence and absence classification, nonwhite in the white and nonwhite classification. We can, in general, relabel the two outcome categories as positive $(+)$ and negative $(-)$. An outcome is *positive* if the primary category is observed and is *negative* if the other category is observed.

It is obvious that in the summary to characterize observations made on a group of people, the number $x$ of positive outcomes is not sufficient; the group size $n$, or total number of observations, should also be recorded. The number $x$ tells us very little and becomes meaningful only after adjusting for the size $n$ of the group; in other words, the two figures $x$ and $n$ are often combined into a *statistic*, called a *proportion*:

$$p = \frac{x}{n}$$

1

The term *statistic* means a summarized figure from observed data. Clearly, $0 \leq p \leq 1$. This proportion $p$ is sometimes expressed as a percentage and is calculated as follows:

$$\text{percent } (\%) = \frac{x}{n}(100)$$

***Example 1.1***   A study published by the Urban Coalition of Minneapolis and the University of Minnesota Adolescent Health Program surveyed 12,915 students in grades 7 through 12 in Minneapolis and St. Paul public schools. The report stated that minority students, about one-third of the group, were much less likely to have had a recent routine physical checkup. Among Asian students, 25.4% said that they had not seen a doctor or a dentist in the last two years, followed by 17.7% of Native Americans, 16.1% of blacks, and 10% of Hispanics. Among whites, it was 6.5%.

*Proportion* is a number used to describe a group of people according to a dichotomous, or binary, characteristic under investigation. It is noted that characteristics with multiple categories can be dichotomized by pooling some categories to form a new one, and the concept of proportion applies. The following are a few illustrations of the use of proportions in the health sciences.

### 1.1.1   Comparative Studies

Comparative studies are intended to show possible differences between two or more groups; Example 1.1 is such a typical comparative study. The survey cited in Example 1.1 also provided the following figures concerning boys in the group who use tobacco at least weekly. Among Asians, it was 9.7%, followed by 11.6% of blacks, 20.6% of Hispanics, 25.4% of whites, and 38.3% of Native Americans.

In addition to surveys that are cross-sectional, as seen in Example 1.1, data for comparative studies may come from different sources; the two fundamental designs being retrospective and prospective. *Retrospective studies* gather past data from selected cases and controls to determine differences, if any, in exposure to a suspected risk factor. These are commonly referred to as *case–control studies*; each study being focused on a particular disease. In a typical case–control study, cases of a specific disease are ascertained as they arise from population-based registers or lists of hospital admissions, and controls are sampled either as disease-free persons from the population at risk or as hospitalized patients having a diagnosis other than the one under study. The advantages of a retrospective study are that it is economical and provides answers to research questions relatively quickly because the cases are already available. Major limitations are due to the inaccuracy of the exposure histories and uncertainty about the appropriateness of the control sample; these problems sometimes hinder retrospective studies and make them less preferred than pro-

**TABLE 1.1**

| Smoking | Shipbuilding | Cases | Controls |
|---------|--------------|-------|----------|
| No      | Yes          | 11    | 35       |
|         | No           | 50    | 203      |
| Yes     | Yes          | 84    | 45       |
|         | No           | 313   | 270      |

spective studies. The following is an example of a retrospective study in the field of occupational health.

*Example 1.2*   A case–control study was undertaken to identify reasons for the exceptionally high rate of lung cancer among male residents of coastal Georgia. Cases were identified from these sources:

(a) Diagnoses since 1970 at the single large hospital in Brunswick
(b) Diagnoses during 1975–1976 at three major hospitals in Savannah
(c) Death certificates for the period 1970–1974 in the area

Controls were selected from admissions to the four hospitals and from death certificates in the same period for diagnoses other than lung cancer, bladder cancer, or chronic lung cancer. Data are tabulated separately for smokers and nonsmokers in Table 1.1. The exposure under investigation, "shipbuilding," refers to employment in shipyards during World War II. By using a separate tabulation, with the first half of the table for nonsmokers and the second half for smokers, we treat *smoking* as a potential confounder. A *confounder* is a factor, an exposure by itself, not under investigation but related to the disease (in this case, lung cancer) and the exposure (shipbuilding); previous studies have linked smoking to lung cancer, and construction workers are more likely to be smokers. The term *exposure* is used here to emphasize that employment in shipyards is a suspected *risk* factor; however, the term is also used in studies where the factor under investigation has beneficial effects.

In an examination of the smokers in the data set in Example 1.2, the numbers of people employed in shipyards, 84 and 45, tell us little because the sizes of the two groups, cases and controls, are different. Adjusting these absolute numbers for the group sizes (397 cases and 315 controls), we have:

1. For the controls,

$$\text{proportion of exposure} = \frac{45}{315}$$

$$= 0.143 \quad \text{or} \quad 14.3\%$$

2. For the cases,

$$\text{proportion of exposure} = \frac{84}{397}$$

$$= 0.212 \quad \text{or} \quad 21.2\%$$

The results reveal different exposure histories: The proportion among cases was higher than that among controls. It is *not* in any way conclusive proof, but it is a good *clue*, indicating a possible relationship between the disease (lung cancer) and the exposure (shipbuilding).

Similar examination of the data for nonsmokers shows that by taking into consideration the numbers of cases and controls, we have the following figures for employment:

1. For the controls,

$$\text{proportion of exposure} = \frac{35}{238}$$

$$= 0.147 \quad \text{or} \quad 14.7\%$$

2. For the cases,

$$\text{proportion of exposure} = \frac{11}{61}$$

$$= 0.180 \quad \text{or} \quad 18.0\%$$

The results also reveal different exposure histories: The proportion among cases was higher than that among controls.

The analyses above also show that the difference between proportions of exposure among smokers, that is,

$$21.2 - 14.3 = 6.9\%$$

is different from the difference between proportions of exposure among non-smokers, which is

$$18.0 - 14.7 = 3.3\%$$

The differences, 6.9% and 3.3%, are *measures* of the strength of the relationship between the disease and the exposure, one for each of the two strata: the two groups of smokers and nonsmokers, respectively. The calculation above shows that the possible effects of employment in shipyards (as a suspected risk factor) are different for smokers and nonsmokers. This difference of differences, if confirmed, is called a *three-term interaction* or *effect modification*, where smok-

**TABLE 1.2**

|  | Population | Cases | Cases per 100,000 |
|---|---|---|---|
| White | 32,930,233 | 2832 | 8.6 |
| Nonwhite | 3,933,333 | 3227 | 82.0 |

ing alters the effect of employment in shipyards as a risk for lung cancer. In that case, *smoking* is not only a confounder, it is an *effect modifier*, which modifies the effects of shipbuilding (on the possibility of having lung cancer).

Another example is provided in the following example concerning glaucomatous blindness.

***Example 1.3***    Data for persons registered blind from glaucoma are listed in Table 1.2.

For these *disease registry data*, direct calculation of a proportion results in a very tiny fraction, that is, the number of cases of the disease per person at risk. For convenience, this is multiplied by 100,000, and hence the result expresses the number of cases per 100,000 people. This data set also provides an example of the use of proportions as disease *prevalence*, which is defined as

$$\text{prevalence} = \frac{\text{number of diseased persons at the time of investigation}}{\text{total number of persons examined}}$$

*Disease prevalence* and related concepts are discussed in more detail in Section 1.2.2.

For blindness from glaucoma, calculations in Example 1.3 reveal a striking difference between the races: The blindness prevalence among nonwhites was over eight times that among whites. The number "100,000" was selected arbitrarily; any power of 10 would be suitable so as to obtain a result between 1 and 100, sometimes between 1 and 1000; it is easier to state the result "82 cases per 100,000" than to say that the prevalence is 0.00082.

### 1.1.2  Screening Tests

Other uses of proportions can be found in the evaluation of screening tests or diagnostic procedures. Following these procedures, clinical observations, or laboratory techniques, people are classified as healthy or as falling into one of a number of disease categories. Such tests are important in medicine and epidemiologic studies and may form the basis of early interventions. Almost all such tests are imperfect, in the sense that healthy persons will occasionally be classified wrongly as being ill, while some people who are really ill may fail to be detected. That is, misclassification is unavoidable. Suppose that each person

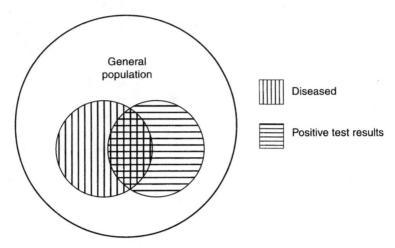

**Figure 1.1**    Graphical display of a screening test.

in a large population can be classified as truly positive or negative for a particular disease; this true diagnosis may be based on more refined methods than are used in the test, or it may be based on evidence that emerges after the passage of time (e.g., at autopsy). For each class of people, diseased and healthy, the test is applied, with the results depicted in Figure 1.1.

The two proportions fundamental to evaluating diagnostic procedures are sensitivity and specificity. *Sensitivity* is the proportion of diseased people detected as positive by the test:

$$\text{sensitivity} = \frac{\text{number of diseased persons who screen positive}}{\text{total number of diseased persons}}$$

The corresponding errors are *false negatives*. *Specificity* is the proportion of healthy people detected as negative by the test:

$$\text{specificity} = \frac{\text{number of healthy persons who screen negative}}{\text{total number of healthy persons}}$$

and the corresponding errors are *false positives*.

Clearly, it is desirable that a test or screening procedure be highly sensitive and highly specific. However, the two types of errors go in opposite directions; for example, an effort to increase sensitivity may lead to more false positives, and vice versa.

***Example 1.4***    A cytological test was undertaken to screen women for cervical cancer. Consider a group of 24,103 women consisting of 379 women whose cervices are abnormal (to an extent sufficient to justify concern with respect to

**TABLE 1.3**

| | Test | | |
| --- | --- | --- | --- |
| True | − | + | Total |
| − | 23,362 | 362 | 23,724 |
| + | 225 | 154 | 379 |

possible cancer) and 23,724 women whose cervices are acceptably healthy. A test was applied and results are tabulated in Table 1.3. (This study was performed with a rather old test and is used here only for illustration.)

The calculations

$$\text{sensitivity} = \frac{154}{379}$$

$$= 0.406 \quad \text{or} \quad 40.6\%$$

$$\text{specificity} = \frac{23,362}{23,724}$$

$$= 0.985 \quad \text{or} \quad 98.5\%$$

show that the test is highly specific (98.5%) but not very sensitive (40.6%); there were more than half (59.4%) false negatives. The implications of the use of this test are:

1. If a woman without cervical cancer is tested, the result would almost surely be negative, *but*
2. If a woman with cervical cancer is tested, the chance is that the disease would go undetected because 59.4% of these cases would lead to false negatives.

Finally, it is important to note that throughout this section, proportions have been defined so that both the numerator and the denominator are counts or frequencies, and the numerator corresponds to a subgroup of the larger group involved in the denominator, resulting in a number between 0 and 1 (or between 0 and 100%). It is straightforward to generalize this concept for use with characteristics having more than two outcome categories; for each category we can define a proportion, and these category-specific proportions add up to 1 (or 100%).

*Example 1.5*  An examination of the 668 children reported living in crack/cocaine households shows 70% blacks, followed by 18% whites, 8% Native Americans, and 4% other or unknown.

### 1.1.3 Displaying Proportions

Perhaps the most effective and most convenient way of presenting data, especially discrete data, is through the use of graphs. Graphs convey the information, the general patterns in a set of data, at a single glance. Therefore, graphs are often easier to read than tables; the most informative graphs are simple and self-explanatory. Of course, to achieve that objective, graphs should be constructed carefully. Like tables, they should be clearly labeled and units of measurement and/or magnitude of quantities should be included. Remember that graphs must tell their own story; they should be complete in themselves and require little or no additional explanation.

***Bar Charts***    Bar charts are a very popular type of graph used to display several proportions for quick comparison. In applications suitable for bar charts, there are several groups and we investigate one binary characteristic. In a bar chart, the various groups are represented along the horizontal axis; they may be arranged alphabetically, by the size of their proportions, or on some other rational basis. A vertical bar is drawn above each group such that the height of the bar is the proportion associated with that group. The bars should be of equal width and should be separated from one another so as not to imply continuity.

***Example 1.6***    We can present the data set on children without a recent physical checkup (Example 1.1) by a bar chart, as shown in Figure 1.2.

***Pie Charts***    Pie charts are another popular type of graph. In applications suitable for pie charts, there is only one group but we want to decompose it into several categories. A pie chart consists of a circle; the circle is divided into

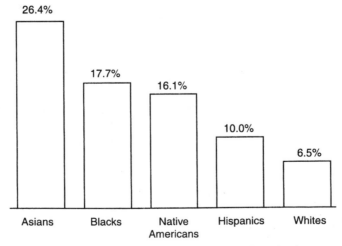

**Figure 1.2**    Children without a recent physical checkup.

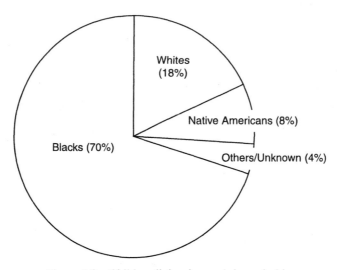

**Figure 1.3**   Children living in crack households.

wedges that correspond to the magnitude of the proportions for various categories. A pie chart shows the differences between the sizes of various categories or subgroups as a decomposition of the total. It is suitable, for example, for use in presenting a budget, where we can easily see the difference between U.S. expenditures on health care and defense. In other words, a bar chart is a suitable graphic device when we have several groups, each associated with a different proportion; whereas a pie chart is more suitable when we have one group that is divided into several categories. The proportions of various categories in a pie chart should add up to 100%. Like bar charts, the categories in a pie chart are usually arranged by the size of the proportions. They may also be arranged alphabetically or on some other rational basis.

*Example 1.7*   We can present the data set on children living in crack households (Example 1.5) by a pie chart as shown in Figure 1.3.

Another example of the pie chart's use is for presenting the proportions of deaths due to different causes.

*Example 1.8*   Table 1.4 lists the number of deaths due to a variety of causes among Minnesota residents for the year 1975. After calculating the proportion of deaths due to each cause: for example,

$$\text{deaths due to cancer} = \frac{6448}{32,686}$$

$$= 0.197 \quad \text{or} \quad 19.7\%$$

we can present the results as in the pie chart shown in Figure 1.4.

**TABLE 1.4**

| Cause of Death | Number of Deaths |
| --- | --- |
| Heart disease | 12,378 |
| Cancer | 6,448 |
| Cerebrovascular disease | 3,958 |
| Accidents | 1,814 |
| Others | 8,088 |
| Total | 32,686 |

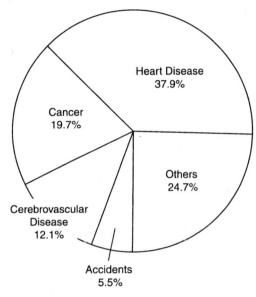

**Figure 1.4**   Causes of death for Minnesota residents, 1975.

*Line Graphs*   A line graph is similar to a bar chart, but the horizontal axis represents time. In the applications most suitable to use line graphs, one binary characteristic is observed repeatedly over time. Different "groups" are consecutive years, so that a line graph is suitable to illustrate how certain proportions change over time. In a line graph, the proportion associated with each year is represented by a point at the appropriate height; the points are then connected by straight lines.

*Example 1.9*   Between the years 1984 and 1987, the crude death rates for women in the United States were as listed in Table 1.5. The change in crude death rate for U.S. women can be represented by the line graph shown in Figure 1.5.

In addition to their use with proportions, line graphs can be used to describe changes in the number of occurrences and with continuous measurements.

**TABLE 1.5**

| Year | Crude Death Rate per 100,000 |
|------|------------------------------|
| 1984 | 792.7 |
| 1985 | 806.6 |
| 1986 | 809.3 |
| 1987 | 813.1 |

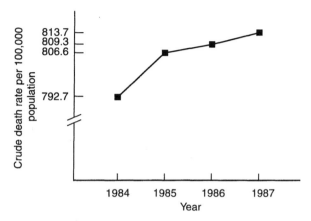

**Figure 1.5**   Death rates for U.S. women, 1984–1987.

***Example 1.10***   The line graph shown in Figure 1.6 displays the trend in rates of malaria reported in the United States between 1940 and 1989 (proportion × 100,000 as above).

## 1.2   RATES

The term *rate* is somewhat confusing; sometimes it is used interchangeably with the term *proportion* as defined in Section 1.1; sometimes it refers to a quantity of a very different nature. In Section 1.2.1, on the *change rate*, we cover this special use, and in the next two Sections, 1.2.2 and 1.2.3, we focus on *rates* used interchangeably with *proportions* as measures of morbidity and mortality. Even when they refer to the same things—measures of morbidity and mortality— there is some degree of difference between these two terms. In contrast to the static nature of proportions, rates are aimed at measuring the occurrences of events during or after a certain time period.

### 1.2.1   Changes

Familiar examples of rates include their use to describe changes after a certain period of time. The *change rate* is defined by

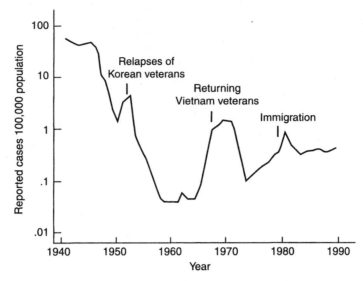

**Figure 1.6**   Malaria rates in the United States, 1940–1989.

$$\text{change rate (\%)} = \frac{\text{new value} - \text{old value}}{\text{old value}} \times 100$$

In general, change rates could exceed 100%. *They are not proportions* (a proportion is a number between 0 and 1 or between 0 and 100%). Change rates are used primarily for description and are not involved in common *statistical analyses*.

***Example 1.11***   The following is a typical paragraph of a *news report*:

A total of 35,238 new AIDS cases was reported in 1989 by the Centers for Disease Control (CDC), compared to 32,196 reported during 1988. The 9% increase is the smallest since the spread of AIDS began in the early 1980s. For example, new AIDS cases were up 34% in 1988 and 60% in 1987. In 1989, 547 cases of AIDS transmissions from mothers to newborns were reported, up 17% from 1988; while females made up just 3971 of the 35,238 new cases reported in 1989; that was an increase of 11% over 1988.

In Example 1.11:

1. The change rate for new AIDS cases was calculated as

$$\frac{35{,}238 - 32{,}196}{32{,}196} \times 100 = 9.4\%$$

(this was *rounded down* to the reported figure of 9% in the news report).

2. For the new AIDS cases transmitted from mothers to newborns, we have

$$17\% = \frac{547 - (1988 \text{ cases})}{1988 \text{ cases}} \times 100$$

leading to

$$1988 \text{ cases} = \frac{547}{1.17}$$
$$= 468$$

(a figure obtainable, as shown above, but usually not reported because of redundancy).

Similarly, the number of new AIDS cases for the year 1987 is calculated as follows:

$$34\% = \frac{32,196 - (1987 \text{ total})}{1987 \text{ total}} \times 100$$

or

$$1987 \text{ total} = \frac{32,196}{1.34}$$
$$= 24,027$$

3. Among the 1989 new AIDS cases, the proportion of females is

$$\frac{3971}{35,238} = 0.113 \quad \text{or} \quad 11.3\%$$

and the proportion of males is

$$\frac{35,238 - 3971}{35,238} = 0.887 \quad \text{or} \quad 88.7\%$$

The proportions of females and males add up to 1.0 or 100%.

### 1.2.2  Measures of Morbidity and Mortality

The field of vital statistics makes use of some special applications of rates, three types of which are commonly mentioned: crude, specific, and adjusted (or standardized). Unlike change rates, these measures are proportions. *Crude rates* are computed for an entire large group or population; they disregard factors such as age, gender, and race. *Specific rates* consider these differences among subgroups or categories of diseases. *Adjusted* or *standardized rates* are used to make valid summary comparisons between two or more groups possessing different age distributions.

The annual *crude death rate* is defined as the number of deaths in a calendar year divided by the population on July 1 of that year (which is usually an estimate); the quotient is often multiplied by 1000 or other suitable power of 10, resulting in a number between 1 and 100 or between 1 and 1000. For example, the 1980 population of California was 23,000,000 (as estimated by July 1) and there were 190,237 deaths during 1980, leading to

$$\text{crude death rate} = \frac{190{,}247}{23{,}000{,}000} \times 1000$$

$$= 8.3 \text{ deaths per 1000 persons per year}$$

The age- and cause-specific death rates are defined similarly.

As for morbidity, the disease prevalence, as defined in Section 1.1, is a proportion used to describe the population at a certain point in time, whereas *incidence* is a rate used in connection with new cases:

$$\text{incidence rate} = \frac{\begin{array}{c}\text{number of persons who developed the disease} \\ \text{over a defined period of time (a year, say)}\end{array}}{\begin{array}{c}\text{number of persons initially without the disease} \\ \text{who were followed for the defined period of time}\end{array}}$$

In other words, the prevalence presents a snapshot of the population's morbidity experience at a certain time point, whereas the incidence is aimed to investigate possible time trends. For example, the 35,238 new AIDS cases in Example 1.11 and the national population without AIDS at the start of 1989 could be combined according to the formula above to yield an incidence of AIDS for the year.

Another interesting use of rates is in connection with *cohort studies*, epidemiological designs in which one enrolls a group of persons and follows them over certain periods of time; examples include occupational mortality studies, among others. The cohort study design focuses on a particular exposure rather than a particular disease as in case–control studies. Advantages of a longitudinal approach include the opportunity for more accurate measurement of exposure history and a careful examination of the time relationships between exposure and any disease under investigation. Each member of a cohort belongs to one of three types of termination:

1. Subjects still alive on the analysis date
2. Subjects who died on a known date within the study period
3. Subjects who are lost to follow-up after a certain date (these cases are a

potential source of bias; effort should be expended on reducing the number of subjects in this category)

The contribution of each member is the length of follow-up time from enrollment to his or her termination. The quotient, defined as the number of deaths observed for the cohort, divided by the total follow-up times (in person-years, say) is the *rate* to characterize the mortality experience of the cohort:

$$\frac{\text{follow-up}}{\text{death rate}} = \frac{\text{number of deaths}}{\text{total person-years}}$$

Rates may be calculated for total deaths and for separate causes of interest, and they are usually multiplied by an appropriate power of 10, say 1000, to result in a single- or double-digit figure: for example, deaths per 1000 months of follow-up. Follow-up death rates may be used to measure the effectiveness of medical treatment programs.

*Example 1.12*    In an effort to provide a complete analysis of the survival of patients with end-stage renal disease (ESRD), data were collected for a sample that included 929 patients who initiated hemodialysis for the first time at the Regional Disease Program in Minneapolis, Minnesota, between January 1, 1976 and June 30, 1982; all patients were followed until December 31, 1982. Of these 929 patients, 257 are diabetics; among the 672 nondiabetics, 386 are classified as low risk (without co-morbidities such as arteriosclerotic heart disease, peripheral vascular disease, chronic obstructive pulmonary, and cancer). Results from these two subgroups are listed in Table 1.6. (Only some summarized figures are given here for illustration; details such as numbers of deaths and total treatment months for subgroups are not included.) For example, for low-risk patients over 60 years of age, there were 38 deaths during 2906 treatment months, leading to

$$\frac{38}{2906} \times 1000 = 13.08 \text{ deaths per 1000 treatment months}$$

**TABLE 1.6**

| Group | Age | Deaths/1000 Treatment Months |
|-------|-----|------------------------------|
| Low-risk | 1–45 | 2.75 |
|  | 46–60 | 6.93 |
|  | 61+ | 13.08 |
| Diabetics | 1–45 | 10.29 |
|  | 46–60 | 12.52 |
|  | 61+ | 22.16 |

**TABLE 1.7**

| Age Group | Alaska Number of Deaths | Persons | Deaths per 100,000 | Florida Number of Deaths | Persons | Deaths per 100,000 |
|---|---|---|---|---|---|---|
| 0–4 | 162 | 40,000 | 405.0 | 2,049 | 546,000 | 375.3 |
| 5–19 | 107 | 128,000 | 83.6 | 1,195 | 1,982,000 | 60.3 |
| 20–44 | 449 | 172,000 | 261.0 | 5,097 | 2,676,000 | 190.5 |
| 45–64 | 451 | 58,000 | 777.6 | 19,904 | 1,807,000 | 1,101.5 |
| 65+ | 444 | 9,000 | 4,933.3 | 63,505 | 1,444,000 | 4,397.9 |
| Total | 1,615 | 407,000 | 396.8 | 91,760 | 8,455,000 | 1,085.3 |

### 1.2.3  Standardization of Rates

Crude rates, as measures of morbidity or mortality, can be used for population description and may be suitable for investigations of their variations over time; however, comparisons of crude rates are often invalid because the populations may be different with respect to an important characteristic such as age, gender, or race (these are potential *confounders*). To overcome this difficulty, an adjusted (or standardized) rate is used in the comparison; the adjustment removes the difference in composition with respect to a confounder.

***Example 1.13***  Table 1.7 provides mortality data for Alaska and Florida for the year 1977.

Example 1.13 shows that the 1977 crude death rate per 100,000 population for Alaska was 396.8 and for Florida was 1085.7, almost a threefold difference. However, a closer examination shows the following:

1. Alaska had higher age-specific death rates for four of the five age groups, the only exception being 45–64 years.
2. Alaska had a higher percentage of its population in the younger age groups.

The findings make it essential to adjust the death rates of the two states in order to make a valid comparison. A simple way to achieve this, called the *direct method*, is to apply to a common standard population, age-specific rates observed from the two populations under investigation. For this purpose, the U.S. population as of the last decennial census is frequently used. The procedure consists of the following steps:

1. The standard population is listed by the same age groups.
2. The expected number of deaths in the standard population is computed

for each age group of each of the two populations being compared. For example, for age group 0–4, the U.S. population for 1970 was 84,416 (per million); therefore, we have:

(a) Alaska rate = 405.0 per 100,000. The expected number of deaths is

$$\frac{(84,416)(405.0)}{100,000} = 341.9$$

$$\simeq 342$$

($\simeq$ means "almost equal to").

(b) Florida rate = 375.3 per 100,000. The expected number of deaths is

$$\frac{(84,416)(375.3)}{100,000} = 316.8$$

$$\simeq 317$$

which is lower than the expected number of deaths for Alaska obtained for the same age group.

3. Obtain the total number of deaths expected.

4. The age-adjusted death rate is

$$\text{adjusted rate} = \frac{\text{total number of deaths expected}}{\text{total standard population}} \times 100,000$$

The calculations are detailed in Table 1.8.

The age-adjusted death rate per 100,000 population for Alaska is 788.6 and for Florida is 770.6. These age-adjusted rates are much closer than as shown by the crude rates, and the adjusted rate for Florida is *lower*. It is important to keep in mind that any population could be chosen as "standard," and because

**TABLE 1.8**

| Age Group | 1970 U.S. Standard Million | Alaska Age-Specific Rate | Alaska Expected Deaths | Florida Age-Specific Rate | Florida Expected Deaths |
|---|---|---|---|---|---|
| 0–4 | 84,416 | 405.0 | 342 | 375.3 | 317 |
| 5–19 | 294,353 | 83.6 | 246 | 60.3 | 177 |
| 20–44 | 316,744 | 261.0 | 827 | 190.5 | 603 |
| 45–64 | 205,745 | 777.6 | 1600 | 1101.5 | 2266 |
| 65+ | 98,742 | 4933.3 | 4871 | 4397.9 | 4343 |
| Total | 1,000,000 | | 7886 | | 7706 |

**TABLE 1.9**

| Age Group | Alaska Population (Used as Standard) | Florida Rate/100,000 | Florida Expected Number of Deaths |
|---|---|---|---|
| 0–4 | 40,000 | 375.3 | 150 |
| 5–19 | 128,000 | 60.3 | 77 |
| 20–44 | 172,000 | 190.5 | 328 |
| 45–64 | 58,000 | 1101.5 | 639 |
| 65+ | 9,000 | 4397.9 | 396 |
| Total | 407,000 | | 1590 |

of this, an adjusted rate is artificial; it does not reflect data from an actual population. The numerical values of the adjusted rates depend in large part on the choice of the standard population. They have real meaning only as relative comparisons.

The advantage of using the U.S. population as the standard is that we can adjust death rates of many states and compare them with each other. Any population could be selected and used as a standard. In Example 1.13 it does not mean that there were only 1 million people in the United States in 1970; it only presents the *age distribution* of 1 million U.S. residents for that year. If all we want to do is to compare Florida with Alaska, we could choose either state as the standard and adjust the death rate of the other; this practice would save half the labor. For example, if we choose Alaska as the standard population, the adjusted death rate for Florida is calculated as shown in Table 1.9. The new adjusted rate,

$$\frac{(1590)(100,000)}{407,000} = 390.7 \text{ per } 100,000$$

is not the same as that obtained using the 1970 U.S. population as the standard (it was 770.6), but it also shows that after age adjustment, the death rate in Florida (390.7 per 100,000) is somewhat lower than that of Alaska (396.8 per 100,000; there is no need for adjustment here because we use Alaska's population as the standard population).

## 1.3    RATIOS

In many cases, such as disease prevalence and disease incidence, proportions and rates are defined very similarly, and the terms *proportions* and *rates* may even be used interchangeably. *Ratio* is a completely different term; it is a computation of the form

$$\text{ratio} = \frac{a}{b}$$

where $a$ and $b$ are *similar quantities* measured from *different groups* or under *different circumstances*. An example is the male/female ratio of smoking rates; such a ratio is positive but may exceed 1.0.

### 1.3.1  Relative Risk

One of the most often used ratios in epidemiological studies is *relative risk*, a concept for the comparison of two groups or populations with respect to a certain unwanted event (e.g., disease or death). The traditional method of expressing it in prospective studies is simply the ratio of the incidence rates:

$$\text{relative risk} = \frac{\text{disease incidence in group 1}}{\text{disease incidence in group 2}}$$

However, the ratio of disease prevalences as well as follow-up death rates can also be formed. Usually, group 2 is under standard conditions—such as nonexposure to a certain risk factor—against which group 1 (exposed) is measured. A relative risk greater than 1.0 indicates harmful effects, whereas a relative risk below 1.0 indicates beneficial effects. For example, if group 1 consists of smokers and group 2 of nonsmokers, we have a *relative risk due to smoking*. Using the data on end-stage renal disease (ESRD) of Example 1.12, we can obtain the relative risks due to diabetes (Table 1.10). All three numbers are greater than 1 (indicating higher mortality for diabetics) and form a decreasing trend with increasing age.

### 1.3.2  Odds and Odds Ratio

The *relative risk*, also called the *risk ratio*, is an important index in epidemiological studies because in such studies it is often useful to measure the *increased* risk (if any) of incurring a particular disease if a certain factor is present. In cohort studies such an index is obtained readily by observing the experience of groups of subjects with and without the factor, as shown above. In a case–control study the data do not present an immediate answer to this type of question, and we now consider how to obtain a useful shortcut solution.

**TABLE 1.10**

| Age Group | Relative Risk |
|-----------|---------------|
| 1–45      | 3.74          |
| 46–60     | 1.81          |
| 61+       | 1.69          |

**TABLE 1.11**

|        | Disease |       |                     |
|--------|---------|-------|---------------------|
| Factor | +       | −     | Total               |
| +      | $A$     | $B$   | $A + B$             |
| −      | $C$     | $D$   | $C + D$             |
| Total  | $A + C$ | $B + D$ | $N = A + B + C + D$ |

Suppose that each subject in a large study, at a particular time, is classified as positive or negative according to some risk factor, and as having or not having a certain disease under investigation. For any such categorization the population may be enumerated in a $2 \times 2$ table (Table 1.11). The entries $A$, $B$, $C$ and $D$ in the table are sizes of the four combinations of disease presence/ absence and factor presence/absence, and the number $N$ at the lower right corner of the table is the total population size. The relative risk is

$$\text{RR} = \frac{A}{A + B} \div \frac{C}{C + D}$$
$$= \frac{A(C + D)}{C(A + B)}$$

In many situations, the number of subjects classified as disease positive is small compared to the number classified as disease negative; that is,

$$C + D \simeq D$$
$$A + B \simeq B$$

and therefore the relative risk can be approximated as follows:

$$\text{RR} \simeq \frac{AD}{BC}$$
$$= \frac{A/B}{C/D}$$
$$= \frac{A/C}{B/D}$$

where the slash denotes division. The resulting ratio, $AD/BC$, is an approximate relative risk, but it is often referred to as an *odds ratio* because:

1. $A/B$ and $C/D$ are the odds in favor of having disease from groups with or without the factor.

2. $A/C$ and $B/D$ are the odds in favor of having been exposed to the factors from groups with or without the disease. These two odds can easily be estimated using case–control data, by using sample frequencies. For example, the odds $A/C$ can be estimated by $a/c$, where $a$ is the number of exposed cases and $c$ the number of nonexposed cases in the sample of cases used in a case–control design.

For the many diseases that are rare, the terms *relative risk* and *odds ratio* are used interchangeably because of the above-mentioned approximation. Of course, it is totally acceptable to draw conclusions on an odds ratio without invoking this approximation for disease that is not rare. The relative risk is an important epidemiological index used to measure seriousness, or the magnitude of the harmful effect of suspected risk factors. For example, if we have

$$RR = 3.0$$

we can say that people exposed have a risk of contracting the disease that is approximately three times the risk of those unexposed. A perfect 1.0 indicates no effect, and beneficial factors result in relative risk values which are smaller than 1.0. From data obtained by a case–control or retrospective study, it is impossible to calculate the relative risk that we want, but if it is reasonable to assume that the disease is rare (prevalence is less than 0.05, say), we can calculate the odds ratio as a stepping stone and use it as an approximate relative risk (we use the notation $\simeq$ for this purpose). In these cases, we interpret the odds ratio calculated just as we would do with the relative risk.

**Example 1.14**    The role of smoking in the etiology of pancreatitis has been recognized for many years. To provide estimates of the quantitative significance of these factors, a hospital-based study was carried out in eastern Massachusetts and Rhode Island between 1975 and 1979. Ninety-eight patients who had a hospital discharge diagnosis of pancreatitis were included in this unmatched case–control study. The control group consisted of 451 patients admitted for diseases other than those of the pancreas and biliary tract. Risk factor information was obtained from a standardized interview with each subject, conducted by a trained interviewer.

Some data for the males are given in Table 1.12. For these data for this example, the approximate relative risks or odds ratios are calculated as follows:

(a)  For ex-smokers,

$$RR_e \simeq \frac{13/2}{80/56}$$

$$= \frac{(13)(56)}{(80)(2)}$$

$$= 4.55$$

**TABLE 1.12**

| Use of Cigarettes | Cases | Controls |
|---|---|---|
| Never | 2 | 56 |
| Ex-smokers | 13 | 80 |
| Current smokers | 38 | 81 |
| Total | 53 | 217 |

[The subscript $e$ in $RR_e$ indicates that we are calculating the relative risk (RR) for ex-smokers.]

(b) For current smokers,

$$RR_c \simeq \frac{38/2}{81/56}$$

$$= \frac{(38)(56)}{(81)(2)}$$

$$= 13.14$$

[The subscript $c$ in $RR_c$ indicates that we are calculating the relative risk (RR) for current smokers.]

In these calculations, the nonsmokers (who never smoke) are used as references. These values indicate that the risk of having pancreatitis for current smokers is approximately 13.14 times the risk for people who never smoke. The effect for ex-smokers is smaller (4.55 times) but is still very high (compared to 1.0, the no-effect baseline for relative risks and odds ratios). In other words, if the smokers were to quit smoking, they would reduce their own risk (from 13.14 times to 4.55 times) but *not* to the normal level for people who never smoke.

### 1.3.3   Generalized Odds for Ordered 2 × k Tables

In this section we provide an interesting generalization of the concept of odds ratios to ordinal outcomes which is sometime used in biomedical research. Readers, especially beginners, may decide to skip it without loss of continuity; if so, corresponding exercises should be skipped accordingly: 1.24(b), 1.25(c), 1.26(b), 1.27(b,c), 1.35(c), 1.38(c), and 1.45(b).

We can see this possible generalization by noting that an odds ratio can be interpreted as an odds for a different event. For example, consider again the same 2 × 2 table as used in Section 1.3.2 (Table 1.11). The number of case–control pairs with different exposure histories is $(AD + BC)$; among them, $AD$ pairs with an exposed case and $BC$ pairs with an exposed control. Therefore $AD/BC$, the odds ratio of Section 1.3.2, can be seen as the odds of finding a pair with an exposed case among discordant pairs (a *discordant pair* is a case–control pair with *different* exposure histories).

**TABLE 1.13**

| Seat Belt | Extent of Injury Received | | | |
|---|---|---|---|---|
| | None | Minor | Major | Death |
| Yes | 75 | 160 | 100 | 15 |
| No | 65 | 175 | 135 | 25 |

The interpretation above of the concept of an odds ratio as an odds can be generalized as follows. The aim here is to present an efficient method for use with ordered $2 \times k$ contingency tables, tables with two rows and $k$ columns having a certain natural ordering. The figure summarized is the generalized odds formulated from the concept of odds ratio. Let us first consider an example concerning the use of seat belts in automobiles. Each accident in this example is classified according to whether a seat belt was used and to the severity of injuries received: none, minor, major, or death (Table 1.13).

To compare the extent of injury from those who used seat belts with those who did not, we can calculate the percent of seat belt users in each injury group that decreases from level "none" to level "death," and the results are:

$$\text{None:} \qquad \frac{75}{75 + 65} = 54\%$$

$$\text{Minor:} \qquad \frac{160}{160 + 175} = 48\%$$

$$\text{Major:} \qquad \frac{100}{100 + 135} = 43\%$$

$$\text{Death:} \qquad \frac{15}{15 + 25} = 38\%$$

What we are seeing here is a *trend* or an *association* indicating that the lower the percentage of seat belt users, the more severe the injury.

We now present the concept of *generalized odds*, a special statistic specifically formulated to measure the strength of such a trend and will use the same example and another one to illustrate its use. In general, consider an ordered $2 \times k$ table with the frequencies shown in Table 1.14.

**TABLE 1.14**

| Row | Column Level | | | | Total |
|---|---|---|---|---|---|
| | 1 | 2 | $\cdots$ | $k$ | |
| 1 | $a_1$ | $a_2$ | $\cdots$ | $a_k$ | $A$ |
| 2 | $b_1$ | $b_2$ | $\cdots$ | $b_k$ | $B$ |
| Total | $n_1$ | $n_2$ | $\cdots$ | $n_k$ | $N$ |

The number of *concordances* is calculated by

$$C = a_1(b_2 + \cdots + b_k) + a_2(b_3 + \cdots + b_k) + \cdots + a_{k-1}b_k$$

(The term *concordance pair* as used above corresponds to a less severe injury for the seat belt user.) The number of *discordances* is

$$D = b_1(a_2 + \cdots + a_k) + b_2(a_3 + \cdots + a_k) + \cdots + b_{k-1}a_k$$

To *measure* the degree of association, we use the index $C/D$ and call it the generalized odds; if there are only two levels of injury, this new index is reduced to the familiar odds ratio. When data are properly arranged, by an a priori hypothesis, the products in the number of concordance pairs $C$ (e.g., $a_1b_2$) go from upper left to lower right, and the products in the number of discordance pairs $D$ (e.g., $b_1a_2$) go from lower left to upper right. In that a priori hypothesis, column 1 is associated with row 1; In the example above, the use of seat belt (yes, first row) is hypothesized to be associated with less severe injury (none, first column). Under this hypothesis, the resulting generalized odds is greater than 1.

**Example 1.15**   For the study above on the use of seat belts in automobiles, we have from the data shown in Table 1.13,

$$C = (75)(175 + 135 + 25) + (160)(135 + 25) + (100)(25)$$
$$= 53{,}225$$
$$D = (65)(160 + 100 + 15) + (175)(100 + 15) + (135)(15)$$
$$= 40{,}025$$

leading to generalized odds of

$$\theta = \frac{C}{D}$$
$$= \frac{53{,}225}{40{,}025}$$
$$= 1.33$$

That is, *given two people with different levels of injury*, the (generalized) odds that the more severely injured person did not wear a seat belt is 1.33. In other words, the people with the more severe injuries would be more likely than the people with less severe injuries to be those who did not use a seat belt.

The following example shows the use of generalized odds in case–control studies with an ordinal risk factor.

**TABLE 1.15**

| Age | Cases | Controls |
|-----|-------|----------|
| 14–17 | 15 | 16 |
| 18–19 | 22 | 25 |
| 20–24 | 47 | 62 |
| 25–29 | 56 | 122 |
| ≥30 | 35 | 78 |

*Example 1.16*   A case–control study of the epidemiology of preterm delivery, defined as one with less than 37 weeks of gestation, was undertaken at Yale–New Haven Hospital in Connecticut during 1977. The study population consisted of 175 mothers of singleton preterm infants and 303 mothers of singleton full-term infants. Table 1.15 gives the distribution of mother's age. We have

$$C = (15)(25 + 62 + 122 + 78) + (22)(62 + 122 + 78)$$
$$+ (47)(122 + 78) + (56)(78)$$
$$= 23{,}837$$
$$D = (16)(22 + 47 + 56 + 35) + (25)(47 + 56 + 35)$$
$$+ (62)(56 + 35) + (122)(35)$$
$$= 15{,}922$$

leading to generalized odds of

$$\theta = \frac{C}{D}$$
$$= \frac{23{,}837}{15{,}922}$$
$$= 1.50$$

This means that the odds that the younger mother has a preterm delivery is 1.5. In other words, the younger mothers would be more likely to have a preterm delivery.

The next example shows the use of generalized odds for contingency tables with more than two rows of data.

*Example 1.17*   Table 1.16 shows the results of a survey in which each subject of a sample of 282 adults was asked to indicate which of three policies he or she favored with respect to smoking in public places. We have

**TABLE 1.16**

| Highest Education Level | Policy Favored | | | Total |
| --- | --- | --- | --- | --- |
| | No Restrictions on Smoking | Smoking Allowed in Designated Areas Only | No Smoking at All | |
| Grade school | 15 | 40 | 10 | 65 |
| High school | 15 | 100 | 30 | 145 |
| College graduate | 5 | 44 | 23 | 72 |
| Total | 35 | 184 | 63 | 300 |

$$C = (15)(100 + 30 + 44 + 23) + (40)(30 + 23) + (100)(23)$$
$$= 8380$$
$$D = (5)(100 + 30 + 40 + 10) + (44)(30 + 10) + (100)(10)$$
$$= 4410$$

leading to generalized odds of

$$\theta = \frac{C}{D}$$
$$= \frac{8380}{4410}$$
$$= 1.90$$

This means that the odds that the more educated person favors more restriction for smoking in public places is 1.90. In other words, people with more education would prefer more restriction on smoking in public places.

### 1.3.4  Mantel–Haenszel Method

In most investigations we are concerned with one primary outcome, such as a disease, and are focusing on one primary (risk) factor, such as an exposure with a possible harmful effect. There are situations, however, where an investigator may want to adjust for a confounder that could influence the outcome of a statistical analysis. A *confounder*, or *confounding variable*, is a variable that may be associated with either the disease or exposure or both. For example, in Example 1.2, a case–control study was undertaken to investigate the relationship between lung cancer and employment in shipyards during World War II among male residents of coastal Georgia. In this case, smoking is a possible counfounder; it has been found to be associated with lung cancer and it may be associated with employment because construction workers are likely to be smokers. Specifically, we want to know:

**TABLE 1.17**

| Exposure | Disease Classification | | Total |
|---|---|---|---|
|  | + | − |  |
| + | $a$ | $b$ | $r_1$ |
| − | $c$ | $d$ | $r_2$ |
| Total | $c_1$ | $c_2$ | $n$ |

(a) Among smokers, whether or not shipbuilding and lung cancer are related

(b) Among nonsmokers, whether or not shipbuilding and lung cancer are related

In fact, the original data were tabulated separately for three smoking levels (nonsmoking, moderate smoking, and heavy smoking); in Example 1.2, the last two tables were combined and presented together for simplicity. Assuming that the confounder, smoking, is not an effect modifier (i.e., smoking does not alter the relationship between lung cancer and shipbuilding), however, we do not want to reach separate conclusions, one at each level of smoking. In those cases, we want to pool data for a combined decision. When both the disease and the exposure are binary, a popular method used to achieve this task is the *Mantel–Haenszel method*. This method provides one single estimate for the common odds ratio and can be summarized as follows:

1. We form $2 \times 2$ tables, one at each level of the confounder.
2. At a level of the confounder, we have the data listed in Table 1.17.

Since we assume that the confounder is not an effect modifier, the odds ratio is constant across its levels. The odds ratio at each level is estimated by $ad/bc$; the Mantel–Haenszel procedure pools data across levels of the confounder to obtain a combined estimate (some kind of weighted average of level-specific odds ratios):

$$OR_{MH} = \frac{\sum ad/n}{\sum bc/n}$$

***Example 1.18*** A case–control study was conducted to identify reasons for the exceptionally high rate of lung cancer among male residents of coastal Georgia as first presented in Example 1.2. The primary risk factor under investigation was employment in shipyards during World War II, and data are tabulated separately for three levels of smoking (Table 1.18).

There are three $2 \times 2$ tables, one for each level of smoking. We begin with the $2 \times 2$ table for nonsmokers (Table 1.19). We have for the nonsmokers

**TABLE 1.18**

| Smoking | Shipbuilding | Cases | Controls |
|---|---|---|---|
| No | Yes | 11 | 35 |
| | No | 50 | 203 |
| Moderate | Yes | 70 | 42 |
| | No | 217 | 220 |
| Heavy | Yes | 14 | 3 |
| | No | 96 | 50 |

**TABLE 1.19**

| Shipbuilding | Cases | Controls | Total |
|---|---|---|---|
| Yes | 11 $(a)$ | 35 $(b)$ | 46 $(r_1)$ |
| No | 50 $(c)$ | 203 $(d)$ | 253 $(r_2)$ |
| Total | 61 $(c_1)$ | 238 $(c_2)$ | 299 $(n)$ |

$$\frac{ad}{n} = \frac{(11)(203)}{299}$$

$$= 7.47$$

$$\frac{bc}{n} = \frac{(35)(50)}{299}$$

$$= 5.85$$

The process is repeated for each of the other two smoking levels. For moderate smokers

$$\frac{ad}{n} = \frac{(70)(220)}{549}$$

$$= 28.05$$

$$\frac{bc}{n} = \frac{(42)(217)}{549}$$

$$= 16.60$$

and for heavy smokers

$$\frac{ad}{n} = \frac{(14)(50)}{163}$$

$$= 4.29$$

$$\frac{bc}{n} = \frac{(3)(96)}{163}$$

$$= 1.77$$

These results are combined to obtain a combined estimate for the common odds ratio:

$$\text{OR}_{\text{MH}} = \frac{7.47 + 28.05 + 4.28}{5.85 + 16.60 + 1.77}$$

$$= 1.64$$

This combined estimate of the odds ratio, 1.64, represents an approximate increase of 64% in lung cancer risk for those employed in the shipbuilding industry.

The following is a similar example aiming at the possible effects of oral contraceptive use on myocardial infarction. The presentation has been shortened, keeping only key figures.

*Example 1.19* A case–control study was conducted to investigate the relationship between myocardial infarction (MI) and oral contraceptive use (OC). The data, stratified by cigarette smoking, are listed in Table 1.20. Application of the Mantel–Haenszel procedure yields the results shown in Table 1.21. The combined odds ratio estimate is

$$\text{OR}_{\text{MH}} = \frac{3.57 + 18.84}{2.09 + 12.54}$$

$$= 1.53$$

representing an approximate increase of 53% in myocardial infarction risk for oral contraceptive users.

**TABLE 1.20**

| Smoking | OC Use | Cases | Controls |
|---------|--------|-------|----------|
| No      | Yes    | 4     | 52       |
|         | No     | 34    | 754      |
| Yes     | Yes    | 25    | 83       |
|         | No     | 171   | 853      |

**TABLE 1.21**

|        | Smoking | |
|--------|---------|-----|
|        | No      | Yes |
| $ad/n$ | 3.57    | 18.84 |
| $bc/n$ | 2.09    | 12.54 |

### 1.3.5   Standardized Mortality Ratio

In a cohort study, the follow-up death rates are calculated and used to describe the mortality experience of the cohort under investigation. However, the observed mortality of the cohort is often compared with that expected from the death rates of the national population (used as *standard* or *baseline*). The basis of this method is the comparison of the observed number of deaths, $d$, from the cohort with the mortality that would have been expected if the group had experienced death rates similar to those of the national population of which the cohort is a part. Let $e$ denote the expected number of deaths; then the comparison is based on the following ratio, called the *standardized mortality ratio:*

$$\text{SMR} = \frac{d}{e}$$

The expected number of deaths is calculated using published national life tables, and the calculation can be approximated as follows:

$$e \simeq \lambda T$$

where $T$ is the total follow-up time (person-years) from the cohort and $\lambda$ the annual death rate (per person) from the referenced population. Of course, the annual death rate of the referenced population changes with age. Therefore, what we actually do in research is more complicated, although based on the same idea. First, we subdivide the cohort into many age groups, then calculate the product $\lambda T$ for each age group using the correct age-specific rate for that group, and add up the results.

*Example 1.20*   Some 7000 British workers exposed to vinyl chloride monomer were followed for several years to determine whether their mortality experience differed from those of the general population. The data in Table 1.22 are for deaths from cancers and are tabulated separately for four groups based on years since entering the industry. This data display shows some interesting features:

1.  For the group with 1–4 years since entering the industry, we have a death rate that is substantially less than that of the general population

**TABLE 1.22**

| Deaths from Cancers | Years Since Entering the Industry | | | | |
|---|---|---|---|---|---|
| | 1–4 | 5–9 | 10–14 | 15+ | Total |
| Observed | 9 | 15 | 23 | 68 | 115 |
| Expected | 20.3 | 21.3 | 24.5 | 60.8 | 126.8 |
| SMR (%) | 44.5 | 70.6 | 94.0 | 111.8 | 90.7 |

(SMR $= 0.445$ or $44.5\%$). This phenomenon, known as the *healthy worker effect*, is probably a consequence of a selection factor whereby workers are necessarily in better health (than people in the general population) at the time of their entry into the workforce.

2. We see an attenuation of the healthy worker effect (i.e., a decreasing trend) with the passage of time, so that the cancer death rates show a slight excess after 15 years. (Vinyl chloride exposures are known to induce a rare form of liver cancer and to increase rates of brain cancer.)

Taking the ratio of two standardized mortality ratios is another way of expressing relative risk. For example, the relative risk of the 15+ years group is 1.58 times the risk of the risk of the 5–9 years group, since the ratio of the two corresponding mortality ratios is

$$\frac{111.8}{70.6} = 1.58$$

Similarly, the risk of the 15+ years group is 2.51 times the risk of the 1–4 years group because the ratio of the two corresponding mortality ratios is

$$\frac{111.8}{44.5} = 2.51$$

## 1.4 NOTES ON COMPUTATIONS

Much of this book is concerned with arithmetic procedures for data analysis, some with rather complicated formulas. In many biomedical investigations, particularly those involving large quantities of data, the analysis (e.g., regression analysis of Chapter 8) gives rise to difficulties in computational implementation. In these investigations it will be necessary to use statistical software specially designed to do these jobs. Most of the calculations described in this book can be carried out readily using statistical packages, and any student or practitioner of data analysis will find the use of such packages essential.

Methods of survival analysis (first half of Chapter 11), for example, and nonparametric methods (Sections 2.4 and 7.4), and of multiple regression analysis (Section 8.2) may best be handled by a specialized package such as SAS; in these sections are included in our examples where they were used. However, students and investigators contemplating use of one of these commercial programs should read the specifications for each program before choosing the options necessary or suitable for any particular procedure. But these sections are exceptions, many calculations described in this book can be carried out readily using Microsoft's Excel, popular software available in every personal computer. Notes on the use of Excel are included in separate sections at the end of each chapter.

A *worksheet* or spreadsheet is a (blank) sheet where you do your work. An Excel *file* holds a stack of worksheets in a *workbook*. You can *name* a sheet, put data on it and *save*; later, *open* and use it. You can *move* or *size* your windows by *dragging* the borders. You can also scroll up and down, or left and right, through an Excel worksheet using the scroll bars on the right side and at the bottom.

An Excel worksheet consists of *grid lines* forming *columns* and *rows*; columns are *lettered* and rows are *numbered*. The intersection of each column and row is a box called a *cell*. Every cell has an *address*, also called a *cell reference*; to refer to a cell, enter the column letter followed by the row number. For example, *the intersection of column C and row 3 is cell C3*. Cells hold numbers, text, or formulas. To refer to a range of cells, enter the cell in the upper left corner of the range followed by a colon (:) and then the lower right corner of the range. For example, A1:B20 refers to the first 20 rows in both columns A and B.

You can click a cell to make it *active* (for use); an active cell is where you enter or edit your data, and it's identified by a *heavy border*. You can also *define or select a range* by left-clicking on the upper leftmost cell and dragging the mouse to the lower rightmost cell. To move around inside a selected range, press Tab or Enter to move forward one cell at a time.

Excel is software designed to handle numbers; so get a project and start typing. Conventionally, files for data analysis use rows for subjects and columns for factors. For example, you conduct a survey using a 10-item questionaire and receive returns from 75 people; your data require a file with 75 rows and 10 columns—not counting labels for columns (factors' names) and rows (subjects' ID). If you made an error, it can be fixed (hit the Del key, which deletes the cell contents). You can change your mind again, deleting the delete, by clicking the *Undo button* (reverse curved arrow). Remember, you can widen your columns by double-clicking their right borders.

The formula bar (near the top, next to an = sign) is a common way to provide the content of an active cell. Excel executes a formula from left to right and performs multiplication (∗) and division (/) before addition (+) and subtraction (−). Parenthenses can/should be used to change the order of calculations. To use formulas (e.g., for data transformation), do it in one of two ways: (1) click the cell you want to fill, then type an = sign followed by the formula in the formula bar (e.g., click C5, then type = A5 + B5); or (2) click the cell you want to fill, then click the *paste function icon*, f∗, which will give you—in a box—a list of Excel functions available for your use.

The *cut and paste* procedure greatly simplifies the typing necessary to form a chart or table or to write numerous formulas. The procedure involves highlighting the cells that contain the information you want to copy, clicking on the cut button (scissors icon) or copy button (two-page icon), selecting the cell(s) in which the information is to be placed, and clicking on the paste button (clipboard and page icon).

The *select and drag* is very efficient for *data transformation*. Suppose that you have the the weight and height of 15 men (weights are in C6:C20 and heights are in D6:D20) and you need their body mass index. You can use the formula bar, for example, clicking E6 and typing $= C6/(D6 \wedge 2)$. The content of E6 now has the body mass index for the first man in your sample, but you do not have to repeat this process 15 times. Notice that when you click on E6, there is a little box in the lower right corner of the cell boundary. If you move the mouse over this box, the cursor changes to a smaller plus sign. If you click on this box, you can then drag the mouse over the remaining cells, and when you release the button, the cells will be filled.

**Bar and Pie Charts**   Forming a bar chart or pie to display proportions is a very simple task. Just click any blank cell to start; you'll be able to move your chart to any location when done. With data ready, click the *ChartWizard* icon (the one with multiple colored bars on the *standard toolbar* near the top). A box appears with choices including bar chart, pie chart, and line chart; the list is on the left side. Choose your chart type and follow instructions. There are many choices, including three dimensions. You can put data and charts side by side for an impressive presentation.

**Rate Standardization**   This is a good problem to practice with Excel: Use it *as a calculator*. Recall this example:

- Florida's rate $= 1085.3$
- Alaska's rate $= 396.8$
- If Florida has Alaska's population (see Table 1.9), you can:
  - Use *formula* to calculate the first number expected.
  - Use *drag and fill* to obtain other numbers expected.
  - Select the last column, then click *Autosum icon* $(\sum)$ to obtain the total number of deaths expected.

**Forming 2 × 2 Tables**   Recall that in a data file you have one row for each subject and one column for each variable. Suppose that *two* of those variables are categorical, say *binary*, and you want to *form a* $2 \times 2$ *table* so you can study their relationship: for example, to calculate an *odds ratio*.

- *Step 0:* Create a (dummy factor), call it, say, *fake*, and fill up that column with "*1*" (you can enter "1" for the first subject, then *select and drag*).
- *Step 1: Activate* a cell (by clicking it), then click *Data* (on the bar above the standard toolbar, near the end); when a box appears, choose *Pivot-Table Report*. Click "*next*" (to indicate that data are here, in Excel, then *highlight* the area containing your data (including variable names on the first row—use the mouse; or you could identify the *range of cells*—say, C5:E28) as a response to a question on *range*. Then click "*next*" to bring in the *PivotTable Wizard*, which shows two groups of things:

(a) A *frame* for a $2 \times 2$ *table* with places identified as *row, column,* and *data*

(b) Names of the factors you chose: say, *exposure, disease,* and *fake*

- *Step 2:* Drag exposure to merge with row (or column), *drag* disease to merge with column (or row), and *drag* fake to merge with data. Then click *Finish*; a $2 \times 2$ *table* appears in the active cell you identified, complete with *cell frequencies, row and column totals,* and *grand total.*

*Note:* If you have another *factor* in addition to exposure and disease available in the data set, even a column for *names* or *IDs*, there is no need to create the *dummy factor*. Complete step 1; then, in step 2, *drag that third factor*, say *ID*, to merge with "*data*" in the frame shown by the *PivotTable Wizard*; it appears as *sum ID*. Click on that item, then choose *count* (to replace *sum*).

## EXERCISES

**1.1**  Self-reported injuries among left- and right-handed people were compared in a survey of 1896 college students in British Columbia, Canada. Of the 180 left-handed students, 93 reported at least one injury, and 619 of the 1716 right-handed students reported at least one injury in the same period. Arrange the data in a $2 \times 2$ table and calculate the proportion of people with at least one injury during the period of observation for each group.

**1.2**  A study was conducted to evaluate the hypothesis that tea consumption and premenstrual syndrome are associated. A group of 188 nursing students and 64 tea factory workers were given questionnaires. The prevalence of premenstrual syndrome was 39% among the nursing students and 77% among the tea factory workers. How many people in each group have premenstrual syndrome? Arrange the data in a $2 \times 2$ table.

**1.3**  The relationship between prior condom use and tubal pregnancy was assessed in a population-based case–control study at Group Health Cooperative of Puget Sound during 1981–1986. The results are shown in Table E1.3. Compute the group size and the proportion of subjects in each group who never used condoms.

**TABLE E1.3**

| Condom Use | Cases | Controls |
|---|---|---|
| Never | 176 | 488 |
| Ever | 51 | 186 |

**1.4**  Epidemic keratoconjunctivitis (EKC) or "shipyard eye" is an acute infectious disease of the eye. A case of EKC is defined as an illness:

- Consisting of redness, tearing, and pain in one or both eyes for more than three days' duration
- Diagnosed as EKC by an ophthalmologist

In late October 1977, one (physician A) of the two ophthalmologists providing the majority of specialized eye care to the residents of a central Georgia county (population 45,000) saw a 27-year-old nurse who had returned from a vacation in Korea with severe EKC. She received symptomatic therapy and was warned that her eye infection could spread to others; nevertheless, numerous cases of an illness similar to hers soon occurred in the patients and staff of the nursing home (nursing home A) where she worked (these people came to physician A for diagnosis and treatment). Table E1.4 provides the exposure history of 22 persons with EKC between October 27, 1977 and January 13, 1978 (when the outbreak stopped after proper control techniques were initiated). Nursing home B, included in this table, is the only other area chronic-care facility. Compute and compare the proportions of cases from the two nursing homes. What would be your conclusion?

**TABLE E1.4**

| Exposure Cohort | Number Exposed | Number of Cases |
|---|---|---|
| Nursing home A | 64 | 16 |
| Nursing home B | 238 | 6 |

**1.5**  In August 1976, tuberculosis was diagnosed in a high school student (index case) in Corinth, Mississippi. Subsequently, laboratory studies revealed that the student's disease was caused by drug-resistant tubercule bacilli. An epidemiologic investigation was conducted at the high school. Table E1.5 gives the rate of positive tuberculin reactions, determined for various groups of students according to degree of exposure to the index case.

**TABLE E1.5**

| Exposure Level | Number Tested | Number Positive |
|---|---|---|
| High | 129 | 63 |
| Low | 325 | 36 |

(a) Compute and compare the proportions of positive cases for the two exposure levels. What would be your conclusion?

(b) Calculate the odds ratio associated with high exposure. Does this result support your conclusion in part (a)?

**1.6** Consider the data taken from a study that attempts to determine whether the use of electronic fetal monitoring (EFM) during labor affects the frequency of cesarean section deliveries. Of the 5824 infants included in the study, 2850 were electronically monitored and 2974 were not. The outcomes are listed in Table E1.6.

**TABLE E1.6**

| | EFM Exposure | | |
| Cesarean Delivery | Yes | No | Total |
| --- | --- | --- | --- |
| Yes | 358 | 229 | 587 |
| No | 2492 | 2745 | 5237 |
| Total | 2850 | 2974 | 5824 |

(a) Compute and compare the proportions of cesarean delivery for the two exposure groups. What would be your conclusion?

(b) Calculate the odds ratio associated with EFM exposure. Does this result support your conclusion in part (a)?

**1.7** A study was conducted to investigate the effectiveness of bicycle safety helmets in preventing head injury. The data consist of a random sample of 793 persons who were involved in bicycle accidents during a one-year period (Table E1.7).

**TABLE E1.7**

| | Wearing Helmet | | |
| Head Injury | Yes | No | Total |
| --- | --- | --- | --- |
| Yes | 17 | 218 | 235 |
| No | 130 | 428 | 558 |
| Total | 147 | 646 | 793 |

(a) Compute and compare the proportions of head injury for the group with helmets versus the group without helmets. What would be your conclusion?

(b) Calculate the odds ratio associated with not using helmet. Does this result support your conclusion in part (a)?

**1.8** A case–control study was conducted in Auckland, New Zealand, to investigate the effects among regular drinkers of alcohol consumption on both nonfatal myocardial infarction and coronary death in the 24 hours after drinking. Data were tabulated separately for men and women (Table E1.8).

**TABLE E1.8**

| | Drink in the Last 24 Hours | Myocardial Infarction | | Coronary Death | |
|---|---|---|---|---|---|
| | | Controls | Cases | Controls | Cases |
| Men | No | 197 | 142 | 135 | 103 |
| | Yes | 201 | 136 | 159 | 69 |
| Women | No | 144 | 41 | 89 | 12 |
| | Yes | 122 | 19 | 76 | 4 |

**(a)** Refer to the myocardial infarction data. Calculate separately for men and women the odds ratio associated with drinking.

**(b)** Compare the two odds ratios in part (a). When the difference is confirmed properly, we have an effect modification.

**(c)** Refer to coronary deaths. Calculte separately for men and women the odds ratio associated with drinking.

**(d)** Compare the two odds ratios in part (c). When the difference is confirmed properly, we have an effect modification.

**1.9** Data taken from a study to investigate the effects of smoking on cervical cancer are stratified by the number of sexual partners (Table E1.9).

**TABLE E1.9**

| Number of Partners | Smoking | Cancer | |
|---|---|---|---|
| | | Yes | No |
| Zero or one | Yes | 12 | 21 |
| | No | 25 | 118 |
| Two or more | Yes | 96 | 142 |
| | No | 92 | 150 |

**(a)** Calculate the odds ratio associated with smoking separately for the two groups, those with zero or one partner and those with two or more partners.

**(b)** Compare the two odds ratios in part (a). When the difference is confirmed properly, we have an effect modification.

**(c)** Assuming that the odds ratios for the two groups, those with zero or one partner and those with two or more partners, are equal (in other words, the number of partners is not an effect modifier), calculate the Mantel–Haenszel estimate of this common odds ratio.

1.10    Table E1.10 provides the proportions of currently married women having an unplanned pregnancy. (Data are tabulated for several different methods of contraception.) Display these proportions in a bar chart.

**TABLE E1.10**

| Method of Contraception | Proportion with Unplanned Pregnancy |
| --- | --- |
| None | 0.431 |
| Diaphragm | 0.149 |
| Condom | 0.106 |
| IUD | 0.071 |
| Pill | 0.037 |

1.11    Table E1.11 summarizes the coronary heart disease (CHD) and lung cancer mortality rates per 1000 person-years by number of cigarettes smoked per day at baseline for men participating in the MRFIT (Multiple Risk Factor Intervention Trial, a very large controlled clinical trial focusing on the relationship between smoking and cardiovascular diseases). For each cause of death, display the rates in a bar chart.

**TABLE E1.11**

|  | Total | CHD Deaths | | Lung Cancer Deaths | |
| --- | --- | --- | --- | --- | --- |
|  |  | $N$ | Rate/1000 yr | $N$ | Rate/1000 yr |
| Never-smokers | 1859 | 44 | 2.22 | 0 | 0 |
| Ex-smokers | 2813 | 73 | 2.44 | 13 | 0.43 |
| Smokers |  |  |  |  |  |
| 1–19 cig./day | 856 | 23 | 2.56 | 2 | 0.22 |
| 20–39 cig./day | 3747 | 173 | 4.45 | 50 | 1.29 |
| ≥40 cig./day | 3591 | 115 | 3.08 | 54 | 1.45 |

1.12    Table E1.12 provides data taken from a study on the association between race and use of medical care by adults experiencing chest pain in the past year. Display the proportions of the three response categories for each group, blacks and whites, in a separate pie chart.

**TABLE E1.12**

| Response | Black | White |
| --- | --- | --- |
| MD seen in past year | 35 | 67 |
| MD seen, not in past year | 45 | 38 |
| MD never seen | 78 | 39 |
| Total | 158 | 144 |

**1.13**  The frequency distribution for the number of cases of pediatric AIDS between 1983 and 1989 is shown in Table E1.13. Display the trend of numbers of cases using a line graph.

**TABLE E1.13**

| Year | Number of Cases | Year | Number of Cases |
|------|-----------------|------|-----------------|
| 1983 | 122             | 1987 | 1,412           |
| 1984 | 250             | 1988 | 2,811           |
| 1985 | 455             | 1989 | 3,098           |
| 1986 | 848             |      |                 |

**1.14**  A study was conducted to investigate the changes between 1973 and 1985 in women's use of three preventive health services. The data were obtained from the National Health Survey; women were divided into subgroups according to age and race. The percentages of women receiving a breast exam within the past two years are given in Table E1.14. Separately for each group, blacks and whites, display the proportions of women receiving a breast exam within the past two years in a bar chart so as to show the relationship between the examination rate and age. Mark the midpoint of each age group on the horizontal axis and display the same data using a line graph.

**TABLE E1.14**

| Age and Race | Breast Exam within Past 2 Years | |
|--------------|------|------|
|              | 1973 | 1985 |
| Total        | 65.5 | 69.6 |
|   Black | 61.7 | 74.8 |
|   White | 65.9 | 69.0 |
| 20–39 years  | 77.5 | 77.9 |
|   Black | 77.0 | 83.9 |
|   White | 77.6 | 77.0 |
| 40–59 years  | 62.1 | 66.0 |
|   Black | 54.8 | 67.9 |
|   White | 62.9 | 65.7 |
| 60–79 years  | 44.3 | 56.2 |
|   Black | 39.1 | 64.5 |
|   White | 44.7 | 55.4 |

**1.15**  Consider the data shown in Table E1.15. Calculate the sensitivity and specificity of x-ray as a screening test for tuberculosis.

**TABLE E1.15**

|  | Tuberculosis | | |
| --- | --- | --- | --- |
| X-ray | No | Yes | Total |
| Negative | 1739 | 8 | 1747 |
| Positive | 51 | 22 | 73 |
| Total | 1790 | 30 | 1820 |

**1.16** Sera from a T-lymphotropic virus type (HTLV-I) risk group (prostitute women) were tested with two commercial "research" enzyme-linked immunoabsorbent assays (EIA) for HTLV-I antibodies. These results were compared with a gold standard, and outcomes are shown in Table E1.16. Calculate and compare the sensitivity and specificity of these two EIAs.

**TABLE E1.16**

|  | Dupont's EIA | | Cellular Product's EIA | |
| --- | --- | --- | --- | --- |
| True | Positive | Negative | Positive | Negative |
| Positive | 15 | 1 | 16 | 0 |
| Negative | 2 | 164 | 7 | 179 |

**1.17** Table E1.17 provides the number of deaths for several leading causes among Minnesota residents for the year 1991.

**TABLE E1.17**

| Cause of Death | Number of Deaths | Rate per 100,000 Population |
| --- | --- | --- |
| Heart disease | 10,382 | 294.5 |
| Cancer | 8,299 | ? |
| Cerebrovascular disease | 2,830 | ? |
| Accidents | 1,381 | ? |
| Other causes | 11,476 | ? |
| Total | 34,368 | ? |

**(a)** Calculate the percent of total deaths from each cause, and display the results in a pie chart.

**(b)** From the death rate (per 100,000 population) for heart disease, calculate the population for Minnesota for the year 1991.

(c) From the result of part (b), fill in the missing death rates (per 100,000 population) in the table.

**1.18** The survey described in Example 1.1, continued in Section 1.1.1, provided percentages of boys from various ethnic groups who use tobacco at least weekly. Display these proportions in a bar chart similar to the one in Figure 1.2.

**1.19** A case–control study was conducted relating to the epidemiology of breast cancer and the possible involvement of dietary fats, vitamins, and other nutrients. It included 2024 breast cancer cases admitted to Roswell Park Memorial Institute, Erie County, New York, from 1958 to 1965. A control group of 1463 was chosen from patients having no neoplasms and no pathology of gastrointestinal or reproductive systems. The primary factors being investigated were vitamins A and E (measured in international units per month). Data for 1500 women over 54 years of age are given in Table E1.19. Calculate the odds ratio associated with a decrease (exposure is low consumption) in ingestion of foods containing vitamin A.

**TABLE E1.19**

| Vitamin A (IU/month) | Cases | Controls |
| --- | --- | --- |
| ≤150,500 | 893 | 392 |
| >150,500 | 132 | 83 |
| Total | 1025 | 475 |

**1.20** Refer to the data set in Table 1.1 (see Example 1.2).
   (a) Calculate the odds ratio associated with employment in shipyards for nonsmokers.
   (b) Calculate the same odds ratio for smokers.
   (c) Compare the results of parts (a) and (b). When the difference is confirmed properly, we have a three-term interaction or effect modification, where smoking alters the effect of employment in shipyards as a risk for lung cancer.
   (d) Assuming that the odds ratios for the two groups, nonsmokers and smokers, are equal (in other words, smoking is not an effect modifier), calculate the Mantel–Haenszel estimate of this common odds ratio.

**1.21** Although cervical cancer is not a major cause of death among American women, it has been suggested that virtually all such deaths are preventable. In an effort to find out who is being screened for the disease, data from the 1973 National Health Interview (a sample of the U.S. population) were used to examine the relationship between Pap testing and some socioeconomic factors. Table E1.21 provides the percentages of women who reported never having had a Pap test. (These are from metropolitan areas.)

**TABLE E1.21**

| Age and Income | White | Black |
|---|---|---|
| 25–44 | | |
|    Poor | 13.0 | 14.2 |
|    Nonpoor | 5.9 | 6.3 |
| 45–64 | | |
|    Poor | 30.2 | 33.3 |
|    Nonpoor | 13.2 | 23.3 |
| 65 and over | | |
|    Poor | 47.4 | 51.5 |
|    Nonpoor | 36.9 | 47.4 |

**(a)** Calculate the odds ratios associated with race (black versus white) among

   **(i)** 25–44 nonpoor

   **(ii)** 45–64 nonpoor

   **(iii)** 65+ nonpoor

Briefly discuss a possible effect modification, if any.

**(b)** Calculate the odds ratios associated with income (poor versus nonpoor) among

   **(i)** 25–44 black

   **(ii)** 45–64 black

   **(iii)** 65+ black

Briefly discuss a possible effect modification, if any.

**(c)** Calculate the odds ratios associated with race (black versus white) among

   **(i)** 65+ poor

   **(ii)** 65+ nonpoor

Briefly discuss a possible effect modification.

**1.22** Since incidence rates of most cancers rise with age, this must always be considered a confounder. Stratified data for an unmatched case–control study are given in Table E1.22. The disease was esophageal cancer among men and the risk factor was alcohol consumption.

   **(a)** Calculate separately for the three age groups the odds ratio associated with *high* alcohol consumption.

**TABLE E1.22**

| Age | Daily Alcohol Consumption | |
|---|---|---|
| | 80+ g | 0–79 g |
| 25–44 | | |
| Cases | 5 | 5 |
| Controls | 35 | 270 |
| 45–64 | | |
| Cases | 67 | 55 |
| Controls | 56 | 277 |
| 65+ | | |
| Cases | 24 | 44 |
| Controls | 18 | 129 |

**(b)** Compare the three odds ratios in part (a). When the difference is confirmed properly, we have an effect modification.

**(c)** Assuming that the odds ratios for the three age groups are equal (in other words, age is not an effect modifier), calculate the Mantel–Haenszel estimate of this common odds ratio.

**1.23** Postmenopausal women who develop endometrial cancer are on the whole heavier than women who do not develop the disease. One possible explanation is that heavy women are more exposed to endogenous estrogens which are produced in postmenopausal women by conversion of steroid precursors to active estrogens in peripheral fat. In the face of varying levels of endogenous estrogen production, one might ask whether the carcinogenic potential of exogenous estrogens would be the same in all women. A case–control study has been conducted to examine the relation among weight, replacement estrogen therapy, and endometrial cancer. The results are shown in Table E1.23.

**TABLE E1.23**

| Weight (kg) | Estrogen Replacement | |
|---|---|---|
| | Yes | No |
| <57 | | |
| Cases | 20 | 12 |
| Controls | 61 | 183 |
| 57–75 | | |
| Cases | 37 | 45 |
| Controls | 113 | 378 |
| >75 | | |
| Cases | 9 | 42 |
| Controls | 23 | 140 |

(a) Calculate separately for the three weight groups the odds ratio associated with estrogen replacement.

(b) Compare the three odds ratios in part (a). When the difference is confirmed properly, we have an effect modification.

(c) Assuming that the odds ratios for the three weight groups are equal (in other words, weight is not an effect modifier), calculate the Mantel–Haenszel estimate of this common odds ratio.

**1.24** The role of menstrual and reproductive factors in the epidemiology of breast cancer has been reassessed using pooled data from three large case–control studies of breast cancer from several Italian regions (Negri et al., 1988). In Table E1.24 data are summarized for age at menopause and age at first live birth.

**TABLE E1.24**

|                          | Cases | Controls |
|--------------------------|-------|----------|
| Age at first live birth  |       |          |
| <22                      | 621   | 898      |
| 22–24                    | 795   | 909      |
| 25–27                    | 791   | 769      |
| ≥28                      | 1043  | 775      |
| Age at menopause         |       |          |
| <45                      | 459   | 543      |
| 45–49                    | 749   | 803      |
| ≥50                      | 1378  | 1167     |

(a) For each of the two factors (age at first live birth and age at menopause), choose the lowest level as the baseline and calculate the odds ratio associated with each other level.

(b) For each of the two factors (age at first live birth and age at menopause), calculate the generalized odds and give your interpretation. How does this result compare with those in part (a)?

**1.25** Risk factors of gallstone disease were investigated in male self-defense officials who received, between October 1986 and December 1990, a retirement health examination at the Self-Defense Forces Fukuoka Hospital, Fukuoka, Japan. Some of the data are shown in Table E1.25.

(a) For each of the three factors (smoking, alcohol, and body mass index), rearrange the data into a 3 × 2 table; the other column is for those without gallstones.

(b) For each of the three 3 × 2 tables in part (a), choose the lowest level as the baseline and calculate the odds ratio associated with each other level.

(c) For each of the three $3 \times 2$ tables in part (a), calculate the generalized odds and give your interpretation. How does this result compare with those in part (b)?

**TABLE E1.25**

| | Number of Men Surveyed | |
|---|---|---|
| Factor | Total | With Number Gallstones |
| Smoking | | |
| Never | 621 | 11 |
| Past | 776 | 17 |
| Current | 1342 | 33 |
| Alcohol | | |
| Never | 447 | 11 |
| Past | 113 | 3 |
| Current | 2179 | 47 |
| Body mass index (kg/m²) | | |
| <22.5 | 719 | 13 |
| 22.5–24.9 | 1301 | 30 |
| ≥25.0 | 719 | 18 |

**1.26**  Data were collected from 2197 white ovarian cancer patients and 8893 white controls in 12 different U.S. case–control studies conducted by various investigators in the period 1956–1986. These were used to evaluate the relationship of invasive epithelial ovarian cancer to reproductive and menstrual characteristics, exogenous estrogen use, and prior pelvic surgeries. Data related to unprotected intercourse and to history of infertility are shown in Table E1.26.

**TABLE E1.26**

| | Cases | Controls |
|---|---|---|
| Duration of unprotected intercourse (years) | | |
| <2 | 237 | 477 |
| 2–9 | 166 | 354 |
| 10–14 | 47 | 91 |
| ≥15 | 133 | 174 |
| History of infertility | | |
| No | 526 | 966 |
| Yes | | |
| No drug use | 76 | 124 |
| Drug use | 20 | 11 |

(a) For each of the two factors (duration of unprotected intercourse and history of infertility, treating the latter as ordinal: no history, history but no drug use, and history with drug use), choose the lowest level as the baseline, and calculate the odds ratio associated with each other level.

(b) For each of the two factors (duration of unprotected intercourse and history of infertility, treating the latter as ordinal: no history, history but no drug use, and history with drug use), calculate the generalized odds and give your interpretation. How does this result compare with those in part (a)?

1.27 Postneonatal mortality due to respiratory illnesses is known to be inversely related to maternal age, but the role of young motherhood as a risk factor for respiratory morbidity in infants has not been explored thoroughly. A study was conducted in Tucson, Arizona, aimed at the incidence of lower respiratory tract illnesses during the first year of life. In this study, over 1200 infants were enrolled at birth between 1980 and 1984. The data shown in Table E1.27 are concerned with wheezing lower respiratory tract illnesses (wheezing LRI: no/yes).

**TABLE E1.27**

| Maternal Age (years) | Boys | | Girls | |
|---|---|---|---|---|
| | No | Yes | No | Yes |
| <21 | 19 | 8 | 20 | 7 |
| 21–25 | 98 | 40 | 128 | 36 |
| 26–30 | 160 | 45 | 148 | 42 |
| >30 | 110 | 20 | 116 | 25 |

(a) For each of the two groups, boys and girls, choose the lowest age group as the baseline and calculate the odds ratio associated with each age group.

(b) For each of the two groups, boys and girls, calculate the generalized odds and give your interpretation. How does this result compare with those in part (a)?

(c) Compare the two generalized odds in part (b) and draw your conclusion.

1.28 An important characteristic of glaucoma, an eye disease, is the presence of classical visual field loss. Tonometry is a common form of glaucoma screening, whereas, for example, an eye is classified as positive if it has an intraocular pressure of 21 mmHg or higher at a single reading. Given the data shown in Table E1.28, calculate the sensitivity and specificity of this screening test.

**TABLE E1.28**

| Field Loss | Test Result | | Total |
|---|---|---|---|
| | Positive | Negative | |
| Yes | 13 | 7 | 20 |
| No | 413 | 4567 | 4980 |

**1.29** From the information in the news report quoted in Example 1.11, calculate:

(a) The number of new AIDS cases for the years 1987 and 1986.

(b) The number of cases of AIDS transmission from mothers to newborns for 1988.

**1.30** In an effort to provide a complete analysis of the survival of patients with end-stage renal disease (ESRD), data were collected for a sample that included 929 patients who initiated hemodialysis for the first time at the Regional Disease Program in Minneapolis, Minnesota between January 1, 1976 and June 30, 1982; all patients were followed until December 31, 1982. Of these 929 patients, 257 are diabetics; among the 672 nondiabetics, 386 are classified as low risk (without comorbidities such as arteriosclerotic heart disease, peripheral vascular disease, chronic obstructive pulmonary, and cancer). For the low-risk ESRD patients, we have the follow-up data shown in Table E1.30 (in addition to those in Example 1.12). Compute the follow-up death rate for each age group and the relative risk for the group "70+" versus "51–60."

**TABLE E1.30**

| Age (years) | Deaths | Treatment Months |
|---|---|---|
| 21–30 | 4 | 1012 |
| 31–40 | 7 | 1387 |
| 41–50 | 20 | 1706 |
| 51–60 | 24 | 2448 |
| 61–70 | 21 | 2060 |
| 70+ | 17 | 846 |

**1.31** Mortality data for the state of Georgia for the year 1977 are given in Table E1.31a.

(a) From this mortality table, calculate the crude death rate for the Georgia.

(b) From Table E1.31 and the data mortality data for Alaska and Florida for the year 1977 (Table 1.7), calculate the age-adjusted death rate for Georgia and compare to those for Alaska and Florida, the

**TABLE E1.31a**

| Age Group | Deaths | Population |
|-----------|--------|------------|
| 0–4   | 2,483  | 424,600   |
| 5–19  | 1,818  | 1,818,000 |
| 20–44 | 3,656  | 1,126,500 |
| 45–64 | 12,424 | 870,800   |
| 65+   | 21,405 | 360,800   |

**TABLE E1.31b**

| Age Group | Population |
|-----------|-----------|
| 0–4   | 84,416  |
| 5–19  | 294,353 |
| 20–44 | 316,744 |
| 45–64 | 205,745 |
| 65+   | 98,742  |
| Total | 1,000,000 |

U.S. population given in Table 1.8, reproduced here in Table E1.31b, being used as the standard.

**(c)** Calculate the age-adjusted death rate for Georgia with the Alaska population serving as the standard population. How does this adjusted death rate compare to the crude death rate of Alaska?

**1.32** Refer to the same set of mortality data as in Exercise 1.31. Calculate and compare the age-adjusted death rates for the states of Alaska and Florida, with the Georgia population serving as the standard population. How do mortality in these two states compare to mortality in the state of Georgia?

**1.33** Some 7000 British workers exposed to vinyl chloride monomer were followed for several years to determine whether their mortality experience differed from those of the general population. In addition to data for deaths from cancers as seen in Example 1.20 (Table 1.23), the study also provided the data shown in Table E1.33 for deaths due to circula-

**TABLE E1.33**

| Deaths | \multicolumn{4}{c}{Years Since Entering the Industry} | Total |
|--------|-----|-----|-------|-----|-------|
|        | 1–4 | 5–9 | 10–14 | 15+ |       |
| Observed | 7    | 25   | 38   | 110   | 180   |
| Expected | 32.5 | 35.6 | 44.9 | 121.3 | 234.1 |

tory disease. Calculate the SMRs for each subgroup and the relative risk for group "15+" versus group "1–4."

**1.34**  A long-term follow-up study of diabetes has been conducted among Pima Indian residents of the Gila River Indian Community of Arizona since 1965. Subjects of this study at least 5 years old and of at least half Pima ancestry were examined approximately every two years; examinations included measurements of height and weight and a number of other factors. Table E1.34 relates diabetes incidence rate (new cases/1000 person-years) to body mass index (a measure of obesity defined as weight/height$^2$). Display these rates by means of a bar chart.

**TABLE E1.34**

| Body Mass Index | Incidence Rate |
|---|---|
| <20 | 0.8 |
| 20–25 | 10.9 |
| 25–30 | 17.3 |
| 30–35 | 32.6 |
| 35–40 | 48.5 |
| ≥40 | 72.2 |

**1.35**  In the course of selecting controls for a study to evaluate the effect of caffeine-containing coffee on the risk of myocardial infarction among women 30–49 years of age, a study noted appreciable differences in coffee consumption among hospital patients admitted for illnesses not known to be related to coffee use. Among potential controls, the coffee consumption of patients who had been admitted to hospital by conditions having an acute onset (such as fractures) was compared to that of patients admitted for chronic disorders (Table E1.35).

**TABLE E1.35**

| Admission by: | Cups of Coffee per Day | | | |
|---|---|---|---|---|
|  | 0 | 1–4 | ≥5 | Total |
| Acute conditions | 340 | 457 | 183 | 980 |
| Chronic conditions | 2440 | 2527 | 868 | 5835 |

(a) Each of the 6815 subjects above is considered as belonging to one of the three groups defined by the number of cups of coffee consumed per day (the three columns). Calculate for each of the three groups the proportion of subjects admitted because of an acute onset. Display these proportions by means of a bar chart.

(b) For those admitted because of their chronic conditions, express their coffee consumption by means of a pie chart.

(c) Calculate the generalized odds and give your interpretation. Exposure is defined as having an acute condition.

**1.36**   In a seroepidemiologic survey of health workers representing a spectrum of exposure to blood and patients with hepatitis B virus (HBV), it was found that infection increased as a function of contact. Table E1.36 provides data for hospital workers with uniform socioeconomic status at an urban teaching hospital in Boston, Massachusetts.

**TABLE E1.36**

| Personnel | Exposure | $n$ | HBV Positive |
|-----------|----------|-----|--------------|
| Physicians | Frequent | 81 | 17 |
| | Infrequent | 89 | 7 |
| Nurses | Frequent | 104 | 22 |
| | Infrequent | 126 | 11 |

(a) Calculate the proportion of HBV-positive workers in each subgroup.

(b) Calculate the odds ratios associated with frequent contacts (compared to infrequent contacts). Do this separately for physicians and nurses.

(c) Compare the two ratios obtained in part (b). A large difference would indicate a three-term interaction or effect modification, where frequent effects are different for physicians and nurses.

(d) Assuming that the odds ratios for the two groups, physicians and nurses, are equal (in other words, type of personnel is not an effect modifier), calculate the Mantel–Haenszel estimate of this common odds ratio.

**1.37**   The results of the Third National Cancer Survey have shown substantial variation in lung cancer incidence rates for white males in Allegheny County, Pennsylvania, which may be due to different smoking rates.

**TABLE E1.37**

| | Lawrenceville | | South Hills | |
|-----|-----|-----|-----|-----|
| Age | $n$ | % | $n$ | % |
| 35–44 | 71 | 54.9 | 135 | 37.0 |
| 45–54 | 79 | 53.2 | 193 | 28.5 |
| 55–64 | 119 | 43.7 | 138 | 21.7 |
| ≥65 | 109 | 30.3 | 141 | 18.4 |
| Total | 378 | 46.8 | 607 | 27.1 |

Table E1.37 gives the percentages of current smokers by age for two study areas.

**(a)** Display the age distribution for Lawrenceville by means of a pie chart.

**(b)** Display the age distribution for South Hills by means of a pie chart. How does this chart compare to the one in part (a)?

**(c)** Display the smoking rates for Lawrenceville and South Hills, side by side, by means of a bar chart.

**1.38** Prematurity, which ranks as the major cause of neonatal morbidity and mortality, has traditionally been defined on the basis of a birth weight under 2500 g. But this definition encompasses two distinct types of infants: infants who are small because they are born early, and infants who are born at or near term but are small because their growth was retarded. *Prematurity* has now been replaced by *low birth weight* to describe the second type, and *preterm* to characterize the first type (babies born before 37 weeks of gestation).

A case–control study of the epidemiology of preterm delivery was undertaken at Yale–New Haven Hospital in Connecticut during 1977. The study population consisted of 175 mothers of singleton preterm infants and 303 mothers of singleton full-term infants. Tables E1.38*a* and E1.38*b* give the distribution of age and socioeconomic status.

**TABLE E1.38*a***

| Age | Cases | Controls |
| --- | --- | --- |
| 14–17 | 15 | 16 |
| 18–19 | 22 | 25 |
| 20–24 | 47 | 62 |
| 25–29 | 56 | 122 |
| ≥30 | 35 | 78 |

**TABLE E1.38*b***

| Socioeconomic Level | Cases | Controls |
| --- | --- | --- |
| Upper | 11 | 40 |
| Upper middle | 14 | 45 |
| Middle | 33 | 64 |
| Lower middle | 59 | 91 |
| Lower | 53 | 58 |
| Unknown | 5 | 5 |

**(a)** Refer to the age data and choose the "≥30" group as the baseline. Calculate the odds ratio associated with every other age group. Is it true, in general, that the younger the mother, the higher the risk?

(b) Refer to the socioeconomic data and choose the "lower" group as the baseline. Calculate the odds ratio associated with every other group. Is it true, in general, that the poorer the mother, the higher the risk?

(c) Refer to the socioeconomic data. Calculate the generalized odds and give your interpretation. Does this support the conclusion in part (b)?

**1.39** *Sudden infant death syndrome* (SIDS), also known as sudden unexplained *death, crib death,* or *cot death,* claims the lives of an alarming number of apparently normal infants every year. In a study at the University of Connecticut School of Medicine, significant associations were found between SIDS and certain demographic characteristics. Significant associations were found between SIDS and certain demographic characteristics. Some of the summarized data are given in Table E1.39. (Expected deaths are calculated using Connecticut infant mortality data for 1974–1976.)

**TABLE E1.39**

|  | Number of Deaths | |
|---|---|---|
|  | Observed | Expected |
| Gender |  |  |
| Male | 55 | 45 |
| Female | 35 | 45 |
| Race |  |  |
| Black | 23 | 11 |
| White | 67 | 79 |

(a) Calculate the standardized mortality ratio (SMR) for each subgroup.

(b) Compare males with females and blacks with whites.

**1.40** Adult male residents of 13 counties in western Washington in whom testicular cancer had been diagnosed during 1977–1983 were interviewed over the telephone regarding their history of genital tract conditions,

**TABLE E1.40**

| Religion | Vasectomy | Cases | Controls |
|---|---|---|---|
| Protestant | Yes | 24 | 56 |
|  | No | 205 | 239 |
| Catholic | Yes | 10 | 6 |
|  | No | 32 | 90 |
| Others | Yes | 18 | 39 |
|  | No | 56 | 96 |

including vasectomy. For comparison, the same interview was given to a sample of men selected from the population of these counties by dialing telephone numbers at random. The data in Table E1.40 are tabulated by religious background. Calculate the odds ratio associated with vasectomy for each religious group. Is there any evidence of an effect modification? If not, calculate the Mantel–Haenszel estimate of the common odds ratio.

**1.41**   The role of menstrual and reproductive factors in the epidemiology of breast cancer has been reassessed using pooled data from three large case–control studies of breast cancer from several Italian regions. The data are summarized in Table E1.24 for age at menopause and age at first live birth. Find a way (or ways) to summarize the data further so as to express the observation that the risk of breast cancer is lower for women with younger ages at first live birth and younger ages at menopause.

**1.42**   In 1979 the U.S. Veterans Administration conducted a health survey of 11,230 veterans. The advantages of this survey are that it includes a large

**TABLE E1.42**

|  | Service in Vietnam | |
|---|---|---|
| Symptom | Yes | No |
| Nightmares | | |
| Yes | 197 | 85 |
| No | 577 | 925 |
| Sleep problems | | |
| Yes | 173 | 160 |
| No | 599 | 851 |
| Troubled memories | | |
| Yes | 220 | 105 |
| No | 549 | 906 |
| Depression | | |
| Yes | 306 | 315 |
| No | 465 | 699 |
| Temper control problems | | |
| Yes | 176 | 144 |
| No | 595 | 868 |
| Life goal association | | |
| Yes | 231 | 225 |
| No | 539 | 786 |
| Omit feelings | | |
| Yes | 188 | 191 |
| No | 583 | 821 |
| Confusion | | |
| Yes | 163 | 148 |
| No | 607 | 864 |

random sample with a high interview response rate, and it was done before the public controversy surrounding the issue of the health effects of possible exposure to Agent Orange. The data in Table E1.42 relate Vietnam service to eight posttraumatic stress disorder symptoms among the 1787 veterans who entered the military service between 1965 and 1975. Calculate the odds ratio for each symptom.

**1.43**  It has been hypothesized that dietary fiber decreases the risk of colon cancer, whereas meats and fats are thought to increase this risk. A large study was undertaken to confirm these hypotheses. Fiber and fat consumptions are classified as low or high, and data are tabulated separately in Table E1.43 for males and females ("low" means below median). For each group (males and females), using "low fat, high fiber" as the baseline, calculate the odds ratio associated with every other dietary combination. Any evidence of an effect modification (interaction between consumption of fat and consumption of fiber)?

**TABLE E1.43**

| | Males | | Females | |
|---|---|---|---|---|
| Diet | Cases | Controls | Cases | Controls |
| Low fat, high fiber | 27 | 38 | 23 | 39 |
| Low fat, low fiber | 64 | 78 | 82 | 81 |
| High fat, high fiber | 78 | 61 | 83 | 76 |
| High fat, low fiber | 36 | 28 | 35 | 27 |

**1.44**  Data are compiled in Table E1.44 from different studies designed to investigate the accuracy of death certificates. The results of 5373 autopsies were compared to the causes of death listed on the certificates. Find a graphical way to display the downward trend of accuracy over time.

**TABLE E1.44**

| | Accurate Certificate | | |
|---|---|---|---|
| Date of Study | Yes | No | Total |
| 1955–1965 | 2040 | 694 | 2734 |
| 1970–1971 | 437 | 203 | 640 |
| 1975–1978 | 1128 | 599 | 1727 |
| 1980 | 121 | 151 | 272 |

**1.45**  A study was conducted to ascertain factors that influence a physician's decision to transfuse a patient. A sample of 49 attending physicians was selected. Each physician was asked a question concerning the frequency with which an unnecessary transfusion was give because another physician suggested it. The same question was asked of a sample of 71 residents. The data are shown in Table E1.45.

**TABLE E1.45**

| | Frequency of Unnecessary Transfusion | | | | |
|---|---|---|---|---|---|
| Type of Physician | Very Frequently (1/week) | Frequently (1/two weeks) | Occasionally (1/month) | Rarely (1/two months) | Never |
| Attending | 1 | 1 | 3 | 31 | 13 |
| Resident | 2 | 13 | 28 | 23 | 5 |

(a) Choose "never" as the baseline and calculate the odds ratio associated with each other frequency and "residency."

(b) Calculate the generalized odds and give your interpretation. Does this result agree with those in part (a)?

**1.46** When a patient is diagnosed as having cancer of the prostate, an important question in deciding on a treatment strategy is whether or not the cancer has spread to the neighboring lymph nodes. The question is so critical in prognosis and treatment that it is customary to operate on the patient (i.e., perform a laparotomy) for the sole purpose of examining the nodes and removing tissue samples to examine under the microscope for evidence of cancer. However, certain variables that can be measured without surgery are predictive of the nodal involvement; and the purpose of the study presented here was to examine the data for 53 prostate cancer patients receiving surgery to determine which of five preoperative variables are predictive of nodal involvement. Table E1.46 presents the complete data set. For each of the 53 patients, there are two continuous independent variables, age at diagnosis and level of serum acid phosphatase ($\times 100$; called "acid"), and three binary variables: x-ray reading, pathology reading (grade) of a biopsy of the tumor obtained by needle before surgery, and a rough measure of the size and location of the tumor (stage) obtained by palpation with the fingers via the rectum. For these three binary independent variables, a value of 1 signifies a positive or more serious state and a 0 denotes a negative or less serious finding. In addition, the sixth column presents the finding at surgery—the primary outcome of interest, which is binary, a value of 1 denoting nodal involvement, and a value of 0 denoting no nodal involvement found at surgery. In this exercise we investigate the effects of the three binary preoperative variables (x-ray, grade, and stage); the effects of the two continuous factors (age and acid phosphatase) will be studied in an exercise in Chapter 2.

(a) Arrange the data on *nodes* and *x-ray* into a 2 × 2 table, calculate the odds ratio associated with x-ray and give your interpretation.

(b) Arrange the data on *nodes* and *grade* into a 2 × 2 table, calculate the odds ratio associated with grade and give your interpretation.

(c) Arrange the data on *nodes* and *stage* into a 2 × 2 table, calculate the odds ratio associated with stage and give your interpretation.

If you use Microsoft's Excel to solve this problem, 2 × 2 tables can be formed using *PivotTable Wizard* in the *Data menu*. An SAS program for part (a), for example, would include these intructions:

```
PROC FREQ;
    TABLES NOTES*XRAY/ OR;
```

**TABLE E1.46   Prostate Cancer Data**

| X-ray | Grade | Stage | Age | Acid | Nodes | X-ray | Grade | Stage | Age | Acid | Nodes |
|---|---|---|---|---|---|---|---|---|---|---|---|
| 0 | 1 | 1 | 64 | 40 | 0 | 0 | 0 | 0 | 60 | 78 | 0 |
| 0 | 0 | 1 | 63 | 40 | 0 | 0 | 0 | 0 | 52 | 83 | 0 |
| 1 | 0 | 0 | 65 | 46 | 0 | 0 | 0 | 1 | 67 | 95 | 0 |
| 0 | 1 | 0 | 67 | 47 | 0 | 0 | 0 | 0 | 56 | 98 | 0 |
| 0 | 0 | 0 | 66 | 48 | 0 | 0 | 0 | 1 | 61 | 102 | 0 |
| 0 | 1 | 1 | 65 | 48 | 0 | 0 | 0 | 0 | 64 | 187 | 0 |
| 0 | 0 | 0 | 60 | 49 | 0 | 1 | 0 | 1 | 58 | 48 | 1 |
| 0 | 0 | 0 | 51 | 49 | 0 | 0 | 0 | 1 | 65 | 49 | 1 |
| 0 | 0 | 0 | 66 | 50 | 0 | 1 | 1 | 1 | 57 | 51 | 1 |
| 0 | 0 | 0 | 58 | 50 | 0 | 0 | 1 | 0 | 50 | 56 | 1 |
| 0 | 1 | 0 | 56 | 50 | 0 | 1 | 1 | 0 | 67 | 67 | 1 |
| 0 | 0 | 1 | 61 | 50 | 0 | 0 | 0 | 1 | 67 | 67 | 1 |
| 0 | 1 | 1 | 64 | 50 | 0 | 0 | 1 | 1 | 57 | 67 | 1 |
| 0 | 0 | 0 | 56 | 52 | 0 | 0 | 1 | 1 | 45 | 70 | 1 |
| 0 | 0 | 0 | 67 | 52 | 0 | 0 | 0 | 1 | 46 | 70 | 1 |
| 1 | 0 | 0 | 49 | 55 | 0 | 1 | 0 | 1 | 51 | 72 | 1 |
| 0 | 1 | 1 | 52 | 55 | 0 | 1 | 1 | 1 | 60 | 76 | 1 |
| 0 | 0 | 0 | 68 | 56 | 0 | 1 | 1 | 1 | 56 | 78 | 1 |
| 0 | 1 | 1 | 66 | 59 | 0 | 1 | 1 | 1 | 50 | 81 | 1 |
| 1 | 0 | 0 | 60 | 62 | 0 | 0 | 0 | 0 | 56 | 82 | 1 |
| 0 | 0 | 0 | 61 | 62 | 0 | 0 | 0 | 1 | 63 | 82 | 1 |
| 1 | 1 | 1 | 59 | 63 | 0 | 1 | 1 | 1 | 65 | 84 | 1 |
| 0 | 0 | 0 | 51 | 65 | 0 | 1 | 0 | 1 | 64 | 89 | 1 |
| 0 | 1 | 1 | 53 | 66 | 0 | 0 | 1 | 0 | 59 | 99 | 1 |
| 0 | 0 | 0 | 58 | 71 | 0 | 1 | 1 | 1 | 68 | 126 | 1 |
| 0 | 0 | 0 | 63 | 75 | 0 | 1 | 0 | 0 | 61 | 136 | 1 |
| 0 | 0 | 1 | 53 | 76 | 0 | | | | | | |

*Note:* This is a very long data file; its electronic copy is available in Web-based form from the author upon request. Summaries would be formed easily using simple computer software.

# 2

# DESCRIPTIVE METHODS FOR CONTINUOUS DATA

A class of measurements or a characteristic on which individual observations or measurements are made is called a *variable*; examples include weight, height, and blood pressure, among others. Suppose that we have a set of numerical values for a variable:

1. If each element of this set may lie only at a few isolated points, we have a *discrete* data set. Examples are race, gender, counts of events, or some sort of artificial grading.
2. If each element of this set may theoretically lie anywhere on the numerical scale, we have a *continuous* data set. Examples are blood pressure, cholesterol level, or time to a certain event such as death.

In Chapter 1 we dealt with the summarization and description of discrete data; in this chapter the emphasis is on continuous measurements.

## 2.1 TABULAR AND GRAPHICAL METHODS

There are various ways of organizing and presenting data; simple tables and graphs, however, are still very effective methods. They are designed to help the reader obtain an intuitive feeling for the data at a glance.

### 2.1.1 One-Way Scatter Plots

One-way scatter plots are the simplest type of graph that can be used to summarize a set of continuous observations. A one-way scatter plot uses a single

393.9                                                                 1242.1

Rate per 100,000 population

**Figure 2.1**  Crude death rates for the United States, 1988.

horizontal axis to display the relative position of each data point. As an example, Figure 2.1 depicts the crude death rates for all 50 states and the District of Columbia, from a low of 393.9 per 100,000 population to a high of 1242.1 per 100,000 population. An advantage of a one-way scatter plot is that since each observation is represented individually, no information is lost; a disadvantage is that it may be difficult to read (and to construct) if values are close to each other.

### 2.1.2  Frequency Distribution

There is no difficulty if the data set is small, for we can arrange those few numbers and write them, say, in increasing order; the result would be sufficiently clear; Figure 2.1 is an example. For fairly large data sets, a useful device for summarization is the formation of a *frequency table* or *frequency distribution*. This is a table showing the number of observations, called *frequency*, within certain ranges of values of the variable under investigation. For example, taking the variable to be the age at death, we have the following example; the second column of the table provides the frequencies.

*Example 2.1*  Table 2.1 gives the number of deaths by age for the state of Minnesota in 1987.

**TABLE 2.1**

| Age | Number of Deaths |
| --- | --- |
| <1 | 564 |
| 1–4 | 86 |
| 5–14 | 127 |
| 15–24 | 490 |
| 25–34 | 667 |
| 35–44 | 806 |
| 45–54 | 1,425 |
| 55–64 | 3,511 |
| 65–74 | 6,932 |
| 75–84 | 10,101 |
| 85+ | 9,825 |
| Total | 34,524 |

If a data set is to be grouped to form a frequency distribution, difficulties should be recognized, and an efficient strategy is needed for better communication. First, there is no clear-cut rule on the number of intervals or classes. With too many intervals, the data are not summarized enough for a clear visualization of how they are distributed. On the other hand, too few intervals are undesirable because the data are oversummarized, and some of the details of the distribution may be lost. In general, between 5 and 15 intervals are acceptable; of course, this also depends on the number of observations, we can and should use more intervals for larger data sets.

The widths of the intervals must also be decided. Example 2.1 shows the special case of mortality data, where it is traditional to show infant deaths (deaths of persons who are born live but die before living one year). Without such specific reasons, intervals generally should be of the same width. This common width $w$ may be determined by dividing the range $R$ by $k$, the number of intervals:

$$w = \frac{R}{k}$$

where the range $R$ is the difference between the smallest and largest in the data set. In addition, a width should be chosen so that it is convenient to use or easy to recognize, such as a multiple of 5 (or 1, for example, if the data set has a narrow range). Similar considerations apply to the choice of the beginning of the first interval; it is a convenient number that is low enough for the first interval to include the smallest observation. Finally, care should be taken in deciding in which interval to place an observation falling on one of the interval boundaries. For example, a consistent rule could be made so as to place such an observation in the interval of which the observation in question is the lower limit.

**Example 2.2**   The following are weights in pounds of 57 children at a day-care center:

| | | | | | | | | | |
|---|---|---|---|---|---|---|---|---|---|
| 68 | 63 | 42 | 27 | 30 | 36 | 28 | 32 | 79 | 27 |
| 22 | 23 | 24 | 25 | 44 | 65 | 43 | 25 | 74 | 51 |
| 36 | 42 | 28 | 31 | 28 | 25 | 45 | 12 | 57 | 51 |
| 12 | 32 | 49 | 38 | 42 | 27 | 31 | 50 | 38 | 21 |
| 16 | 24 | 69 | 47 | 23 | 22 | 43 | 27 | 49 | 28 |
| 23 | 19 | 46 | 30 | 43 | 49 | 12 | | | |

From the data set above we have:

1. The smallest number is 12 and the largest is 79, so that

$$R = 79 - 12$$

$$= 67$$

If five intervals are used, we would have

$$w = \frac{67}{5}$$

$$= 13.4$$

and if 15 intervals are used, we would have

$$w = \frac{67}{15}$$

$$= 4.5$$

Between these two values, 4.5 and 13.6, there are two convenient (or conventional) numbers: 5 and 10. Since the sample size of 57 is not large, a width of 10 should be an apparent choice because it results in fewer intervals (the usual concept of "large" is "100 or more").

2. Since the smallest number is 12, we may begin our first interval at 10. The considerations discussed so far lead to the following seven intervals:

10–19
20–29
30–39
40–49
50–59
60–69
70–79

3. Determining the frequencies or the number of values or measurements for each interval is merely a matter of examining the values one by one and of placing a tally mark beside the appropriate interval. When we do this we have the frequency distribution of the weights of the 57 children (Table 2.2). The temporary column of tallies should be deleted from the final table.

4. An optional but recommended step in the formulation of a frequency distribution is to present the proportion or *relative frequency* in addition to frequency for each interval. These proportions, defined by

$$\text{relative frequency} = \frac{\text{frequency}}{\text{total number of observations}}$$

are shown in Table 2.2 and would be very useful if we need to compare two data sets of different sizes.

**TABLE 2.2**

| Weight Interval (lb) | Tally | Frequency | Relative Frequency (%) |
|---|---|---|---|
| 10–19 | ~~\|\|\|\|~~ | 5 | 8.8 |
| 20–29 | ~~\|\|\|\|~~ ~~\|\|\|\|~~ ~~\|\|\|\|~~ \|\|\|\| | 19 | 33.3 |
| 30–39 | ~~\|\|\|\|~~ ~~\|\|\|\|~~ | 10 | 17.5 |
| 40–49 | ~~\|\|\|\|~~ ~~\|\|\|\|~~ \|\|\| | 13 | 22.8 |
| 50–59 | \|\|\|\| | 4 | 7.0 |
| 60–69 | \|\|\|\| | 4 | 7.0 |
| 70–79 | \|\| | 2 | 3.5 |
| Total | | 57 | 100.0 |

***Example 2.3*** A study was conducted to investigate the possible effects of exercise on the menstrual cycle. From the data collected from that study, we obtained the menarchal age (in years) of 56 female swimmers who began their swimming training after they had reached menarche; these served as controls to compare with those who began their training prior to menarche.

| | | | | | | | |
|---|---|---|---|---|---|---|---|
| 14.0 | 16.1 | 13.4 | 14.6 | 13.7 | 13.2 | 13.7 | 14.3 |
| 12.9 | 14.1 | 15.1 | 14.8 | 12.8 | 14.2 | 14.1 | 13.6 |
| 14.2 | 15.8 | 12.7 | 15.6 | 14.1 | 13.0 | 12.9 | 15.1 |
| 15.0 | 13.6 | 14.2 | 13.8 | 12.7 | 15.3 | 14.1 | 13.5 |
| 15.3 | 12.6 | 13.8 | 14.4 | 12.9 | 14.6 | 15.0 | 13.8 |
| 13.0 | 14.1 | 13.8 | 14.2 | 13.6 | 14.1 | 14.5 | 13.1 |
| 12.8 | 14.3 | 14.2 | 13.5 | 14.1 | 13.6 | 12.4 | 15.1 |

From this data set we have the following:

1. The smallest number is 12.4 and the largest is 16.1, so that

$$R = 16.1 - 12.4$$
$$= 3.7$$

If five intervals are used, we would have

$$w = \frac{3.7}{5}$$
$$= 0.74$$

and if 15 intervals are used, we would have

$$w = \frac{3.7}{15}$$
$$= 0.25$$

**TABLE 2.3**

| Age (years) | Frequency | Relative Frequency (%) |
|---|---|---|
| 12.0–12.4 | 1 | 1.8 |
| 12.5–12.9 | 8 | 14.3 |
| 13.0–13.4 | 5 | 8.9 |
| 13.5–13.9 | 12 | 21.4 |
| 14.0–14.4 | 16 | 28.6 |
| 14.5–14.9 | 4 | 7.1 |
| 15.0–15.4 | 7 | 12.5 |
| 15.5–15.9 | 2 | 3.6 |
| 16.0–16.4 | 1 | 1.8 |
| Total | 56 | 100.0 |

Between these two values, 0.25 and 0.74, 0.5 seems to be a convenient number to use as the width; 0.25 is another choice but it would create many intervals (15) for such a small data set. (Another alternative is to express ages in months and not to deal with decimal numbers.)

2. Since the smallest number is 12.4, we may begin our intervals at 12.0, leading to the following intervals:

> 12.0–12.4
> 12.5–12.9
> 13.0–13.4
> 13.5–13.9
> 14.0–14.4
> 14.5–14.9
> 15.0–15.4
> 15.5–15.9
> 16.0–16.4

3. Count the number of swimmers whose ages belong to each of the nine intervals, and the frequencies, then obtain the frequency distribution of menarchal age of 56 swimmers (Table 2.3), completed with the last column for relative frequencies (expressed as percentages).

### 2.1.3  Histogram and the Frequency Polygon

A convenient way of displaying a frequency table is by means of a histogram and/or a frequency polygon. A *histogram* is a diagram in which:

1. The horizontal scale represents the value of the variable marked at interval boundaries.

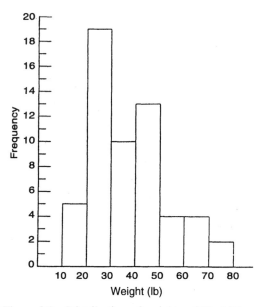

**Figure 2.2** Distribution of weights of 57 children.

2. The vertical scale represents the frequency or relative frequency in each interval (see the exceptions below).

A histogram presents us with a graphic picture of the distribution of measurements. This picture consists of rectangular bars joining each other, one for each interval, as shown in Figure 2.2 for the data set of Example 2.2. If disjoint intervals are used, such as in Table 2.2, the horizontal axis is marked with true boundaries. A *true boundary* is the average of the upper limit of one interval and the lower limit of the next-higher interval. For example, 19.5 serves as the true upper boundary of the first interval and the true lower boundary for the second interval. In cases where we need to compare the shapes of the histograms representing different data sets, or if intervals are of unequal widths, the height of each rectangular bar should represent the density of the interval, where the interval density is defined by

$$\text{density} = \frac{\text{relative frequency (\%)}}{\text{interval width}}$$

The unit for density is *percent per unit* (of measurement): for example, percent per year. If we graph densities on the vertical axis, the relative frequency is represented by the area of the rectangular bar, and the total area under the histogram is 100%. It may always be a good practice to graph densities on the vertical axis with or without having equal class width; when class widths are

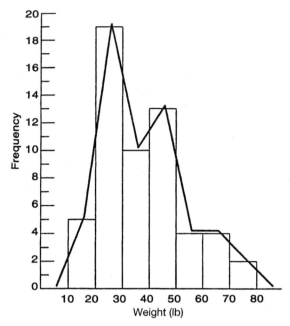

**Figure 2.3**   Distribution of weights of 57 children.

equal, the shape of the histogram looks similar to the graph, with relative frequencies on the vertical axis.

To draw a *frequency polygon*, we first place a dot at the midpoint of the upper base of each rectangular bar. The points are connected with straight lines. At the ends, the points are connected to the midpoints of the previous and succeeding intervals (these are make-up intervals with zero frequency, where widths are the widths of the first and last intervals, respectively). A frequency polygon so constructed is another way to portray graphically the distribution of a data set (Figure 2.3). The frequency polygon can also be shown without the histogram on the same graph.

The frequency table and its graphical relatives, the histogram and the frequency polygon, have a number of applications, as explained below; the first leads to a research question and the second leads to a new analysis strategy.

1. When data are homogeneous, the table and graphs usually show a unimodal pattern with one peak in the middle part. A bimodal pattern might indicate the possible influence or effect of a certain hidden factor or factors.

*Example 2.4*   Table 2.4 provides data on age and percentage saturation of bile for 31 male patients. Using 10% intervals, the data set can be represented by a histogram or a frequency polygon as shown in Figure 2.4. This picture shows

**TABLE 2.4**

| Age | Percent Saturation | Age | Percent Saturation | Age | Percent Saturation |
|-----|-----|-----|-----|-----|-----|
| 23 | 40 | 55 | 137 | 48 | 78 |
| 31 | 86 | 31 | 88 | 27 | 80 |
| 58 | 111 | 20 | 88 | 32 | 47 |
| 25 | 86 | 23 | 65 | 62 | 74 |
| 63 | 106 | 43 | 79 | 36 | 58 |
| 43 | 66 | 27 | 87 | 29 | 88 |
| 67 | 123 | 63 | 56 | 27 | 73 |
| 48 | 90 | 59 | 110 | 65 | 118 |
| 29 | 112 | 53 | 106 | 42 | 67 |
| 26 | 52 | 66 | 110 | 60 | 57 |
| 64 | 88 | | | | |

an apparent bimodal distribution; however, a closer examination shows that among the nine patients with over 100% saturation, eight (or 89%) are over 50 years of age. On the other hand, only four of 22 (or 18%) patients with less than 100% saturation are over 50 years of age. The two peaks in Figure 2.4 might correspond to the two age groups.

2. Another application concerns the symmetry of the distribution as de-picted by the table or its graphs. A symmetric distribution is one in which the distribution has the same shape on both sides of the peak location. If there are more extremely large values, the distribution is then skewed to the right, or *positively* skewed. Examples include family income, antibody level after vaccination, and drug dose to produce a predetermined level of response, among others. It is common that for positively skewed distributions, subsequent statistical analyses should be performed on the log scale: for example, to compute and/or to compare averages of log(dose).

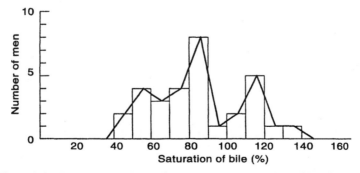

**Figure 2.4**   Frequency polygon for percentage saturation of bile in men.

**TABLE 2.5**

|  | Percent of Families | |
| --- | --- | --- |
| Income ($) | White | Nonwhite |
| 0–14,999 | 13 | 34 |
| 15,000–19,999 | 24 | 31 |
| 20,000–24,999 | 26 | 19 |
| 25,000–34,999 | 28 | 13 |
| 35,000–59,999 | 9 | 3 |
| 60,000 and over | 1 | Negligible |
| Total | 100 | 100 |

*Example 2.5* The distribution of family income for the United States in 1983 by race is shown in Table 2.5. The distribution for nonwhite families is represented in the histogram in Figure 2.5, where the vertical axis represents the density (percent per $1000). It is obvious that the distribution is not symmetric; it is very skewed to the right. In this histogram, we graph the densities on the vertical axis. For example, for the second income interval (15,000–19,999), the relative frequency is 31% and the width of the interval is $5000 (31% per $5000), leading to the density

$$\frac{31}{5000} \times 1000 = 6.2$$

or 6.2% per $1000 (we arbitrarily multiply by 1000, or any power of 10, just to obtain a larger number for easy graphing).

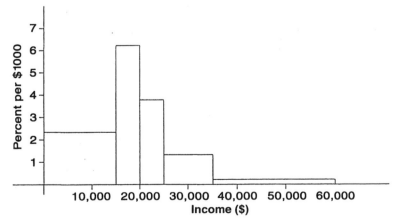

**Figure 2.5**  Income of U.S. nonwhite families, 1983.

**TABLE 2.6**

| Weight Interval (lb) | Frequency | Relative Frequency (%) | Cumulative Relative Frequency (%) |
|---|---|---|---|
| 10–19 | 5 | 8.8 | 8.8 |
| 20–29 | 19 | 33.3 | 42.1 |
| 30–39 | 10 | 17.5 | 59.6 |
| 40–49 | 13 | 22.8 | 82.4 |
| 50–59 | 4 | 7.0 | 89.4 |
| 60–69 | 4 | 7.0 | 96.4 |
| 70–79 | 2 | 3.5 | $99.9 \simeq 100.0$ |
| Total | 57 | 100.0 | |

### 2.1.4   Cumulative Frequency Graph and Percentiles

*Cumulative relative frequency*, or *cumulative percentage*, gives the percentage of persons having a measurement less than or equal to the upper boundary of the class interval. Data from Table 2.2 for the distribution of weights of 57 children are reproduced and supplemented with a column for cumulative relative frequency in Table 2.6. This last column is easy to form; you do it by successively accumulating the relative frequencies of each of the various intervals. In the table the cumulative percentage for the first three intervals is

$$8.8 + 33.3 + 17.5 = 59.6$$

and we can say that 59.6% of the children in the data set have a weight of 39.5 lb or less. Or, as another example, 96.4% of children weigh 69.5 lb or less, and so on.

The cumulative relative frequency can be presented graphically as in Figure 2.6. This type of curve is called a *cumulative frequency graph*. To construct such a graph, we place a point with a horizontal axis marked at the upper class boundary and a vertical axis marked at the corresponding cumulative frequency. Each point represents the cumulative relative frequency and the points are connected with straight lines. At the left end it is connected to the lower boundary of the first interval. If disjoint intervals such as

10–19

20–29

etc. . . .

are used, points are placed at the true boundaries.

The cumulative percentages and their graphical representation, the cumulative frequency graph, have a number of applications. When two cumulative frequency graphs, representing two different data sets, are placed on the same

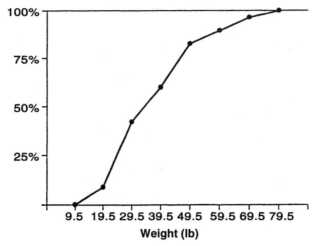

**Figure 2.6**  Cumulative distribution of weights of 57 children.

graph, they provide a rapid visual comparison without any need to compare individual intervals. Figure 2.7 gives such a comparison of family incomes using data of Example 2.5.

The cumulative frequency graph provides a class of important statistics known as *percentiles* or *percentile scores*. The 90th percentile, for example, is the numerical value that exceeds 90% of the values in the data set and is exceeded by only 10% of them. Or, as another example, the 80th percentile is that numerical value that exceeds 80% of the values contained in the data set and is exceeded by 20% of them, and so on. The 50th percentile is commonly

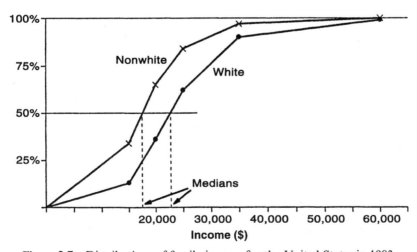

**Figure 2.7**  Distributions of family income for the United States in 1983.

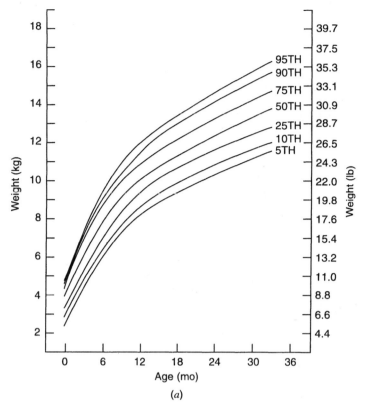

**Figure 2.8**  (*a*) Weight and (*b*) height curves.

called the *median*. In Figure 2.7 the median family income in 1983 for non-whites was about $17,500, compared to a median of about $22,000 for white families. To get the median, we start at the 50% point on the vertical axis and go horizontally until meeting the cumulative frequency graph; the projection of this intersection on the horizontal axis is the median. Other percentiles are obtained similarly.

The cumulative frequency graph also provides an important application in the formation of health norms (see Figure 2.8) for the monitoring of physical progress (weight and height) of infants and children. Here, the same percentiles, say the 90th, of weight or height of groups of different ages are joined by a curve.

***Example 2.6***  Figure 2.9 provides results from a study of Hmong refugees in the Minneapolis–St. Paul area, where each dot represents the average height of five refugee girls of the same age. The graph shows that even though the refugee girls are small, mostly in the lowest 25%, they grow at the same rate as

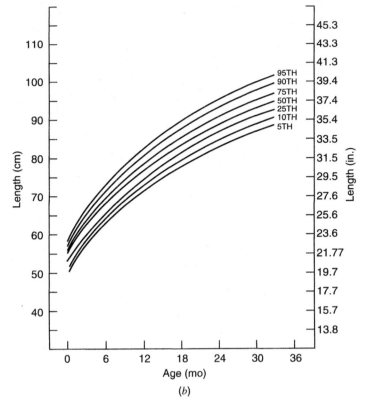

(b)

**Figure 2.8** *(Continued)*

measured by the American standard. However, the pattern changes by the age of 15 years when their average height drops to a level below the 5th percentile. In this example, we plot the *average* height of five girls instead of individual heights; because the Hmong are small, individual heights are likely to be off the chart. This concept of *average* or *mean* is explained in detail in Section 2.2.

### 2.1.5   Stem-and-Leaf Diagrams

A stem-and-leaf diagram is a graphical representation in which the data points are grouped in such a way that we can see the shape of the distribution while retaining the individual values of the data points. This is particularly convenient and useful for smaller data sets. Stem-and-leaf diagrams are similar to frequency tables and histograms, but they also display each and every observation. Data on the weights of children from Example 2.2 are adopted here to illustrate the construction of this simple device. The weights (in pounds) of 57

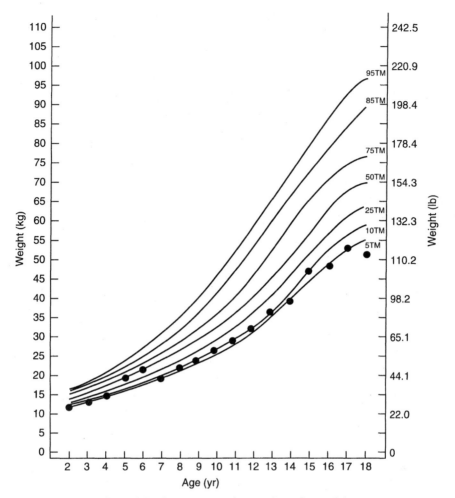

**Figure 2.9**   Mean stature by age for refugee girls.

children at a day-care center are as follows:

| | | | | | | | | | |
|---|---|---|---|---|---|---|---|---|---|
| 68 | 63 | 42 | 27 | 30 | 36 | 28 | 32 | 79 | 27 |
| 22 | 23 | 24 | 25 | 44 | 65 | 43 | 25 | 74 | 51 |
| 36 | 42 | 28 | 31 | 28 | 25 | 45 | 12 | 57 | 51 |
| 12 | 32 | 49 | 38 | 42 | 27 | 31 | 50 | 38 | 21 |
| 16 | 24 | 69 | 47 | 23 | 22 | 43 | 27 | 49 | 28 |
| 23 | 19 | 46 | 30 | 43 | 49 | 12 | | | |

A stem-and-leaf diagram consists of a series of rows of numbers. The number used to label a row is called a *stem*, and the other numbers in the row are

```
1 |            1 | 2 2 2 6 9
2 |            2 | 1 2 2 3 3 3 4 4 5 5 5 7 7 7·7 8 8 8 8
3 |            3 | 0 0 1 1 2 2 6 6 8 8
4 |   ⇒        4 | 2 2 2 3 3 3 4 5 6 7 9 9 9
5 |            5 | 0 0 1 7
6 |            6 | 3 5 8 9
7 |            7 | 4 9
Stems       Stems  Leaves
```

**Figure 2.10**   Typical stem-and-leaf diagram.

called *leaves*. There are no hard rules about how to construct a stem-and-leaf diagram. Generally, it consists of the following steps:

1. Choose some convenient/conventional numbers to serve as stems. The stems chosen are usually the first one or two digits of individual data points.
2. Reproduce the data graphically by recording the digit or digits following the stems as a leaf on the appropriate stem.

If the final graph is turned on its side, it looks similar to a histogram (Figure 2.10). The device is not practical for use with larger data sets because some stems are too long.

## 2.2   NUMERICAL METHODS

Although tables and graphs serve useful purposes, there are many situations that require other types of data summarization. What we need in many applications is the ability to summarize data by means of just a few numerical measures, particularly before inferences or generalizations are drawn from the data. Measures for describing the location (or typical value) of a set of measurements and their variation or dispersion are used for these purposes.

First, let us suppose that we have $n$ measurements in a data set; for example, here is a data set:

$$\{8, 2, 3, 5\}$$

with $n = 4$. We usually denote these numbers as $x_i$'s; thus we have for the example above: $x_1 = 8$, $x_2 = 2$, $x_3 = 3$, and $x_4 = 5$. If we add all the $x_i$'s in the data set above, we obtain 18 as the sum. This addition process is recorded as

$$\sum x = 18$$

where the Greek letter $\Sigma$ is the summation sign. With the summation notation, we are now able to define a number of important summarized measures, starting with the arithmetic average or *mean*.

### 2.2.1    Mean

Given a data set of size $n$,

$$\{x_1, x_2, \ldots, x_n\}$$

the mean of the $x$'s will be denoted by $\bar{x}$ ("x-bar") and is computed by summing all the $x$'s and dividing the sum by $n$. Symbolically,

$$\bar{x} = \frac{\sum x}{n}$$

It is important to know that $\Sigma$ ("sigma") stands for an operation (that of obtaining the sum of the quantities that follow) rather than a quantity itself. For example, considering the data set

$$\{8, 5, 4, 12, 15, 5, 7\}$$

we have

$$n = 7$$
$$\sum x = 56$$

leading to

$$\bar{x} = \frac{56}{7}$$
$$= 8$$

Occasionally, data, especially secondhand data, are presented in the grouped form of a frequency table. In these cases, the mean $\bar{x}$ can be approximated using the formula

$$\bar{x} \simeq \frac{\sum (fm)}{n}$$

where $f$ denotes the frequency (i.e., the number of observations in an interval), $m$ the interval midpoint, and the summation is across the intervals. The midpoint for an interval is obtained by calculating the average of the interval lower true boundary and the upper true boundary. For example, if the first three intervals are

$$10–19$$
$$20–29$$
$$30–39$$

the midpoint for the first interval is

$$\frac{9.5 + 19.5}{2} = 14.5$$

and for the second interval is

$$\frac{19.5 + 29.5}{2} = 24.5$$

This process for calculation of the mean $\bar{x}$ using Table 2.3 is illustrated in Table 2.7.

$$\bar{x} \simeq \frac{2086.5}{57}$$
$$= 36.6 \text{ lb}$$

(If individual weights were used, we would have $\bar{x} = 36.7$ lb.)

Of course, the mean $\bar{x}$ obtained from this technique with a frequency table is different from the $\bar{x}$ using individual or raw data. However, the process saves some computational labor, and the difference between the results, $\bar{x}$'s, is very small if the data set is large and the interval width is small.

As indicated earlier, a characteristic of some interest is the symmetry or lack of symmetry of a distribution, and it is recommended that for very positively

**TABLE 2.7**

| Weight Interval | Frequency, $f$ | Interval Midpoint, $m$ | $fm$ |
|---|---|---|---|
| 10–19 | 5 | 14.5 | 72.5 |
| 20–29 | 19 | 24.5 | 465.5 |
| 30–39 | 10 | 34.5 | 345.0 |
| 40–49 | 13 | 44.5 | 578.5 |
| 50–59 | 4 | 54.5 | 218.0 |
| 60–69 | 4 | 64.5 | 258.0 |
| 70–79 | 2 | 74.5 | 149.0 |
| Total | 57 | | 2086.5 |

**TABLE 2.8**

| $x$ | $\ln x$ |
|---|---|
| 8 | 2.08 |
| 5 | 1.61 |
| 4 | 1.39 |
| 12 | 2.48 |
| 15 | 2.71 |
| 7 | 1.95 |
| 28 | 3.33 |
| 79 | 15.55 |

skewed distributions, analyses are commonly done on the log scale. After obtaining a mean on the log scale, we should take the antilog to return to the original scale of measurement; the result is called the *geometric mean* of the $x$'s. The effect of this process is to minimize the influences of extreme observations (very large numbers in the data set). For example, considering the data set

$$\{8, 5, 4, 12, 15, 7, 28\}$$

with one unusually large measurement, we have Table 2.8, with natural logs presented in the second column. The mean is

$$\bar{x} = \frac{79}{7}$$

$$= 11.3$$

while on the log scale we have

$$\frac{\sum \ln x}{n} = \frac{15.55}{7}$$

$$= 2.22$$

leading to a geometric mean of 9.22, which is less affected by the large measurements. Geometric mean is used extensively in microbiological and serological research, in which distributions are often skewed positively.

*Example 2.7* In some studies the important number is the time to an event, such as death; it is called the *survival time*. The term *survival time* is conventional even though the primary event could be nonfatal, such as a relapse or the appearance of the first disease symptom. Similar to cases of income and antibody level, the distributions of survival times are positively skewed; therefore, data are often summarized using the median or geometric mean. The following is a typical example.

The remission times of 42 patients with acute leukemia were reported from a clinical trial undertaken to assess the ability of the drug 6-mercaptopurine (6-MP) to maintain remission. Each patient was randomized to receive either 6-MP or placebo. The study was terminated after one year; patients have different follow-up times because they were enrolled sequentially at different times. Times to relapse in weeks for the 21 patients in the placebo group were

$$1, 1, 2, 2, 3, 4, 4, 5, 5, 8, 8, 8, 8, 11, 11, 12, 12, 15, 17, 22, 23$$

The mean is

$$\bar{x} = \frac{\sum x}{n}$$
$$= 8.67 \text{ weeks}$$

and on the log scale we have

$$\frac{\sum \ln x}{n} = 1.826$$

leading to a geometric mean of 6.21, which, in general, is less affected by the large measurements.

### 2.2.2    Other Measures of Location

Another useful measure of location is the *median*. If the observations in the data set are arranged in increasing or decreasing order, the median is the middle observation, which divides the set into equal halves. If the number of observations $n$ is odd, there will be a unique median, the $\frac{1}{2}(n + 1)$th number from either end in the ordered sequence. If $n$ is even, there is strictly no middle observation, but the median is defined by convention as the average of the two middle observations, the $(\frac{1}{2}n)$th and $\frac{1}{2}(n + 1)$th from either end. In Section 2.1 we showed a quicker way to get an approximate value for the median using the cumulative frequency graph (see Figure 2.6).

The two data sets $\{8, 5, 4, 12, 15, 7, 28\}$ and $\{8, 5, 4, 12, 15, 7, 49\}$, for example, have different means but the same median, 8. Therefore, the advantage of the median as a measure of location is that it is less affected by extreme observations. However, the median has some disadvantages in comparison with the mean:

1. It takes no account of the precise magnitude of most of the observations and is therefore less efficient than the mean because it wastes information.
2. If two groups of observations are pooled, the median of the combined group cannot be expressed in terms of the medians of the two component

groups. However, the mean can be so expressed. If component groups are of sizes $n_1$ and $n_2$ and have means $\bar{x}_1$ and $\bar{x}_2$ respectively, the mean of the combined group is

$$\bar{x} = \frac{n_1\bar{x}_1 + n_2\bar{x}_2}{n_1 + n_2}$$

3. In large data sets, the median requires more work to calculate than the mean and is not much use in the elaborate statistical techniques (it is still useful as a descriptive measure for skewed distributions).

A third measure of location, the *mode*, was introduced briefly in Section 2.1.3. It is the value at which the frequency polygon reaches a peak. The mode is not used widely in analytical statistics, other than as a descriptive measure, mainly because of the ambiguity in its definition, as the fluctuations of small frequencies are apt to produce spurious modes. For these reasons, in the remainder of the book we focus on a single measure of location, the mean.

### 2.2.3   Measures of Dispersion

When the mean $\bar{x}$ of a set of measurements has been obtained, it is usually a matter of considerable interest to measure the degree of variation or dispersion around this mean. Are the $x$'s all rather close to $\bar{x}$, or are some of them dispersed widely in each direction? This question is important for purely descriptive reasons, but it is also important because the measurement of dispersion or variation plays a central part in the methods of statistical inference described in subsequent chapters.

An obvious candidate for the measurement of dispersion is the *range R*, defined as the difference between the largest value and the smallest value, which was introduced in Section 2.1.3. However, there are a few difficulties about use of the range. The first is that the value of the range is determined by only two of the original observations. Second, the interpretation of the range depends in a complicated way on the number of observations, which is an undesirable feature.

An alternative approach is to make use of *deviations* from the mean, $x - \bar{x}$; it is obvious that the greater the variation in the data set, the larger the magnitude of these deviations will tend to be. From these deviations, the *variance $s^2$* is computed by squaring each deviation, adding them, and dividing their sum by one less than $n$:

$$s^2 = \frac{\sum(x - \bar{x})^2}{n - 1}$$

The use of the divisor $(n - 1)$ instead of $n$ is clearly not very important when $n$ is large. It is more important for small values of $n$, and its justification will be explained briefly later in this section. The following should be noted:

1. It would be no use to take the mean of deviations because

$$\sum(x - \bar{x}) = 0$$

2. Taking the mean of the absolute values, for example

$$\frac{\sum|x - \bar{x}|}{n}$$

is a possibility. However, this measure has the drawback of being difficult to handle mathematically, and we do not consider it further in this book.

The variance $s^2$ ($s$ squared) is measured in the square of the units in which the $x$'s are measured. For example, if $x$ is the time in seconds, the variance is measured in seconds squared ($\sec^2$). It is convenient, therefore, to have a measure of variation expressed in the same units as the $x$'s, and this can be done easily by taking the square root of the variance. This quantity is the *standard deviation*, and its formula is

$$s = \sqrt{\frac{\sum(x - \bar{x})^2}{n - 1}}$$

Consider again the data set

$$\{8, 5, 4, 12, 15, 5, 7\}$$

Calculation of the variance $s^2$ and standard deviation $s$ is illustrated in Table 2.9.

In general, this calculation process is likely to cause some trouble. If the mean is not a "round" number, say $\bar{x} = 10/3$, it will need to be rounded off,

**TABLE 2.9**

| $x$ | $x - \bar{x}$ | $(x - \bar{x})^2$ |
|---|---|---|
| 8 | 0 | 0 |
| 5 | −3 | 9 |
| 4 | −4 | 16 |
| 12 | 4 | 16 |
| 15 | 7 | 49 |
| 5 | −3 | 9 |
| 7 | −1 | 1 |
| $\sum x = 56$ | | $\sum(x - \bar{x})^2 = 100$ |
| $n = 7$ | | $s^2 = 100/6 = 16.67$ |
| $\bar{x} = 8$ | | $s = \sqrt{16.67} = 4.08$ |

**TABLE 2.10**

| $x$ | $x^2$ |
|-----|-------|
| 8 | 64 |
| 5 | 25 |
| 4 | 16 |
| 12 | 144 |
| 15 | 225 |
| 5 | 25 |
| 7 | 49 |
| 56 | 548 |

and errors arise in the subtraction of this figure from each $x$. This difficulty can easily be overcome by using the following shortcut formula for the variance:

$$s^2 = \frac{\sum x^2 - (\sum x)^2/n}{n - 1}$$

Our earlier example is reworked in Table 2.10, yielding identical results.

$$s^2 = \frac{(548) - (56)^2/7}{6}$$

$$= 16.67$$

When data are presented in the grouped form of a frequency table, the variance is calculated using the following modified shortcut formula:

$$s^2 \simeq \frac{\sum fm^2 - (\sum fm)^2/n}{n - 1}$$

where $f$ denotes an interval frequency, $m$ the interval midpoint calculated as in Section 2.2.2 and the summation is across the intervals. This approximation is illustrated in Table 2.11.

$$s^2 \simeq \frac{89,724.25 - (2086.5)^2/57}{56}$$

$$= 238.35$$

$$s \simeq 15.4 \text{ lb}$$

(If individual weights were used, we would have $s = 15.9$ lb.)

It is often not clear to beginners why we use $(n - 1)$ instead of $n$ as the denominator for $s^2$. This number, $n - 1$, called the *degrees of freedom*, repre-

**TABLE 2.11**

| Weight Interval | $f$ | $m$ | $m^2$ | $fm$ | $fm^2$ |
|---|---|---|---|---|---|
| 10–19 | 5 | 14.5 | 210.25 | 72.5 | 1,051.25 |
| 20–29 | 19 | 24.5 | 600.25 | 465.5 | 11,404.75 |
| 30–39 | 10 | 34.5 | 1,190.25 | 345.0 | 11,902.50 |
| 40–49 | 13 | 44.5 | 1,980.25 | 578.5 | 25,743.25 |
| 50–59 | 4 | 54.5 | 2,970.25 | 218.0 | 11,881.00 |
| 60–69 | 4 | 64.5 | 4,160.25 | 258.0 | 16,641.00 |
| 70–79 | 2 | 74.5 | 5,550.25 | 149.0 | 11,100.50 |
| Total | 57 | | | 2,086.5 | 89,724.25 |

sent the amount of information contained in the sample. The real explanation for $n - 1$ is hard to present at the level of this book; however, it may be seen this way. What we are trying to do with $s$ is to provide a measure of variability, a measure of the average gap or average distance between numbers in the sample, and there are $n - 1$ gaps between $n$ numbers. When $n = 2$ there is only one gap or distance beween the two numbers, and when $n = 1$, there is no variability to measure.

Finally, it is occasionally useful to describe the variation by expressing the standard deviation as a proportion or percentage of the mean. The resulting measure,

$$CV = \frac{s}{\bar{x}} \times 100\%$$

is called the *coefficient of variation*. It is an *index*, a dimensionless quantity because the standard deviation is expressed in the same units as the mean and could be used to compare the difference in variation between two types of measurements. However, its use is rather limited and we will not present it at this level.

### 2.2.4  Box Plots

The *box plot* is a graphical representation of a data set that gives a visual impression of location, spread, and the degree and direction of skewness. It also allows for the identification of outliers. Box plots are similar to one-way scatter plots in that they require a single horizontal axis; however, instead of plotting every observation, they display a summary of the data. A box plot consists of the following:

1. A central box extends from the 25th to the 75th percentiles. This box is divided into two compartments at the median value of the data set. The relative sizes of the two halves of the box provide an indication of the

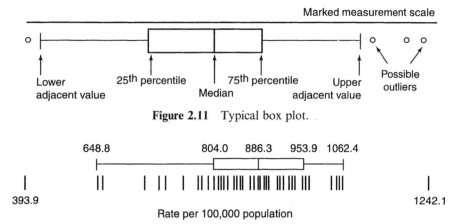

**Figure 2.11**  Typical box plot.

Rate per 100,000 population

**Figure 2.12**  Crude death rates for the United States, 1988: a combination of one-way scatter and box plots.

distribution symmetry. If they are approximately equal, the data set is roughly symmetric; otherwise, we are able to see the degree and direction of skewness (Figure 2.11).

2. The line segments projecting out from the box extend in both directions to the *adjacent values*. The adjacent values are the points that are 1.5 times the length of the box beyond either quartile. All other data points outside this range are represented individually by little circles; these are considered to be outliers or extreme observations that are not typical of the rest of the data.

Of course, it is possible to combine a one-way scatter plot and a box plot so as to convey an even greater amount of information (Figure 2.12). There are other ways of constructing box plots; for example, one may make it vertically or divide it into different levels of outliers.

## 2.3   SPECIAL CASE OF BINARY DATA

Observations or measurements may be made on different scales. If each element of a data set may lie at only a few isolated points, we have a *discrete* data set. A special case of discrete data are binary data, where each outcome has only two possible values; examples are gender and an indication of whether a treatment is a success or a failure. If each element of this set may theoretically lie anywhere on a numerical scale, we have a *continuous* data set; examples are blood pressure and cholesterol level. Chapter 1 deals with the summarization and description of discrete data, especially binary data; the primary statistic was proportion. In this chapter the emphasis so far has been on continuous measurements, where, for example, we learn to form sample mean and use it as a measure of location, a *typical* value representing the data set. In addition, the *variance* and/or *standard deviation* is formed and used to measure the degree of variation or dispersion of data around the mean. In this short section we will see that binary data can be treated as a special case of continuous data.

Many outcomes can be classified as belonging to one of two possible categories: presence and absence, nonwhite and white, male and female, improved and not improved. Of course, one of these two categories is usually identified as being of primary interest; for example, presence in the presence and absence classification, or nonwhite in the white and nonwhite classification. We can, in general, relabel the two outcome categories as positive $(+)$ and negative $(-)$. An outcome is positive if the primary category is observed and is negative if the other category is observed. The proportion is defined as in Chapter 1:

$$p = \frac{x}{n}$$

where $x$ is the number of positive outcomes and $n$ is the sample size. However, it can also be expressed as

$$p = \frac{\sum x_i}{n}$$

where $x_i$ is "1" if the $i$th outcome is positive and "0" otherwise. In other words, a sample proportion can be viewed as a special case of sample means where data are coded as 0 or 1. But what do we mean by *variation* or *dispersion*, and how do we measure it?

Let us write out the variance $s^2$ using the shortcut formula of Section 2.2 but with the denominator $n$ instead of $n - 1$ (this would make little difference because we almost always deal with *large* samples of binary data):

$$s = \sqrt{\frac{\sum x_i^2 - (\sum x_i)^2 / n}{n}}$$

Since $x_i$ is binary, with "1" if the $i$th outcome is positive and "0" otherwise, we have

$$x_i^2 = x_i$$

and therefore,

$$s^2 = \frac{\sum x_i - (\sum x_i)^2 / n}{n}$$
$$= \frac{\sum x_i}{n} \left( 1 - \frac{\sum x_i}{n} \right)$$
$$= p(1 - p)$$

In other words, the statistic $p(1 - p)$ can be used in place of $s^2$ as a measure of *variation*; the logic can be seen as follows. First, the quantity $p(1 - p)$, with $0 \leq p \leq 1$, attains its maximum value when $p = 0.5$. For example,

$$(0.1)(0.9) = 0.09$$
$$\vdots$$
$$(0.4)(0.6) = 0.24$$
$$(0.5)(0.5) = 0.25$$
$$(0.6)(0.4) = 0.24$$
$$\vdots$$
$$(0.9)(0.1) = 0.09$$

The values of $p(1 - p)$ are greatest in the vicinity of $p = 0.5$ and decrease as we go toward both ends (0 and 1) of the range of $p$. If we are performing a coin-tossing experiment or conducting an election; the result would be most *unpredictable* when the chance to obtain the outcome wanted is in the vicinity of $p = 0.5$. In other words, the quantity $p(1 - p)$ is a suitable statistic to measure the *volatility, dispersion*, and *variation*. The corresponding statistic for standard deviation is $\sqrt{p(1 - p)}$.

## 2.4  COEFFICIENTS OF CORRELATION

Methods discussed in this chapter have been directed to the analyses of data where a single continuous measurement was made on each element of a sample. However, in many important investigations we may have two measurements made: where the sample consists of pairs of values and the research objective is concerned with the association between these variables. For example, what is the relationship between a mother's weight and her baby's weight? In Section 1.3 we were concerned with the association between dichotomous variables. For example, if we want to investigate the relationship between a disease and a certain risk factor, we could calculate an odds ratio to represent the strength of the relationship. In this section we deal with continuous measurements, and the method is referred to as *correlation analysis*. Correlation is a concept that carries the common colloquial implication of association, such as "height and weight are correlated." The statistical procedure will give the word a technical meaning; we can actually calculate a number that tells the *strength* of the association.

When dealing with the relationship between two continuous variables, we first have to distinguish between a deterministic relationship and a statistical relationship. For a *deterministic relationship*, values of the two variables are related through an exact mathematical formula. For example, consider the

**TABLE 2.12**

| $x$ (oz) | $y$ (%) | $x$ (oz) | $y$ (%) |
| --- | --- | --- | --- |
| 112 | 63 | 81 | 120 |
| 111 | 66 | 84 | 114 |
| 107 | 72 | 118 | 42 |
| 119 | 52 | 106 | 72 |
| 92 | 75 | 103 | 90 |
| 80 | 118 | 94 | 91 |

relationship between hospital cost and number of days in hospital. If the costs are $100 for admission and $150 per day, we can easily calculate the total cost given the number of days in hospital, and if any set of data is plotted, say cost versus number of days, all data points fall perfectly on a straight line. Unlike a deterministic relationship, a *statistical relationship* is not perfect. In general, the points do not fall perfectly on any line or curve.

Table 2.12 gives the values for the birth weight ($x$) and the increase in weight between days 70 and 100 of life, expressed as a percentage of the birth weight ($y$) for 12 infants. If we let each pair of numbers ($x$, $y$) be represented by a dot in a diagram with the $x$'s on the horizontal axis, we have Figure 2.13. The dots do not fall perfectly on a straight line, but rather, scatter around a line, very typical for statistical relationships. Because of this scattering of dots, the diagram is called a *scatter diagram*. The positions of the dots provide some information about the direction as well as the strength of the association under the investigation. If they tend to go from lower left to upper right, we have a positive association; if they tend to go from upper left to lower right, we have a negative association. The relationship becomes weaker and weaker as the dis-

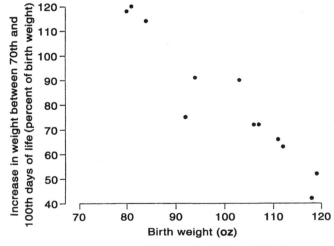

**Figure 2.13**  Scatter diagram for birth-weight data.

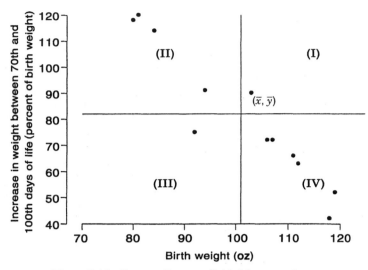

**Figure 2.14**  Scatter diagram divided into quadrants.

tribution of the dots clusters less closely around the line, and becomes virtually no correlation when the distribution approximates a circle or oval (the method is ineffective for measuring a relationship that is not linear).

### 2.4.1  Pearson's Correlation Coefficient

Consider the scatter diagram shown in Figure 2.14, where we have added a vertical and a horizontal line through the point $(\bar{x}, \bar{y})$ and label the four quarters as I, II, III, and IV. It can be seen that

- In quarters I and III,

$$(x - \bar{x})(y - \bar{y}) > 0$$

so that for positive association, we have

$$\sum (x - \bar{x})(y - \bar{y}) > 0$$

Furthermore, this sum is large for stronger relationships because most of the dots, being closely clustered around the line, are in these two quarters.
- Similarly, in quarters II and IV,

$$(x - \bar{x})(y - \bar{y}) < 0$$

leading to

$$\sum (x - \bar{x})(y - \bar{y}) < 0$$

for negative association.

With proper standardization, we obtain

$$r = \frac{\sum (x - \bar{x})(y - \bar{y})}{\sqrt{[\sum (x - \bar{x})^2][\sum (y - \bar{y})^2]}}$$

so that

$$-1 \leq r \leq 1$$

This statistic, $r$, called the *correlation coefficient*, is a popular measure for the strength of a statistical relationship; here is a shortcut formula:

$$r = \frac{\sum xy - (\sum x)(\sum y)/n}{\sqrt{[\sum x^2 - (\sum x)^2/n][\sum y^2 - (\sum y)^2/n]}}$$

Meanningful interpretation of the correlation coefficient $r$ is rather complicated at this level. We will revisit the topic in Chapter 8 in the context of regression analysis, a statistical method that is closely connected to correlation. Generally:

- Values near 1 indicate a strong positive association.
- Values near $-1$ indicate a strong negative association.
- Values around 0 indicate a weak association.

Interpretation of $r$ should be made cautiously, however. It is true that a scatter plot of data that results in a correlation number of $+1$ or $-1$ has to lie in a perfectly straight line. But a correlation of 0 doesn't mean that there is no association; it means that there is no *linear* association. You can have a correlation near 0 and yet have a very strong association, such as the case when the data fall neatly on a sharply bending curve.

***Example 2.8*** Consider again the birth-weight problem described earlier in this section. We have the data given in Table 2.13. Using the five totals, we obtain

$$r = \frac{94{,}322 - [(1207)(975)]/12}{\sqrt{[123{,}561 - (1207)^2/12][86{,}487 - (975)^2/12]}}$$

$$= -0.946$$

indicating a very strong negative association.

**TABLE 2.13**

| $x$ | $y$ | $x^2$ | $y^2$ | $xy$ |
|---|---|---|---|---|
| 112 | 63 | 12,544 | 3,969 | 7,056 |
| 111 | 66 | 12,321 | 4,356 | 7,326 |
| 107 | 72 | 11,449 | 5,184 | 7,704 |
| 119 | 52 | 14,161 | 2,704 | 6,188 |
| 92 | 75 | 8,464 | 5,625 | 6,900 |
| 80 | 118 | 6,400 | 13,924 | 9,440 |
| 81 | 120 | 6,561 | 14,400 | 9,720 |
| 84 | 114 | 7,056 | 12,996 | 9,576 |
| 118 | 42 | 13,924 | 1,764 | 4,956 |
| 106 | 72 | 11,236 | 5,184 | 7,632 |
| 103 | 90 | 10,609 | 8,100 | 9,270 |
| 94 | 91 | 8,836 | 8,281 | 8,554 |
| 1,207 | 975 | 123,561 | 86,487 | 94,322 |

The following example presents a problem with similar data structure where the target of investigation is a possible relationship between a woman's age and her systolic blood pressure.

***Example 2.9***    The data in Table 2.14 represent systolic blood pressure readings on 15 women. We set up a work table (Table 2.15) as in Example 2.8. Using these totals, we obtain

$$r = \frac{146,260 - [(984)(2193)]/15}{\sqrt{[67,954 - (984)^2/15][325,889 - (2193)^2/15]}}$$

$$= 0.566$$

indicating a moderately positive association.

**TABLE 2.14**

| Age $(x)$ | SBP $(y)$ | Age $(x)$ | SBP $(y)$ |
|---|---|---|---|
| 42 | 130 | 85 | 162 |
| 46 | 115 | 72 | 158 |
| 42 | 148 | 64 | 155 |
| 71 | 100 | 81 | 160 |
| 80 | 156 | 41 | 125 |
| 74 | 162 | 61 | 150 |
| 70 | 151 | 75 | 165 |
| 80 | 156 | | |

**TABLE 2.15**

| $x$ | $y$ | $x^2$ | $y^2$ | $xy$ |
|---|---|---|---|---|
| 42 | 130 | 1,764 | 16,900 | 5,460 |
| 46 | 115 | 2,116 | 13,225 | 5,290 |
| 42 | 148 | 1,764 | 21,904 | 6,216 |
| 71 | 100 | 5,041 | 10,000 | 7,100 |
| 80 | 156 | 6,400 | 24,336 | 12,480 |
| 74 | 162 | 5,476 | 26,224 | 11,988 |
| 70 | 151 | 4,900 | 22,801 | 10,570 |
| 80 | 156 | 6,400 | 24,336 | 12,480 |
| 85 | 162 | 7,225 | 26,224 | 13,770 |
| 72 | 158 | 5,184 | 24,964 | 11,376 |
| 64 | 155 | 4,096 | 24,025 | 9,920 |
| 81 | 160 | 6,561 | 25,600 | 12,960 |
| 41 | 125 | 1,681 | 15,625 | 5,125 |
| 61 | 150 | 3,721 | 22,500 | 9,150 |
| 75 | 165 | 5,625 | 27,225 | 12,375 |
| 984 | 2,193 | 67,954 | 325,889 | 146,260 |

### 2.4.2 Nonparametric Correlation Coefficients

Suppose that the data set consists of $n$ pairs of observations $\{(x_i, y_i)\}$, expressing a possible relationship between two continuous variables. We characterize the strength of such a relationship by calculating the coefficient of correlation:

$$r = \frac{\sum(x - \bar{x})(y - \bar{y})}{\sqrt{[\sum(x - \bar{x})^2][\sum(y - \bar{y})^2]}}$$

called *Pearson's correlation coefficient*. Like other common statistics, such as the mean $\bar{x}$ and the standard deviation $s$, the correlation coefficient $r$ is very sensitive to extreme observations. We may be interested in calculating a measure of association that is more robust with respect to outlying values. There are not one but two nonparametric procedures: Spearman's rho and Kendall's tau rank correlations.

*Spearman's Rho* Spearman's rank correlation is a direct nonparametric counterpart of Pearson's correlation coefficient. To perform this procedure, we first arrange the $x$ values from smallest to largest and assign a rank from 1 to $n$ for each value; let $R_i$ be the rank of value $x_i$. Similarly, we arrange the $y$ values from smallest to largest and assign a rank from 1 to $n$ for each value; let $S_i$ be the rank of value $y_i$. If there are tied observations, we assign an average rank, averaging the ranks that the tied observations take jointly. For example, if the second and third measurements are equal, they are both assigned 2.5 as their

**TABLE 2.16**

| Birth Weight | | Increase in Weight | | | |
|---|---|---|---|---|---|
| x (oz) | Rank R | y (%) | Rank S | R − S | $(R - S)^2$ |
| 112 | 10 | 63 | 3 | 7 | 49 |
| 111 | 9 | 66 | 4 | 5 | 25 |
| 107 | 8 | 72 | 5.5 | 2.5 | 6.25 |
| 119 | 12 | 52 | 2 | 10 | 100 |
| 92 | 4 | 75 | 7 | −3 | 9 |
| 80 | 1 | 118 | 11 | −10 | 100 |
| 81 | 2 | 120 | 12 | −10 | 100 |
| 84 | 3 | 114 | 10 | −7 | 49 |
| 118 | 11 | 42 | 1 | 10 | 100 |
| 106 | 7 | 72 | 5.5 | 1.5 | 2.25 |
| 103 | 6 | 90 | 8 | −2 | 4 |
| 94 | 5 | 91 | 9 | −416 | |
| | | | | | 560.50 |

common rank. The next step is to replace, in the formula of Pearson's correlation coefficient $r$, $x_i$ by its rank $R_i$ and $y_i$ by its rank $S_i$. The result is *Spearman's rho*, a popular rank correlation:

$$\rho = \frac{\sum(R_i - \bar{R})(S_i - \bar{S})}{\sqrt{[\sum(R_i - \bar{R})^2][\sum(S_i - \bar{S})^2]}}$$

$$= 1 - \frac{6\sum(R_i - S_i)^2}{n(n^2 - 1)}$$

The second expression is simpler and easier to use.

***Example 2.10***   Consider again the birth-weight problem of Example 2.8. We have the data given in Table 2.16. Substituting the value of $\sum(R_i - S_i)^2$ into the formula for rho $(\rho)$, we obtain

$$\rho = 1 - \frac{(6)(560.5)}{(12)(143)}$$

$$= -0.96$$

which is very close to the value of $r$ (−0.946) obtained in Example 2.8. This closeness is true when there are few or no extreme observations.

***Kendall's Tau***   Unlike Spearman's rho, the other rank correlation—Kendall's tau $\tau$—is defined and calculated very differently, even though the two correla-

**TABLE 2.17**

|   |   |   |   |   |   |   |   |   |   |   |   |   | Total |
|---|---|---|---|---|---|---|---|---|---|---|---|---|---|
| $x$ | 80 | 81 | 84 | 92 | 94 | 103 | 106 | 107 | 111 | 112 | 118 | 119 | |
| $y$ | 118 | 120 | 114 | 75 | 91 | 90 | 72 | 72 | 66 | 63 | 42 | 52 | |
| $C$ | 1 | 0 | 0 | 2 | 0 | 0 | 0 | 0 | 0 | 0 | 0 | 0 | 3 |
| $D$ | 10 | 10 | 9 | 6 | 7 | 6 | 4 | 4 | 3 | 2 | 0 | 0 | 61 |

tions often yield similar numerical results. The birth-weight problem of Example 2.8 is adapted to illustrate this method with the following steps:

1. The $x$ and $y$ values are presented in two rows; the $x$ values in the first row are arranged from smallest to largest.
2. For each $y$ value in the second row, we count
   (a) The number of larger $y$ values to its right (third row). The sum of these is denoted by $C$.
   (b) The number of smaller $y$ values to its right (fourth row). The sum of these is denoted by $D$.
   $C$ and $D$ are the numbers of concordant and discordant pairs.
3. Kendall's rank correlation is defined by

$$\tau = \frac{C - D}{\frac{1}{2}n(n-1)}$$

***Example 2.11***   For the birth-weight problem above, we have the data given in Table 2.17. The value of Kendall's tau is

$$\tau = \frac{3 - 61}{\frac{1}{2}(12)(11)}$$
$$= -0.88$$

*Note:* A SAS program would include these instructions:

```
PROC CORR PEARSON SPEARMAN KENDALL;
VAR BWEIGHT INCREASE;
```

If there are more than two variable names listed, the CORR procedure will compute correlation coefficients between all pairs of variables.

## 2.5   NOTES ON COMPUTATIONS

In Section 1.4 we covered basic techniques for Microsoft's Excel: how to open/form a spreadsheet, save, and retrieve it. Topics included data-entry steps such as *select and drag*, use of *formula bar*, and bar and pie charts. In this short section we focus on continuous data, covering topics such as the construction of histograms, basic descriptive statistics, and correlation analysis.

**Histograms**   With a frequency table ready, click the *ChartWizard* icon (the one with multiple colored bars on the *standard toolbar* near the top). A box appears with choices (as when you learned to form a bar chart or pie chart); select the column chart type. Then click on *next*.

- For the data range, highlight the frequency column. This can be done by clicking on the first observation and dragging the mouse to the last observation. Then click on *next*.
- To remove the gridlines, click on the gridline tab and uncheck the box. To remove the legend, you can do the same using the legend tab. Now click *finish*.
- The problem is that there are still gaps. To remove these, double-click on a bar of the graph and a new set of options should appear. Click on the options tab and change the gap width from 150 to 0.

### Descriptive Statistics

- First, click the cell you want to fill, then click the *paste function icon*, f*, which will give you—in a box—a list Excel functions available for your use.
- The item you need in this list is *Statistical*; upon hitting this, a new list appears with *function names*, each for a statistical procedure.
- The following are procedures/names we learn in this chapter (alphabetically): AVERAGE: provides the sample mean, GEOMEAN: provides the geometric mean, MEDIAN: provides the sample median, STDEV: provides the standard deviation, and VAR: provides the variance. In each case you can obtain only one statistic at a time. First, you have to enter the *range* containing your sample: for example, D6:D20 (you can see what you are entering on the *formula bar*). The computer will return with a numerical value for the statistic requested in a *preselected* cell.

### Pearson's Coefficient of Correlation

- First, click the cell you want to fill, then click the *paste function icon*, f*, which will give you—in a box—a list of Excel functions available for your use.
- The item you need in this list is *Statistical*; upon hitting this, a new list appears with *function names*, each for a statistical procedure. Click *CORREL*, for correlation.

- In the new box, move the cursor to fill in the X and Y ranges in the two rows marked *Array 1* and *Array 2*. The computer will return with numerical value for the statistic requested, Pearson's correlation coefficient $r$, in a *preselected* cell.

## EXERCISES

**2.1**   Table E2.1 gives the values of serum cholesterol levels for 1067 U.S. men aged 25 to 34 years.

**TABLE E2.1**

| Cholesterol Level (mg/100 mL) | Number of Men |
| --- | --- |
| 80–119 | 13 |
| 120–159 | 150 |
| 160–199 | 442 |
| 200–239 | 299 |
| 240–279 | 115 |
| 280–319 | 34 |
| 320–399 | 14 |
| Total | 1067 |

**(a)** Plot the histogram, frequency polygon, and cumulative frequency graph.

**(b)** Find, approximately, the median using your cumulative frequency graph.

**2.2**   Table E2.2 provides the relative frequencies of blood lead concentrations for two groups of workers in Canada, one examined in 1979 and the other in 1987.

**TABLE E2.2**

| Blood Lead (μg/dL) | Relative Frequency (%) | |
| --- | --- | --- |
| | 1979 | 1987 |
| 0–19 | 11.5 | 37.8 |
| 20–29 | 12.1 | 14.7 |
| 30–39 | 13.9 | 13.1 |
| 40–49 | 15.4 | 15.3 |
| 50–59 | 16.5 | 10.5 |
| 60–69 | 12.8 | 6.8 |
| 70–79 | 8.4 | 1.4 |
| 80+ | 9.4 | 0.4 |

(a) Plot the histogram and frequency polygon for each year on separate graphs.

(b) Plot the two cumulative frequency graphs in one figure.

(c) Find and compare the medians.

2.3    A study on the effects of exercise on the menstrual cycle provides the following ages (years) of menarche (the beginning of menstruation) for 96 female swimmers who began training prior to menarche.

| | | | | | | | |
|---|---|---|---|---|---|---|---|
| 15.0 | 17.1 | 14.6 | 15.2 | 14.9 | 14.4 | 14.7 | 15.3 |
| 13.6 | 15.1 | 16.2 | 15.9 | 13.8 | 15.0 | 15.4 | 14.9 |
| 14.2 | 16.5 | 13.2 | 16.8 | 15.3 | 14.7 | 13.9 | 16.1 |
| 15.4 | 14.6 | 15.2 | 14.8 | 13.7 | 16.3 | 15.1 | 14.5 |
| 16.4 | 13.6 | 14.8 | 15.5 | 13.9 | 15.9 | 16.0 | 14.6 |
| 14.0 | 15.1 | 14.8 | 15.0 | 14.8 | 15.3 | 15.7 | 14.3 |
| 13.9 | 15.6 | 15.4 | 14.6 | 15.2 | 14.8 | 13.7 | 16.3 |
| 15.1 | 14.5 | 13.6 | 15.1 | 16.2 | 15.9 | 13.8 | 15.0 |
| 15.4 | 14.9 | 16.2 | 15.9 | 13.8 | 15.0 | 15.4 | 14.9 |
| 14.2 | 16.5 | 13.4 | 16.5 | 14.8 | 15.1 | 14.9 | 13.7 |
| 16.2 | 15.8 | 15.4 | 14.7 | 14.3 | 15.2 | 14.6 | 13.7 |
| 14.9 | 15.8 | 15.1 | 14.6 | 13.8 | 16.0 | 15.0 | 14.6 |

(a) Form a frequency distribution including relative frequencies and cumulative relative frequencies.

(b) Plot the frequency polygon and cumulative frequency graph.

(c) Find the median and 95th percentile.

2.4    The following are the menarchal ages (in years) of 56 female swimmers who began training after they had reached menarche.

| | | | | | | | |
|---|---|---|---|---|---|---|---|
| 14.0 | 16.1 | 13.4 | 14.6 | 13.7 | 13.2 | 13.7 | 14.3 |
| 12.9 | 14.1 | 15.1 | 14.8 | 12.8 | 14.2 | 14.1 | 13.6 |
| 14.2 | 15.8 | 12.7 | 15.6 | 14.1 | 13.0 | 12.9 | 15.1 |
| 15.0 | 13.6 | 14.2 | 13.8 | 12.7 | 15.3 | 14.1 | 13.5 |
| 15.3 | 12.6 | 13.8 | 14.4 | 12.9 | 14.6 | 15.0 | 13.8 |
| 13.0 | 14.1 | 13.8 | 14.2 | 13.6 | 14.1 | 14.5 | 13.1 |
| 12.8 | 14.3 | 14.2 | 13.5 | 14.1 | 13.6 | 12.4 | 15.1 |

(a) Form a frequency distribution using the same age intervals as in Exercise 2.3. (These intervals may be different from those in Example 2.3.)

(b) Display on the same graph two cumulative frequency graphs, one for the group trained before menarche and one for the group trained after menarche. Compare these two graphs and draw your conclusion.

(c) Find the median and 95th percentile and compare to the results of Exercise 2.3.

**2.5** The following are the daily fat intake (grams) of a group of 150 adult males.

| | | | | | | | | | |
|---|---|---|---|---|---|---|---|---|---|
| 22 | 62 | 77 | 84 | 91 | 102 | 117 | 129 | 137 | 141 |
| 42 | 56 | 78 | 73 | 96 | 105 | 117 | 125 | 135 | 143 |
| 37 | 69 | 82 | 93 | 93 | 100 | 114 | 124 | 135 | 142 |
| 30 | 77 | 81 | 94 | 97 | 102 | 119 | 125 | 138 | 142 |
| 46 | 89 | 88 | 99 | 95 | 100 | 116 | 121 | 131 | 152 |
| 63 | 85 | 81 | 94 | 93 | 106 | 114 | 127 | 133 | 155 |
| 51 | 80 | 88 | 98 | 97 | 106 | 119 | 122 | 134 | 151 |
| 52 | 70 | 76 | 95 | 107 | 105 | 117 | 128 | 144 | 150 |
| 68 | 79 | 82 | 96 | 109 | 108 | 117 | 120 | 147 | 153 |
| 67 | 75 | 76 | 92 | 105 | 104 | 117 | 129 | 148 | 164 |
| 62 | 85 | 77 | 96 | 103 | 105 | 116 | 132 | 146 | 168 |
| 53 | 72 | 72 | 91 | 102 | 101 | 128 | 136 | 143 | 164 |
| 65 | 73 | 83 | 92 | 103 | 118 | 127 | 132 | 140 | 167 |
| 68 | 75 | 89 | 95 | 107 | 111 | 128 | 139 | 148 | 168 |
| 68 | 79 | 82 | 96 | 109 | 108 | 117 | 130 | 147 | 153 |

(a) Form a frequency distribution, including relative frequencies and cumulative relative frequencies.

(b) Plot the frequency polygon and investigate the symmetry of the distribution.

(c) Plot the cumulative frequency graph and find the 25th and 75th percentiles. Also calculate the *midrange* = 75th percentile − 25th percentile. (This is another good descriptive measure of variation; it is similar to the *range* but is less affected by extreme observations.)

**2.6** Refer to the data on daily fat intake in Exercise 2.5.

(a) Calculate the mean using raw data.

(b) Calculate, approximately, the mean using the frequency table obtained in Exercise 2.5.

**2.7** Using the income data of Example 2.5:

(a) Plot the histogram for the white families. Does it have the same shape as that for nonwhite families shown in Figure 2.5?

(b) Plot and compare the two cumulative frequency graphs, whites versus nonwhites, confirming the results shown in Figure 2.7.

(c) Compute, approximately, the means of the two groups and compare the results to the medians referred in Section 2.1.4.

**2.8** Refer to the percentage saturation of bile for the 31 male patients in Example 2.4.

(a) Compute the mean, variance, and standard deviation.

(b) The frequency polygon of Figure 2.3 is based on the grouping (arbitrary choices) shown in Table E2.8. Plot the cumulative frequency graph and obtain, approximately, the median from this graph. How does the answer compare to the exact median (the sixteenth largest saturation percentage)?

**TABLE E2.8**

| Interval (%) | Frequency | Interval (%) | Frequency |
|---|---|---|---|
| 40–49 | 2 | 100–109 | 2 |
| 50–59 | 4 | 110–119 | 5 |
| 60–69 | 3 | 120–129 | 1 |
| 70–79 | 4 | 130–139 | 1 |
| 80–89 | 8 | 140–149 | 0 |
| 90–99 | 1 | | |

**2.9** The study cited in Example 2.4 also provided data (percentage saturation of bile) for 29 women. These percentages were

| | | | | | |
|---|---|---|---|---|---|
| 65 | 58 | 52 | 91 | 84 | 107 |
| 86 | 98 | 35 | 128 | 116 | 84 |
| 76 | 146 | 55 | 75 | 73 | 120 |
| 89 | 80 | 127 | 82 | 87 | 123 |
| 142 | 66 | 77 | 69 | 76 | |

(a) Form a frequency distribution using the same intervals as in Example 2.4 and Exercise 2.8.

(b) Plot in the same graph and compare the two frequency polygons and cumulative frequency graphs: men and women.

(c) Compute the mean, variance, and standard deviation using these new data for women and compare the results to those for men in Exercise 2.8.

(d) Compute and compare the two coefficients of variation, men versus women.

**2.10** The following frequency distribution was obtained for the preoperational percentage hemoglobin values of a group of subjects from a village where there has been a malaria eradication program (MEP):

| Hemoglobin (%) | 30–39 | 40–49 | 50–59 | 60–69 | 70–79 | 80–89 | 90–99 |
|---|---|---|---|---|---|---|---|
| Frequency | 2 | 7 | 14 | 10 | 8 | 2 | 2 |

The results in another group was obtained after MEP:

| 43 | 63 | 63 | 75 | 95 | 75 | 80 | 48 | 62 | 71 | 76 | 90 |
|----|----|----|-----|----|----|----|----|----|----|----|----|
| 51 | 61 | 74 | 103 | 93 | 82 | 74 | 65 | 63 | 53 | 64 | 67 |
| 80 | 77 | 60 | 69  | 73 | 76 | 91 | 55 | 65 | 69 | 84 | 78 |
| 50 | 68 | 72 | 89  | 75 | 57 | 66 | 79 | 85 | 70 | 59 | 71 |
| 87 | 67 | 72 | 52  | 35 | 67 | 99 | 81 | 97 | 74 | 61 | 62 |

(a) Form a frequency distribution using the same intervals as in the first table.

(b) Plot in the same graph and compare the two cumulative frequency graphs: before and after the malaria eradication program.

**2.11**  In a study of water pollution, a sample of mussels was taken and lead concentration (milligrams per gram dry weight) was measured from each one. The following data were obtained:

$$\{113.0, 140.5, 163.3, 185.7, 202.5, 207.2\}$$

Calculate the mean $\bar{x}$, variance $s^2$, and standard deviation $s$.

**2.12**  Consider the data taken from a study that examines the response to ozone and sulfur dioxide among adolescents suffering from asthma. The following are measurements of forced expiratory volume (liters) for 10 subjects:

$$\{3.50, 2.60, 2.75, 2.82, 4.05, 2.25, 2.68, 3.00, 4.02, 2.85\}$$

Calculate the mean $\bar{x}$, variance $s^2$, and standard deviation $s$.

**2.13**  The percentage of ideal body weight was determined for 18 randomly selected insulin-dependent diabetics. The outcomes (%) are

| 107 | 119 | 99  | 114 | 120 | 104 | 124 | 88 | 114 |
|-----|-----|-----|-----|-----|-----|-----|----|-----|
| 116 | 101 | 121 | 152 | 125 | 100 | 114 | 95 | 117 |

Calculate the mean $\bar{x}$, variance $s^2$, and standard deviation $s$.

**2.14**  A study on birth weight provided the following data (in ounces) on 12 newborns:

$$\{112, 111, 107, 119, 92, 80, 81, 84, 118, 106, 103, 94\}$$

Calculate the mean $\bar{x}$, variance $s^2$, and standard deviation $s$.

**2.15** The following are the activity values (micromoles per minute per gram of tissue) of a certain enzyme measured in the normal gastric tissue of 35 patients with gastric carcinoma:

| | | | | | | |
|---|---|---|---|---|---|---|
| 0.360 | 1.189 | 0.614 | 0.788 | 0.273 | 2.464 | 0.571 |
| 1.827 | 0.537 | 0.374 | 0.449 | 0.262 | 0.448 | 0.971 |
| 0.372 | 0.898 | 0.411 | 0.348 | 1.925 | 0.550 | 0.622 |
| 0.610 | 0.319 | 0.406 | 0.413 | 0.767 | 0.385 | 0.674 |
| 0.521 | 0.603 | 0.533 | 0.662 | 1.177 | 0.307 | 1.499 |

Calculate the mean $\bar{x}$, variance $s^2$, and standard deviation $s$.

**2.16** The data shown in Table 2.14 represent systolic blood pressure readings on 15 women (see Example 2.9). Calculate the mean $\bar{x}$, variance $s^2$, and standard deviation $s$ for systolic blood pressure and for age.

**2.17** The ages (in days) at time of death for samples of 11 girls and 16 boys who died of sudden infant death syndrome are shown in Table E2.17. Calculate the mean $\bar{x}$, variance $s^2$, and standard deviation $s$ for each group.

**TABLE E2.17**

| Females | Males | |
|---|---|---|
| 53 | 46 | 115 |
| 56 | 52 | 133 |
| 60 | 58 | 134 |
| 60 | 59 | 175 |
| 78 | 77 | 175 |
| 87 | 78 | |
| 102 | 80 | |
| 117 | 81 | |
| 134 | 84 | |
| 160 | 103 | |
| 277 | 114 | |

**2.18** A study was conducted to investigate whether oat bran cereal helps to lower serum cholesterol in men with high cholesterol levels. Fourteen men were placed randomly on a diet that included either oat bran or cornflakes; after two weeks, their low-density lipoprotein (LDL) cholesterol levels were recorded. Each man was then switched to the alternative diet. After a second two-week period, the LDL cholesterol level of each person was recorded again. The data are shown in Table E2.18. Calculate the LDL difference (cornflake—oat bran) for each of the 14 men, then the mean $\bar{x}$, variance $s^2$, and standard deviation $s$ for this sample of differences.

**TABLE E2.18**

| Subject | Cornflakes | Oat Bran |
|---------|------------|----------|
| | LDL (mmol/L) | |
| 1 | 4.61 | 3.84 |
| 2 | 6.42 | 5.57 |
| 3 | 5.40 | 5.85 |
| 4 | 4.54 | 4.80 |
| 5 | 3.98 | 3.68 |
| 6 | 3.82 | 2.96 |
| 7 | 5.01 | 4.41 |
| 8 | 4.34 | 3.72 |
| 9 | 3.80 | 3.49 |
| 10 | 4.56 | 3.84 |
| 11 | 5.35 | 5.26 |
| 12 | 3.89 | 3.73 |
| 13 | 2.25 | 1.84 |
| 14 | 4.24 | 4.14 |

**2.19**  An experiment was conducted at the University of California–Berkeley to study the effect of psychological environment on the anatomy of the brain. A group of 19 rats was randomly divided into two groups. Twelve animals in the treatment group lived together in a large cage furnished with playthings that were changed daily; animals in the control group lived in isolation with no toys. After a month, the experimental animals were killed and dissected. Table E2.19 gives the cortex weights (the thinking part of the brain) in milligrams. Calculate separately for each group the mean $\bar{x}$, variance $s^2$, and standard deviation $s$ of the cortex weight.

**TABLE E2.19**

| Treatment | | Control |
|-----------|-----|---------|
| 707 | 696 | 669 |
| 740 | 712 | 650 |
| 745 | 708 | 651 |
| 652 | 749 | 627 |
| 649 | 690 | 656 |
| 676 | | 642 |
| 699 | | 698 |

**2.20**  Ozone levels around Los Angeles have been measured as high as 220 parts per billion (ppb). Concentrations this high can cause the eyes to burn and are a hazard to both plant and animal life. But what about other cities? The following are data (ppb) on the ozone level obtained in a forested area near Seattle, Washington:

$$
\begin{array}{ccccc}
160 & 165 & 170 & 172 & 161 \\
176 & 163 & 196 & 162 & 160 \\
162 & 185 & 167 & 180 & 168 \\
163 & 161 & 167 & 173 & 162 \\
169 & 164 & 179 & 163 & 178
\end{array}
$$

**(a)** Calculate $\bar{x}$, $s^2$, and $s$; compare the mean to that of Los Angeles.

**(b)** Calculate the coefficient of variation.

**2.21** The systolic blood pressures (in mmHg) of 12 women between the ages of 20 and 35 were measured before and after administration of a newly developed oral contraceptive (Table E2.21). Focus on the column of differences in the table; calculate the mean $\bar{x}$, variance $s^2$, and standard deviation $s$.

**TABLE E2.21**

| Subject | Before | After | After–Before Difference, $d_i$ |
|---------|--------|-------|-------------------------------|
| 1       | 122    | 127   | 5   |
| 2       | 126    | 128   | 2   |
| 3       | 132    | 140   | 8   |
| 4       | 120    | 119   | −1  |
| 5       | 142    | 145   | 3   |
| 6       | 130    | 130   | 0   |
| 7       | 142    | 148   | 6   |
| 8       | 137    | 135   | −2  |
| 9       | 128    | 129   | 1   |
| 10      | 132    | 137   | 5   |
| 11      | 128    | 128   | 0   |
| 12      | 129    | 133   | 4   |

**2.22** A group of 12 hemophiliacs, all under 41 years of age at the time of HIV seroconversion, were followed from primary AIDS diagnosis until death (ideally, we should take as a starting point the time at which a person contracts AIDS rather than the time at which the patient is diagnosed, but this information is unavailable). Survival times (in months) from diagnosis until death of these hemophiliacs were: 2, 3, 6, 6, 7, 10, 15, 15, 16, 27, 30, and 32. Calculate the mean, geometric mean, and median.

**2.23** Suppose that we are interested in studying patients with systemic cancer who subsequently develop a brain metastasis; our ultimate goal is to prolong their lives by controlling the disease. A sample of 23 such patients, all of whom were treated with radiotherapy, were followed from the first day of their treatment until recurrence of the original tumor. Recurrence is defined as the reappearance of a metastasis in exactly the same site, or in the case of patients whose tumor never completely disappeared, enlargement of the original lesion. Times to recurrence (in weeks) for the 23 patients were: 2, 2, 2, 3, 4, 5, 5, 6, 7, 8, 9, 10, 14, 14, 18, 19, 20, 22, 22, 31, 33, 39, and 195. Calculate the mean, geometric mean, and median.

**2.24** A laboratory investigator interested in the relationship between diet and the development of tumors divided 90 rats into three groups and fed them with low-fat, saturated-fat, and unsaturated-fat diets, respectively. The rats were all the same age and species and were in similar physical condition. An identical amount of tumor cells was injected into a foot pad of each rat. The tumor-free time is the time from injection of tumor cells to the time that a tumor develops. All 30 rats in the unsaturated-fat diet group developed tumors; tumor-free times (in days) were: 112, 68, 84, 109, 153, 143, 60, 70, 98, 164, 63, 63, 77, 91, 91, 66, 70, 77, 63, 66, 66, 94, 101, 105, 108, 112, 115, 126, 161, and 178. Calculate the mean, geometric mean, and median.

**2.25** The Data shown in Table E2.25 are from a study that compares adolescents who have bulimia to healthy adolescents with similar body compositions and levels of physical activity. The table provides measures of daily caloric intake (kcal/kg) for random samples of 23 bulimic adolescents and 15 healthy ones.

**TABLE E2.25**

| Bulimic Adolescents | | | Healthy Adolescents | |
| --- | --- | --- | --- | --- |
| 15.9 | 17.0 | 18.9 | 30.6 | 40.8 |
| 16.0 | 17.6 | 19.6 | 25.7 | 37.4 |
| 16.5 | 28.7 | 21.5 | 25.3 | 37.1 |
| 18.9 | 28.0 | 24.1 | 24.5 | 30.6 |
| 18.4 | 25.6 | 23.6 | 20.7 | 33.2 |
| 18.1 | 25.2 | 22.9 | 22.4 | 33.7 |
| 30.9 | 25.1 | 21.6 | 23.1 | 36.6 |
| 29.2 | 24.5 | | 23.8 | |

(a) Calculate and compare the means.

(b) Calculate and compare the variances.

**2.26** Two drugs, amantadine (A) and rimantadine (R), are being studied for use in combating the influenza virus. A single 100-mg dose is administered orally to healthy adults. The response variable is the time (minutes) required to reach maximum concentration. Results are shown in Table E2.26.

**TABLE E2.26**

| Drug A | | | Drug R | | |
|---|---|---|---|---|---|
| 105 | 123 | 124 | 221 | 227 | 280 |
| 126 | 108 | 134 | 261 | 264 | 238 |
| 120 | 112 | 130 | 250 | 236 | 240 |
| 119 | 132 | 130 | 230 | 246 | 283 |
| 133 | 136 | 142 | 253 | 273 | 516 |
| 145 | 156 | 170 | 256 | 271 | |
| 200 | | | | | |

(a) Calculate and compare the means.

(b) Calculate and compare the variances and standard deviations.

(c) Calculate and compare the medians.

**2.27** Data are shown in Table E2.27 for two groups of patients who died of acute myelogenous leukemia. Patients were classified into the two groups according to the presence or absence of a morphologic characteristic of white cells. Patients termed AG positive were identified by the presence of Auer rods and/or significant granulature of the leukemic cells in the bone marrow at diagnosis. For AG-negative patients, these factors were absent. Leukemia is a cancer characterized by an overproliferation of white blood cells; the higher the white blood count (WBC), the more severe the disease.

**TABLE E2.27**

| AG Positive, $N = 17$ | | AG Negative, $N = 16$ | |
|---|---|---|---|
| WBC | Survival Time (weeks) | WBC | Survival Time (weeks) |
| 2,300 | 65 | 4,400 | 56 |
| 750 | 156 | 3,000 | 65 |
| 4,300 | 100 | 4,000 | 17 |
| 2,600 | 134 | 1,500 | 7 |
| 6,000 | 16 | 9,000 | 16 |
| 10,500 | 108 | 5,300 | 22 |
| 10,000 | 121 | 10,000 | 3 |
| 17,000 | 4 | 19,000 | 4 |
| 5,400 | 39 | 27,000 | 2 |
| 7,000 | 143 | 28,000 | 3 |
| 9,400 | 56 | 31,000 | 8 |
| 32,000 | 26 | 26,000 | 4 |
| 35,000 | 22 | 21,000 | 3 |
| 100,000 | 1 | 79,000 | 30 |
| 100,000 | 1 | 100,000 | 4 |
| 52,000 | 5 | 100,000 | 43 |
| 100,000 | 65 | | |

(a) Calculate separately for each group (AG positive and AG negative) the mean $\bar{x}$, variance $s^2$, and standard deviation $s$ for survival time.

(b) Calculate separately for each group (AG positive and AG negative) the mean, geometric mean, and median for white blood count.

**2.28** Refer to the data on systolic blood pressure (mmHg) of 12 women in Exercise 2.21. Calculate Pearson's correlation coefficient, Kendall's tau, and Spearman's rho rank correlation coefficients representing the strength of the relationship between systolic blood pressures measured before and after administration of the oral contraceptive.

**2.29** The following are the heights (measured to the nearest 2 cm) and weights (measured to the nearest kilogram) of 10 men and 10 women. For the men,

| Height | 162 | 168 | 174 | 176 | 180 | 180 | 182 | 184 | 186 | 186 |
|--------|-----|-----|-----|-----|-----|-----|-----|-----|-----|-----|
| Weight | 65  | 65  | 84  | 63  | 75  | 76  | 82  | 65  | 80  | 81  |

and for the women,

| Height | 152 | 156 | 158 | 160 | 162 | 162 | 164 | 164 | 166 | 166 |
|--------|-----|-----|-----|-----|-----|-----|-----|-----|-----|-----|
| Weight | 52  | 50  | 47  | 48  | 52  | 55  | 55  | 56  | 60  | 60  |

(a) Draw a scatter diagram, for men and women separately, to show the association, if any, between height and weight.

(b) Calculate Pearson's correlation coefficient, Kendall's tau, and Spearman's rho rank correlation coefficients of height and weight for men and women separately.

**2.30** Table E2.30 gives the net food supply ($x$, the number of calories per person per day) and the infant mortality rate ($y$, the number of infant

**TABLE E2.30**

| Country | $x$ | $y$ | Country | $x$ | $y$ |
|---------|-----|-----|---------|-----|-----|
| Argentina | 2730 | 98.8 | Iceland | 3160 | 42.4 |
| Australia | 3300 | 39.1 | India | 1970 | 161.6 |
| Austria | 2990 | 87.4 | Ireland | 3390 | 69.6 |
| Belgium | 3000 | 83.1 | Italy | 2510 | 102.7 |
| Burma | 1080 | 202.1 | Japan | 2180 | 60.6 |
| Canada | 3070 | 67.4 | Netherlands | 3010 | 37.4 |
| Chile | 2240 | 240.8 | New Zealand | 3260 | 32.2 |
| Cuba | 2610 | 116.8 | Sweden | 3210 | 43.3 |
| Egypt | 2450 | 162.9 | U.K. | 3100 | 55.3 |
| France | 2880 | 66.1 | U.S. | 3150 | 53.2 |
| Germany | 2960 | 63.3 | Uruguay | 2380 | 94.1 |

deaths per 1000 live births) for certain selected countries before World War II.

**(a)** Draw a scatter diagram to show the association, if any, between between the average daily number of calories per person and the infant mortality rate.

**(b)** Calculate Pearson's correlation coefficient, Kendall's tau, and Spearman's rho rank correlation coefficients.

**2.31** In an assay of heparin, a standard preparation is compared with a test preparation by observing the log clotting times ($y$, in seconds) of blood containing different doses of heparin ($x$ is the log dose) (Table E2.31). Replicate readings are made at each dose level.

**TABLE E2.31**

| Log Clotting Times | | | | |
|---|---|---|---|---|
| Standard | | Test | | Log Dose |
| 1.806 | 1.756 | 1.799 | 1.763 | 0.72 |
| 1.851 | 1.785 | 1.826 | 1.832 | 0.87 |
| 1.954 | 1.929 | 1.898 | 1.875 | 1.02 |
| 2.124 | 1.996 | 1.973 | 1.982 | 1.17 |
| 2.262 | 2.161 | 2.140 | 2.100 | 1.32 |

**(a)** Draw a scatter diagram to show the association, if any, between between the log clotting times and log dose separately for the standard preparation and the test preparation. Do they appear to be linear?

**(b)** Calculate Pearson's correlation coefficient for log clotting times and log dose separately for the standard preparation and the test preparation. Do they appear to be different?

**2.32** When a patient is diagnosed as having cancer of the prostate, an important question in deciding on treatment strategy for the patient is whether or not the cancer has spread to the neighboring lymph nodes. The question is so critical in prognosis and treatment that it is customary to operate on the patient (i.e., perform a laparotomy) for the sole purpose of examining the nodes and removing tissue samples to examine under the microscope for evidence of cancer. However, certain variables that can be measured without surgery are predictive of the nodal involvement; and the purpose of the study presented here was to examine the data for 53 prostate cancer patients receiving surgery, to determine which of five preoperative variables are predictive of nodal involvement. Table E2.32 presents the complete data set. For each of the 53 patients, there are two continuous independent variables (i.e., preoperative factors), age at diagnosis and level of serum acid phosphatase ($\times 100$; called "acid"), and three binary variables: x-ray reading, pathology reading (grade) of a biopsy of the tumor obtained by needle before surgery, and a rough measure of the size and location of the tumor (stage) obtained by palpation with the fingers via the rectum. For these three binary independent variables, a value of 1 signifies a positive or more serious state and a 0 denotes a negative or less serious finding. In addition, the sixth column presents the finding at surgery—the primary outcome of interest, which is binary, a value of 1 denoting nodal involvement, and a value of 0 denoting no nodal involvement found at surgery.

**TABLE E2.32**

| X-ray | Grade | Stage | Age | Acid | Nodes | X-ray | Grade | Stage | Age | Acid | Nodes |
|---|---|---|---|---|---|---|---|---|---|---|---|
| 0 | 1 | 1 | 64 | 40 | 0 | 0 | 0 | 0 | 60 | 78 | 0 |
| 0 | 0 | 1 | 63 | 40 | 0 | 0 | 0 | 0 | 52 | 83 | 0 |
| 1 | 0 | 0 | 65 | 46 | 0 | 0 | 0 | 1 | 67 | 95 | 0 |
| 0 | 1 | 0 | 67 | 47 | 0 | 0 | 0 | 0 | 56 | 98 | 0 |
| 0 | 0 | 0 | 66 | 48 | 0 | 0 | 0 | 1 | 61 | 102 | 0 |
| 0 | 1 | 1 | 65 | 48 | 0 | 0 | 0 | 0 | 64 | 187 | 0 |
| 0 | 0 | 0 | 60 | 49 | 0 | 1 | 0 | 1 | 58 | 48 | 1 |
| 0 | 0 | 0 | 51 | 49 | 0 | 0 | 0 | 1 | 65 | 49 | 1 |
| 0 | 0 | 0 | 66 | 50 | 0 | 1 | 1 | 1 | 57 | 51 | 1 |
| 0 | 0 | 0 | 58 | 50 | 0 | 0 | 1 | 0 | 50 | 56 | 1 |
| 0 | 1 | 0 | 56 | 50 | 0 | 1 | 1 | 0 | 67 | 67 | 1 |
| 0 | 0 | 1 | 61 | 50 | 0 | 0 | 0 | 1 | 67 | 67 | 1 |
| 0 | 1 | 1 | 64 | 50 | 0 | 0 | 1 | 1 | 57 | 67 | 1 |
| 0 | 0 | 0 | 56 | 52 | 0 | 0 | 1 | 1 | 45 | 70 | 1 |
| 0 | 0 | 0 | 67 | 52 | 0 | 0 | 0 | 1 | 46 | 70 | 1 |
| 1 | 0 | 0 | 49 | 55 | 0 | 1 | 0 | 1 | 51 | 72 | 1 |
| 0 | 1 | 1 | 52 | 55 | 0 | 1 | 1 | 1 | 60 | 76 | 1 |
| 0 | 0 | 0 | 68 | 56 | 0 | 1 | 1 | 1 | 56 | 78 | 1 |
| 0 | 1 | 1 | 66 | 59 | 0 | 1 | 1 | 1 | 50 | 81 | 1 |
| 1 | 0 | 0 | 60 | 62 | 0 | 0 | 0 | 0 | 56 | 82 | 1 |
| 0 | 0 | 0 | 61 | 62 | 0 | 0 | 0 | 1 | 63 | 82 | 1 |
| 1 | 1 | 1 | 59 | 63 | 0 | 1 | 1 | 1 | 65 | 84 | 1 |
| 0 | 0 | 0 | 51 | 65 | 0 | 1 | 0 | 1 | 64 | 89 | 1 |
| 0 | 1 | 1 | 53 | 66 | 0 | 0 | 1 | 0 | 59 | 99 | 1 |
| 0 | 0 | 0 | 58 | 71 | 0 | 1 | 1 | 1 | 68 | 126 | 1 |
| 0 | 0 | 0 | 63 | 75 | 0 | 1 | 0 | 0 | 61 | 136 | 1 |
| 0 | 0 | 1 | 53 | 76 | 0 | | | | | | |

*Note:* This is a very long data file; its electronic copy, in a Web-based form, is available from the author upon request.

In Exercise 1.46, we investigated the effects of the three binary pre-operative variables (x-ray, grade, and stage); in this exercise, we focus on the effects of the two continuous factors (age and acid phosphatase). The 53 patients are divided into two groups by the finding at surgery, a group with nodal involvement and a group without (denoted by 1 or 0 in the sixth column). For each group and for each of the two factors age at diagnosis and level of serum acid phosphatase, calculate the mean $\bar{x}$, variance $s^2$, and standard deviation $s$.

**2.33** Refer to the data on cancer of the prostate in Exercise 2.32. Investigate the relationship between age at diagnosis and level of serum acid phosphatase by calculating Pearson's correlation coefficient and draw your conclusion. Repeat this analysis, but analyze the data separately for the two groups, the group with nodal involvement and the group without. Does the nodal involvement seem to have any effect on the strength of this relationship?

**2.34** A study was undertaken to examine the data for 44 physicians working for an emergency department at a major hospital so as to determine which of a number of factors are related to the number of complaints received during the preceding year. In addition to the number of complaints, data available consist of the number of visits—which serves as the *size* for the observation unit, the physician—and four other factors under investigation. Table E2.34 presents the complete data set. For each of the 44 physicians there are two continuous explanatory factors, revenue (dollars per hour) and workload at the emergency service (hours), and two binary variables, gender (female/male) and residency training in emergency services (no/yes). Divide the number of complaints by the number of visits and use this ratio (number of complaints per visit) as the primary *outcome* or *endpoint X*.

(a) For each of the two binary factors, gender (female/male) and residency training in emergency services (no/yes), which divide the 44 physicians into two subgroups—say, men and women—calculate the mean $\bar{x}$ and standard deviation $s$ for the endpoint $X$.

(b) Investigate the relationship between the outcome, number of complaints per visit, and each of two continuous explanatory factors, revenue (dollars per hour) and workload at the emergency service (hours), by calculating Pearson's correlation coefficient, and draw your conclusion.

(c) Draw a scatter diagram to show the association, if any, between the number of complaints per visit and the workload at the emergency service. Does it appear to be linear?

**TABLE E2.34**

| No. of Visits | Complaint | Residency | Gender | Revenue | Hours |
|---|---|---|---|---|---|
| 2014 | 2 | Y | F | 263.03 | 1287.25 |
| 3091 | 3 | N | M | 334.94 | 1588.00 |
| 879 | 1 | Y | M | 206.42 | 705.25 |
| 1780 | 1 | N | M | 226.32 | 1005.50 |
| 3646 | 11 | N | M | 288.91 | 1667.25 |
| 2690 | 1 | N | M | 275.94 | 1517.75 |
| 1864 | 2 | Y | M | 295.71 | 967.00 |
| 2782 | 6 | N | M | 224.91 | 1609.25 |
| 3071 | 9 | N | F | 249.32 | 1747.75 |
| 1502 | 3 | Y | M | 269.00 | 906.25 |
| 2438 | 2 | N | F | 225.61 | 1787.75 |
| 2278 | 2 | N | M | 212.43 | 1480.50 |
| 2458 | 5 | N | M | 211.05 | 1733.50 |
| 2269 | 2 | N | F | 213.23 | 1847.25 |
| 2431 | 7 | N | M | 257.30 | 1433.00 |
| 3010 | 2 | Y | M | 326.49 | 1520.00 |
| 2234 | 5 | Y | M | 290.53 | 1404.75 |
| 2906 | 4 | N | M | 268.73 | 1608.50 |
| 2043 | 2 | Y | M | 231.61 | 1220.00 |
| 3022 | 7 | N | M | 241.04 | 1917.25 |
| 2123 | 5 | N | F | 238.65 | 1506.25 |
| 1029 | 1 | Y | F | 287.76 | 589.00 |
| 3003 | 3 | Y | F | 280.52 | 1552.75 |
| 2178 | 2 | N | M | 237.31 | 1518.00 |
| 2504 | 1 | Y | F | 218.70 | 1793.75 |
| 2211 | 1 | N | F | 250.01 | 1548.00 |
| 2338 | 6 | Y | M | 251.54 | 1446.00 |
| 3060 | 2 | Y | M | 270.52 | 1858.25 |
| 2302 | 1 | N | M | 247.31 | 1486.25 |
| 1486 | 1 | Y | F | 277.78 | 933.75 |
| 1863 | 1 | Y | M | 259.68 | 1168.25 |
| 1661 | 0 | N | M | 260.92 | 877.25 |
| 2008 | 2 | N | M | 240.22 | 1387.25 |
| 2138 | 2 | N | M | 217.49 | 1312.00 |
| 2556 | 5 | N | M | 250.31 | 1551.50 |
| 1451 | 3 | Y | F | 229.43 | 973.75 |
| 3328 | 3 | Y | M | 313.48 | 1638.25 |
| 2927 | 8 | N | M | 293.47 | 1668.25 |
| 2701 | 8 | N | M | 275.40 | 1652.75 |
| 2046 | 1 | Y | M | 289.56 | 1029.75 |
| 2548 | 2 | Y | M | 305.67 | 1127.00 |
| 2592 | 1 | N | M | 252.35 | 1547.25 |
| 2741 | 1 | Y | F | 276.86 | 1499.25 |
| 3763 | 10 | Y | M | 308.84 | 1747.50 |

*Note:* This is a very long data file; its electronic copy, in a Web-based form, is available from the author upon request.

**2.35**  There have been times when the city of London, England, experienced periods of dense fog. Table E2.35 shows such data for a very severe 15-day period, including the number of deaths in each day ($y$), the mean atmospheric smoke ($x_1$, in mg/m$^3$), and the mean atmospheric sulfur dioxide content ($x_2$, in ppm).

**TABLE E2.35**

| Number of Deaths | Smoke | Sulfur Dioxide |
|---|---|---|
| 112 | 0.30 | 0.09 |
| 140 | 0.49 | 0.16 |
| 143 | 0.61 | 0.22 |
| 120 | 0.49 | 0.14 |
| 196 | 2.64 | 0.75 |
| 294 | 3.45 | 0.86 |
| 513 | 4.46 | 1.34 |
| 518 | 4.46 | 1.34 |
| 430 | 1.22 | 0.47 |
| 274 | 1.22 | 0.47 |
| 255 | 0.32 | 0.22 |
| 236 | 0.29 | 0.23 |
| 256 | 0.50 | 0.26 |
| 222 | 0.32 | 0.16 |
| 213 | 0.32 | 0.16 |

(a) Calculate Pearson's correlation coefficient for y and $x_1$.

(b) Calculate Pearson's correlation coefficient for y and $x_2$.

# 3

# PROBABILITY AND
# PROBABILITY MODELS

## 3.1 PROBABILITY

Most of Chapter 1 dealt with proportions. A proportion is defined to represent the relative size of the portion of a population with a certain (binary) characteristic. For example, *disease prevalence* is the proportion of a population with a disease. Similarly, we can talk about the proportion of positive reactors to a certain screening test, the proportion of males in colleges, and so on. A proportion is used as a descriptive measure for a target population with respect to a binary or dichotomous characteristic. It is a number between 0 and 1 (or 100%); the larger the number, the larger the subpopulation with the chacteristic [e.g., 70% male means *more* males (than 50%)].

Now consider a population with certain binary characteristic. A *random selection* is defined as one in which each person has an equal *chance* of being selected. What is the *chance* that a person with the characteristic will be selected (e.g., the chance of selecting, say, a diseased person)? The answer depends on the size of the subpopulation to which he or she belongs (i.e., the proportion). The larger the proportion, the higher the chance (of such a person being selected). That *chance* is measured by the proportion, a number between 0 and 1, called the *probability*. *Proportion* measures size; it is a descriptive statistic. *Probability* measures chance. When we are concerned about the outcome (still *uncertain* at this stage) with a random selection, a proportion (static, no action) becomes a probability (action about to be taken). Think of this simple example about a box containing 100 marbles, 90 of them red and the other 10 blue. If the question is: "Are there red marbles in the box?", someone who saw the box's contents would answer "90%." But if the question is: "If I take one marble at random, do you think I would have a red one?", the answer would be "90% chance." The first 90% represents a proportion; the second 90% indicates the probability. In addition, if we keep taking random selections (called *repeated sampling*), the *accumulated long-term relative frequency* with which an event occurs (i.e., characteristic to be observed) is *equal* to the proportion of the subpopulation with that characteristic. Because of this observation, *proportion* and *probability* are sometimes used interchangingly. In the following sections we deal with the concept of *probability* and some simple applications in making health decisions.

### 3.1.1  Certainty of Uncertainty

Even science is uncertain. Scientists are sometimes wrong. They arrive at different conclusions in many different areas: the effects of a certain food ingredient or of low-level radioactivity, the role of fats in diets, and so on. Many studies are inconclusive. For example, for decades surgeons believed that a radical mastectomy was the only treatment for breast cancer. More recently, carefully designed clinical trials showed that less drastic treatments seem equally effective.

Why is it that science is not always certain? Nature is complex and full of *unexplained biological variability*. In addition, almost all methods of observation and experiment are imperfect. Observers are subject to human bias and error. Science is a continuing story; subjects vary; measurements fluctuate. Biomedical science, in particular, contains controversy and disagreement; with the best of intentions, biomedical data—medical histories, physical examinations, interpretations of clinical tests, descriptions of symptoms and diseases— are somewhat inexact. But most important of all, we always have to deal with incomplete information: It is either impossible, or too costly, or too time consuming, to study the entire population; we often have to rely on information gained from a *sample*—that is, a subgroup of the population under investigation. So some uncertainty almost always prevails. Science and scientists cope with uncertainty by using the concept of *probability*. By calculating probabilities, they are able to describe what has happened and predict what should happen in the future under similar conditions.

### 3.1.2  Probability

The target population of a specific research effort is the entire set of subjects at which the research is aimed. For example, in a screening for cancer in a community, the target population will consist of all persons in that community who are at risk for the disease. For one cancer site, the target population might be all women over the age of 35; for another site, all men over the age of 50.

The *probability* of an event, such as a screening test being positive, in a target population is defined as the *relative frequency* (i.e., *proportion*) with which the event occurs in that target population. For example, the probability of having a disease is the disease prevalence. For another example, suppose that out of $N = 100,000$ persons of a certain target population, a total of 5500 are positive reactors to a certain screening test; then the probability of being positive, denoted by Pr(positive), is

$$\text{Pr(positive)} = \frac{5500}{100,000}$$

$$= 0.055 \quad \text{or} \quad 5.5\%$$

A probability is thus a descriptive measure for a target population with respect to a certain event of interest. It is a number between 0 and 1 (or zero and 100%); the larger the number, the larger the subpopulation. For the case of continuous measurement, we have the probability of being within a certain interval. For example, the probability of a serum cholesterol level between 180 and 210 (mg/100 mL) is the proportion of people in a certain target population who have cholesterol levels falling between 180 and 210 (mg/100 mL). This is measured, in the context of the histogram of Chapter 2, by the area of a rectangular bar for the class (180–210). Now of critical importance in the interpretation of probability is the concept of random sampling so as to associate the concept of probability with uncertainty and chance.

Let the size of the target population be $N$ (usually, a very large number), a sample is any subset—say, $n$ in number ($n < N$)—of the target population. Simple random sampling from the target population is sampling so that every possible sample of size $n$ has an equal chance of selection. For simple random sampling:

1. Each individual draw is uncertain with respect to any event or characteristic under investigation (e.g., having a disease), *but*
2. In repeated sampling from the population, the accumulated long-run relative frequency with which the event occurs is the population relative frequency of the event.

The physical process of random sampling can be carried out as follows (or in a fashion logically equivalent to the following steps).

1. A list of all $N$ subjects in the population is obtained. Such a list is termed a *frame* of the population. The subjects are thus available to an arbitrary numbering (e.g., from 000 to $N = 999$). The frame is often based on a directory (telephone, city, etc.) or on hospital records.
2. A tag is prepared for each subject carrying a number $1, 2, \ldots, N$.
3. The tags are placed in a receptacle (e.g., a box) and mixed thoroughly.
4. A tag is drawn blindly. The number on the tag then identifies the subject from the population; this subject becomes a member of the sample.

Steps 2 to 4 can also be implemented using a table of random numbers (Appendix A). Arbitrarily pick a three-digit column (or four-digit column if the population size is large), and a number selected arbitrarily in that column serves to identify the subject from the population. In practice, this process has been computerized.

We can now link the concepts of probability and random sampling as follows. In the example of cancer screening in a community of $N = 100,000$ persons, the calculated probability of 0.055 is interpreted as: "The probability of a randomly drawn person from the target population having a positive test result is 0.055 or 5.5%." The rationale is as follows. On an initial draw, the subject chosen may or may not be a positive reactor. However, if this process—of randomly drawing one subject at a time from the population—is repeated over and over again a large number of times, the accumulated long-run relative frequency of positive receptors in the sample will approximate 0.055.

### 3.1.3   Statistical Relationship

The data from the cancer screening test of Example 1.4 are reproduced here as Table 3.1. In this design, each member of the population is characterized by two variables: the test result $X$ and the true disease status $Y$. Following our definition above, the probability of a positive test result, denoted $Pr(X = +)$, is

$$Pr(X = +) = \frac{516}{24,103}$$

$$= 0.021$$

and the probability of a negative test result, denoted $Pr(X = -)$, is

$$Pr(X = -) = \frac{23,587}{24,103}$$

$$= 0.979$$

and similarly, the probabilities of having $(Y = +)$ and not having $(Y = -)$ the disease are given by

**TABLE 3.1**

| Disease, $Y$ | Test Result, $X$ | | Total |
|---|---|---|---|
| | $+$ | $-$ | |
| $+$ | 154 | 225 | 379 |
| $-$ | 362 | 23,362 | 23,724 |
| Total | 516 | 23,587 | 24,103 |

$$\Pr(Y = +) = \frac{379}{24,103}$$

$$= 0.015$$

and

$$\Pr(Y = -) = \frac{23,724}{24,103}$$

$$= 0.985$$

Note that the sum of the probabilities for each variable is unity:

$$\Pr(X = +) + \Pr(X = -) = 1.0$$
$$\Pr(Y = +) + \Pr(Y = -) = 1.0$$

This is an example of the *addition rule* of probabilities for mutually exclusive events: One of the two events $(X = +)$ or $(X = -)$ is certain to be true for a person selected randomly from the population.

Further, we can calculate the *joint probabilities*. These are the probabilities for two events—such as having the disease *and* having a positive test result—occurring simultaneously. With two variables, $X$ and $Y$, there are four conditions of outcomes and the associated joint probabilities are

$$\Pr(X = +, Y = +) = \frac{154}{24,103}$$

$$= 0.006$$

$$\Pr(X = +, Y = -) = \frac{362}{24,103}$$

$$= 0.015$$

$$\Pr(X = -, Y = +) = \frac{225}{24,103}$$

$$= 0.009$$

and

$$\Pr(X = -, Y = -) = \frac{23,362}{24,103}$$

$$= 0.970$$

The second of the four joint probabilities, 0.015, represents the probability of a person drawn randomly from the target population having a positive test result but being healthy (i.e., a *false positive*). These joint probabilities and the

**TABLE 3.2**

|  | $X$ | | |
|---|---|---|---|
| $Y$ | $+$ | $-$ | Total |
| $+$ | 0.006 | 0.009 | 0.015 |
| $-$ | 0.015 | 0.970 | 0.985 |
| Total | 0.021 | 0.979 | 1.00 |

marginal probabilities above, calculated separately for $X$ and $Y$, are summarized and displayed in Table 3.2. Observe that the four cell probabilities add to unity [i.e., one of the four events $(X = +, Y = +)$ or $(X = +, Y = -)$ or $(X = -, Y = +)$ or $(X = -, Y = -)$ is certain to be true for a randomly selected individual from the population]. Also note that the joint probabilities in each row (or column) add up to the *marginal* or *univariate probability* at the margin of that row (or column). For example,

$$\Pr(X = +, Y = +) + \Pr(X = -, Y = +) = \Pr(Y = +)$$

$$= 0.015$$

We now consider a third type of probability. For example, the *sensitivity* is expressible as

$$\text{sensitivity} = \frac{154}{379}$$

$$= 0.406$$

calculated for the event $(X = +)$ using the subpopulation having $(Y = +)$. That is, of the total number of 379 persons with cancer, the proportion with a positive test result, is 0.406 or 40.6%. This number, denoted by $\Pr(X = + \mid Y = +)$, is called a *conditional probability* ($Y = +$ being the condition) and is related to the other two types of probability:

$$\Pr(X = + \mid Y = +) = \frac{\Pr(X = +, Y = +)}{\Pr(Y = +)}$$

or

$$\Pr(X = +, Y = +) = \Pr(X = + \mid Y = +) \Pr(Y = +)$$

Clearly, we want to distinguish this conditional probability, $\Pr(X = + \mid Y = +)$, from the *marginal probability*, $\Pr(X = +)$. If they are equal,

$$\Pr(X = + \mid Y = +) = \Pr(X = +)$$

the two events $(X = +)$ and $(Y = +)$ are said to be *independent* (because the condition $Y = +$ does not change the probability of $X = +$) and we have the *multiplication rule* for probabilities of independent events:

$$Pr(X = +, Y = +) = Pr(X = +) \, Pr(Y = +)$$

If the two events are not independent, they have a statistical relationship or we say that they are *statistically associated*. For the screening example above,

$$Pr(X = +) = 0.021$$

$$Pr(X = + \mid Y = +) = 0.406$$

clearly indicating a strong statistical relationship [because $Pr(X = + \mid Y = +) \neq Pr(X = +)$]. Of course, it makes sense to have a strong statistical relationship here; otherwise, the screening is useless. However, it should be emphasized that a statistical association does not necessarily mean that there is a cause and an effect. Unless a relationship is so strong and repeated so constantly that the case is overwhelming, a statistical relationship, especially one observed from a sample (because the totality of population information is rarely available), is only a clue, meaning that more study or confirmation is needed.

It should be noted that there are several different ways to check for the presence of a statistical relationship.

1. *Calculation of the odds ratio.* When $X$ and $Y$ are independent, or not associated statistically, the odds ratio equals 1. Here we refer to the odds ratio value for the population; this value is defined as

$$\text{odds ratio} = \frac{Pr(X = + \mid Y = +)/(Pr(X = - \mid Y = +))}{Pr(X = + \mid Y = -)/(Pr(X = - \mid Y = -))}$$

and can be expressed, equivalently, in terms of the joint probabilities as

$$\text{odds ratio} = \frac{Pr(X = +, Y = +) \, Pr(X = -, Y = -)}{Pr(X = +, Y = -) \, Pr(X = -, Y = +)}$$

and the example above yields

$$OR = \frac{(0.006)(0.970)}{(0.015)(0.009)}$$

$$= 43.11$$

clearly indicating a statistical relationship.

2. *Comparison of conditional probability and unconditional (or marginal) probability*: for example, $\Pr(X = + \mid Y = +)$ versus $\Pr(X = +)$.
3. *Comparison of conditional probabilities*: for example, $\Pr(X = + \mid Y = +)$ versus $\Pr(X = + \mid Y = -)$. The screening example above yields

$$\Pr(X = + \mid Y = +) = 0.406$$

whereas

$$\Pr(X = + \mid Y = -) = \frac{362}{23,724}$$

$$= 0.015$$

again clearly indicating a statistical relationship. It should also be noted that we illustrate the concepts using data from a cancer screening test but that these concepts apply to any cross-classification of two binary factors or variables. The primary aim is to determine whether a statistical relationship is present; Exercise 3.1, for example, deals with relationships between health services and race.

The next two sections present some applications of those simple probability rules introduced in Section 3.1.2: the problem of *when* to use screening tests and the problem of *how* to measure agreement.

### 3.1.4   Using Screening Tests

We have introduced the concept of *conditional probability*. However, it is important to distinguish the two conditional probabilities, $\Pr(X = + \mid Y = +)$ and $\Pr(Y = + \mid X = +)$. In Example 1.4, reintroduced in Section 3.1.3, we have

$$\Pr(X = + \mid Y = +) = \frac{154}{379}$$

$$= 0.406$$

whereas

$$\Pr(Y = + \mid X = +) = \frac{154}{516}$$

$$= 0.298$$

Within the context of screening test evaluation:

**TABLE 3.3**

| Population A | | | | Population B | | |
|---|---|---|---|---|---|---|
| | X | | | | X | |
| Y | + | − | Y | + | − |
| + | 45,000 | 5,000 | + | 9,000 | 1,000 |
| − | 5,000 | 45,000 | − | 9,000 | 81,000 |

1. $\Pr(X = + \mid Y = +)$ and $\Pr(X = - \mid Y = -)$ are the sensitivity and specificity, respectively.
2. $\Pr(Y = + \mid X = +)$ and $\Pr(Y = - \mid X = -)$ are called the *positive predictivity* and *negative predictivity*.

With positive predictivity (or *positive predictive value*), the question is: Given that the test $X$ suggests cancer, what is the probability that, in fact, cancer is present? Rationales for these predictive values are that a test passes through several stages. Initially, the original test idea occurs to a researcher. It must then go through a developmental stage. This may have many aspects (in biochemistry, microbiology, etc.) one of which is in biostatistics: trying the test out on a pilot population. From this developmental stage, the efficiency of the test is characterized by its sensitivity and specificity. An efficient test will then go through an applicational stage with an actual application of $X$ to a target population; and here we are concerned with its predictive values. The simple example given in Table 3.3 shows that unlike sensitivity and specificity, the positive and negative predictive values depend not only on the efficiency of the test but also on the disease prevalence of the target population. In both cases, the test is 90% sensitive and 90% specific. However:

1. Population A has a prevalence of 50%, leading to a positive predictive value of 90%.
2. Population B has a prevalence of 10%, leading to a positive predictive value of 50%.

The conclusion is clear: If a test—even a highly sensitive and highly specific one—is applied to a target population in which the disease prevalence is low (e.g., population screening for a rare disease), the positive predictive value is low. (How does this relate to an important public policy: Should we conduct random testing for AIDS?)

In the actual application of a screening test to a target population (the applicational stage), data on the disease status of individuals are not available (otherwise, screening would not be needed). However, disease prevalences are often available from national agencies and health surveys. Predictive values are then calculated from

$$\frac{\text{positive}}{\text{predictivity}} = \frac{(\text{prevalence})(\text{sensitivity})}{(\text{prevalence})(\text{sensitivity}) + (1 - \text{prevalence})(1 - \text{specificity})}$$

and

$$\frac{\text{negative}}{\text{predictivity}} = \frac{(1 - \text{prevalence})(\text{specificity})}{(1 - \text{prevalence})(\text{specificity}) + (\text{prevalence})(1 - \text{sensitivity})}$$

These formulas, called *Bayes' theorem*, allow us to calculate the predictive values without having data from the application stage. All we need are the disease prevalence (obtainable from federal health agencies) and sensitivity and specificity; these were obtained after the developmental stage. It is not too hard to prove these formulas using the addition and multiplication rules of probability. For example, we have

$$\Pr(Y = + \,|\, X = +) = \frac{\Pr(X = +, Y = +)}{\Pr(X = +)}$$

$$= \frac{\Pr(X = +, Y = +)}{\Pr(X = +, Y = +) + \Pr(X = +, Y = -)}$$

$$= \frac{\Pr(Y = +)\,\Pr(X = + \,|\, Y = +)}{\Pr(Y = +)\,\Pr(X = + \,|\, Y = +) + \Pr(Y = -)\,\Pr(X = + \,|\, Y = -)}$$

$$= \frac{\Pr(Y = +)\,\Pr(X = + \,|\, Y = +)}{\Pr(Y = +)\,\Pr(X = + \,|\, Y = +) + [1 - \Pr(Y = +)][1 - \Pr(X = - \,|\, Y = -)]}$$

which is the first equation for positive predictivity. You can also see, instead of going through formal proofs, our illustration of their validity using the population B data above:

1. Direct calculation of positive predictivity yields

$$\frac{9000}{18,000} = 0.5$$

2. Use of prevalence, sensitivity, and specificity yields

$$\frac{(\text{prevalence})(\text{sensitivity})}{(\text{prevalence})(\text{sensitivity}) + (1 - \text{prevalence})(1 - \text{specificity})}$$

$$= \frac{(0.1)(0.9)}{(0.1)(0.9) + (1 - 0.1)(1 - 0.9)}$$

$$= 0.5$$

**TABLE 3.4**

|  | Observer 2 | | |
| Observer 1 | Category 1 | Category 2 | Total |
| --- | --- | --- | --- |
| Category 1 | $n_{11}$ | $n_{12}$ | $n_{1+}$ |
| Category 2 | $n_{21}$ | $n_{22}$ | $n_{2+}$ |
| Total | $n_{+1}$ | $n_{+2}$ | $n$ |

### 3.1.5  Measuring Agreement

Many research studies rely on an observer's judgment to determine whether a disease, a trait, or an attribute is present or absent. For example, results of ear examinations will certainly have effects on a comparison of competing treatments for ear infection. Of course, the basic concern is the issue of reliability. Sections 1.1.2 and 3.1.4 dealt with an important aspect of reliability, the validity of the assessment. However, to judge a method's validity, an exact method for classification, or *gold standard*, must be available for the calculation of sensitivity and specificity. When an exact method is *not* available, reliability can only be judged *indirectly* in terms of *reproducibility*; the most common way for doing that is measuring the agreement between examiners.

For simplicity, assume that each of two observers independently assigns each of $n$ items or subjects to one of two categories. The sample may then be enumerated in a $2 \times 2$ table (Table 3.4) or in terms of the cell probabilities (Table 3.5). Using these frequencies, we can define:

1. An overall proportion of *concordance*:

$$C = \frac{n_{11} + n_{22}}{n}$$

2. Category-specific proportions of concordance:

$$C_1 = \frac{2n_{11}}{2n_{11} + n_{12} + n_{21}}$$

$$C_2 = \frac{2n_{22}}{2n_{22} + n_{12} + n_{21}}$$

**TABLE 3.5**

|  | Observer 2 | | |
| Observer 1 | Category 1 | Category 2 | Total |
| --- | --- | --- | --- |
| Category 1 | $p_{11}$ | $p_{12}$ | $p_{1+}$ |
| Category 2 | $p_{21}$ | $p_{22}$ | $p_{2+}$ |
| Total | $p_{+1}$ | $p_{+2}$ | 1.0 |

The distinction between concordance and association is that for two responses to be associated perfectly, we require only that we can predict the category on one response from the category of the other response, while for two responses to have a perfect concordance, they must fall into the identical category. However, the proportions of concordance, overall or category-specific, do not measure agreement. Among other reasons, they are affected by the marginal totals. One possibility is to compare the overall concordance,

$$\theta_1 = \sum_i p_{ii}$$

where $p$'s are the proportions in the second $2 \times 2$ table above, with the *chance concordance*,

$$\theta_2 = \sum_i p_{i+} p_{+i}$$

which occurs if the row variable is independent of the column variable, because if two events are independent, the probability of their joint occurrence is the product of their individual or marginal probabilities (the multiplication rule). This leads to a measure of agreement,

$$\kappa = \frac{\theta_1 - \theta_2}{1 - \theta_2}$$

called the *kappa statistic*, $0 \leq \kappa \leq 1$, which can be expressed as

$$\kappa = \frac{2(n_{11}n_{22} - n_{12}n_{21})}{n_{1+}n_{+2} + n_{+1}n_{2+}}$$

and the following are guidelines for the evaluation of kappa in clinical research:

$$\kappa > 0.75: \quad \text{excellent reproducibility}$$
$$0.40 \leq \kappa \leq 0.75: \quad \text{good reproducibility}$$
$$0 \leq \kappa < 0.40: \quad \text{marginal/poor reproducibility}$$

In general, reproducibility that is not good indicates the need for multiple assessment.

**Example 3.1**  Two nurses perform ear examinations, focusing on the color of the eardrum (tympanic membrane); each independently assigns each of 100 ears to one of two categories: (a) normal or gray, or (b) not normal (white, pink, orange, or red). The data are shown in Table 3.6. The result,

**TABLE 3.6**

|  | Nurse 2 | | |
| Nurse 1 | Normal | Not Normal | Total |
| --- | --- | --- | --- |
| Normal | 35 | 10 | 45 |
| Not normal | 20 | 35 | 55 |
| Total | 55 | 45 | 100 |

$$\kappa = \frac{(2)[(35)(35) - (20)(10)]}{(45)(45) + (55)(55)}$$

$$= 0.406$$

indicates that the agreement is barely acceptable.

It should also be pointed out that:

1. The kappa statistic, as a measure for agreement, can also be used when there are more than two categories for classification:

$$\kappa = \frac{\sum_i p_{ii} - \sum_i p_{i+}p_{+i}}{1 - \sum_i p_{i+}p_{+i}}$$

2. We can form category-specific kappa statistics (e.g., with two categories); we have

$$\kappa_1 = \frac{p_{11} - p_{1+}p_{+1}}{1 - p_{1+}p_{+1}}$$

$$\kappa_2 = \frac{p_{22} - p_{2+}p_{+2}}{1 - p_{2+}p_{+2}}$$

3. The major problem with kappa is that it approaches zero (even with a high degree of agreement) if the prevalence is near 0 or near 1.

## 3.2  NORMAL DISTRIBUTION

### 3.2.1  Shape of the Normal Curve

The histogram of Figure 2.3 is reproduced here as Figure 3.1 (for numerical details, see Table 2.2). A close examination shows that in general, the relative frequencies (or densities) are greatest in the vicinity of the intervals 20–29, 30–39, and 40–49 and decrease as we go toward both extremes of the range of measurements.

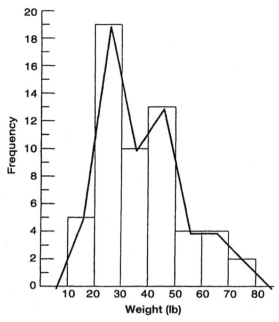

**Figure 3.1** Distribution of weights of 57 children.

Figure 3.1 shows a distribution based on a total of 57 children; the frequency distribution consists of intervals with a width of 10 lb. Now imagine that we increase the number of children to 50,000 and decrease the width of the intervals to 0.01 lb. The histogram would now look more like the one in Figure 3.2, where the step to go from one rectangular bar to the next is very small. Finally, suppose that we increase the number of children to 10 million and decrease the width of the interval to 0.00001 lb. You can now imagine a histogram with bars having practically no widths and thus the steps have all but disappeared. If we continue to increase the size of the data set and decrease the interval width, we eventually arrive at a smooth curve superimposed on the histogram of Figure 3.2 called a *density curve*. You may already have heard about the *normal distribution*; it is described as being a bell-shaped distribution, sort of like a handlebar moustache, similar to Figure 3.2. The name may suggest that most dis-

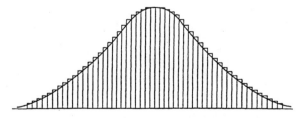

**Figure 3.2** Histogram based on a large data set of weights.

tributions in nature are normal. Strictly speaking, that is false. Even more strictly speaking, they *cannot be exactly normal.* Some, such as heights of adults of a particular gender and race, are amazingly close to normal, *but never exactly.*

The normal distribution is extremely useful in statistics, but for a very different reason—not because it occurs in nature. Mathematicians proved that for samples that are "big enough," values of their sample means, $\bar{x}'s$ (including sample proportions as a special case), are approximately distributed as normal, even if the samples are taken from really strangely shaped distributions. This important result is called the *central limit theorem.* It is as important to statistics as the understanding of germs is to the understanding of disease. Keep in mind that "normal" is just a name for this curve; if an attribute is not distributed normally, it does not imply that it is "abnormal." Many statistics texts provide statistical procedures for finding out whether a distribution is normal, but they are beyond the scope of this book.

From now on, to distinguish samples from populations (a sample is a subgroup of a population), we adopt the set of notations defined in Table 3.7. Quantities in the second column ($\mu$, $\sigma^2$, and $\pi$) are parameters representing numerical properties of populations; $\mu$ and $\sigma^2$ for continuously measured information and $\pi$ for binary information. Quantities in the first column ($\bar{x}$, $s^2$, and $p$) are statistics representing summarized information from samples. Parameters are fixed (constants) but unknown, and each statistic can be used as an estimate for the parameter listed in the same row of the foregoing table. For example, $\bar{x}$ is used as an estimate of $\mu$; this topic is discussed in more detail in Chapter 4. A major problem in dealing with statistics such as $\bar{x}$ and $p$ is that if we take a different sample—even using the same sample size—values of a statistic change from sample to sample. The central limit theorem tells us that if sample sizes are fairly large, values of $\bar{x}$ (or $p$) in repeated sampling have a very nearly normal distribution. Therefore, to handle variability due to *chance,* so as to be able to declare—for example—that a certain observed difference is more than would occur by chance but is real, we first have to learn how to calculate probabilities associated with *normal curves.*

The term *normal curve,* in fact, refers not to one curve but to a family of curves, each characterized by a mean $\mu$ and a variance $\sigma^2$. In the special case where $\mu = 0$ and $\sigma^2 = 1$, we have the *standard normal curve.* For a given $\mu$ and

**TABLE 3.7**

| Quantity | Notation | |
|---|---|---|
| | Sample | Population |
| Mean | $\bar{x}$ (x-bar) | $\mu$ (mu) |
| Variance | $s^2$ (s squared) | $\sigma^2$ (sigma squared) |
| Standard deviation | $s$ | $\sigma$ |
| Proportion | $p$ | $\pi$ (pi) |

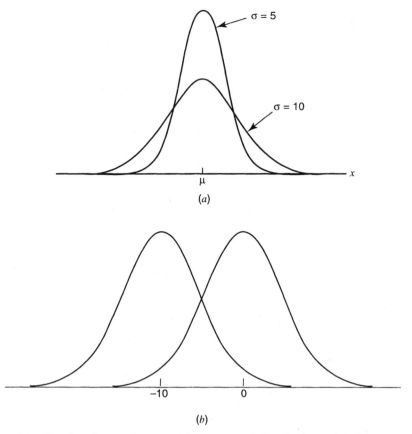

**Figure 3.3**  Family of normal curves: (*a*) two normal distributions with the same mean but different variances; (*b*) two normal distributions with the same variance but different means.

a given $\sigma^2$, the curve is bell-shaped with the tails dipping down to the baseline. In theory, the tails get closer and closer to the baseline but never touch it, proceeding to infinity in either direction. In practice, we ignore that and work within practical limits.

The peak of the curve occurs at the mean $\mu$ (which for this special distribution is also median and mode), and the height of the curve at the peak depends, inversely, on the variance $\sigma^2$. Figure 3.3 shows some of these curves.

### 3.2.2  Areas under the Standard Normal Curve

A variable that has a normal distribution with mean $\mu = 0$ and variance $\sigma^2 = 1$ is called the *standard normal variate* and is commonly designated by the letter $Z$. As with any continuous variable, probability calculations here are always

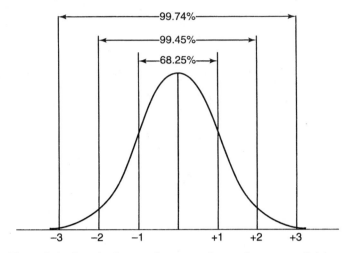

**Figure 3.4**    Standard normal curve and some important divisions.

concerned with finding the probability that the variable assumes any value in an interval between two specific points $a$ and $b$. The probability that a continuous variable assumes a value between two points $a$ and $b$ is the area under the graph of the density curve between $a$ and $b$; the vertical axis of the graph represents the densities as defined in Chapter 2. The total area under any such curve is unity (or 100%), and Figure 3.4 shows the standard normal curve with some important divisions. For example, about 68% of the area is contained within $\pm 1$:

$$\Pr(-1 < z < 1) = 0.6826$$

and about 95% within $\pm 2$:

$$\Pr(-2 < z < 2) = 0.9545$$

More areas under the standard normal curve have been computed and are available in tables, one of which is our Appendix B. The entries in the table of Appendix B give the area under the standard normal curve between the mean ($z = 0$) and a specified positive value of $z$. Graphically, it is represented by the shaded region in Figure 3.5.

Using the table of Appendix B and the symmetric property of the standard normal curve, we show how some other areas are computed. [With access to some computer packaged program, these can be obtained easily; see Section 3.5. However, we believe that these practices do add to the learning, even though they may no longer be needed.]

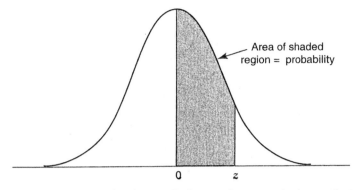

**Figure 3.5**    Area under the standard normal curve as in Appendix B.

***How to Read the Table in Appendix B***    The entries in Appendix B give the area under the standard normal curve between zero and a positive value of $z$. Suppose that we are interested in the area between $z = 0$ and $z = 1.35$ (numbers are first rounded off to two decimal places). To do this, first find the row marked with 1.3 in the left-hand column of the table, and then find the column marked with .05 in the top row of the table $(1.35 = 1.30 + 0.05)$. Then looking in the body of the table, we find that the "1.30 row" and the ".05 column" intersect at the value .4115. This number, 0.4115, is the desired area between $z = 0$ and $z = 1.35$. A portion of Appendix B relating to these steps is shown in Table 3.8. Another example: The area between $z = 0$ and $z = 1.23$ is 0.3907; this value is found at the intersection of the "1.2 row" and the ".03 column" of the table.

Inversely, given the area between zero and some positive value $z$, we can find that value of $z$. Suppose that we are interested in a $z$ value such that the area between zero and $z$ is 0.20. To find this $z$ value, we look into the body of the table to find the tabulated area value nearest to 0.20, which is .2019. This number is found at the intersection of the ".5 row" and the ".03 column." Therefore, the desired $z$ value is 0.53 $(0.53 = 0.50 + 0.03)$.

**TABLE 3.8**

| $z$ | .00 | .01 | .02 | .03 | .04 | .05 | etc... |
|-----|-----|-----|-----|-----|-----|-----|--------|
| .0 | | | | | | | |
| .1 | | | | | | | |
| .2 | | | | | | | |
| ⋮ | | | | | | | |
| 1.3 | | | | | | | .4115 |
| ⋮ | | | | | | | |

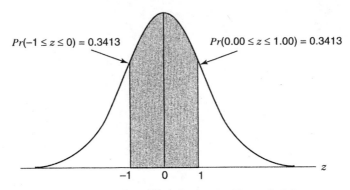

$Pr(-1 \leq z \leq 0) = 0.3413$    $Pr(0.00 \leq z \leq 1.00) = 0.3413$

$-1 \quad 0 \quad 1$    $z$

**Figure 3.6**    Graphical display for Example 3.2.

***Example 3.2***    What is the probability of obtaining a $z$ value between $-1$ and 1? We have

$$Pr(-1 \leq z \leq 1) = Pr(-1 \leq z \leq 0) + Pr(0 \leq z \leq 1)$$
$$= 2 \times Pr(0 \leq z \leq 1)$$
$$= (2)(0.3413)$$
$$= 0.6826$$

which confirms the number listed in Figure 3.4. This area is shown graphically in Figure 3.6.

***Example 3.3***    What is the probability of obtaining a $z$ value of at least 1.58? We have

$$Pr(z \geq 1.58) = 0.5 - Pr(0 \leq z \leq 1.58)$$
$$= 0.5 - 0.4429$$
$$= 0.0571$$

and this probability is shown in Figure 3.7.

***Example 3.4***    What is the probability of obtaining a $z$ value of $-0.5$ or larger? We have

$$Pr(z \geq -0.5) = Pr(-0.5 \leq z \leq 0) + Pr(0 \leq z)$$
$$= Pr(0 \leq z \leq 0.5) + Pr(0 \leq z)$$
$$= 0.1915 + 0.5$$
$$= 0.6915$$

and this probability is shown in Figure 3.8.

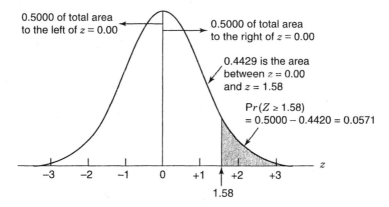

**Figure 3.7**   Graphical display for Example 3.3.

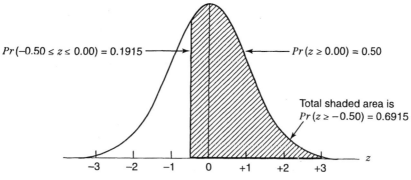

**Figure 3.8**   Graphical display for Example 3.4.

***Example 3.5***   What is the probability of obtaining a $z$ value between 1.0 and 1.58? We have

$$\Pr(1.0 \le z \le 1.58) = \Pr(0 \le z \le 1.58) - \Pr(0 \le z \le 1.0)$$
$$= 0.4429 - 0.3413$$
$$= 0.1016$$

and this probability is shown in Figure 3.9.

***Example 3.6***   Find a $z$ value such that the probability of obtaining a larger $z$ value is only 0.10. We have

$$\Pr(z \ge ?) = 0.10$$

and this is illustrated in Figure 3.10. Scanning the table in Appendix B, we find

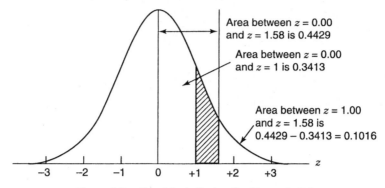

**Figure 3.9**   Graphical display for Example 3.5.

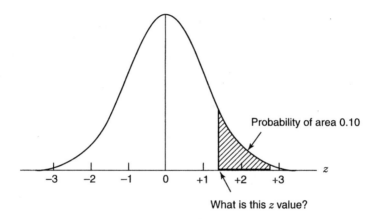

**What is this $z$ value?**

**Figure 3.10**   Graphical display for Example 3.6.

.3997 (area between 0 and 1.28), so that

$$\Pr(z \geq 1.28) = 0.5 - \Pr(0 \leq z \leq 1.28)$$

$$= 0.5 - 0.3997$$

$$\simeq 0.10$$

In terms of the question asked, there is approximately a 0.1 probability of obtaining a $z$ value of 1.28 or larger.

### 3.2.3   Normal Distribution as a Probability Model

The reason we have been discussing the standard normal distribution so extensively with many examples is that probabilities for all normal distributions are computed using the standard normal distribution. That is, when we have a normal distribution with a given mean $\mu$ and a given standard deviation $\sigma$ (or variance $\sigma^2$), we answer probability questions about the distribution by first converting (or standardizing) to the standard normal:

$$z = \frac{x - \mu}{\sigma}$$

Here we interpret the $z$ value (or $z$ *score*) as the number of standard deviations from the mean.

***Example 3.7***   If the total cholesterol values for a certain target population are approximately normally distributed with a mean of 200 (mg/100 mL) and a standard deviation of 20 (mg/100 mL), the probability that a person picked at random from this population will have a cholesterol value greater than 240 (mg/100 mL) is

$$\Pr(x \geq 240) = \Pr\left(\frac{x - 200}{20} \geq \frac{240 - 200}{20}\right)$$
$$= \Pr(z \geq 2.0)$$
$$= 0.5 - \Pr(z \leq 2.0)$$
$$= 0.5 - 0.4772$$
$$= 0.0228 \quad \text{or} \quad 2.28\%$$

***Example 3.8***   Figure 3.11 is a model for hypertension and hypotension (*Journal of the American Medical Association*, 1964), presented here as a simple illustration on the use of the normal distribution; acceptance of the model itself is not universal.

Data from a population of males were collected by age as shown in Table 3.9. From this table, using Appendix B, systolic blood pressure limits for each group can be calculated (Table 3.10). For example, the highest healthy limit for the 20–24 age group is obtained as follows:

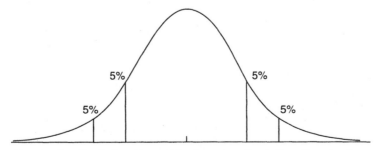

Hypotensive  Borderline  Normal blood pressure  Borderline  Hypertensive

**Figure 3.11**   Graphical display of a hypertension model.

**TABLE 3.9**

| Age (Years) | Systolic Blood Pressure (mmHg) | |
|---|---|---|
| | Mean | Standard Deviation |
| 16 | 118.4 | 12.17 |
| 17 | 121.0 | 12.88 |
| 18 | 119.8 | 11.95 |
| 19 | 121.8 | 14.99 |
| 20–24 | 123.9 | 13.74 |
| 25–29 | 125.1 | 12.58 |
| 30–34 | 126.1 | 13.61 |
| 35–39 | 127.1 | 14.20 |
| 40–44 | 129.0 | 15.07 |
| 45–54 | 132.3 | 18.11 |
| 55–64 | 139.8 | 19.99 |

**TABLE 3.10**

| Age | Hypotension if below: | Lowest Healthy | Highest Healthy | Hypertension if above: |
|---|---|---|---|---|
| 16 | 98.34 | 102.80 | 134.00 | 138.46 |
| 17 | 99.77 | 104.49 | 137.51 | 142.23 |
| 18 | 100.11 | 104.48 | 135.12 | 139.49 |
| 19 | 97.10 | 102.58 | 141.02 | 146.50 |
| 20–24 | ? | ? | ? | ? |
| 25–29 | ? | ? | ? | ? |
| 30–34 | 103.67 | 108.65 | 143.55 | 148.53 |
| 35–39 | 103.70 | 108.90 | 145.30 | 150.50 |
| 40–44 | 104.16 | 109.68 | 148.32 | 153.84 |
| 45–54 | 102.47 | 109.09 | 155.41 | 162.03 |
| 55–64 | 106.91 | 114.22 | 165.38 | 172.74 |

$$\Pr(x \geq ?) = 0.10$$

$$= \Pr\left(\frac{x - 123.9}{13.74} \geq \frac{? - 123.9}{13.74}\right)$$

and from Example 3.5 we have

$$1.28 = \frac{? - 123.9}{13.74}$$

leading to

$$? = 123.9 + (1.28)(13.74)$$

$$= 141.49$$

## 3.3  PROBABILITY MODELS FOR CONTINUOUS DATA

In Section 3.2 we treated the family of normal curves very informally because it was intended to reach more students and readers for whom mathematical formulas may not be very relevant. In this section we provide some supplementary information that may be desirable for those who may be more interested in the fundamentals of biostatistical inference.

A class of measurements or a characteristic on which individual observations or measurements are made is called a *variable*. If values of a variable may theoretically lie anywhere on a numerical scale, we have a *continuous variable*; examples include weight, height, and blood pressure, among others. We saw in Section 3.2 that each continuous variable is characterized by a smooth *density curve*. Mathematically, a curve can be characterized by an equation of the form

$$y = f(x)$$

called a *probability density function*, which includes one or several parameters; the total area under a density curve is 1.0. The probability that the variable assumes any value in an interval between two specific points $a$ and $b$ is given by

$$\int_a^b f(x)\, dx$$

The probability density function for the family of normal curves, sometimes referred to as the *Gaussian distribution*, is given by

$$f(x) = \frac{1}{\sigma\sqrt{2\pi}} \exp\left[-\frac{1}{2}\left(\frac{x-\mu}{\sigma}\right)^2\right] \qquad \text{for } -\infty < x < \infty$$

The meaning and significance of the parameters $\mu$ and $\sigma/\sigma^2$ have been discussed in Section 3.2; $\mu$ is the mean, $\sigma^2$ is the variance, and $\sigma$ is the standard deviation. When $\mu = 1$ and $\sigma^2 = 1$, we have the *standard normal distribution*. The numerical values listed in Appendix B are those given by

$$\int_0^z \frac{1}{\sqrt{2\pi}} \exp\left[-\frac{1}{2}(x)^2\right] dx$$

The normal distribution plays an important role in statistical inference because:

1. Many real-life distributions are approximately normal.
2. Many other distributions can be almost normalized by appropriate data transformations (*e.g.*, taking the log). When log $X$ has a normal distribution, $X$ is said to have a *lognormal distribution*.

3. As a sample size increases, the means of samples drawn from a population of any distribution will approach the normal distribution. This theorem, when stated rigorously, is known as the central limit theorem (more details in Chapter 4).

In addition to the normal distribution (Appendix B), topics introduced in subsequent chapters involve three other continuous distributions:

- The $t$ distribution (Appendix C)
- The chi-square distribution (Appendix D)
- The $F$ distribution (Appendix E)

The $t$ distribution is similar to the standard normal distribution in that it is unimodal, bell-shaped, and symmetrical; extends infinitely in either direction; and has a *mean* of zero. This is a family of curves, each indexed by a number called *degrees of freedom* (df). Given a sample of continuous data, the degrees of freedom measure the quantity of information available in a data set that can be used for estimating the population variance $\sigma^2$ (i.e., $n - 1$, the denominator of $s^2$). The $t$ curves have "thicker" tails than those of the standard normal curve; their variance is slightly greater than $1$ $[= \mathrm{df}/(\mathrm{df} - 2)]$. However, the area under each curve is still equal to unity (or 100%). Areas under a curve from the right tail, shown by the shaded region, are listed in Appendix C; the $t$ distribution for infinite degrees of freedom is precisely equal to the standard normal distribution. This equality is readily seen by examining the column marked, say, "Area = .025." The last row (infinite df) shows a value of 1.96, which can be verified using Appendix B.

Unlike the normal and $t$ distributions, the chi-square and $F$ distributions are concerned with nonnegative attributes and will be used only for certain "tests" in Chapter 6 (chi-square distribution) and Chapter 7 ($F$ distribution). Similar to the case of the $t$ distribution, the formulas for the probability distribution functions of the chi-square and $F$ distributions are rather complex mathematically and are not presented here. Each chi-square distribution is indexed by a number called the *degrees of freedom $r$*. We refer to it as the chi-square distribution with $r$ degrees of freedom; its mean and variance are $r$ and $2r$, respectively. An $F$ distribution is indexed by 2 degrees of freedom $(m, n)$.

## 3.4   PROBABILITY MODELS FOR DISCRETE DATA

Again, a class of measurements or a characteristic on which individual observations or measurements are made is called a *variable*. If values of a variable may lie at only a few isolated points, we have a *discrete variable*; examples include race, gender, or some sort of artificial grading. Topics introduced in subsequent chapters involve two of these discrete distributions: the binomial distribution and the Poisson distribution.

### 3.4.1  Binomial Distribution

In Chapter 1 we discussed cases with dichotomous outcomes such as male–female, survived–not survived, infected–not infected, white–nonwhite, or simply positive–negative. We have seen that such data can be summarized into proportions, rates, and ratios. In this section we are concerned with the probability of a compound event: the occurrence of $x$ (positive) outcomes ($0 \le x \le n$) in $n$ trials, called a *binomial probability*. For example, if a certain drug is known to cause a side effect 10% of the time and if five patients are given this drug, what is the probability that four or more experience the side effect?

Let $S$ denote a side-effect outcome and $N$ an outcome without side effects. The process of determining the chance of $x$ $S$'s in $n$ trials consists of listing all the possible mutually exclusive outcomes, calculating the probability of each outcome using the multiplication rule (where the trials are assumed to be independent), and then combining the probability of all those outcomes that are compatible with the desired results using the addition rule. With five patients there are 32 mutually exclusive outcomes, as shown in Table 3.11.

Since the results for the five patients are independent, the multiplication rule produces the probabilities shown for each combined outcome. For example:

- The probability of obtaining an outcome with four $S$'s and one $N$ is

$$(0.1)(0.1)(0.1)(0.1)(1 - 0.1) = (0.1)^4(0.9)$$

- The probability of obtaining all five $S$'s is

$$(0.1)(0.1)(0.1)(0.1)(0.1) = (0.1)^5$$

Since the event "all five with side effects" corresponds to only one of the 32 outcomes above and the event "four with side effects and one without" pertains to five of the 32 outcomes, each with probability $(0.1)^4(0.9)$, the addition rule yields a probability

$$(0.1)^5 + (5)(0.1)^4(0.9) = 0.00046$$

for the compound event that "four or more have side effects." In general, the binomial model applies when each trial of an experiment has two possible outcomes (often referred to as "failure" and "success" or "negative" and "positive"; one has a success when the primary outcome is observed). Let the probabilities of failure and success be, respectively, $1 - \pi$ and $\pi$, and we "code" these two outcomes as 0 (zero successes) and 1 (one success). The experiment consists of $n$ repeated trials satisfying these assumptions:

1. The $n$ trials are all independent.
2. The parameter $\pi$ is the same for each trial.

**TABLE 3.11**

| First Patient | Second Patient | Third Patient | Fourth Patient | Fifth Patient | Probability | Number of Patients having Side Effects |
|---|---|---|---|---|---|---|
| S | S | S | S | S | $(0.1)^5$ | → 5 |
| S | S | S | S | N | $(0.1)^4(0.9)$ | → 4 |
| S | S | S | N | S | $(0.1)^4(0.9)$ | → 4 |
| S | S | S | N | N | $(0.1)^3(0.9)^2$ | 3 |
| S | S | N | S | S | $(0.1)^4(0.9)$ | → 4 |
| S | S | N | S | N | $(0.1)^3(0.9)^2$ | 3 |
| S | S | N | N | S | $(0.1)^3(0.9)^2$ | 3 |
| S | S | N | N | N | $(0.1)^2(0.9)^3$ | 2 |
| S | N | S | S | S | $(0.1)^4(0.9)$ | → 4 |
| S | N | S | S | N | $(0.1)^3(0.9)^2$ | 3 |
| S | N | S | N | S | $(0.1)^3(0.9)^2$ | 3 |
| S | N | S | N | N | $(0.1)^2(0.9)^3$ | 2 |
| S | N | N | S | S | $(0.1)^3(0.9)^2$ | 3 |
| S | N | N | S | N | $(0.1)^2(0.9)^3$ | 2 |
| S | N | N | N | S | $(0.1)^2(0.9)^3$ | 2 |
| S | N | N | N | N | $(0.1)(0.9)^4$ | 1 |
| N | S | S | S | S | $(0.1)^4(0.9)$ | → 4 |
| N | S | S | S | N | $(0.1)^3(0.9)^2$ | 3 |
| N | S | S | N | S | $(0.1)^3(0.9)^2$ | 3 |
| N | S | S | N | N | $(0.1)^2(0.9)^3$ | 2 |
| N | S | N | S | S | $(0.1)^3(0.9)^2$ | 3 |
| N | S | N | S | N | $(0.1)^2(0.9)^3$ | 2 |
| N | S | N | N | S | $(0.1)^2(0.9)^3$ | 2 |
| N | S | N | N | N | $(0.1)(0.9)^4$ | 1 |
| N | N | S | S | S | $(0.1)^3(0.9)^2$ | 3 |
| N | N | S | S | N | $(0.1)^2(0.9)^3$ | 2 |
| N | N | S | N | S | $(0.1)^2(0.9)^3$ | 2 |
| N | N | S | N | N | $(0.1)(0.9)^4$ | 1 |
| N | N | N | S | S | $(0.1)^2(0.9)^3$ | 2 |
| N | N | N | S | N | $(0.1)(0.9)^4$ | 1 |
| N | N | N | N | S | $(0.1)(0.9)^4$ | 1 |
| N | N | N | N | N | $(0.9)^5$ | 0 |

The model is concerned with the total number of successes in $n$ trials as a random variable, denoted by $X$. Its probability density function is given by

$$\Pr(X = x) = \binom{n}{x}\pi^x(1 - \pi)^{n-x} \qquad \text{for } x = 0, 1, 2, \ldots, n$$

where $\binom{n}{x}$ is the number of combinations of $x$ objects selected from a set of $n$

objects,

$$\binom{n}{x} = \frac{n!}{x!\,(n-x)!}$$

and $n!$ is the product of the first $n$ integers. For example,

$$3! = (1)(2)(3)$$

The mean and variance of the binomial distribution are

$$\mu = n\pi$$

$$\sigma^2 = n\pi(1 - \pi)$$

and when the number of trials $n$ is from moderate to large ($n > 25$, say), we we approximate the binomial distribution by a normal distribution and answer probability questions by first converting to a standard normal score:

$$z = \frac{x - n\pi}{\sqrt{n\pi(1 - \pi)}}$$

where $\pi$ is the probability of having a positive outcome from a single trial. For example, for $\pi = 0.1$ and $n = 30$, we have

$$\mu = (30)(0.1)$$

$$= 3$$

$$\sigma^2 = (30)(0.1)(0.9)$$

$$= 2.7$$

so that

$$\Pr(x \geq 7) \simeq \Pr\left(z \geq \frac{7-3}{\sqrt{2.7}}\right)$$

$$= \Pr(z \geq 2.43)$$

$$= 0.0075$$

In other words, if the true probability for having the side effect is 10%, the probability of having seven or more of 30 patients with the side effect is less than 1% ($= 0.0075$).

### 3.4.2  Poisson Distribution

The next discrete distribution that we consider is the Poisson distribution, named after a French mathematician. This distribution has been used extensively in health science to model the distribution of the number of occurrences $x$ of some random event in an interval of time or space, or some volume of matter. For example, a hospital administrator has been studying daily emergency admissions over a period of several months and has found that admissions have averaged three per day. He or she is then interested in finding the probability that no emergency admissions will occur on a particular day. The *Poisson distribution* is characterized by its probability density function:

$$\Pr(X = x) = \frac{\theta^x e^{-\theta}}{x!} \qquad \text{for } x = 0, 1, 2, \ldots$$

It turns out, interestingly enough, that for a Poisson distribution the variance is equal to the mean, the parameter $\theta$ above. Therefore, we can answer probability questions by using the formula for the Poisson density above or by converting the number of occurrences $x$ to the standard normal score, provided that $\theta \geq 10$:

$$z = \frac{x - \theta}{\sqrt{\theta}}$$

In other words, we can approximate a Poisson distribution by a normal distribution with mean $\theta$ if $\theta$ is at least 10.

Here is another example involving the Poisson distribution. The infant mortality rate (IMR) is defined as

$$\text{IMR} = \frac{d}{N}$$

for a certain target population during a given year, where $d$ is the number of deaths during the first year of life and $N$ is the total number of live births. In the studies of IMRs, $N$ is conventionally assumed to be fixed and $d$ to follow a Poisson distribution.

***Example 3.9***  For the year 1981 we have the following data for the New England states (Connecticut, Maine, Massachusetts, New Hampshire, Rhode Island, and Vermont):

$$d = 1585$$

$$N = 164,200$$

For the same year, the national infant mortality rate was 11.9 (per 1000 live

births). If we apply the national IMR to the New England states, we would
have

$$\theta = (11.9)(164.2)$$

$$\simeq 1954 \text{ infant deaths}$$

Then the event of having as few as 1585 infant deaths would occur with a
probability

$$\Pr(d \leq 1585) = \Pr\left(z \leq \frac{1585 - 1954}{\sqrt{1954}}\right)$$

$$= \Pr(z \leq -8.35)$$

$$\simeq 0$$

The conclusion is clear: Either we observed an extremely improbable event, or
infant mortality in the New England states is lower than the national average.
The rate observed for the New England states was 9.7 deaths per 1000 live
births.

## 3.5   BRIEF NOTES ON THE FUNDAMENTALS

In this section we provide some brief notes on the foundation of some methods
in previous sections. Readers, especially beginners, may decide to skip it without loss of continuity.

### 3.5.1   Mean and Variance

As seen in Sections 3.3 and 3.4, a probability density function $f$ is defined so
that:

(a) $f(k) = \Pr(X = k)$ in the discrete case
(b) $f(x)\,dx = \Pr(x \leq X \leq x + dx)$ in the continuous case

For a continuous distribution, such as the normal distribution, the mean $\mu$
and variance $\sigma^2$ are calculated from:

(a) $\mu = \int xf(x)\,dx$
(b) $\sigma^2 = \int (x - \mu)^2 f(x)\,dx$

For a discrete distribution, such as the binomial distribution or Poisson distribution, the mean $\mu$ and variance $\sigma^2$ are calculated from:

(a) $\mu = \sum xf(x)$

(b) $\sigma^2 = \sum (x - \mu)^2 f(x)$

For example, we have for the binomial distribution,

$$\mu = np$$

$$\sigma^2 = np(1 - p)$$

and for the Poisson distribution,

$$\mu = \theta$$

$$\sigma^2 = \theta$$

### 3.5.2   Pair-Matched Case–Control Study

Data from epidemiologic studies may come from various sources, the two fundamental designs being *retrospective* and *prospective* (or *cohort*). Retrospective studies gather past data from selected cases (diseased individuals) and controls (nondiseased individuals) to determine differences, if any, in the exposure to a suspected risk factor. They are commonly referred to as *case–control studies*. Cases of a specific disease, such as lung cancer, are ascertained as they arise from population-based disease registers or lists of hospital admissions, and controls are sampled either as disease-free persons from the population at risk or as hospitalized patients having a diagnosis other than the one under investigation. The advantages of a case–control study are that it is economical and that it is possible to answer research questions relatively quickly because the cases are already available. Suppose that each person in a large population has been classified as exposed or not exposed to a certain factor, and as having or not having some disease. The population may then be enumerated in a $2 \times 2$ table (Table 3.12), with entries being the proportions of the total population.

Using these proportions, the association (if any) between the factor and the disease could be measured by the ratio of risks (or relative risk) of being disease positive for those with or without the factor:

**TABLE 3.12**

| Factor | Disease | | Total |
|---|---|---|---|
| | + | − | |
| + | $P_1$ | $P_3$ | $P_1 + P_3$ |
| − | $P_2$ | $P_4$ | $P_2 + P_4$ |
| Total | $P_1 + P_2$ | $P_3 + P_4$ | 1 |

$$\text{relative risk} = \frac{P_1}{P_1 + P_3} \div \frac{P_2}{P_2 + P_4}$$

$$= \frac{P_1(P_2 + P_4)}{P_2(P_1 + P_3)}$$

since in many (although not all) situations, the proportions of subjects classified as disease positive will be small. That is, $P_1$ is small in comparison with $P_3$, and $P_2$ will be small in comparison with $P_4$. In such a case, the relative risk is almost equal to

$$\theta = \frac{P_1 P_4}{P_2 P_3}$$

$$= \frac{P_1/P_3}{P_2/P_4}$$

the odds ratio of being disease positive, or

$$= \frac{P_1/P_2}{P_3/P_4}$$

the odds ratio of being exposed. This justifies the use of an odds ratio to determine differences, if any, in the exposure to a suspected risk factor.

As a technique to control confounding factors in a designed study, individual cases are matched, often one to one, to a set of controls chosen to have similar values for the important confounding variables. The simplest example of pair-matched data occurs with a single binary exposure (e.g., smoking versus nonsmoking). The data for outcomes can be represented by a $2 \times 2$ table (Table 3.13) where $(+, -)$ denotes (exposed, unexposed).

For example, $n_{10}$ denotes the number of pairs where the case is exposed, but the matched control is unexposed. The most suitable statistical model for making inferences about the odds ratio $\theta$ is to use the conditional probability of the number of exposed cases among the discordant pairs. Given $n = n_{10} + n_{01}$ being fixed, it can be seen that $n_{10}$ has $B(n, p)$, where

$$p = \frac{\theta}{1 + \theta}$$

**TABLE 3.13**

| Control | Case | |
|---|---|---|
| | $+$ | $-$ |
| $+$ | $n_{11}$ | $n_{01}$ |
| $-$ | $n_{10}$ | $n_{00}$ |

The proof can be presented briefly as follows. Denoting by

$$\lambda_1 = 1 - \psi_1 \qquad (0 \le \lambda_1 \le 1)$$

$$\lambda_0 = 1 - \psi_0 \qquad (0 \le \lambda_0 \le 1)$$

the exposure probabilities for cases and controls, respectively, the probability of observing a case–control pair with only the case exposed is $\lambda_1\psi_0$, while that of observing a pair where only the control is exposed is $\psi_1\lambda_0$. Hence the conditional probability of observing a pair of the former variety, given that it is discordant, is

$$P = \frac{\lambda_1\psi_0}{\lambda_1\psi_0 + \psi_1\lambda_0}$$

$$= \frac{\lambda_1\psi_0/\psi_1\lambda_0}{\lambda_1\psi_0/\psi_1\lambda_0 + 1}$$

$$= \frac{(\lambda_1/\psi_1)/(\lambda_0/\psi_0)}{(\lambda_1/\psi_1)/(\lambda_0/\psi_0) + 1}$$

$$= \frac{\theta}{\theta + 1}$$

a function of the odds ratio $\theta$ only.

## 3.6   NOTES ON COMPUTATIONS

In Sections 1.4 and 2.5 we covered basic techniques for Microsoft's Excel: how to open/form a spreadsheet, save it, retrieve it, and perform certain descriptive statistical tasks. Topics included data-entry steps, such as *select and drag*, use of *formula bar*, bar and pie charts, histograms, calculations of descritive statistics such as mean and standard deviation, and calculation of a coefficient of correlation. In this short section we focus on probability models related to the calculation of areas under density curves, especially normal curves and $t$ curves.

*Normal Curves*   The first two steps are the same as in obtaining descriptive statistics (but no data are needed now): (1) click the *paste function icon*, f*, and (2) click *Statistical*. Among the functions available, two are related to *normal curves*: NORMDIST and NORMINV. Excel provides needed information for any normal distribution, not just the standard normal distribution as in Appendix B. Upon selecting either one of the two functions above, a box appears asking you to provide (1) the mean $\mu$, (2) the standard deviation $\sigma$, and (3) in the last row, marked *cumulative*, to enter *TRUE* (there is a choice *FALSE*, but you do not need that). The answer will appear in a *preselected* cell.

- NORMDIST gives the *area under the normal curve* (with mean and variance provided) all the way from the far-left side (minus infinity) to the value *x that you have to specify*. For example, if you specify $\mu = 0$ and $\sigma = 1$, the return is the area under the standard normal curve up to the point specified (which is the same as the number from Appendix B *plus 0.5*).

- NORMINV performs the inverse process, where *you* provide the area under the normal curve (a number between 0 and 1), together with the mean $\mu$ and standard deviation $\sigma$, and requests the point *x* on the horizontal axis so that the area under that normal curve from the far-left side (minus infinity) to the value *x* is equal to the number provided between 0 and 1. For example, if you put in $\mu = 0$, $\sigma - 1$, and probability = 0.975, the return is 1.96; unlike Appendix B, if you want a number in the right tail of the curve, the input probability should be a number greater than 0.5.

***The t Curves: Procedures TDIST and TINV***    We want to learn how to find the areas under the normal curves so that we can determine the *p values* for statistical tests (a topic starting in Chapter 5). Another popular family in this category is the *t distributions*, which begin with the same first two steps: (1) click the *paste function icon*, f\*, and (2) click *Statistical*. Among the functions available, two related to the *t distributions* are TDIST and TINV. Similar to the case of NORMDIST and NORMINV, TDIST gives the *area under the t curve*, and TINV performs the inverse process where you provide the area under the curve and request point *x* on the horizontal axis. In each case you have to provide the *degrees of freedom*. In addition, in the last row, marked with *tails*, enter:

- (Tails=) *1* if you want *one-sided*
- (Tails=) *2* if you want *two-sided*

(More details on the concepts of one- and two-sided areas are given in Chapter 5.) For example:

- *Example 1:* If you enter (x=) *2.73*, (deg freedom=) *18*, and, (Tails=) *1*, you're requesting the area under a *t* curve with 18 degrees of freedom and *to the right* of 2.73 (i.e., right tail); the answer is 0.00687.
- *Example 2:* If you enter (x=) *2.73*, (deg freedom=) *18*, and (Tails=) *2*, you're requesting the area under a *t* curve with 18 degrees of freedom and *to the right* of 2.73 and *to the left* of −2.73 (i.e., both right and left tails); the answer is 0.01374, which is twice the previous answer of 0.00687.

## EXERCISES

**3.1**    Although cervical cancer is not a leading cause of death among women in the United States, it has been suggested that virtually all such deaths are preventable (5166 American women died from cervical cancer in 1977). In an effort to find out who is being or not being screened for cervical cancer (Pap testing), data were collected from a certain community (Table E3.1). Is there a statistical relationship here? (Try a few different methods: calculation of odds ratio, comparison of conditional and unconditional probabilities, and comparison of conditional probabilities.)

**TABLE E3.1**

| | Race | | |
|---|---|---|---|
| Pap Test | White | Black | Total |
| No | 5,244 | 785 | 6,029 |
| Yes | 25,117 | 2,348 | 27,465 |
| Total | 30,361 | 3,133 | 33,494 |

**3.2** In a study of intraobserver variability in assessing cervical smears, 3325 slides were screened for the presence or absence of abnormal squamous cells. Each slide was screened by a particular observer and then re-screened six months later by the same observer. The results are shown in Table E3.2. Is there a statistical relationship between first screening and second screening? (Try a few different methods as in Exercise 3.1.)

**TABLE E3.2**

| | Second Screening | | |
|---|---|---|---|
| First Screening | Present | Absent | Total |
| Present | 1763 | 489 | 2252 |
| Absent | 403 | 670 | 1073 |
| Total | 2166 | 1159 | 3325 |

**3.3** From the intraobserver variability study above, find:

**(a)** The probability that abnormal squamous cells were found to be absent in both screenings.

**(b)** The probability of an absence in the second screening given that abnormal cells were found in the first screening.

**(c)** The probability of an abnormal presence in the second screening given that no abnormal cells were found in the first screening.

**(d)** The probability that the screenings disagree.

**3.4** Given the screening test of Example 1.4, where

$$\text{sensitivity} = 0.406$$

$$\text{specificity} = 0.985$$

calculate the positive predictive values when the test is applied to the following populations:

Population A: 80% prevalence

Population B: 25% prevalence

**3.5** Consider the data shown in Table E3.5 on the use of x-ray as a screening test for tuberculosis:

**TABLE E3.5**

| X-ray | Tuberculosis | |
|---|---|---|
| | No | Yes |
| Negative | 1739 | 8 |
| Positive | 51 | 22 |
| Total | 1790 | 30 |

**(a)** Calculate the sensitivity and specificity.

**(b)** Find the disease prevalence.

**(c)** Calculate the positive predictive value both directly and indirectly using Bayes' theorem.

**3.6** From the sensitivity and specificity of x-rays found in Exercise 3.5, compute the positive predictive value corresponding to these prevalences: 0.2, 0.4, 0.6, 0.7, 0.8, and 0.9. Can we find a prevalence when the positive predictive value is preset at 0.80 or 80%?

**3.7** Refer to the standard normal distribution. What is the probability of obtaining a $z$ value of:

**(a)** At least 1.25?

**(b)** At least $-0.84$?

**3.8** Refer to the standard normal distribution. What is the probability of obtaining a $z$ value:

**(a)** Between $-1.96$ and 1.96?

**(b)** Between 1.22 and 1.85?

**(c)** Between $-0.84$ and 1.28?

**3.9** Refer to the standard normal distribution. What is the probability of obtaining $z$ value:

**(a)** Less than 1.72?

**(b)** Less than $-1.25$?

**3.10** Refer to the standard normal distribution. Find a $z$ value such that the probability of obtaining a larger $z$ value is:

(a) 0.05.

(b) 0.025.

(c) 0.20.

**3.11** Verify the numbers in the first two rows of Table 3.10 in Example 3.8; for example, show that the lowest healthy systolic blood pressure for 16-year-old boys is 102.8.

**3.12** Complete Table 3.10 in Example 3.8 at the question marks.

**3.13** Medical research has concluded that people experience a common cold roughly two times per year. Assume that the time between colds is normally distributed with a mean of 160 days and a standard deviation of 40 days.

(a) What is the probability of going 200 or more days between colds? Of going 365 or more days?

(b) What is the probability of getting a cold within 80 days of a previous cold?

**3.14** Assume that the test scores for a large class are normally distributed with a mean of 74 and a standard deviation of 10.

(a) Suppose that you receive a score of 88. What percent of the class received scores higher than yours?

(b) Suppose that the teacher wants to limit the number of A grades in the class to no more than 20%. What would be the lowest score for an A?

**3.15** Intelligence test scores, referred to as *intelligence quotient* or IQ scores, are based on characteristics such as verbal skills, abstract reasoning power, numerical ability, and spatial visualization. If plotted on a graph, the distribution of IQ scores approximates a normal curve with a mean of about 100. An IQ score above 115 is considered superior. Studies of "intellectually gifted" children have generally defined the lower limit of their IQ scores at 140; approximately 1% of the population have IQ scores above this limit. (Based on Biracree, 1984.)

(a) Find the standard deviation of this distribution.

(b) What percent are in the "superior" range of 115 or above?

(c) What percent of the population have IQ scores of 70 or below?

**3.16** IQ scores for college graduates are normally distributed with a mean of 120 (as compared to 100 for the general population) with a standard deviation of 12. What is the probability of randomly selecting a graduate student with an IQ score:

(a) Between 110 and 130?

(b) Above 140?

(c) Below 100?

**3.17** Suppose it is known that the probability of recovery for a certain disease is 0.4. If 35 people are stricken with the disease, what is the probability that:

**(a)** 25 or more will recover?

**(b)** Fewer than five will recover?

(Use the normal approximation.)

**3.18** A study found that for 60% of the couples who have been married 10 years or less, both spouses work. A sample of 30 couples who have been married 10 years or less are selected from marital records available at a local courthouse. We are interested in the number of couples in this sample in which both spouses work. What is the probability that this number is:

**(a)** 20 or more?

**(b)** 25 or more?

**(c)** 10 or fewer?

(Use the normal approximation.)

**3.19** Many samples of water, all the same size, are taken from a river suspected of having been polluted by irresponsible operators at a sewage treatment plant. The number of coliform organisms in each sample was counted; the average number of organisms per sample was 15. Assuming the number of organisms to be Poisson distributed, find the probability that:

**(a)** The next sample will contain at least 20 organisms.

**(b)** The next sample will contain no more than five organisms.

**3.20** For the year 1981 (see Example 3.9), we also have the following data for the South Atlantic states (Delaware, Florida, Georgia, Maryland, North and South Carolina, Virginia, and West Virginia, and the District of Columbia):

$$d = 7643 \text{ infant deaths}$$

$$N = 550,300 \text{ live births}$$

Find the infant mortality rate, and compare it to the national average using the method of Example 3.9.

**3.21** For a $t$ curve with 20 df, find the areas:

**(a)** To the left of 2.086 and of 2.845.

**(b)** To the right of 1.725 and of 2.528.

**(c)** Beyond $\pm 2.086$ and beyond $\pm 2.845$.

**3.22** For a chi-square distribution with 2 df, find the areas:

    **(a)** To the right of 5.991 and of 9.210.

    **(b)** To the right of 6.348.

    **(c)** Between 5.991 and 9.210.

**3.23** For an $F$ distribution with two numerator df's and 30 denominator df's, find the areas:

    **(a)** To the right of 3.32 and of 5.39.

    **(b)** To the right of 2.61.

    **(c)** Between 3.32 and 5.39.

**3.24** In a study of intraobserver variability in assessing cervical smears, 3325 slides were screened for the presence or absence of abnormal squamous cells. Each slide was screened by a particular observer and then re-screened six months later by the same observer. The results are shown in Table E3.2. Calculate the kappa statistic representing the agreement between the two screenings.

**3.25** Ninety-eight heterosexual couples, at least one of whom was HIV-infected, were enrolled in an HIV transmission study and interviewed about sexual behavior. Table E3.25 provides a summary of condom use reported by heterosexual partners. How strongly do the couples agree?

**TABLE E3.25**

| Woman | Man | | Total |
| --- | --- | --- | --- |
| | Ever | Never | |
| Ever | 45 | 6 | 51 |
| Never | 7 | 40 | 47 |
| Total | 52 | 46 | 98 |

# 4

# ESTIMATION OF PARAMETERS

The entire process of statistical design and analysis can be described briefly as follows. The target of a scientist's investigation is a population with certain characteristic of interest: for example, a man's systolic blood pressure or his cholesterol level, or whether a leukemia patient responds to an investigative drug. A numerical characteristic of a target population is called a *parameter*: for example, the population mean $\mu$ (average SBP) or the population proportion $\pi$ (a drug's response rate). Generally, it would be too time consuming or too costly to obtain the totality of population information in order to learn about the parameter(s) of interest. For example, there are millions of men to survey in a target population, and the value of the information may not justify the high cost. Sometimes the population does not even exist. For example, in the case of an investigative drug for leukemia, we are interested in *future* patients as well as present patients. To deal with the problem, the researcher may decide to take a sample or to conduct a small phase II clinical trial. Chapters 1 and 2 provide methods by which we can learn about data from the sample or samples. We learned how to organize data, how to summarize data, and how to present them. The topic of probability in Chapter 3 sets the framework for dealing with uncertainties. By this point the researcher is ready to draw inferences about the population of interest based on what he or she learned from his or her sample(s). Depending on the research's objectives, we can classify inferences into two categories: one in which we want to estimate the value of a parameter, for example the response rate of a leukemia investigative drug, and one where we want to compare the parameters for two subpopulations using statistical tests of significance. For example, we want to know whether men have higher cholesterol levels, on average, than women. In

147

this chapter we deal with the first category and the statistical procedure called *estimation*. It is extremely useful, one of the most useful procedures of statistics. The word *estimate* actually has a language problem, the opposite of the language problem of statistical "tests" (the topic of Chapter 5). The colloquial meaning of the word *test* carries the implication that statistical tests are especially objective, no-nonsense procedures that reveal the truth. Conversely, the colloquial meaning of the word *estimate* is that of guessing, perhaps off the top of the head and uninformed, not to be taken too seriously. It is used by car body repair shops, which "estimate" how much it will cost to fix a car after an accident. The estimate in that case is actually a bid from a for-profit business establishment seeking your trade. In our case, the word *estimation* is used in the usual sense that provides a "substitute" for an unknown truth, but it isn't that bad a choice of word once you understand *how* to do it. But it is important to make it clear that statistical estimation is no less objective than any other formal statistical procedure; statistical estimation requires calculations and tables just as statistical testing does. In addition, it is very important to differentiate formal statistical estimation from ordinary guessing. In formal statistical estimation, we can determine the *amount of uncertainty* (and so the error) in the estimate. How often have you heard of someone making a guess and then giving you a number measuring the "margin of error" of the guess? That's what statistical estimation does. It gives you the best guess and then tells you how "wrong" the guess could be, in quite precise terms. Certain media, sophisticated newspapers in particular, are starting to educate the public about statistical estimation. They do it when they report the results of polls. They say things like, "74% of the voters disagree with the governor's budget proposal," and then go on to say that the margin error is plus or minus 3%. What they are saying is that whoever conducted the poll is claiming to have polled about 1000 people chosen at random and that statistical estimation theory tells us to be 95% certain that if *all* the voters were polled, their disagreement percentage would be discovered to be within 3% of 74%. In other words, it's very unlikely that the 74% is off the mark by more than 3%; the truth is almost certainly between 71 and 77%. In subsequent sections of this chapter we introduce the strict interpretation of these *confidence intervals*.

## 4.1 BASIC CONCEPTS

A class of measurements or a characteristic on which individual observations or measurements are made is called a *variable* or *random variable*. The value of a random variable varies from subject to subject; examples include weight, height, blood pressure, or the presence or absence of a certain habit or practice, such as smoking or use of drugs. The distribution of a random variable is often assumed to belong to a certain family of distributions, such as binomial, Poisson, or normal. This assumed family of distributions is specified or indexed by one or several parameters, such as a population mean $\mu$ or a population proportion $\pi$. It is usually either impossible, too costly, or too time consuming to obtain the entire population data on any variable in order to learn about a parameter involved in its distribution. Decisions in health science are thus often made using a small sample of a population. The problem for a decision maker is to decide on the basis of data the estimated value of a parameter, such as the population mean, as well as to provide certain ideas concerning errors associated with that estimate.

### 4.1.1    Statistics as Variables

A parameter is a numerical property of a population; examples include population mean $\mu$ and population proportion $\pi$. The corresponding quantity obtained from a sample is called a *statistic*; examples of statistics include the sample mean $\bar{x}$ and sample proportion $p$. Statistics help us draw inferences or conclusions about population parameters. After a sample has already been obtained, the value of a statistic—for example, the sample mean $\bar{x}$—is known and fixed; however, if we take a different sample, we almost certainly have a different numerical value for that same statistic. In this repeated sampling context, a statistic is looked upon as a variable that takes different values from sample to sample.

### 4.1.2    Sampling Distributions

The distribution of values of a statistic obtained from repeated samples of the same size from a given population is called the *sampling distribution* of that statistic.

*Example 4.1*    Consider a population consisting of six subjects (this small size is impractical, but we need something small enough to use as an illustration here). Table 4.1 gives the subject names (for identification) and values of a variable under investigation (e.g., 1 for a smoker and 0 for a nonsmoker). In this case the population mean $\mu$ (also the population proportion $\pi$ for this very special dichotomous variable) is 0.5 ($= 3/6$). We now consider all *possible* samples, without replacement, of size 3; none or some or all subjects in each sample have

**TABLE 4.1**

| Subject | Value |
|---------|-------|
| A | 1 |
| B | 1 |
| C | 1 |
| D | 0 |
| E | 0 |
| F | 0 |

**TABLE 4.2**

| Samples | Number of Samples | Value of Sample Mean, $\bar{x}$ |
|---|---|---|
| (D, E, F) | 1 | 0 |
| (A, D, E), (A, D, F), (A, E, F) (B, D, E), (B, D, F), (B, E, F) (C, D, E), (C, D, F), (C, E, F) | 9 | $\frac{1}{3}$ |
| (A, B, D), (A, B, E), (A, B, F) (A, C, D), (A, C, E), (A, C, F) (B, C, D), (B, C, E), (B, C, F) | 9 | $\frac{2}{3}$ |
| (A, B, C) | 1 | 1 |
| Total | 20 | |

value "1," the remaining, "0." Table 4.2 represents the sampling distribution of the sample mean.

This sampling distribution gives us a few interesting properties:

1. Its mean (i.e., the mean of *all possible* sample means) is

$$\frac{(1)(0) + (9)\left(\frac{1}{3}\right) + (9)\left(\frac{2}{3}\right) + (1)(1)}{20} = 0.5$$

which is the same as the mean of the original distribution. Because of this, we say that the sample mean (sample proportion) is an *unbiased estimator* for the population mean (population proportion). In other words, if we use the sample mean (sample proportion) to estimate the population mean (population proportion), we are *correct on the average*.

2. If we form a bar graph for this sampling distribution (Figure 4.1). It shows a shape somewhat similar to that of a symmetric, bell-shaped nor-

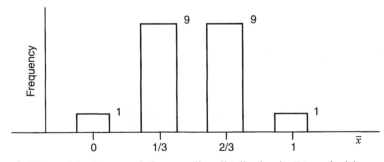

**Figure 4.1**   Bar graph for sampling distribution in Example 4.1.

**TABLE 4.3**

| Samples | Number of Samples | Value of Sample Mean, $\bar{x}$ |
|---|---|---|
| (A, D, E, F), (B, D, E, F), (C, D, E, F) | 3 | 0.25 |
| (A, B, D, E), (A, B, D, F), (A, B, E, F) (A, C, D, E), (A, C, D, F), (A, C, E, F) (B, C, D, E), (B, C, D, F), (B, C, E, F) | 9 | 0.50 |
| (A, B, C, D), (A, B, C, E), (A, B, C, F) | 3 | 0.75 |
| Total | 15 | |

mal curve. This resemblance is much clearer with real populations and larger sample sizes.

We now consider the same population and all possible samples of size $n = 4$. Table 4.3 represents the new sampling distribution. It can be seen that we have a different sampling distribution because the sample size is different. However, we still have both above-mentioned properties:

1. Unbiasedness of the sample mean:

$$\frac{(3)(0.25) + (9)(0.50) + (3)(0.75)}{15} = 0.5$$

2. Normal shape of the sampling distribution (bar graph; Figure 4.2).
3. In addition, we can see that the variance of the new distribution is smaller. The two faraway values of $\bar{x}$, 0 and 1, are no longer possible; new values—0.25 and 0.75—are closer to the mean 0.5, and the majority (nine samples) have values that are right at the sampling distribution mean. The major reason for this is that the new sampling distribution is associated with a larger sample size, $n = 4$, compared to $n = 3$ for the previous sampling distribution.

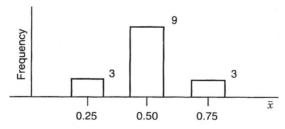

**Figure 4.2**   Normal shape of sampling distribution for Example 4.1.

### 4.1.3   Introduction to Confidence Estimation

*Statistical inference* is a procedure whereby inferences about a population are made on the basis of the results obtained from a sample drawn from that population. Professionals in health science are often interested in a parameter of a certain population. For example, a health professional may be interested in knowing what proportion of a certain type of person, treated with a particular drug, suffers undesirable side effects. The process of estimation entails calculating, from the data of a sample, some statistic that is offered as an estimate of the corresponding parameter of the population from which the sample was drawn.

A point estimate is a single numerical value used to estimate the corresponding population parameter. For example, the sample mean is a point estimate for the population mean, and the sample proportion is a point estimate for the population proportion. However, having access to the data of a sample and a knowledge of statistical theory, we can do more than just providing a point estimate. The sampling distribution of a statistic—if available—would provide information on biasedness/unbiasedness (several statistics, such as $\bar{x}$, $p$, and $s^2$, are unbiased) and variance.

Variance is important; a small variance for a sampling distribution indicates that most possible values for the statistic are close to each other, so that a particular value is more likely to be reproduced. In other words, the variance of a sampling distribution of a statistic can be used as a measure of precision or reproducibility of that statistic; the smaller this quantity, the better the statistic as an estimate of the corresponding parameter. The square root of this variance is called the *standard error* of the statistic; for example, we will have the standard error of the sample mean, or $SE(\bar{x})$; the standard error of the sample proportion, $SE(p)$; and so on. It is the same quantity, but we use the term *standard deviation* for measurements and the term *standard error* when we refer to the standard deviation of a statistic. In the next few sections we introduce a process whereby the point estimate and its standard error are combined to form an interval estimate or *confidence interval*. A confidence interval consists of two numerical values, defining an interval which, with a specified degree of confidence, we believe includes the parameter being estimated.

## 4.2   ESTIMATION OF MEANS

The results of Example 4.1 are not coincidences but are examples of the characteristics of sampling distributions in general. The key tool here is the *central limit theorem*, introduced in Section 3.2.1, which may be summarized as follows: Given any population with mean $\mu$ and variance $\sigma^2$, the sampling distribution of $\bar{x}$ will be approximately normal with mean $\mu$ and variance $\sigma^2/n$ when the sample size $n$ is large (of course, the larger the sample size, the better the approximation; in practice, $n = 25$ or more could be considered adequately large). This means that we have the two properties

$$\mu_{\bar{x}} = \mu$$

$$\sigma_{\bar{x}}^2 = \frac{\sigma^2}{n}$$

as seen in Example 4.1.

The following example shows how good $\bar{x}$ is as an estimate for the population $\mu$ even if the sample size is as small as 25. (Of course, it is used only as an illustration; in practice, $\mu$ and $\sigma^2$ are unknown.)

***Example 4.2***    Birth weights obtained from deliveries over a long period of time at a certain hospital show a mean $\mu$ of 112 oz and a standard deviation $\sigma$ of 20.6 oz. Let us suppose that we want to compute the probability that the mean birth weight from a sample of 25 infants will fall between 107 and 117 oz (i.e., the estimate is off the mark by no more than 5 oz). The central limit theorem is applied and it indicates that $\bar{x}$ follows a normal distribution with mean

$$\mu_{\bar{x}} = 112$$

and variance

$$\sigma_{\bar{x}}^2 = \frac{(20.6)^2}{25}$$

or standard error

$$\sigma_{\bar{x}} = 4.12$$

It follows that

$$\Pr(107 \leq \bar{x} \leq 117) = \Pr\left(\frac{107 - 112}{4.12} \leq z \leq \frac{117 - 112}{4.12}\right)$$

$$= \Pr(-1.21 \leq z \leq 1.21)$$

$$= (2)(0.3869)$$

$$= 0.7738$$

In other words, if we use the mean of a sample of size $n = 25$ to estimate the population mean, about 80% of the time we are correct within 5 oz; this figure would be 98.5% if the sample size were 100.

### 4.2.1    Confidence Intervals for a Mean

Similar to what was done in Example 4.2, we can write, for example,

$$\Pr\left[-1.96 \leq \frac{\bar{x} - \mu}{\sigma/\sqrt{n}} \leq 1.96\right] = (2)(0.475)$$

$$= 0.95$$

This statement is a consequence of the central limit theorem, which indicates that for a large sample size $n$, $\bar{x}$ is a random variable (in the context of repeated sampling) with a normal sampling distribution in which

$$\mu_{\bar{x}} = \mu$$

$$\sigma_{\bar{x}}^2 = \sigma^2/n$$

The quantity inside brackets in the equation above is equivalent to

$$\bar{x} - 1.96\sigma/\sqrt{n} \leq \mu \leq \bar{x} + 1.96\sigma/\sqrt{n}$$

All we need to do now is to select a random sample, calculate the numerical value of $\bar{x}$ and its standard error with $\sigma$ replaced by sample variance $s$, $s/\sqrt{n}$, and substitute these values to form the endpoints of the interval

$$\bar{x} \pm 1.96s/\sqrt{n}$$

In a specific numerical case this will produce two numbers,

$$a = \bar{x} - 1.96s/\sqrt{n}$$

and

$$b = \bar{x} + 1.96s/\sqrt{n}$$

and we have the interval

$$a \leq \mu \leq b$$

But here we run into a logical problem. We are sampling from a fixed population. We are examining values of a random variable obtained by selecting a random sample from that fixed population. The random variable has a distribution with mean $\mu$ that we wish to estimate. Since the population and the distribution of the random variable we are investigating are fixed, it follows that the parameter $\mu$ is fixed. The quantities $\mu$, $a$, and $b$ are all fixed (after the sample has been obtained); then we cannot assert that the probability that $\mu$ lies

**TABLE 4.4**

| Degree of Confidence | Coefficient |
|---|---|
| 99% | 2.576 |
| → 95% | 1.960 |
| 90% | 1.645 |
| 80% | 1.282 |

between $a$ and $b$ is 0.95. In fact, either $\mu$ lies in $(a, b)$ or it does not, and it is not correct to assign a probability to the statement (even the truth remains unknown).

The difficulty here arises at the point of substitution of the numerical values observed for $\bar{x}$ and its standard error. The random variation in $\bar{x}$ is variation from sample to sample in the context of repeated sampling. When we substitute $\bar{x}$ and its standard error $s/\sqrt{n}$ by their numerical values resulting in interval $(a, b)$, it is understood that the repeated sampling process could produce many different intervals of the same form:

$$\bar{x} \pm 1.96\,\text{SE}(\bar{x})$$

About 95% of these intervals would actually include $\mu$. Since we have only one of these possible intervals, the interval $(a, b)$ from our sample, we say we are 95% confident that $\mu$ lies between these limits. The interval $(a, b)$ is called a *95% confidence interval* for $\mu$, and the figure "95" is called the *degree of confidence* or *confidence level*.

In forming confidence intervals, the degree of confidence is determined by the investigator of a research project. Different investigators may prefer different confidence intervals; the coefficient to be multiplied with the standard error of the mean should be determined accordingly. A few typical choices are given in Table 4.4; 95% is the most conventional.

Finally, it should be noted that since the standard error is

$$\text{SE}(\bar{x}) = s/\sqrt{n}$$

the width of a confidence interval becomes narrower as sample size increases, and the process above is applicable only to large samples ($n > 25$, say). In the next section we show how to handle smaller samples (there is nothing magic about "25"; see the note at the end of Section 4.2.2).

*Example 4.3* For the data on percentage saturation of bile for 31 male patients of Example 2.4:

> 40, 86, 111, 86, 106, 66, 123, 90, 112, 52, 88, 137, 88, 88,
> 65, 79, 87, 56, 110, 106, 110, 78, 80, 47, 74, 58, 88, 73, 118,
> 67, 57

we have

$$n = 31$$

$$\bar{x} = 84.65$$

$$s = 24.00$$

leading to a standard error

$$SE(\bar{x}) = \frac{24.00}{\sqrt{31}}$$

$$= 4.31$$

and a 95% confidence interval for the population mean:

$$84.65 \pm (1.96)(4.31) = (76.2, 93.1)$$

(The resulting interval is wide, due to a large standard deviation as observed from the sample, $s = 24.0$, reflecting heterogeneity of sample subjects.)

### 4.2.2  Uses of Small Samples

The procedure for confidence intervals in Section 4.2.1 is applicable only to large samples (say, $n > 25$). For smaller samples, the results are still valid if the population variance $\sigma^2$ is known and standard error is expressed as $\sigma/\sqrt{n}$. However, $\sigma^2$ is almost always unknown. When $\sigma$ is unknown, we can estimate it by $s$ but the procedure has to be modified by changing the coefficient to be multiplied by the standard error to accommodate the error in estimating $\sigma$ by $s$; how much larger the coefficient is depends on how much information we have in estimating $\sigma$ (by $s$), that is, the sample size $n$.

Therefore, instead of taking coefficients from the standard normal distribution table (numbers such as 2.576, 1.960, 1.645, and 1.282 for degrees of confidence 99%, 95%, 90%, and 80%), we will use corresponding numbers from the $t$ curves where the quantity of information is indexed by the degree of freedom (df $= n - 1$). The figures are listed in Appendix C; the column to read is the one with the correct normal coefficient on the bottom row (marked with df $= \infty$). See, for example, Table 4.5 for the case where the degree of confidence is 0.95. *(For better results, it is always a good practice to use the t table regardless of sample size because coefficients such as 1.96 are only for very large sample sizes.)*

***Example 4.4***   In an attempt to assess the physical condition of joggers, a sample of $n = 25$ joggers was selected and maximum volume oxygen ($VO_2$) uptake was measured, with the following results:

TABLE 4.5

| df | $t$ Coefficient (percentile) |
|---|---|
| 5 | 2.571 |
| 10 | 2.228 |
| 15 | 2.131 |
| 20 | 2.086 |
| 24 | 2.064 |
| $\to \infty$ | 1.960 |

$$\bar{x} = 47.5 \text{ mL/kg}$$

$$s = 4.8 \text{ mL/kg}$$

$$\text{SE}(\bar{x}) = \frac{4.8}{\sqrt{25}}$$

$$= 0.96$$

From Appendix C we find that the $t$ coefficient with 24 df for use with a 95% confidence interval is 2.064, leading to a 95% confidence interval for the population mean $\mu$ (this is the population of joggers' $VO_2$ uptake) of

$$47.5 \pm (2.064)(0.96) = (45.5, 49.5)$$

***Example 4.5***  In addition to the data in Example 4.4, we have data from a second sample consisting of 26 nonjoggers which were summarized into these statistics:

$$n_2 = 26$$

$$\bar{x}_2 = 37.5 \text{ mL/kg}$$

$$s_2 = 5.1 \text{ mL/kg}$$

$$\text{SE}(\bar{x}) = \frac{5.1}{\sqrt{26}}$$

$$= 1.0$$

From Appendix C we find that the $t$ coefficient with 25 df for use with a 95% confidence interval is 2.060, leading to a 95% confidence interval for the population mean $\mu$ (this is the population of joggers' $VO_2$ uptake) of

$$37.5 \pm (2.060)(1.0) = (35.4, 39.6)$$

### 4.2.3 Evaluation of Interventions

In efforts to determine the effect of a risk factor or an intervention, we may want to estimate the *difference of means*: say, between the population of cases and the population of controls. However, we choose not to present the methodology with much detail at this level—with one exception, the case of matched design or before-and-after intervention, where each experimental unit serves as its own control. This design makes it possible to control for confounding variables that are difficult to measure (e.g. environmental exposure) and therefore difficult to adjust at the analysis stage. The main reason to include this method here, however, is because we treat the data as one sample and the aim is still estimating the (population) mean. That is, data from matched or before-and-after experiments should not be considered as coming from two independent samples. The procedure is to reduce the data to a one-sample problem by computing before-and-after (or control-and-case) differences for each subject (or pairs of matched subjects). By doing this with paired observations, we get a set of differences that can be handled as a single sample problem. The mean to be estimated, using the sample of differences, represents the effects of the intervention (or the effects of of the disease) under investigation.

*Example 4.6*    The systolic blood pressures of 12 women between the ages of 20 and 35 were measured before and after administration of a newly developed oral contraceptive. Given the data in Table 4.6, we have from the column of differences, the $d_i$'s,

$$n = 12$$
$$\sum d_i = 31$$
$$\sum d_i^2 = 185$$

**TABLE 4.6**

| Subject | Systolic Blood Pressure (mmHg) | | After–Before Difference, $d_i$ | $d_i^2$ |
|---|---|---|---|---|
| | Before | After | | |
| 1 | 122 | 127 | 5 | 25 |
| 2 | 126 | 128 | 2 | 4 |
| 3 | 132 | 140 | 8 | 64 |
| 4 | 120 | 119 | −1 | 1 |
| 5 | 142 | 145 | 3 | 9 |
| 6 | 130 | 130 | 0 | 0 |
| 7 | 142 | 148 | 6 | 36 |
| 8 | 137 | 135 | −2 | 4 |
| 9 | 128 | 129 | 1 | 1 |
| 10 | 132 | 137 | 5 | 25 |
| 11 | 128 | 128 | 0 | 0 |
| 12 | 129 | 133 | 4 | 16 |

leading to

$$\bar{d} = \text{average difference}$$

$$= \frac{31}{12}$$

$$= 2.58 \text{ mmHg}$$

$$s^2 = \frac{185 - (31)^2/12}{11}$$

$$= 9.54$$

$$s = 3.09$$

$$SE(\bar{d}) = \frac{3.09}{\sqrt{12}}$$

$$= 0.89$$

With a degree of confidence of 0.95 the $t$ coefficient from Appendix C is 2.201, for 11 degrees of freedom, so that a 95% confidence interval for the mean difference is

$$2.58 \pm (2.201)(0.89) = (0.62, 4.54)$$

This means that the "after" mean is larger than the "before" mean, an increase of between 0.62 and 4.54.

In many other interventions, or in studies to determine possible effects of a risk factor, it may not be possible to employ matched design. The comparison of means is based on data from two independent samples. The process of estimating the *difference of means* is summarized briefly as follows:

1. Data are summarized separately to obtain

$$\text{sample 1:} \quad n_1, \bar{x}_1, s_1^2$$
$$\text{sample 2:} \quad n_2, \bar{x}_2, s_2^2$$

2. The standard error of the difference of means is given by

$$SE(\bar{x}_1 - \bar{x}_2) = \sqrt{\frac{s_1^2}{n_1} + \frac{s_2^2}{n_2}}$$

3. Finally, a 95% confidence interval for the difference of population means, $\mu_1 - \mu_2$, can be calculated from the formula

$$(\bar{x}_1 - \bar{x}_2) \pm (\text{coefficient}) SE(\bar{x}_1 - \bar{x}_2)$$

where the coefficient is 1.96 if $n_1 + n_2$ is large; otherwise, a $t$ coefficient is used with approximately

$$\mathrm{df} = n_1 + n_2 - 2$$

## 4.3   ESTIMATION OF PROPORTIONS

The sample proportion is defined as in Chapter 1:

$$p = \frac{x}{n}$$

where $x$ is the number of positive outcomes and $n$ is the sample size. However, the proportion $p$ can also be viewed as a sample mean $\bar{x}$, where $x_i$ is 1 if the $i$th outcome is positive and 0 otherwise:

$$p = \frac{\sum x_i}{n}$$

Its standard error is still derived using the same process:

$$\mathrm{SE}(p) = \frac{s}{\sqrt{n}}$$

with the standard deviation $s$ given as in Section 2.3:

$$s = \sqrt{p(1 - p)}$$

In other words, the standard error of the sample proportion is calculated from

$$\mathrm{SE}(p) = \sqrt{\frac{p(1 - p)}{n}}$$

To state it more formally, the central limit theorem implies that the sampling distribution of $p$ will be approximately normal when the sample size $n$ is large; the mean and variance of this sampling distribution are

$$\mu_p = \pi$$

and

$$\sigma_p^2 = \frac{\pi(1 - \pi)}{n}$$

respectively, where $\pi$ is the population proportion.

***Example 4.7***  Suppose that the true proportion of smokers in a community is known to be in the vicinity of $\pi = 0.4$, and we want to estimate it using a sample of size $n = 100$. The central limit theorem indicates that $p$ follows a normal distribution with mean

$$\mu_p = 0.40$$

and variance

$$\sigma_p^2 = \frac{(0.4)(0.6)}{100}$$

or standard error

$$\sigma_p = 0.049$$

Suppose that we want our estimate to be correct within $\pm 3\%$; it follows that

$$\Pr(0.37 \le p \le 0.43) = \Pr\left(\frac{0.37 - 0.40}{0.049} \le z \le \frac{0.43 - 0.40}{0.049}\right)$$

$$= \Pr(-0.61 \le z \le 0.61)$$

$$= (2)(0.2291)$$

$$= 0.4582 \quad \text{or} \quad \text{approximately } 46\%$$

That means if we use the proportion of smokers from a sample of $n = 100$ to estimate the true proportion of smokers, only about 46% of the time are we correct within $\pm 3\%$; this figure would be 95.5% if the sample size is raised to $n = 1000$. What we learn from this example is that compared to the case of continuous data in Example 4.2, it should take a much larger sample to have a good estimate of a proportion such as a disease prevalence or a drug side effect.

From this sampling distribution of the sample proportion, in the context of repeated sampling, we have an approximate 95% confidence interval for a population proportion $\pi$:

$$p \pm 1.96 SE(p)$$

where, again, the standard error of the sample proportion is calculated from

$$SE(p) = \sqrt{\frac{p(1-p)}{n}}$$

There are no easy ways for small samples; this is applicable only to larger samples ($n > 25$, $n$ should be much larger for a narrow intervals; procedures for small samples are rather complicated and are not covered in this book).

**Example 4.8**    Consider the problem of estimating the prevalence of malignant melanoma in 45- to 54-year-old women in the United States. Suppose that a random sample of $n = 5000$ women is selected from this age group and $x = 28$ are found to have the disease. Our point estimate for the prevalence of this disease is

$$p = \frac{28}{5000}$$
$$= 0.0056$$

Its standard error is

$$SE(p) = \sqrt{\frac{(0.0056)(1 - 0.0056)}{5000}}$$
$$= 0.0011$$

Therefore, a 95% confidence interval for the prevalence $\pi$ of malignant melanoma in 45- to 54-year-old women in the United States is given by

$$0.0056 \pm (1.96)(0.0011) = (0.0034, 0.0078)$$

**Example 4.9**    A public health official wishes to know how effective health education efforts are regarding smoking. Of $n_1 = 100$ males sampled in 1965 at the time of the release of the Surgeon General's Report on the health consequences of smoking, $x_1 = 51$ were found to be smokers. In 1980 a second random sample of $n_2 = 100$ males, gathered similarly, indicated that $x_2 = 43$ were smokers. Application of the method above yields the following 95% confidence intervals for the smoking rates:

(a) In 1965, the estimated rate was

$$p_1 = \frac{51}{100}$$
$$= 0.51$$

with its standard error

$$SE(p_1) = \sqrt{\frac{(0.51)(1 - 0.51)}{100}}$$
$$= 0.05$$

leading to a 95% confidence interval of

$$0.51 \pm (1.96)(0.05) = (0.41, 0.61)$$

(b) In 1980, the estimated rate was

$$p_2 = \frac{43}{100}$$
$$= 0.43$$

with its standard error

$$SE(p_2) = \sqrt{\frac{(0.43)(1 - 0.43)}{100}}$$
$$= 0.05$$

leading to a 95% confidence interval of

$$0.43 \pm (1.96)(0.05) = (0.33, 0.53)$$

It can be seen that the two confidence intervals, one for 1965 and one for 1980, are both quite long and overlapsed, even though the estimated rates show a decrease of 8% in smoking rate, because the sample sizes are rather small.

*Example 4.10* A study was conducted to look at the effects of oral contraceptives (OC) on heart disease in women 40–44 years of age. It is found that among $n_1 = 5000$ current OC users, 13 develop a myocardial infarction (MI) over a three-year period, while among $n_2 = 10,000$ non-OC users, seven develop an MI over a three-year period. Application of the method described above yields the following 95% confidence intervals for the MI rates:

(a) For OC users, the estimated rate was

$$p_1 = \frac{13}{5000}$$
$$= 0.0026$$

with its standard error

$$SE(p_2) = \sqrt{\frac{(0.0026)(1 - 0.0026)}{5000}}$$
$$= 0.0007$$

leading to a 95% confidence interval of

$$0.0026 \pm (1.96)(0.0007) = (0.0012, 0.0040)$$

(b) For non-OC users, the estimated rate was

$$p_2 = \frac{7}{10,000}$$
$$= 0.0007$$

with its standard error

$$SE(p_2) = \sqrt{\frac{(0.0007)(1 - 0.0007)}{10,000}}$$
$$= 0.0003$$

leading to a 95% confidence interval of

$$0.0007 \pm (1.96)(0.0003) = (0.0002, 0.0012)$$

It can be seen that the two confidence intervals, one for OC users and one for non-OC users, do not overlap, a strong indication that the two population MI rates are probably not the same.

In many trials for interventions, or in studies to determine possible effects of a risk factor, the comparison of proportions is based on data from two independent samples. However, the process of constructing two confidence intervals separately, one from each sample, as mentioned briefly at the end of the last few examples, is not efficient. The reason is that the *overall confidence* level may no longer be, say, 95% as intended because the process involves *two* separate inferences; possible errors may add up. The estimation of the *difference of proportions* should be formed using the following formula (for a 95% confidence interval):

$$(p_1 - p_2) \pm (1.96)SE(p_1 - p_2)$$

where

$$SE(p_1 - p_2) = \sqrt{\frac{p_1(1 - p_1)}{n_1} + \frac{p_2(1 - p_2)}{n_2}}$$

## 4.4  ESTIMATION OF ODDS RATIOS

So far we have relied heavily on the central limit theorem in forming confidence intervals for the means (Section 4.2) and the proportions (Section 4.3). The central limit theorem stipulates that as a sample size increases, the means of samples drawn from a population of any distribution will approach the normal distribution; and a proportion can be seen as a special case of the means. Even when the sample sizes are not large, since many real-life distributions are approximately normal, we still can form confidence intervals for the means (see Section 4.2.2 on the uses of small samples).

Besides the mean and the proportion, we have had two other statistics of interest, the *odds ratio* and Pearson's *coefficient of correlation*. However, the method used to form confidence intervals for the means and proportions does not apply directly to the case of these two new parameters. The sole reason is that they do not have the backing of the central limit theorem. The sampling distributions of the (sample) odds ratio and (sample) coefficient of correlation are positively skewed. Fortunately, these sampling distributions can be almost normalized by an appropriate data transformation: in these cases, by taking the logarithm. Therefore, we learn to form confidence intervals on the log scale; then taking antilogs of the two endpoints, a method used in Chapter 2 to obtain the *geometric mean*. In this section we present in detail such a method for the calculation of confidence intervals for odds ratios.

Data from a case–control study, for example, may be summarized in a $2 \times 2$ table (Table 4.7). We have:

(a)  The odds that a case was exposed is

$$\text{odds for cases} = \frac{a}{b}$$

(b)  The odds that a control was exposed is

$$\text{odds for controls} = \frac{c}{d}$$

Therefore, the (observed) odds ratio from the samples is

$$\text{OR} = \frac{a/b}{c/d}$$

$$= \frac{ad}{bc}$$

**TABLE 4.7**

| Exposure | Cases | Controls |
|----------|-------|----------|
| Exposed | $a$ | $c$ |
| Unexposed | $b$ | $d$ |

Confidence intervals are derived from the normal approximation to the sampling distribution of ln(OR) with variance

$$\text{Var}[\ln(\text{OR})] \simeq \frac{1}{a} + \frac{1}{b} + \frac{1}{c} + \frac{1}{d}$$

(ln is *logarithm to base e*, or the *natural logarithm*.) Consequently, an approximate 95% confidence interval, on the log scale, for odds ratio is given by

$$\ln \frac{ad}{bc} \pm 1.96\sqrt{\frac{1}{a} + \frac{1}{b} + \frac{1}{c} + \frac{1}{d}}$$

A 95% confidence interval for the odds ratio under investigation is obtained by *exponentiating* (the reverse log operation or antilog) the two endpoints:

$$\ln \frac{ad}{bc} - 1.96\sqrt{\frac{1}{a} + \frac{1}{b} + \frac{1}{c} + \frac{1}{d}}$$

and

$$\ln \frac{ad}{bc} + 1.96\sqrt{\frac{1}{a} + \frac{1}{b} + \frac{1}{c} + \frac{1}{d}}$$

**Example 4.11**   The role of smoking in pancreatitis has been recognized for many years; the data shown in Table 4.8 are from a case–control study carried out in eastern Massachusetts and Rhode Island in 1975–1979 (see Example 1.14). We have

(a) For ex-smokers, compared to those who never smoked,

$$\text{OR} = \frac{(13)(56)}{(80)(2)}$$

$$= 4.55$$

and a 95% confidence interval for the population odds ratio on the log scale is from

**TABLE 4.8**

| Use of Cigarettes | Cases | Controls |
|---|---|---|
| Current smokers | 38 | 81 |
| Ex-smokers | 13 | 80 |
| Never | 2 | 56 |

$$\ln 4.55 - 1.96\sqrt{\frac{1}{13} + \frac{1}{56} + \frac{1}{80} + \frac{1}{2}} = -0.01$$

to

$$\ln 4.55 + 1.96\sqrt{\frac{1}{13} + \frac{1}{56} + \frac{1}{80} + \frac{1}{2}} = 3.04$$

and hence the corresponding 95% confidence interval for the population odds ratio is $(0.99, 20.96)$.

(b) For current smokers, compared to those who never smoked,

$$OR = \frac{(38)(56)}{(81)(2)}$$

$$= 13.14$$

and a 95% confidence interval for the population odds ratio on the log scale is from

$$\ln 13.14 - 1.96\sqrt{\frac{1}{38} + \frac{1}{56} + \frac{1}{81} + \frac{1}{2}} = 1.11$$

to

$$\ln 13.14 + 1.96\sqrt{\frac{1}{38} + \frac{1}{56} + \frac{1}{81} + \frac{1}{2}} = 4.04$$

and hence the corresponding 95% confidence interval for the population odds ratio is $(3.04, 56.70)$.

***Example 4.12*** Toxic shock syndrome (TSS) is a disease first recognized in the 1980s, characterized by sudden onset of high fever ($>102°F$), vomiting, diarrhea, rapid progression to hypotension, and in most cases, shock. Because of the striking association with menses, several studies have been undertaken to look at various practices associated with the menstrual cycle. In a study by the Centers for Disease Control, 30 of 40 TSS cases and 30 of 114 controls who used a single brand of tampons used the Rely brand. Data are presented in a $2 \times 2$ table (Table 4.9). We have

$$OR = \frac{(30)(84)}{(10)(30)}$$

$$= 8.4$$

and a 95% confidence interval for the population odds ratio on the log scale is from

**TABLE 4.9**

| Brand | Cases | Controls |
|-------|-------|----------|
| Rely  | 30    | 30       |
| Others | 10   | 84       |
| Total | 40    | 114      |

$$\ln 8.4 - 1.96\sqrt{\frac{1}{30} + \frac{1}{10} + \frac{1}{30} + \frac{1}{84}} = 1.30$$

to

$$\ln 8.4 + 1.96\sqrt{\frac{1}{30} + \frac{1}{10} + \frac{1}{30} + \frac{1}{84}} = 2.96$$

and hence the corresponding 95% confidence interval for the population odds ratio is $(3.67, 19.30)$, indicating a very high risk elevation for Rely users.

## 4.5  ESTIMATION OF CORRELATION COEFFICIENTS

Similar to the case of the odds ratio, the sampling distribution of Pearson's coefficient of correlation is also positively skewed. After completing our descriptive analysis (Section 2.4), information about a possible relationship between two continuous factors are sufficiently contained in two statistics: the number of pairs of data $n$ (sample size) and Pearson's coefficient of correlation $r$ (which is a number between 0 and 1). Confidence intervals are then derived from the normal approximation to the sampling distribution of

$$z = \frac{1}{2} \ln \frac{1+r}{1-r}$$

with variance of approximately

$$\text{Var}(z) = \frac{1}{n-3}$$

Consequently, an approximate 95% confidence for the correlation coefficient interval, on this newly transformed scale, for Pearson's correlation coefficient is given by

$$z \pm 1.96\sqrt{\frac{1}{n-3}}$$

A 95% confidence interval $(r_l, r_u)$ for the coefficient of correlation under investigation is obtained by transforming the two endpoints,

$$z_l = z - 1.96\sqrt{\frac{1}{n-3}}$$

and

$$z_u = z + 1.96\sqrt{\frac{1}{n-3}}$$

as follows to obtain the lower endpoint,

$$r_l = \frac{\exp(2z_l) - 1}{\exp(2z_l) + 1}$$

and the upper endpoint of the confidence interval for the population coefficient of correlation,

$$r_u = \frac{\exp(2z_u) - 1}{\exp(2z_u) + 1}$$

(In these formulas, "exp" is the exponentiation, or antinatural log, operation.)

**Example 4.13**   The data shown in Table 4.10 represent systolic blood pressure readings on 15 women. The descriptive analysis in Example 2.9 yields $r = 0.566$ and we have

$$z = \frac{1}{2}\ln\frac{1+0.566}{1-0.566}$$
$$= 0.642$$

$$z_l = 0.642 - 1.96\sqrt{\frac{1}{12}}$$
$$= 0.076$$

**TABLE 4.10**

| Age ($x$) | SBP ($y$) | Age ($x$) | SBP ($y$) |
|-----------|-----------|-----------|-----------|
| 42 | 130 | 80 | 156 |
| 46 | 115 | 74 | 162 |
| 42 | 148 | 70 | 151 |
| 71 | 100 | 80 | 156 |
| 80 | 156 | 41 | 125 |
| 74 | 162 | 61 | 150 |
| 70 | 151 | 75 | 165 |
| 80 | 156 | | |

$$z_u = 0.642 + 1.96\sqrt{\frac{1}{12}}$$

$$= 1.207$$

$$r_l = \frac{\exp(0.152) - 1}{\exp(0.152) + 1}$$

$$= 0.076$$

$$r_u = \frac{\exp(2.414) - 1}{\exp(2.414) + 1}$$

$$= 0.836$$

or a 95% confidence interval for the population coefficient of correlation of $(0.076, 0.836)$, indicating a positive association between a woman's age and her systolic blood pressure; that is, older women are likely to have higher systolic blood pressure (it is just a coincidence that $z_l = r_l$).

***Example 4.14*** Table 4.11 gives the values for the birth weight $(x)$ and the increase in weight between days 70 and 100 of life, expressed as a percentage of the birth weight $(y)$ for 12 infants. The descriptive analysis in Example 2.8 yields $r = -0.946$ and we have

$$z = \frac{1}{2} \ln \frac{1 - 0.946}{1 + 0.946}$$

$$= -1.792$$

$$z_l = 0.014 - 1.96\sqrt{\frac{1}{9}}$$

$$= -2.446$$

$$z_u = 0.014 + 1.96\sqrt{\frac{1}{9}}$$

$$= -1.139$$

**TABLE 4.11**

| $x$ (oz) | $y$ (%) | $x$ (oz) | $y$ (%) |
|---|---|---|---|
| 112 | 63 | 81 | 120 |
| 111 | 66 | 84 | 114 |
| 107 | 72 | 118 | 42 |
| 119 | 52 | 106 | 72 |
| 92 | 75 | 103 | 90 |
| 80 | 118 | 94 | 91 |

$$r_l = \frac{\exp(-4.892) - 1}{\exp(-4.892) + 1}$$

$$= -0.985$$

$$r_u = \frac{\exp(-2.278) - 1}{\exp(-2.278) + 1}$$

$$= -0.814$$

or a 95% confidence interval for the population coefficient of correlation of $(-0.985, -0.814)$, indicating a very strong negative association between a baby's birth weight and his or her increase in weight between days 70 and 100 of life; that is, smaller babies are likely to grow faster during that period (that may be why, at three months, most babies look the same size!).

## 4.6   BRIEF NOTES ON THE FUNDAMENTALS

Problems in the biological and health sciences are formulated mathematically by considering the data that are to be used for making a decision as the observed values of a certain random variable $X$. The distribution of $X$ is assumed to belong to a certain family of distributions specified by one or several parameters; examples include the normal distribution, the binomial distribution, and the Poisson distribution, among others. The magnitude of a parameter often represents the effect of a risk or environmental factor and knowing its value, even approximately, would shed some light on the impact of such a factor. The problem for decision makers is to decide on the basis of the data which members of the family could represent the distribution of $X$, that is, to predict or estimate the value of the primary parameter $\theta$.

***Maximum Likelihood Estimation***   The likelihood function $L(x; \theta)$ for random sample $\{x_i\}$ of size $n$ from the probability density function (pdf) $f(x; \theta)$ is

$$L(x; \theta) = \prod_{i=1}^{n} f(x_i; \theta)$$

The maximum likelihood estimator (MLE) of $\theta$ is the value $\hat{\theta}$ for which $L(x; \theta)$ is maximized. Calculus suggests setting the derivative of $L$ with respect to $\theta$ equal to zero and solving the resulting equation. We can obtain, for example:

1. For a binomial distribution,

$$L(x; p) = \binom{n}{x} p^x (1 - p)^{n-x}$$

leading to

$$\hat{p} = \frac{x}{n} \quad \text{(sample proportion)}$$

2. For the Poisson distribution,

$$L(x; \theta) = \prod_{i=1}^{n} \frac{\theta^{x_i} e^{-\theta}}{x_i!}$$

leading to

$$\hat{\theta} = \frac{\sum x_i}{n}$$

$$= \bar{x} \quad \text{(sample mean)}$$

3. For the normal distribution,

$$L(x; \mu, \theta) = \prod_{i=1}^{n} \frac{1}{\sigma\sqrt{2\pi}} \exp\left[-\frac{(x_i - \mu)^2}{2\sigma^2}\right]$$

leading to

$$\hat{\mu} = \bar{x}$$

**Matched Case–Control Studies**  Pair-matched case–control studies with a binary risk factor were introduced near the end of Chapter 3. To control confounding factors in this design, individual diseased cases are matched one to one to a set of controls, or disease-free persons, chosen to have similar values for important confounding variables. The data are represented by a $2 \times 2$ table (Table 4.12) where $(+, -)$ denotes (exposed, unexposed) categories. Let $\theta$ be the odds ratio associated with the exposure under investigation; $n_{10}$ was shown to have the binomial distribution $B(n, p)$, where

**TABLE 4.12**

|  | Case | |
|---|---|---|
| Control | + | − |
| + | $n_{11}$ | $n_{01}$ |
| − | $n_{10}$ | $n_{00}$ |

$$n = n_{10} + n_{01}$$

$$p = \frac{\theta}{\theta + 1}$$

This corresponds to the following likelihood function:

$$L(x; \theta) = \binom{n_{10} + n_{01}}{n_{10}} \left(\frac{\theta}{\theta + 1}\right)^{n_{10}} \left(\frac{1}{\theta + 1}\right)^{n_{01}}$$

leading to a simple point estimate for the odds ratio $\hat{\theta} = n_{10}/n_{01}$.

## 4.7   NOTES ON COMPUTATIONS

All the computations for confidence intervals can be put together using a calculator, even though some are quite tedious, especially the confidence intervals for odds ratios and coefficients of correlation. Descriptive statistics such as mean $\bar{x}$ and standard deviation $s$ can be obtained with the help of Excel (see Section 2.5). Standard normal and $t$ coefficients can also be obtained with the help of Excel (see Section 3.5). If you try the first two usual steps: (1) click the *paste function icon*, f*, and (2) click *Statistical,* among the functions available you will find CONFIDENCE, which is intended for use in forming confidence intervals. But it is not worth the effort; the process is only for 95% confidence intervals for the mean using a large sample (with coefficient 1.96), and you still need to enter the sample mean, standard deviation, and sample size.

## EXERCISES

**4.1**   Consider a population consisting of four subjects, $A$, $B$, $C$, and $D$. The values for a random variable $X$ under investigation are given in Table E4.1. Form the sampling distribution for the sample mean of size $n = 2$ and verify that $\mu_{\bar{x}} = \mu$. Then repeat the process with sample size of $n = 3$.

**TABLE E4.1**

| Subject | Value |
|---------|-------|
| $A$ | 1 |
| $B$ | 1 |
| $C$ | 0 |
| $D$ | 0 |

**4.2**   The body mass index $(kg/m^2)$ is calculated by dividing a person's weight by the square of his or her height and is used as a measure of the extent to which the person is overweight. Suppose that the distribution of the body mass index for men has a standard deviation of $\sigma = 3$ kg/m$^2$, and we wish to estimate the mean $\mu$ using a sample of size $n = 49$. Find the probability that we would be correct within 1 kg/m$^2$.

**4.3**    Self-reported injuries among left- and right-handed people were compared in a survey of 1896 college students in British Columbia, Canada. Of the 180 left-handed students, 93 reported at least one injury, and 619 of the 1716 right-handed students reported at least one injury in the same period. Calculate the 95% confidence interval for the proportion of students with at least one injury for each of the two subpopulations, left- and right-handed students.

**4.4**    A study was conducted to evaluate the hypothesis that tea consumption and premenstrual syndrome are associated. A total of 188 nursing students and 64 tea factory workers were given questionnaires. The prevalence of premenstrual syndrome was 39% among the nursing students and 77% among the tea factory workers. Calculate the 95% confidence interval for the prevalence of premenstrual syndrome for each of the two populations, nursing students and tea factory workers.

**4.5**    A study was conducted to investigate drinking problems among college students. In 1983, a group of students were asked whether they had ever driven an automobile while drinking. In 1987, after the legal drinking age was raised, a different group of college students were asked the same question. The results are given in Table E4.5. Calculate, separately for 1983 and 1987, the 95% confidence interval for the proportion of students who had driven an automobile while drinking.

**TABLE E4.5**

| Drove While Drinking | Year | | Total |
|---|---|---|---|
| | 1983 | 1987 | |
| Yes | 1250 | 991 | 2241 |
| No | 1387 | 1666 | 3053 |
| Total | 2637 | 2657 | 5294 |

**4.6**    In August 1976, tuberculosis was diagnosed in a high school student (INDEX CASE) in Corinth, Mississippi. Subsequently, laboratory studies revealed that the student's disease was caused by drug-resistant tubercule bacilli. An epidemiologic investigation was conducted at the high school. Table E4.6 gives the rates of positive tuberculin reaction, determined for various groups of students according to degree of exposure to the index case. Calculate the 95% confidence interval for the rate of positive tuberculin reaction separately for each of the two subpopulations, those with high exposure and those with low exposure.

**TABLE E4.6**

| Exposure Level | Number Tested | Number Positive |
|---|---|---|
| High | 129 | 63 |
| Low | 325 | 36 |

**4.7** The prevalence rates of hypertension among adult (ages 18–74) white and black Americans were measured in the second National Health and Nutrition Examination Survey, 1976–1980. Prevalence estimates (and their standard errors) for women are given in Table E4.7. Calculate and compare the 95% confidence intervals for the proportions of the two groups, balcks and whites, and draw appropriate conclusion. Do you need the sample sizes to do your calculations?

**TABLE E4.7**

|  | $p$ (%) | SE($p$) |
|---|---|---|
| Whites | 25.3 | 0.9 |
| Blacks | 38.6 | 1.8 |

**4.8** Consider the data given in Table E4.8. Calculate the 95% confidence intervals for the sensitivity and specificity of x-ray as a screening test for tuberculosis.

**TABLE E4.8**

| X-ray | Tuberculosis | | Total |
|---|---|---|---|
|  | No | Yes |  |
| Negative | 1739 | 8 | 1747 |
| Positive | 51 | 22 | 73 |
| Total | 1790 | 30 | 1820 |

**4.9** Sera from a T-lymphotropic virus type (HTLV-I) risk group (prostitute women) were tested with two commercial research enzyme-linked immunoabsorbent assays (EIA) for HTLV-I antibodies. These results were compared with a gold standard, and the outcomes are shown in Table E4.9. Calculate the 95% confidence intervals for the sensitivity and specificity separately for the two EIAs.

**TABLE E4.9**

| True | Dupont's EIA | | Cellular Product's EIA | |
|---|---|---|---|---|
| | Positive | Negative | Positive | Negative |
| Positive | 15 | 1 | 16 | 0 |
| Negative | 2 | 164 | 7 | 179 |

**4.10** In a seroepidemiologic survey of health workers representing a spectrum of exposure to blood and patients with hepatitis B virus (HBV), it was found that infection increased as a function of contact. Table E4.10 provides data for hospital workers with uniform socioeconomic status at an urban teaching hospital in Boston, Massachusetts. Calculate the 95% confidence intervals for the proportions of HBV-positive workers in each subpopulation.

**TABLE E4.10**

| Personnel | Exposure | $n$ | HBV Positive |
|---|---|---|---|
| Physicians | Frequent | 81 | 17 |
| | Infrequent | 89 | 7 |
| Nurses | Frequent | 104 | 22 |
| | Infrequent | 126 | 11 |

**4.11** Consider the data taken from a study attempting to determine whether the use of electronic fetal monitoring (EFM) during labor affects the frequency of cesarean section deliveries. Of the 5824 infants included in the study, 2850 were electronically monitored and 2974 were not. The outcomes are given in Table E4.11.

**TABLE E4.11**

| Cesarean Delivery | EFM Exposure | | Total |
|---|---|---|---|
| | Yes | No | |
| Yes | 358 | 229 | 587 |
| No | 2492 | 2745 | 5237 |
| Total | 2850 | 2974 | 5824 |

**(a)** Use the data from the group without EFM exposure, calculate the 95% confidence interval for the proportion of cesarean delivery.

**(b)** Calculate the 95% confidence interval for the odds ratio representing the relationship between EFM exposure and cesarean delivery.

**4.12** A study was conducted to investigate the effectiveness of bicycle safety helmets in preventing head injury. The data consist of a random sample of 793 persons who were involved in bicycle accidents during a one-year period (Table E4.12).

**TABLE E4.12**

| | Wearing Helmet | | |
|---|---|---|---|
| Head Injury | Yes | No | Total |
| Yes | 17 | 218 | 235 |
| No | 130 | 428 | 558 |
| Total | 147 | 646 | 793 |

**(a)** Use the data from the group without helmets, calculate the 95% confidence interval for the proportion of head injury.

**(b)** Calculate the 95% confidence interval for the odds ratio representing the relationship between use (or nonuse) of a helmet and head injury.

**4.13** A case–control study was conducted in Auckland, New Zealand, to investigate the effects of alcohol consumption on both nonfatal myocardial infarction and coronary death in the 24 hours after drinking, among regular drinkers. Data were tabulated separately for men and women (Table E4.13).

**TABLE E4.13**

| | Drink in the Last 24 hours | Myocardial Infarction | | Coronary Death | |
|---|---|---|---|---|---|
| | | Controls | Cases | Controls | Cases |
| Men | No | 197 | 142 | 135 | 103 |
| | Yes | 201 | 136 | 159 | 69 |
| Women | No | 144 | 41 | 89 | 12 |
| | Yes | 122 | 19 | 76 | 4 |

**(a)** Refer to the myocardial infarction data and calculate separately for men and women the 95% confidence interval for the odds ratio associated with drinking.

**(b)** Refer to coronary death data and calculate separately for men and women the 95% confidence interval for the odds ratio associated with drinking.

**(c)** From the results in parts (a) and/or (b), is there any indication that *gender* may act as an effect modifier?

**4.14** Adult male residents of 13 counties of western Washington in whom testicular cancer had been diagnosed during 1977–1983 were interviewed over the telephone regarding both their history of genital tract conditions and possible vasectomy. For comparison, the same interview was given to a sample of men selected from the population of these counties by dialing telephone numbers at random. The data, tabulated by religious background, are given in Table E4.14. Calculate the 95% confidence interval for the odds ratio associated with vasectomy for each religious group. Is there any evidence of an effect modification?

**TABLE E4.14**

| Religion | Vasectomy | Cases | Controls |
|---|---|---|---|
| Protestant | Yes | 24 | 56 |
|  | No | 205 | 239 |
| Catholic | Yes | 10 | 6 |
|  | No | 32 | 90 |
| Others | Yes | 18 | 39 |
|  | No | 56 | 96 |

**4.15** A case–control study was conducted relating to the epidemiology of breast cancer and the possible involvement of dietary fats, along with vitamins and other nutrients. It included 2024 breast cancer cases who were admitted to Roswell Park Memorial Institute, Erie County, New York, from 1958 to 1965. A control group of 1463 was chosen from patients having no neoplasms and no pathology of gastrointestinal or reproductive systems. The primary factors being investigated were vitamins A and E (measured in international units per month). The data listed in Table E4.15 are for 1500 women over 54 years of age.

**TABLE E4.15**

| Vitamin A (IU/month) | Cases | Controls |
|---|---|---|
| $\leq 150{,}500$ | 893 | 392 |
| $> 150{,}500$ | 132 | 83 |
| Total | 1025 | 475 |

(a) Calculate the 95% confidence interval for the proportion among the controls who consumed at least 150,500 international units of vitamin A per month.

(b) Calculate the 95% confidence interval for the odds ratio associated with vitamin A deficiency.

**4.16** A study was undertaken to investigate the effect of bloodborne environmental exposures on ovarian cancer from an assessment of consumption of coffee, tobacco, and alcohol. Study subjects consist of 188 women in the San Francisco Bay area with epithelial ovarian cancers diagnosed in

1983–1985, and 539 control women. Of the 539 controls, 280 were hospitalized women without overt cancer, and 259 were chosen from the general population by random telephone dialing. Data for coffee consumption are summarized in Table E4.16. Calculate the odds ratio and its 95% confidence interval for:

**TABLE E4.16**

| Coffee Drinkers | Cases | Hospital Controls | Population Controls |
|---|---|---|---|
| No | 11 | 31 | 26 |
| Yes | 177 | 249 | 233 |

**(a)** Cases versus hospital controls.

**(b)** Cases versus population controls.

**4.17** Postneonatal mortality due to respiratory illnesses is known to be inversely related to maternal age, but the role of young motherhood as a risk factor for respiratory morbidity in infants has not been explored thoroughly. A study was conducted in Tucson, Arizona, aimed at the incidence of lower respiratory tract illnesses during the first year of life. In this study, over 1200 infants were enrolled at birth between 1980 and 1984. The data shown in Table E4.17 are concerned with wheezing lower respiratory tract illnesses (wheezing LRI): no/yes. Using ">30" as the baseline, calculate the odds ratio and its 95% confidence interval for each other maternal age group.

**TABLE E4.17**

| Maternal Age (years) | Boys | | Girls | |
|---|---|---|---|---|
| | No | Yes | No | Yes |
| <21 | 19 | 8 | 20 | 7 |
| 21–25 | 98 | 40 | 128 | 36 |
| 26–30 | 160 | 45 | 148 | 42 |
| >30 | 110 | 20 | 116 | 25 |

**4.18** Data were collected from 2197 white ovarian cancer patients and 8893 white controls in 12 different U.S. case–control studies conducted by various investigators in the period 1956–1986 (*American Journal of Epidemiology*, 1992). These were used to evaluate the relationship of invasive epithelial ovarian cancer to reproductive and menstrual characteristics, exogenous estrogen use, and prior pelvic surgeries. Part of the data is shown in Table E4.18.

**TABLE E4.18**

|  | Cases | Controls |
|---|---|---|
| Duration of unprotected intercourse (years) |  |  |
| <2 | 237 | 477 |
| 2–9 | 166 | 354 |
| 10–14 | 47 | 91 |
| ≥15 | 133 | 174 |
| History of infertility |  |  |
| No | 526 | 966 |
| Yes |  |  |
| No drug use | 76 | 124 |
| Drug use | 20 | 11 |

**(a)** For the first factor in Table E4.18, using "<2" as the baseline calculate the odds ratio and its 95% confidence interval for each other level of exposure.

**(b)** For the second factor, using "no history of infertility" as the baseline, calculate the odds ratio and its 95% confidence interval for each group with a history of infertility.

**4.19** Consider the following measurements of forced expiratory volume (liters) for 10 subjects taken from a study that examines the response to ozone and sulfur dioxide among adolescents suffering from asthma:

$$\{3.50, 2.60, 2.75, 2.82, 4.05, 2.25, 2.68, 3.00, 4.02, 2.85\}$$

Calculate the 95% confidence interval for the (population) mean of forced expiratory volume (liters).

**4.20** The percentage of ideal body weight was determined for 18 randomly selected insulin-dependent diabetics. The outcomes (%) were

$$\begin{array}{ccccccccc} 107 & 119 & 99 & 114 & 120 & 104 & 124 & 88 & 114 \\ 116 & 101 & 121 & 152 & 125 & 100 & 114 & 95 & 117 \end{array}$$

Calculate the 95% confidence interval for the (population) mean of the percentage of ideal body weight.

**4.21** A study on birth weight provided the following data (in ounces) on 12 newborns:

$$\{112, 111, 107, 119, 92, 80, 81, 84, 118, 106, 103, 94\}$$

Calculate the 95% confidence interval for the (population) mean of the birth weight.

**4.22** The ages (in days) at time of death for samples of 11 girls and 16 boys who died of sudden infant death syndrome are shown in Table E4.22. Calculate separately for boys and girls the 95% confidence interval for the (population) mean of age (in days) at time of death.

**TABLE E4.22**

| Females | Males | |
|---------|-------|------|
| 53 | 46 | 115 |
| 56 | 52 | 133 |
| 60 | 58 | 134 |
| 60 | 59 | 175 |
| 78 | 77 | 175 |
| 87 | 78 | |
| 102 | 80 | |
| 117 | 81 | |
| 134 | 84 | |
| 160 | 103 | |
| 277 | 114 | |

**4.23** A study was conducted to investigate whether oat bran cereal helps to lower serum cholesterol in men with high cholesterol levels. Fourteen men were randomly placed on a diet that included either oat bran or cornflakes; after two weeks, their low-density-lipoprotein (LDL) cholesterol levels were recorded. Each man was then switched to the alternative diet. After a second two-week period, the LDL cholesterol level of each person was recorded again. The data are shown in Table E4.23. Calcu-

**TABLE E4.23**

| Subject | LDL (mmol/L) | |
|---------|-----------|----------|
| | Cornflakes | Oat Bran |
| 1 | 4.61 | 3.84 |
| 2 | 6.42 | 5.57 |
| 3 | 5.40 | 5.85 |
| 4 | 4.54 | 4.80 |
| 5 | 3.98 | 3.68 |
| 6 | 3.82 | 2.96 |
| 7 | 5.01 | 4.41 |
| 8 | 4.34 | 3.72 |
| 9 | 3.80 | 3.49 |
| 10 | 4.56 | 3.84 |
| 11 | 5.35 | 5.26 |
| 12 | 3.89 | 3.73 |
| 13 | 2.25 | 1.84 |
| 14 | 4.24 | 4.14 |

late the 95% confidence interval for the (population) mean difference of the LDL cholesterol level (mmol/L; cornflakes—oat bran).

**4.24** An experiment was conducted at the University of California–Berkeley to study the psychological environment effect on the anatomy of the brain. A group of 19 rats was randomly divided into two groups. Twelve animals in the treatment group lived together in a large cage, furnished with playthings that were changed daily; animals in the control group lived in isolation with no toys. After a month, the experimental animals were killed and dissected. Table E4.24 gives the cortex weights (the thinking part of the brain) in milligrams. Calculate separately for each treatment the 95% confidence interval for the (population) mean of the cortex weight. How do the means compare?

**TABLE E4.24**

| Treatment | | Control |
|---|---|---|
| 707 | 696 | 669 |
| 740 | 712 | 650 |
| 745 | 708 | 651 |
| 652 | 749 | 627 |
| 649 | 690 | 656 |
| 676 | | 642 |
| 699 | | 698 |

**4.25** The systolic blood pressures (mmHg) of 12 women between the ages of 20 and 35 were measured before and after administration of a newly developed oral contraceptive (Table E4.25).

**TABLE E4.25**

| Subject | Before | After | After–Before Difference, $d_i$ |
|---|---|---|---|
| 1 | 122 | 127 | 5 |
| 2 | 126 | 128 | 2 |
| 3 | 132 | 140 | 8 |
| 4 | 120 | 119 | −1 |
| 5 | 142 | 145 | 3 |
| 6 | 130 | 130 | 0 |
| 7 | 142 | 148 | 6 |
| 8 | 137 | 135 | −2 |
| 9 | 128 | 129 | 1 |
| 10 | 132 | 137 | 5 |
| 11 | 128 | 128 | 0 |
| 12 | 129 | 133 | 4 |

(a) Calculate the 95% confidence interval for the mean systolic blood pressure *change*. Does the oral contraceptive seem to change the mean systolic blood pressure?

(b) Calculate a 95% confidence interval for Pearson's correlation coefficient representing a possible relationship between systolic blood pressures measured before and after the administration of oral contraceptive. What does it mean that these measurements are correlated (if confirmed)?

**4.26**  Suppose that we are interested in studying patients with systemic cancer who subsequently develop a brain metastasis; our ultimate goal is to prolong their lives by controlling the disease. A sample of 23 such patients, all of whom were treated with radiotherapy, were followed from the first day of their treatment until recurrence of the original tumor. Recurrence is defined as the reappearance of a metastasis in exactly the same site, or in the case of patients whose tumor never completely disappeared, enlargement of the original lesion. Times to recurrence (in weeks) for the 23 patients were: 2, 2, 2, 3, 4, 5, 5, 6, 7, 8, 9, 10, 14, 14, 18, 19, 20, 22, 22, 31, 33, 39, and 195. First, calculate the 95% confidence interval for the mean time to recurrence on the log scale; then convert the endpoints to *weeks*.

**4.27**  An experimental study was conducted with 136 five-year-old children in four Quebec schools to investigate the impact of simulation games designed to teach children to obey certain traffic safety rules. The transfer of learning was measured by observing children's reactions to a quasi-real-life model of traffic risks. The scores on the transfer of learning for the control and attitude/behavior simulation game groups are summarized in Table E4.27. Find and compare the 95% confidence intervals for the means of the two groups, and draw an appropriate conclusion.

**TABLE E4.27**

| Summarized Data | Control | Simulation Game |
|---|---|---|
| $n$ | 30 | 33 |
| $\bar{x}$ | 7.9 | 10.1 |
| $s$ | 3.7 | 2.3 |

**4.28**  The body mass index is calculated by dividing a person's weight by the square of his or her height (it is used as a measure of the extent to which the person is overweight). A sample of 58 men, selected (retrospectively) from a large group of middle-aged men who later developed diabetes mellitus, yields $\bar{x} = 25.0$ kg/m$^2$ and $s = 2.7$ kg/m$^2$.

(a) Calculate a 95% confidence interval for the mean of this subpopulation.

**(b)** If it is known that the average body mass index for middle-aged men who do not develop diabetes is 24.0 kg/m², what can you say about the relationship between body mass index and diabetes in middle-aged men?

**4.29** A study was undertaken to clarify the relationship between heart disease and occupational carbon disulfide exposure along with another important factor, elevated diastolic blood pressure (DBP), in a data set obtained from a 10-year prospective follow-up of two cohorts of over 340 male industrial workers in Finland. Carbon disulfide is an industrial solvent that is used all over the world in the production of viscose rayon fibers. Table E4.29 gives the mean and standard deviation (SD) of serum cholesterol (mg/100 mL) among exposed and nonexposed cohorts, by diastolic blood pressure (DBP). Compare serum cholesterol levels between exposed and nonexposed cohorts at each level of DBP by calculating the two 95% confidence intervals for the means (exposed and nonexposed groups).

**TABLE E4.29**

| DBP (mmHg) | Exposed | | | Nonexposed | | |
|---|---|---|---|---|---|---|
| | $n$ | Mean | SD | $n$ | Mean | SD |
| <95 | 205 | 220 | 50 | 271 | 221 | 42 |
| 95–100 | 92 | 227 | 57 | 53 | 236 | 46 |
| ≥100 | 20 | 233 | 41 | 10 | 216 | 48 |

**4.30** Refer to the data on cancer of the prostate in Exercise 2.39, and calculate the 95% confidence interval for Pearson's correlation between age and the level of serum acid phosphatase.

**4.31** Table E4.31 give the net food supply ($x$, the number of calories per person per day) and the infant mortality rate ($y$, the number of infant

**TABLE E4.31**

| Country | $x$ | $y$ | Country | $x$ | $y$ |
|---|---|---|---|---|---|
| Argentina | 2730 | 98.8 | Iceland | 3160 | 42.4 |
| Australia | 3300 | 39.1 | India | 1970 | 161.6 |
| Austria | 2990 | 87.4 | Ireland | 3390 | 69.6 |
| Belgium | 3000 | 83.1 | Italy | 2510 | 102.7 |
| Burma | 1080 | 202.1 | Japan | 2180 | 60.6 |
| Canada | 3070 | 67.4 | Netherlands | 3010 | 37.4 |
| Chile | 2240 | 240.8 | New Zealand | 3260 | 32.2 |
| Cuba | 2610 | 116.8 | Sweden | 3210 | 43.3 |
| Egypt | 2450 | 162.9 | U.K. | 3100 | 55.3 |
| France | 2880 | 66.1 | U.S. | 3150 | 53.2 |
| Germany | 2960 | 63.3 | Uruguay | 2380 | 94.1 |

deaths per 1000 live births) for certain selected countries before World War II. Calculate the 95% confidence interval for Pearson's correlation coefficient between the net food supply and the infant mortality rate.

**4.32** In an assay of heparin, a standard preparation is compared with a test preparation by observing the log clotting times ($y$, in seconds) of blood containing different doses of heparin ($x$ is the log dose) (Table E4.32). Replicate readings are made at each dose level. Calculate separately for the standard preparation and the test preparation the 95% confidence interval for Pearson's correlation coefficient between the log clotting times and log dose.

**TABLE E4.32**

| Log Clotting Times | | | | |
|---|---|---|---|---|
| Standard | | Test | | Log Dose |
| 1.806 | 1.756 | 1.799 | 1.763 | 0.72 |
| 1.851 | 1.785 | 1.826 | 1.832 | 0.87 |
| 1.954 | 1.929 | 1.898 | 1.875 | 1.02 |
| 2.124 | 1.996 | 1.973 | 1.982 | 1.17 |
| 2.262 | 2.161 | 2.140 | 2.100 | 1.32 |

**4.33** There have been times when the city of London, England, experienced periods of dense fog. Table E4.33 shows such data for a 15-day very severe period which include the number of deaths each day ($y$), the mean atmospheric smoke ($x_1$, in mg/m$^3$), and the mean atmospheric sulfur dioxide content ($x_2$, in ppm). Calculate the 95% confidence interval for Pearson's correlation coefficient between the number of deaths ($y$) and

**TABLE E4.33**

| Number of Deaths | Smoke | Sulfur Dioxide |
|---|---|---|
| 112 | 0.30 | 0.09 |
| 140 | 0.49 | 0.16 |
| 143 | 0.61 | 0.22 |
| 120 | 0.49 | 0.14 |
| 196 | 2.64 | 0.75 |
| 294 | 3.45 | 0.86 |
| 513 | 4.46 | 1.34 |
| 518 | 4.46 | 1.34 |
| 430 | 1.22 | 0.47 |
| 274 | 1.22 | 0.47 |
| 255 | 0.32 | 0.22 |
| 236 | 0.29 | 0.23 |
| 256 | 0.50 | 0.26 |
| 222 | 0.32 | 0.16 |
| 213 | 0.32 | 0.16 |

the mean atmospheric smoke ($x_1$, in mg/m$^3$), and between the number of deaths ($y$) and the mean atmospheric sulfur dioxide content ($x_2$), respectively.

**4.34**   The following are the heights (measured to the nearest 2 cm) and the weights (measured to the nearest kilogram) of 10 men:

| Height | 162 | 168 | 174 | 176 | 180 | 180 | 182 | 184 | 186 | 186 |
|--------|-----|-----|-----|-----|-----|-----|-----|-----|-----|-----|
| Weight | 65  | 65  | 84  | 63  | 75  | 76  | 82  | 65  | 80  | 81  |

and 10 women:

| Height | 152 | 156 | 158 | 160 | 162 | 162 | 164 | 164 | 166 | 166 |
|--------|-----|-----|-----|-----|-----|-----|-----|-----|-----|-----|
| Weight | 52  | 50  | 47  | 48  | 52  | 55  | 55  | 56  | 60  | 60  |

Calculate, separately for men and women, the 95% confidence interval for Pearson's correlation coefficient between height and weight. Is there any indication that the two population correlation coefficients are different.

**4.35**   Data are shown in Table E4.35 for two groups of patients who died of acute myelogenous leukemia. Patients were classified into the two groups according to the presence or absence of a morphologic characteristic of white cells. Patients termed AG positive were identified by the presence

**TABLE E4.35**

| AG Positive, $N = 17$ | | AG Negative, $N = 16$ | |
|---|---|---|---|
| WBC | Survival Time (weeks) | WBC | Survival Time (weeks) |
| 2,300   | 65  | 4,400   | 56 |
| 750     | 156 | 3,000   | 65 |
| 4,300   | 100 | 4,000   | 17 |
| 2,600   | 134 | 1,500   | 7  |
| 6,000   | 16  | 9,000   | 16 |
| 10,500  | 108 | 5,300   | 22 |
| 10,000  | 121 | 10,000  | 3  |
| 17,000  | 4   | 19,000  | 4  |
| 5,400   | 39  | 27,000  | 2  |
| 7,000   | 143 | 28,000  | 3  |
| 9,400   | 56  | 31,000  | 8  |
| 32,000  | 26  | 26,000  | 4  |
| 35,000  | 22  | 21,000  | 3  |
| 100,000 | 1   | 79,000  | 30 |
| 100,000 | 1   | 100,000 | 4  |
| 52,000  | 5   | 100,000 | 43 |
| 100,000 | 65  |         |    |

of Auer rods and/or significant granulature of the leukemic cells in the bone marrow at diagnosis. For AG-negative patients these factors were absent. Leukemia is a cancer characterized by an overproliferation of white blood cells; the higher the white blood count (WBC), the more severe the disease. Calculate separately for the AG-positive and AG-negative patients the 95% confidence interval for Pearson's correlation coefficient between survival time and white blood count (both on a log scale). Is there any indication that the two population correlation coefficients are different?

# 5

# INTRODUCTION TO STATISTICAL TESTS OF SIGNIFICANCE

This chapter covers the most used and yet most misunderstood statistical procedures, called *tests* or *tests of significance*. The reason for the misunderstanding is simple: language. The colloquial meaning of the word *test* is one of no-nonsense objectivity. Students take tests in school, hospitals draw blood to be sent to laboratories for tests, and automobiles are tested by the manufacturer for performance and safety. It is thus natural to think that statistical tests are the "objective" procedures to use on data. The truth is that statistical tests are no more or less objective than any other statistical procedure, such as confidence estimation (Chapter 4).

Statisticians have made the problem worse by using the word *significance*, another word that has a powerful meaning in ordinary, colloquial language: *importance*. Statistical tests that result in significance are naturally misunderstood by the public to mean that the findings or results are important. That's not what statisticians mean; it only means that, for example, the difference they hypothesized was *real*.

Statistical tests are commonly seriously misinterpreted by nonstatisticians, but the misinterpretations are very natural. It is very natural to look at data and ask whether there is "anything going on" or whether it is just a bunch of meaningless numbers that can't be interpreted. Statistical tests appeal to investigators and readers of research for a reason in addition to the aforementioned reasons of language confusion. Statistical tests are appealing because they seem to make a *decision*; they are attractive because they say "yes" or "no." There is comfort in using a procedure that gives definitive *answers* from confusing data.

One way of explaining statistical tests is to use criminal court procedures as a metaphor. In criminal court, the accused is "presumed innocent" until "proved guilty beyond all reasonable doubt." This framework of presumed innocence has nothing whatsoever to do with anyone's personal belief as to the innocence or guilt of the defendant. Sometimes everybody, including the jury, the judge, and even the defendant's attorney, think the defendant is guilty. The rules and procedures of the criminal court must be followed, however. There may be a mistrial, or a hung jury, or the arresting officer forgot to read the defendant his or her rights. Any number of things can happen to save the guilty from a conviction. On the other hand, an innocent defendant is sometimes convicted by overwhelming circumstantial evidence. Criminal courts occasionally make mistakes, sometimes releasing the guilty and sometimes convicting the innocent. Statistical tests are like that. Sometimes, statistical significance is attained when nothing is going on, and sometimes, no statistical significance is attained when something very important is going on.

Just as in the courtroom, everyone would like statistical tests to make mistakes as infrequently as possible. Actually, the mistake rate of one of two possible mistakes made by statistical tests has usually been chosen (arbitrarily) to be 5% or 1%. The kind of mistake referred to here is the mistake of attaining statistical significance when there is actually nothing going on, just as the mistake of convicting the innocent in a trial by jury. This mistake is called a type I mistake or *type I error*. Statistical tests are often constructed so that type I errors occur 5% or 1% of the time. There is no custom regarding the rate of *type II errors*, however. A type II error is the mistake of not getting statistical significance when there is something going on, just as the mistake of releasing the guilty in a trial by jury. The rate of type II mistakes is dependent on several factors. One of the factors is *how much* is going on, just as the severity of the crime in a trial by jury. If there is a lot going on, one is less likely to make type II errors. Another factor is the amount of variability ("noise") there is in the data, just as in the quality of evidence available in a trial by jury. A lot of variability makes type II errors more likely. Yet another factor is the size of the study, just as the amount of evidence in a trial by jury. There are more type II errors in small studies than there are in large ones. Type II errors are rare in really huge studies but quite common in small studies.

There is a very important, subtle aspect of statistical tests, based on the aforementioned three things that make type II errors very improbable. Since really huge studies virtually guarantee getting statistical significance if there is even the slightest amount going on, such studies result in statistical significance when the *amount* that is going on is of no practical importance. In this case, statistical significance is attained in the face of no practical significance. On the other hand, small studies can result in statistical *nonsignificance* when something of great practical importance is going on. The conclusion is that the attainment of statistical significance in a study is just as affected by extraneous factors as it is by practical importance. It is essential to learn that statistical significance is not synonymous with practical importance.

## 5.1    BASIC CONCEPTS

From the introduction of sampling distributions in Chapter 4, it was clear that the value of a sample mean is influenced by:

1. The population $\mu$, because

$$\mu_{\bar{x}} = \mu$$

2. Chance; $\bar{x}$ and $\mu$ are almost never identical. The variance of the sampling distribution is

$$\sigma_{\bar{x}}^2 = \frac{\sigma^2}{n}$$

a combined effect of natural variation in the population ($\sigma^2$) and sample size $n$.

Therefore, when an observed value $\bar{x}$ is far from a hypothesized value of $\mu$ (e.g., mean high blood pressures for a group of oral contraceptive users compared to a typical average for women in the same age group), a natural question would be: Was it just due to chance, or something else? To deal with questions such as this, statisticians have invented the concept of *hypothesis tests*, and these tests have become widely used statistical techniques in the health sciences. In fact, it is almost impossible to read a research article in public health or medical sciences without running across hypothesis tests!

### 5.1.1    Hypothesis Tests

When a health investigator seeks to understand or explain something, for example the effect of a toxin or a drug, he or she usually formulates his or her research question in the form of a *hypothesis*. In the statistical context, a hypothesis is a statement about a distribution (e.g., "the distribution is normal") or its underlying parameter(s) (e.g., "$\mu = 10$"), or a statement about the relationship between probability distributions (e.g., "there is no statistical relationship") or its parameters (e.g., "$\mu_1 = \mu_2$"—equality of population means). The hypothesis to be tested is called the *null hypothesis* and will be denoted by $H_0$; it is usually stated in the null form, indicating no difference or no relationship between distributions or parameters, similar to the constitutional guarantee that the accused is presumed innocent until proven guilty. In other words, under the null hypothesis, an observed difference (like the one between sample means $\bar{x}_1$ and $\bar{x}_2$ for sample 1 and sample 2, respectively) just reflects chance variation. A *hypothesis test* is a decision-making process that examines a set or sets of data, and on the basis of expectation under $H_0$, leads to a decision as to whether or not to reject $H_0$. An *alternative hypothesis*, which we denote by $H_A$, is a hypothesis that in some sense contradicts the null hypothesis $H_0$, just as the *charge* by the prosecution in a trial by jury. Under $H_A$, the observed difference is real (e.g., $\bar{x}_1 \neq \bar{x}_2$ not by chance but because $\mu_1 \neq \mu_2$). A null hypothesis is rejected if and only if there is sufficiently strong evidence from the data to support its alternative—the names are somewhat unsettling, because the alternative hypothesis is, for a health investigator, the one that he or she usually wants to prove. (The null hypothesis is just a dull explanation of the findings–in terms of chance variation!) However, these are entrenched statistical terms and will be used as standard terms for the rest of this book.

Why is hypothesis testing important? Because in many circumstances we merely wish to know whether a certain proposition is true or false. The process of hypothesis tests provides a framework for making decisions on an *objective* basis, by weighing the relative merits of different hypotheses, rather than on a *subjective* basis by simply looking at the numbers. Different people can form different opinions by looking at data (confounded by chance variation or sampling errors), but a hypothesis test provides a standardized decision-making process that will be consistent for all people. The mechanics of the tests vary with the hypotheses and measurement scales (Chapters 6, 7, and 8), but the general philosophy and foundation is common and is discussed in some detail in this chapter.

### 5.1.2 Statistical Evidence

A null hypothesis is often concerned with a parameter or parameters of population(s). However, it is often either impossible, or too costly or time consuming, to obtain the entire population data on any variable in order to see whether or not a null hypothesis is true. Decisions are thus made using sample data. Sample data are summarized into a statistic or statistics that are used to estimate the parameter(s) involved in the null hypothesis. For example, if a null hypothesis is about $\mu$ (e.g., $H_0$: $\mu = 10$), a good place to look for information about $\mu$ is $\bar{x}$. In that context, the statistic $\bar{x}$ is called a *test statistic*; a test statistic can be used to measure the difference between the data (i.e., the numerical value of $\bar{x}$ obtained from the sample) and what is expected if the null hypothesis is true (i.e., "$\mu = 10$"). However, this evidence is statistical evidence; it varies from sample to sample (in the context of repeated sampling). It is a variable with a specific sampling distribution. The observed value is thus usually converted to a standard unit: the number of standard errors away from a hypothesized value. At this point, the logic of the test can be seen more clearly. It is an argument by contradiction, designed to show that the null hypothesis will lead to a less acceptable conclusion (an almost impossible event—some event that occurs with near-zero probability) and must therefore be rejected. In other words, the difference between the data and what is expected on the null hypothesis would be very difficult—even absurd—to explain as a chance variation; it makes you want to abandon (or reject) the null hypothesis and believe in the alternative hypothesis because it is more plausible.

**TABLE 5.1**

|  | Decision | |
|---|---|---|
| Truth | $H_0$ Is Not Rejected | $H_0$ Is Rejected |
| $H_0$ is true | Correct decision | Type I error |
| $H_0$ is false | Type II error | Correct decision |

### 5.1.3 Errors

Since a null hypothesis $H_0$ may be true or false and our possible decisions are whether to reject or not to reject it, there are four possible outcomes or combinations. Two of the four outcomes are correct decisions:

1. Not rejecting a true $H_0$
2. Rejecting a false $H_0$

but there are also two possible ways to commit an error:

1. *Type I:* A true $H_0$ is rejected.
2. *Type II:* A false $H_0$ is not rejected.

These possibilities are shown in Table 5.1.

The general aim in hypothesis testing is to keep $\alpha$ and $\beta$, the probabilities—in the context of repeated sampling—of types I and II respectively, as small as possible. However, if resources are limited, this goal requires a compromise because these actions are contradictory (e.g., a decision to decrease the size of $\alpha$ will increase the size of $\beta$, and vice versa). Conventionally, we fix $\alpha$ at some specific conventional level—say, 0.05 or 0.01—and $\beta$ is controlled through the use of sample size(s).

*Example 5.1*  Suppose that the national smoking rate among men is 25% and we want to study the smoking rate among men in the New England states. Let $\pi$ be the proportion of New England men who smoke. The null hypothesis that the smoking prevalence in New England is the same as the national rate is expressed as

$$H_0: \pi = 0.25$$

Suppose that we plan to take a sample of size $n = 100$ and use this decision-making rule:

$$\text{If } p \le 0.20, \ H_0 \text{ is rejected,}$$

where $p$ is the proportion obtained from the sample.

(a) Alpha ($\alpha$) is defined as the probability of wrongly rejecting a true null hypothesis, that is,

$$\alpha = \Pr(p \leq 0.20, \text{ given that } \pi = 0.25)$$

Since $n = 100$ is large enough for the central limit theorem to apply, the sampling distribution of $p$ is approximately normal with mean and variance, under $H_0$, given by

$$\mu_p = \pi$$
$$= 0.25$$
$$\sigma_p^2 = \frac{\pi(1 - \pi)}{n}$$
$$= (0.043)^2$$

respectively. Therefore, for this decision-making rule,

$$\alpha = \Pr\left(z \leq \frac{0.20 - 0.25}{0.043}\right)$$
$$= \Pr(z \leq -1.16)$$
$$= 0.123 \quad \text{or} \quad 12.3\%$$

Of course, we can make this smaller (as small as we wish) by changing the decision-making rule; however, that action will increase the value of $\beta$ (or the probability of a type II error).

(b) Suppose that the truth is

$$H_A: \pi = 0.15$$

Beta ($\beta$) is defined as the probability of not rejecting a false $H_0$, that is,

$$\beta = \Pr(p > 0.20; \text{ knowing that } \pi = 0.15)$$

Again, an application of the central limit theorem indicates that the sampling distribution of $p$ is approximately normal with mean

$$\mu_p = 0.15$$

and variance

$$\sigma_p^2 = \frac{(0.15)(0.85)}{100}$$
$$= (0.036)^2$$

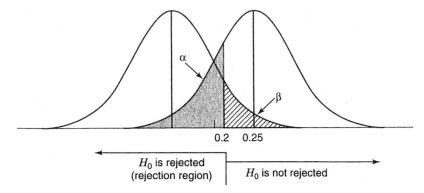

0.2   0.25

$H_0$ is rejected
(rejection region)        $H_0$ is not rejected

**Figure 5.1**   Graphical display of type I and type II errors.

Therefore,

$$\beta = \Pr\left(z \geq \frac{0.20 - 0.15}{0.036}\right)$$

$$= \Pr(z \geq 1.39)$$

$$= 0.082 \quad \text{or} \quad 8.2\%$$

The results above can be represented graphically as shown in Figure 5.1. It can be seen that $\beta$ depends on a specific alternative (e.g., $\beta$ is larger for $H_A: \pi = 0.17$ or any alternative hypothesis that specifies a value of $\pi$ that is farther away from 0.25) and from Figure 5.1, if we change the decision-making rule by using a smaller *cut point*, we would decrease $\alpha$ but increase $\beta$.

## 5.2   ANALOGIES

To reinforce some of the definitions or terms that we have encountered, we consider in this section two analogies: trials by jury and medical screening tests.

### 5.2.1   Trials by Jury

Statisticians and statistics users may find a lot in common between a court trial and a statistical test of significance. In a criminal court, the jury's duty is to evaluate the evidence of the prosecution and the defense to determine whether a defendant is guilty or innocent. By use of the judge's instructions, which provide guidelines for their reaching a decision, the members of the jury can arrive at one of two verdicts: guilty or not guilty. Their decision may be correct or they could make one of two possible errors: convict an innocent person or free a criminal. The analogy between statistics and trials by jury goes as follows:

Test of significance ↔ Court trial

Null hypothesis ↔ "Every defendant is innocent until proved guilty"

Research design ↔ Police investigation

Data/test statistics ↔ Evidence/exhibits

Statistical principles ↔ Judge's instruction

Statistical decision ↔ Verdict

Type I error ↔ Conviction of an innocent defendant

Type II error ↔ Acquittal of a criminal

This analogy clarifies a very important concept: When a null hypothesis is not rejected, it does not necessarily lead to its acceptance, because a "not guilty" verdict is just an indication of "lack of evidence," and "innocence" is one of the possibilities. That is, when a difference is not statistically significant, there are still two possibilities:

1. The null hypothesis is true.
2. The null hypothesis is false, but there is not enough evidence from sample data to support its rejection (i.e., sample size is too small).

### 5.2.2  Medical Screening Tests

Another analogy of hypothesis testing can be found in the application of screening tests or diagnostic procedures. Following these procedures, clinical observations, or laboratory techniques, people are classified as healthy or as having a disease. Of course, these tests are imperfect: Healthy persons will occasionally be classified wrongly as being ill, while some who are ill may fail to be detected. The analogy between statistical tests and screening tests goes briefly as follows:

$$type\ I\ error \leftrightarrow false\ positives$$
$$type\ II\ error \leftrightarrow false\ negatives$$

so that

$$\alpha = 1 - specificity$$
$$\beta = 1 - sensitivity$$

### 5.2.3  Common Expectations

The medical care system, with its high visibility and remarkable history of achievements, has been perceived somewhat naively by the general public as a

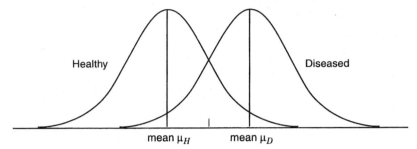

**Figure 5.2**    Graphical display of a translational model of diseases.

perfect remedy factory. Medical tests are expected to diagnose correctly any disease (and physicians are expected to treat and cure all diseases effectively!). Another common misconception is the assumption that all tests, regardless of the disease being tested for, are equally accurate. People are shocked to learn that a test result is wrong (of course, the psychological effects could be devastating). Another analogy between tests of significance and screening tests exists here: Statistical tests are also expected to provide a correct decision!

In some medical cases such as infections, the presence or absence of bacteria and viruses is easier to confirm correctly. In other cases, such as the diagnosis of diabetes by a blood sugar test, the story is different. One very simple model for these situations would be to assume that the variable $X$ (e.g., the sugar level in blood) on which the test is based is distributed with different means for the healthy and diseased subpopulations (Figure 5.2).

It can be seen from Figure 5.2 that errors are unavoidable, especially when the two means, $\mu_H$ and $\mu_D$, are close. The same is true for statistical tests of significance; when the null hypothesis $H_0$ is not true, it could be a little wrong or very wrong. For example, for

$$H_0: \mu = 10$$

the truth could be $\mu = 12$ or $\mu = 50$. If $\mu = 50$, type II errors would be less likely, and if $\mu = 12$, type II errors are more likely.

## 5.3   SUMMARIES AND CONCLUSIONS

To perform a hypothesis test, we take the following steps:

1. Formulate a null hypothesis and an alternative hypothesis. This would follow our research question, providing an explanation of what we want to prove in terms of chance variation; the statement resulting from our research question forms our alternative hypothesis.
2. Design the experiment and obtain data.

3. Choose a test statistic. This choice depends on the null hypothesis as well as the measurement scale.

4. Summarize findings and state appropriate conclusions.

This section involves the final step of the process outlined above.

### 5.3.1  Rejection Region

The most common approach is the formation of a decision rule. All possible values of the chosen test statistic (in the repeated sampling context) are divided into two regions. The region consisting of values of the test statistic for which the null hypothesis $H_0$ is rejected is called the *rejection region*. The values of the test statistic comprising the rejection region are those values that are less likely to occur if the null hypothesis is true, and the decision rule tells us to reject $H_0$ if the value of the test statistic that we compute from our sample(s) is one of the values in this region. For example, if a null hypothesis is about $\mu$, say

$$H_0: \mu = 10$$

then a good place to look for a test statistic for $H_0$ is $\bar{x}$, and it is obvious that $H_0$ should be rejected if $\bar{x}$ is far away from "10", the hypothesized value of $\mu$. Before we proceed, a number of related concepts should be made clear:

***One-Sided versus Two-Sided Tests***    In the example above, a vital question is: Are we interested in the deviation of $\bar{x}$ from 10 in one or both directions? If we are interested in determining whether $\mu$ is significantly *different* from 10, we would perform a two-sided test and the rejection region would be as shown in Figure 5.3a. On the other hand, if we are interested in whether $\mu$ is significantly *larger* than 10, we would perform a one-sided test and the rejection region would be as shown in Figure 5.3b.

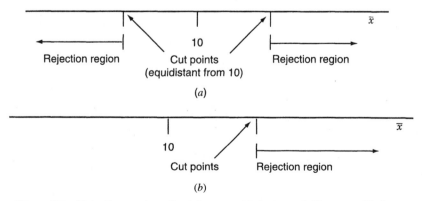

**Figure 5.3**  Rejection regions for (a) a two-sided test and (b) a one-sided test.

A one-sided test is indicated for research questions like these: Is a new drug *superior* to a standard drug? Does the air pollution *exceed* safe limits? Has the death rate been *reduced* for those who quit smoking? A two-sided test is indicated for research questions like these: Is there a *difference* between the cholesterol levels of men and women? Does the mean age of a target population *differ* from that of the general population?

***Level of Significance***    The decision as to which values of the test statistic go into the rejection region, or as to the location of the cut point, is made on the basis of the desired level of type I error $\alpha$ (also called the *size* of the test). A computed value of the test statistic that falls in the rejection region is said to be *statistical significant*. Common choices for $\alpha$, the level of significance, are 0.01, 0.05, and 0.10; the 0.05 or 5% level is especially popular.

***Reproducibility***    Here we aim to clarify another misconception about hypothesis tests. A very simple and common situation for hypothesis tests is that the test statistic, for example the sample mean $\bar{x}$, is normally distributed with different means under the null hypothesis $H_0$ and alternative hypothesis $H_A$. A one-sided test could be represented graphically as shown in Figure 5.4. It should now be clear that a statistical conclusion is not guaranteed to be reproducible. For example, if the alternative hypothesis is true and the mean of the distribution of the test statistic (see Figure 5.4) is right at the cut point, the probability would be 50% to obtain a test statistic inside the rejection region.

### 5.3.2   *p* Values

Instead of saying that an observed value of the test statistic is significant (i.e., falling into the rejection region for a given choice of $\alpha$) or is not significant, many writers in the research literature prefer to report findings in terms of *p values*. The *p* value is the probability of getting values of the test statistic as extreme as, or more extreme than, that observed if the null hypothesis is true. For the example above of

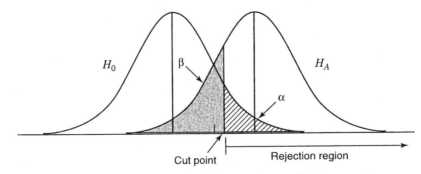

**Figure 5.4**   Graphical display of a one-sided test.

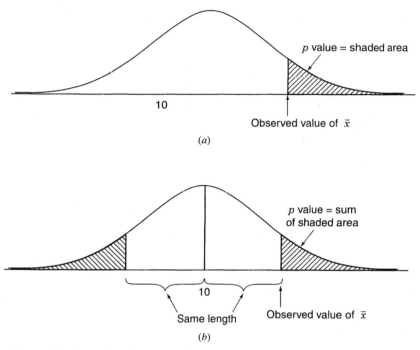

**Figure 5.5**  Graphical display of (a) a one-sided test and (b) a two-sided test.

$$H_0\colon \mu = 10$$

if the test is a one-sided test, we would have the $p$ value shown in Figure 5.5a, and if the test is a two-sided test, we have the $p$ value shown in Figure 5.5b. The curve in these graphs represents the sampling distribution of $\bar{x}$ if $H_0$ is true.

Compared to the approach of choosing a level of significance and formulating a decision rule, the use of the $p$-value criterion would be as follows:

1. If $p < \alpha$, $H_0$ is rejected.
2. if $p \geq \alpha$, $H_0$ is not rejected.

However, the reporting of $p$ values as part of the results of an investigation is more informative to readers than statements such as "the null hypothesis is rejected at the 0.05 level of significance" or "the results were not significant at the 0.05 level." Reporting the $p$ value associated with a test lets the reader know how common or how rare is the computed value of the test statistic given that $H_0$ is true. In other words, the $p$ value can be used as a *measure* of the compatibility between the data (reality) and a null hypothesis (theory); the smaller the $p$ value, the less compatible the theory and the reality. A compromise between the two approaches would be to report both in statements such as

**TABLE 5.2**

| $p$ Value | Interpretation |
|---|---|
| $p > 0.10$ | Result is not significant |
| $0.05 < p < 0.10$ | Result is marginally significant |
| $0.01 < p < 0.05$ | Result is significant |
| $p < 0.01$ | Result is highly significant |

"the difference is statistically significant ($p < 0.05$)." In doing so, researchers generally agree on the conventional terms listed in Table 5.2.

Finally, it should be noted that the difference between means, for example, although statistically significant, may be so small that it has little health consequence. In other words, the result may be *statistically significant* but may not be *practically significant*.

**Example 5.2** Suppose that the national smoking rate among men is 25% and we want to study the smoking rate among men in the New England states. The null hypothesis under investigation is

$$H_0: \pi = 0.25$$

Of $n = 100$ males sampled, $x = 15$ were found to be smokers. Does the proportion $\pi$ of smokers in New England states *differ* from that in the nation?

Since $n = 100$ is large enough for the central limit theorem to apply, it indicates that the sampling distribution of the sample proportion $p$ is approximately normal with mean and variance under $H_0$:

$$\mu_p = 0.25$$

$$\sigma_p^2 = \frac{(0.25)(1 - 0.25)}{100}$$

$$= (0.043)^2$$

The value of $p$ observed from our sample is

$$\frac{15}{100} = 0.15$$

representing a difference of 0.10 from the hypothesized value of 0.25. The $p$ value is defined as the probability of getting a value of the test statistic as extreme as, or more extreme than, that observed if the null hypothesis is true. This is represented graphically as shown in Figure 5.6.

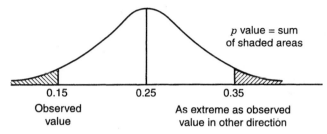

**Figure 5.6**  Graphical display of the $p$ value for Example 5.2.

Therefore,

$$p \text{ value} = \Pr(p \leq 0.15 \text{ or } p \geq 0.35)$$
$$= 2 \times \Pr(p \geq 0.35)$$
$$= 2 \times \Pr\left(z \geq \frac{0.35 - 0.25}{0.043}\right)$$
$$= 2 \times \Pr(z \geq 2.33)$$
$$= (2)(0.5 - 0.4901)$$
$$\simeq 0.02$$

In other words, with the data given, the difference between the national smoking rate and the smoking rate of New England states is statistically significant ($p < 0.05$).

### 5.3.3  Relationship to Confidence Intervals

Suppose that we consider a hypothesis of the form

$$H_0: \mu = \mu_0$$

where $\mu_0$ is a known hypothesized value. A two-sided hypothesis test for $H_0$ is related to confidence intervals as follows:

1. If $\mu_0$ is not included in the 95% confidence interval for $\mu$, $H_0$ should be rejected at the 0.05 level. This is represented graphically as shown in Figure 5.7.
2. If $\mu_0$ is included in the 95% confidence interval for $\mu$, $H_0$ should not be rejected at the 0.05 level (Figure 5.8).

*Example 5.3*  Consider the hypothetical data set in Example 5.2. Our point estimate of smoking prevalence in New England is

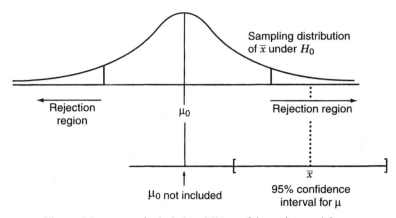

**Figure 5.7**   $\mu_0$ not included at 95% confidence interval for $\mu$.

$$p = \frac{15}{100}$$

$$= 0.15$$

Its standard error is

$$\text{SE}(p) = \sqrt{\frac{(0.15)(1 - 0.15)}{100}}$$

$$= 0.036$$

Therefore, a 95% confidence interval for the New England states smoking rate $\pi$ is given by

$$0.15 \pm (1.96)(0.036) = (0.079, 0.221)$$

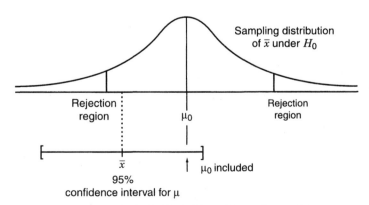

**Figure 5.8**   $\mu_0$ included in 95% confidence interval for $\mu$.

It is noted that the national rate of 0.25 is *not included* in that confidence interval.

## 5.4  BRIEF NOTES ON THE FUNDAMENTALS

A statistical hypothesis is a statement about a probability distribution or its underlying parameter(s), or a statement about the relationship between probability distributions and their parameters. If the hypothesis specifies a probability density function (pdf) completely, it is called *simple*; otherwise, it is *composite*. The hypothesis to be tested, the *null hypothesis*, is denoted by $H_0$; it is always stated in the null form, indicating no difference or no relationship between distributions or parameters. A statistical test is a decision-making process that examines a set or sets of sample data and on the basis of expectation under $H_0$ leads to a decision as to whether or not to reject $H_0$. An alternative hypothesis, which we denote by $H_A$, is a hypothesis that in some sense contradicts the null hypothesis $H_0$. A null hypothesis is rejected if and only if there is sufficiently strong evidence from the data to support its alternative.

### 5.4.1  Type I and Type II Errors

Since a null hypothesis may be true or false and our possible decisions are whether to reject or not reject it, there are four possible combinations. Two of the four combinations are correct decisions, but there are two possible ways to commit an error:

1. *Type I:* A true $H_0$ is wrongly rejected.
2. *Type II:* A false $H_0$ is not rejected.

The probability of a type I error is usually denoted by $\alpha$ and is commonly referred to as the *significance level* of a test. The probability of a type II error (for a specific alternative $H_A$) is denoted by $\beta$ and $(1 - \beta)$ is called the *power* of the test. The general aim in hypothesis testing is to use statistical tests that make $\alpha$ and $\beta$ as small as possible. This goal requires a compromise because these actions are contradictory; in a statistical data analysis, we fix $\alpha$ at some specific conventional level—say, 0.01 or 0.05—and use the test that minimizes $\beta$ or, equivalently, maximizes the power.

### 5.4.2  More about Errors and *p* Values

Given that type I error is more important or clear-cut, statistical tests are formed according to the *Neyman–Pearson framework*. Begin by specifying a small number $\alpha > 0$ such that probabilities of type I error greater than $\alpha$ are undesirable. Then restrict attention to tests which have the probability of rejection under $H_0$ less than or equal to $\alpha$; such tests are said to to have *level of significance* $\alpha$; the values 0.01 and 0.05 are common choices for $\alpha$. Since a test of level *alpha* is also of level $\alpha' > \alpha$, it is convenient to give a name to the smallest level of significance of a test. This quantity is called the *size* of the test. In general, if we have a test statistic $T$ and use a critical value $t$, our test has size $\alpha(t)$, which depends on the value of $t$, given by

$$\alpha(t) = \Pr(T \geq t)$$

The problem is that different people faced with the same testing problem may have different criteria of size; investigator A may be satisfied to reject the null hypothesis using a test of size 0.05, while investigator B insists on using size 0.01. If the two investigators can agree on a common test statistic $T$, this difficulty may be overcome by reporting the outcome of the investigation in terms of the *observed size* or $p$ value of the test. In other words, the $p$ value is $\alpha$ ($t =$ observed $T$). This quantity is a statistic that is defined as the smallest level of significance at which an investigator using $T$ would rejuect on the basis of the outcome observed. That is, if the investigator's critical value corresponds to a test of size less than the $p$ value, the null hypothesis is rejected. If the null hypothesis is true and the investigation repeated, $p$ values are distributed uniformly on the interval $[0, 1]$.

## EXERCISES

**5.1**  For each part, state the null $(H_0)$ and alternative $(H_A)$ hypotheses.

(a) Has the average community level of suspended particulates for the month of August exceeded 30 $\mu g/cm^3$?

(b) Does the mean age of onset of a certain acute disease for schoolchildren differ from 11.5?

(c) A psychologist claims that the average IQ of a sample of 60 children is significantly above the normal IQ of 100.

(d) Is the average cross-sectional area of the lumen of coronary arteries for men, ages 40–59, less than 31.5% of the total arterial cross section?

(e) Is the mean hemoglobin level of high-altitude workers different from 16 $g/cm^3$?

(f) Does the average speed of 50 cars as checked by radar on a particular highway differ from 55 mph?

**5.2**  The distribution of diastolic blood pressures for the population of female diabetics between the ages of 30 and 34 has an unknown mean $\mu$ and a standard deviation of $\sigma = 9$ mmHg. It may be useful to physicians to know whether the mean $\mu$ of this population is equal to the mean diastolic blood pressure of the general population of females of this age group, which is 74.5 mmHg. What are the null hypothesis and alternative hypothesis for this test?

**5.3**  *E. canis* infection is a tick-borne disease of dogs that is sometimes contracted by humans. Among infected humans, the distribution of white blood cell counts has an unknown mean $\mu$ and a standard deviation $\sigma$. In the general population the mean white blood count is 7250 per $mm^3$. It is believed that persons infected with *E. canis* must on average have a lower white blood cell count. What is the null hypothesis for the test? Is this a one- or two-sided alternative?

**5.4**  It is feared that the smoking rate in young females has increased in the last several years. In 1985, 38% of the females in the 17- to 24-year age group were smokers. An experiment is to be conducted to gain evidence to support the increase contention. Set up the appropriate null and alternative hypotheses. Explain in a practical sense what, if anything, has occurred if a type I or type II error has been committed.

**5.5**  A group of investigators wishes to explore the relationship between the use of hair dyes and the development of breast cancer in females. A group of 1000 beauticians 40–49 years of age is identified and followed for five years. After five years, 20 new cases of breast cancer have occurred. Assume that breast cancer incidence over this time period for average American women in this age group is 7/1000. We wish to test the hypothesis that using hair dyes increases the risk of breast cancer. Is a one- or two-sided test appropriate here? Compute the *p* value for your choice.

**5.6**  Height and weight are often used in epidemiological studies as possible predictors of disease outcomes. If the people in the study are assessed in a clinic, heights and weights are usually measured directly. However, if the people are interviewed at home or by mail, a person's self-reported height and weight are often used instead. Suppose that we conduct a study on 10 people to test the comparability of these two methods. Data from these 10 people were obtained using both methods on each person. What is the criterion for the comparison? What is the null hypothesis? Should a two- or a one-sided test be used here?

**5.7**  Suppose that 28 cancer deaths are noted among workers exposed to asbestos in a building materials plant from 1981 to 1985. Only 20.5 cancer deaths are expected from statewide mortality rates. Suppose that we want to know if there is a significant excess of cancer deaths among these workers. What is the null hypothesis? Is a one- or two-sided test appropriate here?

**5.8**  A food frequency questionnaire was mailed to 20 subjects to assess the intake of various food groups. The sample standard deviation of vitamin C intake over the 20 subjects was 15 (exclusive of vitamin C supplements). Suppose that we know from using an in-person diet interview method in an earlier large study that the standard deviation is 20. Formulate the null and alternative hypotheses if we want to test for any differences between the standard deviations of the two methods.

**5.9**    In Example 5.1 it was assumed that the national smoking rate among men is 25%. A study is to be conducted for New England states using a sample size $n = 100$ and the decision rule

$$\text{If } p \leq 0.20, \ H_0 \text{ is rejected}$$

where $H_0$ is

$$H_0\colon \pi = 0.25$$

where $\pi$ and $p$ are population and sample proportions, respectively, for New England states. Is this a one- or a two-tailed test?

**5.10**    In Example 5.1, with the rule

$$\text{If } p \leq 0.20, \ H_0 \text{ is rejected}$$

it was found that the probabilities of type I and type II errors are

$$\alpha = 0.123$$
$$\beta = 0.082$$

for $H_A\colon \pi = 0.15$. Find $\alpha$ and $\beta$ if the rule is changed to

$$\text{If } p \leq 0.18, \ H_0 \text{ is rejected}$$

How does this change affect $\alpha$ and $\beta$ values?

**5.11**    Answer the questions in Exercise 5.10 for the decision rule

$$\text{If } p \leq 0.22, \ H_0 \text{ is rejected}$$

**5.12**    Recalculate the $p$ value in Example 5.2 if it was found that 18 (instead of 15) men in a sample of $n = 100$ are smokers.

**5.13**    Calculate the 95% confidence interval for $\pi$ using the sample in Exercise 5.12 and compare the findings to the testing results of Exercise 5.12.

**5.14**    Plasma glucose levels are used to determine the presence of diabetes. Suppose that the mean log plasma glucose concentration (mg/dL) in 35- to 44-year-old is 4.86 with standard deviation 0.54. A study of 100 sedentary persons in this age group is planned to test whether they have higher levels of plasma glucose than the general population.

**(a)** Set up the null and alternative hypotheses.

**(b)** If the real increase is 0.1 log unit, what is the power of such a study if a two-sided test is to be used with $\alpha = 0.05$?

**5.15** Suppose that we are interested in investigating the effect of race on level of blood pressure. The mean and standard deviation of systolic blood pressure among 25- to 34-year-old white males were reported as 128.6 mmHg and 11.1 mmHg, respectively, based on a very large sample. Suppose that the actual mean for black males in the same age group is 135 mmHg. What is the power of the test (two-sided, $\alpha = 0.05$) if $n = 100$ and we assume that the variances are the same for whites and blacks?

# 6

# COMPARISON OF POPULATION PROPORTIONS

In this chapter we present basic inferential methods for categorical data, especially the analysis of two-way contingency tables. Let $X_1$ and $X_2$ denote two categorical variables, $X_1$ having $I$ levels and $X_2$ having $J$ levels, thus $IJ$ combinations of classifications. We display the data in a rectangular table having $I$ rows for the categories of $X_1$ and $J$ columns for the categories of $X_2$; the $IJ$ cells represent the $IJ$ combinations of outcomes. When the cells contain frequencies of outcomes, the table is called a *contingency table* or *cross-classified table*, also referred to as a $I$ by $J$ or $I \times J$ table. Most topics in this chapter are devoted to the analyses of these two-way tables; however, before we can get there, let's start with the simplest case: that of a one-sample problem with binary data.

## 6.1 ONE-SAMPLE PROBLEM WITH BINARY DATA

In this type of problem, we have a sample of binary data $(n, x)$ with $n$ being an adequately large sample size and $x$ the number of positive outcomes among the $n$ observations, and we consider the null hypothesis

$$H_0: \pi = \pi_0$$

where $\pi_0$ is a fixed and known number between 0 and 1: for example,

$$H_0: \pi = 0.25$$

$\pi_0$ is often a standardized or referenced figure, for example, the effect of a standardized drug or therapy, or the national smoking rate (where the national sample is often large enough so as to produce negligible sampling error in $\pi_0$). Or we could be concerned with a research question such as: Does the side effect (of a certain drug) exceed regulated limit $\pi_0$? In Exercise 5.5 we compared the incidence of breast cancer among female beauticians (who are frequently exposed to the use of hair dyes) versus a standard level of 7/1000 (for five years) for "average" American women. The figure 7/1000 is $\pi_0$ for that example.

In a typical situation, the null hypothesis of a statistical test is concerned with a parameter; the parameter in this case is the proportion $\pi$. Sample data are summarized into a statistic used to estimate the parameter under investigation. Since the parameter under investigation is the proportion $\pi$, our focus in this case is the sample proportion $p$. In general, a statistic is itself a variable with a specific sampling distribution (in the context of repeated sampling). Our statistic in this case is the sample proportion $p$; the corresponding sampling distribution is obtained easily by invoking the *central limit theorem*. With large sample size and assuming that the null hypothesis $H_0$ is true, it is the normal distribution with mean and variance given by

$$\mu_p = \pi_0$$

$$\sigma_p^2 = \frac{\pi_0(1 - \pi_0)}{n}$$

respectively. From this sampling distribution, the observed value of the sample proportion can be converted to a standard unit: the number of standard errors away from the hypothesized value of $\pi_0$. In other words, to perform a test of significance for $H_0$, we proceed with the following steps:

1. Decide whether a one- or a two-sided test is appropriate.
2. Choose a level of significance $\alpha$, a common choice being 0.05.
3. Calculate the $z$ score

$$z = \frac{p - \pi_0}{\sqrt{\pi_0(1 - \pi_0)/n}}$$

4. From the table for the standard normal distribution (Appendix B) and the choice of $\alpha$ (e.g., $\alpha = 0.05$), the rejection region is determined by:
   (a) For a one-sided test:

   $$z \leq -1.65 \quad \text{for } H_A: \pi < \pi_0$$

   $$z \geq 1.65 \quad \text{for } H_A: \pi > \pi_0$$

   (b) For a two-sided test or $H_A: \pi \neq \pi_0$:

   $$z \leq -1.96 \quad \text{or} \quad z \geq 1.96$$

*Example 6.1*   A group of investigators wish to explore the relationship between the use of hair dyes and the development of breast cancer in women. A sample of $n = 1000$ female beauticians 40–49 years of age is identified and followed for five years. After five years, $x = 20$ new cases of breast cancer have occurred. It is known that breast cancer incidence over this time period for average American women in this age group is $\pi_0 = 7/1000$. We wish to test the hypothesis that using hair dyes *increases* the risk of breast cancer (a one-sided alternative). We have:

1. A one-sided test with

$$H_A: \pi > \frac{7}{1000}$$

2. Using the conventional choice of $\alpha = 0.05$ leads to the rejection region $z > 1.65$.
3. From the data,

$$p = \frac{20}{1000}$$
$$= 0.02$$

leading to a "$z$ score" of:

$$z = \frac{0.02 - 0.007}{\sqrt{(0.007)(0.993)/1000}}$$
$$= 4.93$$

(i.e., the observed proportion $p$ is 4.93 standard errors away from the hypothesized value of $\pi_0 = 0.007$).
4. Since the computed $z$ score falls into the rejection region $(4.93 > 1.65)$, the null hypothesis is rejected at the 0.05 level chosen. In fact, the difference is very highly significant $(p < 0.001)$.

## 6.2   ANALYSIS OF PAIR-MATCHED DATA

The method presented in this section applies to cases where each subject or member of a group is observed twice for the presence or absence of a certain characteristic (e.g., at admission to and discharge from a hospital), or matched pairs are observed for the presence or absence of the same characteristic. A popular application is an epidemiological design called a *pair-matched case–control study*. In case–control studies, cases of a specific disease are ascertained as they arise from population-based registers or lists of hospital admissions,

**TABLE 6.1**

|  | Control | |
|---|---|---|
| Case | + | − |
| + | $a$ | $b$ |
| − | $c$ | $d$ |

and controls are sampled either as disease-free persons from the population at risk or as hospitalized patients having a diagnosis other than the one under investigation. As a technique to control confounding factors, individual cases are matched, often one to one, to controls chosen to have similar values for confounding variables such as age, gender, and race.

For pair-matched data with a single binary exposure (e.g., smoking versus nonsmoking), data can be represented by a $2 \times 2$ table (Table 6.1) where $(+, -)$ denotes the (exposed, nonexposed) outcome. In this $2 \times 2$ table, $a$ denotes the number of pairs with two exposed members, $b$ denotes the number of pairs where the case is exposed but the matched control is unexposed, $c$ denotes the number of pairs where the case is unexposed but the matched control is exposed, and $d$ denotes the number of pairs with two unexposed members. The analysis of pair-matched data with a single binary exposure can be seen, heuristically, as follows. What we really want to do is to compare the incidence of exposure among the cases versus the controls; the parts of the data showing no difference, the number $a$ of pairs with two exposed members, and the number $d$ of pairs with two unexposed members would contribute nothing as evidence in such a comparison. The comparison therefore relies solely on two other frequencies, $b$ and $c$; under the null hypothesis that the exposure has nothing to do with the disease, we *expect* that $b = c$ or $b/(b + c) = 0.5$. In other words, the analysis of pair-matched data with a single binary exposure can be seen as a special case of the one-sample problem with binary of Section 6.1 with $n = b + c$, $x = b$, and $\pi_0 = 0.5$. Recall the form of the test statistic of Section 6.1; we have

$$z = \frac{p - \pi_0}{\sqrt{\pi_0(1 - \pi_0)/(b + c)}}$$

$$= \frac{[b/(b + c)] - \frac{1}{2}}{\sqrt{\left(\frac{1}{2}\right)\left(1 - \frac{1}{2}\right)/(b + c)}}$$

$$= \frac{b - c}{\sqrt{b + c}}$$

The decision is based on the standardized $z$ score and referring to the percentiles of the standard normal distribution or, in the two-sided form, the square of the statistic above, denoted by

$$X^2 = \frac{(b-c)^2}{b+c}$$

and the test is known as *McNemar's chi-square*. If the test is one-sided, $z$ is used and the null hypothesis is rejected at the 0.05 level when

$$z \geq 1.65$$

If the test is two-sided, $X^2$ is used and the null hypothesis is rejected at the 0.05 level when

$$X^2 \geq 3.84$$

[It should be noted that $3.84 = (1.96)^2$, so that $X^2 \geq 3.84$ is equivalent to $z \leq -1.96$ or $z \geq 1.96$.]

***Example 6.2***    It has been noted that metal workers have an increased risk for cancer of the internal nose and paranasal sinuses, perhaps as a result of exposure to cutting oils. Therefore, a study was conducted to see whether this particular exposure also increases the risk for squamous cell carcinoma of the scrotum.

Cases included all 45 squamous cell carcinomas of the scrotum diagnosed in Connecticut residents from 1955 to 1973, as obtained from the Connecticut Tumor Registry. Matched controls were selected for each case based on age at death (within eight years), year of death (within three years), and number of jobs as obtained from combined death certificate and directory sources. An occupational indicator of metal worker (yes/no) was evaluated as the possible risk factor in this study; results are shown in Table 6.2. We have, for a one-tailed test,

$$z = \frac{26 - 5}{\sqrt{26 + 5}}$$
$$= 3.77$$

indicating a very highly significant increase of risk associated with the exposure ($p < 0.001$).

**TABLE 6.2**

|  | Controls | |
| --- | --- | --- |
| Cases | Yes | No |
| Yes | 2 | 26 |
| No | 5 | 12 |

**TABLE 6.3**

| Widowed Men | Married Men | |
| --- | --- | --- |
| | Dead | Alive |
| Dead | 2 | 292 |
| Alive | 210 | 700 |

*Example 6.3*   A study in Maryland identified 4032 white persons, enumerated in a unofficial 1963 census, who became widowed between 1963 and 1974. These people were matched, one to one, to married persons on the basis of race, gender, year of birth, and geography of residence. The matched pairs were followed to a second census in 1975. The overall male mortality is shown in Table 6.3. An application of McNemar's chi-square test (two-sided) yields

$$X^2 = \frac{(292 - 210)^2}{292 + 210}$$
$$= 13.39$$

It can be seen that the null hypothesis of equal mortality should be rejected at the 0.05 level (13.39 > 3.84).

## 6.3   COMPARISON OF TWO PROPORTIONS

Perhaps the most common problem involving categorical data is the comparison of two proportions. In this type of problem we have two independent samples of binary data $(n_1, x_1)$ and $(n_2, x_2)$ where the $n$'s are adequately large sample sizes that may or may not be equal. The $x$'s are the numbers of "positive" outcomes in the two samples, and we consider the null hypothesis

$$H_0: \pi_1 = \pi_2$$

expressing the equality of the two population proportions.

To perform a test of significance for $H_0$, we proceed with the following steps:

1. Decide whether a one-sided test, say,

$$H_A: \pi_2 > \pi_1$$

or a two-sided test,

$$H_A: \pi_1 \neq \pi_2$$

is appropriate.

2. Choose a significance level $\alpha$, a common choice being 0.05.
3. Calculate the $z$ score

$$z = \frac{p_2 - p_1}{\sqrt{p(1-p)(1/n_1 + 1/n_2)}}$$

where $p$ is the pooled proportion, defined by

$$p = \frac{x_1 + x_2}{n_1 + n_2}$$

an estimate of the common proportion under $H_0$.

4. Refer to the table for standard normal distribution (Appendix B) for selecting a cut point. For example, if the choice of $\alpha$ is 0.05, the rejection region is determined by:
   (a) For the one-sided alternative $H_A$: $\pi_2 > \pi_1$, $z \geq 1.65$.
   (b) For the one-sided alternative $H_A$: $\pi_2 < \pi_1$, $z \leq -1.65$.
   (c) For the two-sided alternative $H_A$: $\pi_1 \neq \pi_2$, $z \leq -1.96$ or $z \geq 1.96$.

What we are doing here follows the same format used in previous sections.

- The basic term of $p_2 - p_1$ *measures the difference* between the two samples,
- Its expected hypothesized value (i.e., under $H_0$) is zero.
- The denominator of $z$ is the standard error of $p_2 - p_1$, a measure of how good $p_2 - p_1$ is as an estimate of $\pi_2 - \pi_1$.
- Therefore, $z$ measures the number of standard errors that $p_2 - p_1$, the *evidence*, is away from its hypothesized value.

In the two-sided form, the square of the $z$ score, denoted $X^2$, is more often used. The test is referred to as the *chi-square test*. The test statistic can also be obtained using the shortcut formula

$$X^2 = \frac{(n_1 + n_2)[x_1(n_2 - x_2) - x_2(n_1 - x_1)]^2}{n_1 n_2 (x_1 + x_2)(n_1 + n_2 - x_1 - x_2)}$$

and the null hypothesis is rejected at the 0.05 level when

$$X^2 \geq 3.84$$

It should be noted that in the general case, with data in a $2 \times 2$ table (Table 6.4), the chi-square statistic above is simply

$$X^2 = \frac{(a+b+c+d)(ad-bc)^2}{(a+c)(b+d)(a+b)(c+d)}$$

its denominator being the product of the four marginal totals.

**TABLE 6.4**

| Factor | Sample 1 | Sample 2 | Total |
|---|---|---|---|
| Present | $a$ | $c$ | $a + c$ |
| Absent | $b$ | $d$ | $b + d$ |
| Sample size | $n_1 = a + b$ | $n_2 = c + d$ | $N = a + b + c + d$ |

**Example 6.4** A study was conducted to see whether an important public health intervention would significantly reduce the smoking rate among men. Of $n_1 = 100$ males sampled in 1965 at the time of the release of the Surgeon General's report on the health consequences of smoking, $x_1 = 51$ were found to be smokers. In 1980 a second random sample of $n_2 = 100$ males, similarly gathered, indicated that $x_2 = 43$ were smokers.

An application of the method above yields

$$p = \frac{51 + 43}{100 + 100}$$

$$= 0.47$$

$$z = \frac{0.51 - 0.43}{\sqrt{(0.47)(0.53)\left(\frac{1}{100} + \frac{1}{100}\right)}}$$

$$= 1.13$$

It can be seen that the rate observed was reduced from 51% to 43%, but the reduction is not statistically significant at the 0.05 level ($z = 1.13 < 1.65$).

**Example 6.5** An investigation was made into fatal poisonings of children by two drugs which were among the leading causes of such deaths. In each case, an inquiry was made as to how the child had received the fatal overdose and responsibility for the accident was assessed. Results are shown in Table 6.5. We have the proportions of cases for which the child is responsible:

$$p_A = \frac{8}{8 + 31}$$

$$= 0.205 \quad \text{or} \quad 20.5\%$$

$$p_B = \frac{12}{12 + 19}$$

$$= 0.387 \quad \text{or} \quad 38.7\%$$

**TABLE 6.5**

| | Drug A | Drug B |
|---|---|---|
| Child responsible | 8 | 12 |
| Child not responsible | 31 | 19 |

suggesting that they are not the same and that a child seems more prone to taking drug B than drug A. However, the chi-square statistic

$$X^2 = \frac{(39 + 31)[(8)(19) - (31)(12)]^2}{(39)(31)(20)(50)}$$

$$= 2.80 \quad (<3.84; \ \alpha = 0.05)$$

shows that the difference is not statistically significant at the 0.05 level.

***Example 6.6***   In Example 1.2, a case–control study was conducted to identify reasons for the exceptionally high rate of lung cancer among male residents of coastal Georgia. The primary risk factor under investigation was employment in shipyards during World War II, and Table 6.6 provides data for non-smokers. We have for the cases,

$$p_2 = \frac{11}{61}$$

$$= 0.180$$

and for the controls,

$$p_1 = \frac{35}{238}$$

$$= 0.147$$

An application of the procedure yields a pooled proportion of

$$p = \frac{11 + 35}{61 + 238}$$

$$= 0.154$$

leading to

$$z = \frac{0.180 - 0.147}{\sqrt{(0.154)(0.846)\left(\frac{1}{61} + \frac{1}{238}\right)}}$$

$$= 0.64$$

**TABLE 6.6**

| Shipbuilding | Cases | Controls |
|---|---|---|
| Yes | 11 | 35 |
| No | 50 | 203 |

It can be seen that the rate of employment for the cases (18.0%) was higher than that for the controls (14.7%), but the difference is not statistically significant at the 0.05 level ($z = 0.64 < 1.65$).

***Example 6.7***   The role of smoking in the etiology of pancreatitis has been recognized for many years. To provide estimates of the quantitative significance of these factors, a hospital-based study was carried out in eastern Massachusetts and Rhode Island between 1975 and 1979. Ninety-eight patients who had a hospital discharge diagnosis of pancreatitis were included in this unmatched case–control study. The control group consisted of 451 patients admitted for diseases other than those of the pancreas and biliary tract. Risk factor information was obtained from a standardized interview with each subject, conducted by a trained interviewer.

Some data for the males are shown in Table 6.7. With currently smoking being the exposure, we have for the cases,

$$p_2 = \frac{38}{53}$$

$$- 0.717$$

and for the controls,

$$p_1 = \frac{81}{217}$$

$$= 0.373$$

An application of the procedure yields a pooled proportion of

$$p = \frac{38 + 81}{53 + 217}$$

$$= 0.441$$

leading to

$$z = \frac{0.717 - 0.373}{\sqrt{(0.441)(0.559)\left(\frac{1}{53} + \frac{1}{217}\right)}}$$

$$= 4.52$$

**TABLE 6.7**

| Use of Cigarettes | Cases | Controls |
|---|---|---|
| Current smokers | 38 | 81 |
| Never or ex-smokers | 15 | 136 |
| Total | 53 | 217 |

It can be seen that the proportion of smokers among the cases (71.7%) was higher than that for the controls (37.7%) and the difference is highly statistically significant ($p < 0.001$).

## 6.4   MANTEL–HAENSZEL METHOD

We are often interested only in investigating the relationship between two binary variables (e.g., a disease and an exposure); however, we have to control for confounders. A confounding variable is a variable that may be associated with either the disease or exposure or both. For example, in Example 1.2, a case–control study was undertaken to investigate the relationship between lung cancer and employment in shipyards during World War II among male residents of coastal Georgia. In this case, smoking is a confounder; it has been found to be associated with lung cancer and it may be associated with employment because construction workers are likely to be smokers. Specifically, we want to know:

- Among smokers, whether or not shipbuilding and lung cancer are related
- Among nonsmokers, whether or not shipbuilding and lung cancer are related

The underlying question is the question concerning conditional independence between lung cancer and shipbuilding; however, we do not want to reach separate conclusions, one at each level of smoking. Assuming that the confounder, smoking, is not an effect modifier (i.e., smoking does not alter the relationship between lung cancer and shipbuilding), we want to pool data for a combined decision. When both the disease and the exposure are binary, a popular method to achieve this task is the Mantel–Haenszel method. The process can be summarized as follows:

1. We form $2 \times 2$ tables, one at each level of the confounder.
2. At a level of the confounder, we have the frequencies as shown in Table 6.8.

**TABLE 6.8**

| Exposure | Disease Classification | | Total |
| --- | --- | --- | --- |
| | + | − | |
| + | $a$ | $b$ | $r_1$ |
| − | $c$ | $d$ | $r_2$ |
| Total | $c_1$ | $c_2$ | $n$ |

Under the null hypothesis and fixed marginal totals, cell $(1, 1)$ frequency $a$ is distributed with mean and variance:

$$E_0(a) = \frac{r_1 c_1}{n}$$

$$\mathrm{Var}_0(a) = \frac{r_1 r_2 c_1 c_2}{n^2(n-1)}$$

and the Mantel–Haenszel test is based on the $z$ statistic:

$$z = \frac{\sum a - \sum(r_1 c_1/n)}{\sqrt{\sum(r_1 r_2 c_1 c_2/(n^2(n-1)))}}$$

where the summation $(\sum)$ is across levels of the confounder. Of course, one can use the square of the $z$ score, a *chi-square test* at one degree of freedom, for two-sided alternatives.

When the test above is statistically significant, the association between the disease and the exposure is *real*. Since we assume that the confounder is not an effect modifier, the odds ratio is constant across its levels. The odds ratio at each level is estimated by $ad/bc$; the Mantel–Haenszel procedure pools data across levels of the confounder to obtain a combined estimate:

$$\mathrm{OR}_{\mathrm{MH}} = \frac{\sum(ad/n)}{\sum(bc/n)}$$

***Example 6.8***   A case–control study was conducted to identify reasons for the exceptionally high rate of lung cancer among male residents of coastal Georgia. The primary risk factor under investigation was employment in shipyards during World War II, and data are tabulated separately in Table 6.9 for three levels of smoking. There are three $2 \times 2$ tables, one for each level of smoking; in Example 1.1, the last two tables were combined and presented together for simplicity.

We begin with the $2 \times 2$ table for nonsmokers (Table 6.10). We have, for the nonsmokers,

**TABLE 6.9**

| Smoking | Shipbuilding | Cases | Controls |
|---------|--------------|-------|----------|
| No | Yes | 11 | 35 |
| | No | 50 | 203 |
| Moderate | Yes | 70 | 42 |
| | No | 217 | 220 |
| Heavy | Yes | 14 | 3 |
| | No | 96 | 50 |

**TABLE 6.10**

| Shipbuilding | Cases | Controls | Total |
|---|---|---|---|
| Yes | 11 $(a)$ | 35 $(b)$ | 46 $(r_1)$ |
| No | 50 $(c)$ | 203 $(d)$ | 253 $(r_2)$ |
| Total | 61 $(c_1)$ | 238 $(c_2)$ | 299 $(n)$ |

$$a = 11$$

$$\frac{r_1 c_1}{n} = \frac{(46)(61)}{299}$$

$$= 9.38$$

$$\frac{r_1 r_2 c_1 c_2}{n^2(n-1)} = \frac{(46)(253)(61)(238)}{(299)^2(298)}$$

$$= 6.34$$

$$\frac{ad}{n} = \frac{(11)(203)}{299}$$

$$= 7.47$$

$$\frac{bc}{n} = \frac{(35)(50)}{299}$$

$$= 5.85$$

The process is repeated for each of the other two smoking levels. For moderate smokers,

$$a = 70$$

$$\frac{r_1 c_1}{n} = \frac{(112)(287)}{549}$$

$$= 58.55$$

$$\frac{r_1 r_2 c_1 c_2}{n^2(n-1)} = \frac{(112)(437)(287)(262)}{(549)^2(548)}$$

$$= 22.28$$

$$\frac{ad}{n} = \frac{(70)(220)}{549}$$

$$= 28.05$$

$$\frac{bc}{n} = \frac{(42)(217)}{549}$$

$$= 16.60$$

and for heavy smokers,

$$a = 14$$

$$\frac{r_1 c_1}{n} = \frac{(17)(110)}{163}$$

$$= 11.47$$

$$\frac{r_1 r_2 c_1 c_2}{n^2(n-1)} = \frac{(17)(146)(110)(53)}{(163)^2(162)}$$

$$= 3.36$$

$$\frac{ad}{n} = \frac{(14)(50)}{163}$$

$$= 4.29$$

$$\frac{bc}{n} = \frac{(3)(96)}{163}$$

$$- 1.77$$

These results are combined to obtain the $z$ score:

$$z = \frac{(11 - 9.38) + (70 - 58.55) + (14 - 11.47)}{\sqrt{6.34 + 22.28 + 3.36}}$$

$$= 2.76$$

and a $z$ score of 2.76 yields a one-tailed $p$ value of 0.0029, which is beyond the 1% level. This result is stronger than those for tests at each level because it is based on more information, where all data at all three smoking levels are used. The combined odds ratio estimate is

$$OR_{MH} = \frac{7.47 + 28.05 + 4.29}{5.85 + 16.60 + 1.77}$$

$$= 1.64$$

representing an approximate increase of 64% in lung cancer risk for those employed in the shipbuilding industry.

*Note:* An SAS program would include these instructions:

```
DATA;
INPUT SMOKE SHIP CANCER COUNT;
CARDS;
1 1 1 11
1 1 2 35
```

```
1 2 1 50
1 2 2 203
2 1 1 70
2 1 2 42
2 2 1 217
2 2 2 220
3 1 1 14
3 1 2 3
3 2 1 96
3 2 2 50;
PROC FREQ;
WEIGHT COUNT;
TABLES SMOKE*SHIP*CANCER/CMH;
```

The result is given in a chi-square form ($X^2 = 7.601$, $p = 0.006$); CMH stands for the Cochran–Mantel–Haenszel statistic; noting that $(2.76)^2 = 7.601$.

***Example 6.9*** A case–control study was conducted to investigate the relationship between myocardial infarction (MI) and oral contraceptive use (OC). The data, stratified by cigarette smoking, are given in Table 6.11. An application of the Mantel–Haenszel procedure yields the results shown in Table 6.12. The combined $z$ score is

**TABLE 6.11**

| Smoking | OC Use | Cases | Controls |
|---|---|---|---|
| No | Yes | 4 | 52 |
|  | No | 34 | 754 |
| Yes | Yes | 25 | 83 |
|  | No | 171 | 853 |

**TABLE 6.12**

|  | Smoking | |
|---|---|---|
|  | No | Yes |
| $a$ | 4 | 25 |
| $\dfrac{r_1 c_1}{n}$ | 2.52 | 18.70 |
| $\dfrac{r_1 r_2 c_1 c_2}{n^2(n-1)}$ | 2.25 | 14.00 |
| $\dfrac{ad}{n}$ | 3.57 | 18.84 |
| $\dfrac{bc}{n}$ | 2.09 | 12.54 |

$$z = \frac{(4 - 2.52) + (25 - 18.70)}{\sqrt{2.25 + 14.00}}$$

$$= 1.93$$

which is significant at the 5% level (one-sided). The combined odds ratio estimate is

$$\text{OR}_{\text{MH}} = \frac{3.57 + 18.84}{2.09 + 12.54}$$

$$= 1.53$$

representing an approximate increase of 53% in myocardial infarction for oral contraceptive users.

## 6.5  INFERENCES FOR GENERAL TWO-WAY TABLES

Data forming two-way contingency tables do not necessarily come from two binomial samples. There may be more than two binomial samples to compare. They may come from two independent samples, but the endpoint may have more than two categories. They may come from a survey (i.e., *one* sample), but data are cross-tabulated based on two binary factors of interest (so we still have a $2 \times 2$ table as in the comparison of two proportions).

Consider the general case of an $I \times J$ table: say, resulting from a survey of size $n$. Let $X_1$ and $X_2$ denote two categorical variables, $X_1$ having $I$ levels and $X_2$ having $J$ levels; there are $IJ$ combinations of classifications. The $IJ$ cells represent the $IJ$ combinations of classifications; their probabilities are $\{\pi_{ij}\}$, where $\pi_{ij}$ denotes the probability that the outcome $(X_1, X_2)$ falls in the cell in row $i$ and column $j$. When two categorical variables forming the two-way table are independent, all $\pi_{ij} = \pi_{i+}\pi_{+j}$. This is the *multiplication rule* for probabilities of independence events introduced in Chapter 3; here $\pi_{i+}$ and $\pi_{+j}$ are the two marginal or univariate probabilities. The estimate of $\pi_{ij}$ under this condition is

$$\widehat{\pi_{ij}} = \widehat{\pi_{i+}} \, \widehat{\pi_{+j}}$$

$$= p_{i+}p_{+j}$$

$$= \frac{x_{i+}}{n} \frac{x_{+j}}{n}$$

$$= \frac{x_{i+}x_{+j}}{n^2}$$

where the $x$'s are the observed frequencies. Under the assumption of independence, we would have in cell $(i, j)$:

$$e_{ij} = n\widehat{\pi_{ij}}$$

$$= \frac{x_{i+}x_{+j}}{n}$$

$$= \frac{(\text{row total})(\text{column total})}{\text{sample size}}$$

The $\{e_{ij}\}$ are called *estimated expected frequencies*, the frequencies we expect to have under the null hypothesis of independence. They have the same marginal totals as those of the data observed. In this problem we do not compare proportions (because we have only one sample), what we really want to see is if the two factors or variables $X_1$ and $X_2$ are *related*; the task we perform is a *test for independence*. We achieve that by comparing the observed frequencies, the $x$'s, versus those expected under the null hypothesis of independence, the expected frequencies $e$'s. This needed comparison is done through Pearson's chi-quare statistic:

$$\chi^2 = \sum_{i,j} \frac{(x_{ij} - e_{ij})^2}{e_{ij}}$$

For large samples, $X^2$ has approximately a chi-square distribution with degrees of freedom under the null hypothesis of independence,

$$df = (I - 1)(J - 1)$$

with greater values lead to a rejection of $H_0$.

***Example 6.10***  In 1979 the U.S. Veterans Administration conducted a health survey of 11,230 veterans. The advantages of this survey are that it includes a large random sample with a high interview response rate, and it was done before the public controversy surrounding the issue of the health effects of possible exposure to Agent Orange. The data shown in Table 6.13 relate Vietnam service to having sleep problems among the 1787 veterans who entered the military service between 1965 and 1975. We have

**TABLE 6.13**

| | Service in Vietnam | | |
|---|---|---|---|
| Sleep Problems | Yes | No | Total |
| Yes | 173 | 160 | 333 |
| No | 599 | 851 | 1450 |
| Total | 772 | 1011 | 1783 |

$$e_{11} = \frac{(333)(772)}{1783}$$

$$= 144.18$$

$$e_{12} = 333 - 144.18$$

$$= 188.82$$

$$e_{21} = 772 - 144.18$$

$$= 627.82$$

$$e_{22} = 1011 - 188.82$$

$$= 822.18$$

leading to

$$X^2 = \frac{(173 - 144.18)^2}{144.18} + \frac{(160 - 188.82)^2}{188.82} + \frac{(599 - 627.82)^2}{627.82} + \frac{(851 - 822.18)^2}{822.18}$$

$$= 12.49$$

This statistic, at 1 df, indicate a significant correlation ($p < 0.001$) relating Vietnam service to having sleep problems among the veterans. It is interesting to note that we needed to calculate only *one* expected frequency, $e_{11}$; this explains the *one* degree of freedom we used.

***Example 6.11*** Table 6.14 shows the results of a survey in which each subject of a sample of 300 adults was asked to indicate which of three policies they favored with respect to smoking in public places. The numbers in parentheses are expected frequencies. An application of Pearson's chi-quare test, at 6 degrees of freedom, yields

$$X^2 = \frac{(5 - 8.75)^2}{8.75} + \frac{(44 - 46)^2}{46} + \cdots + \frac{(10 - 4.5)^2}{4.5}$$

$$= 22.57$$

**TABLE 6.14**

| Highest Education Level | Policy Favored | | | | |
|---|---|---|---|---|---|
| | No Restrictions on Smoking | Smoking Allowed in Designated Areas Only | No Smoking at All | No Opinion | Total |
| College graduate | 5 (8.75) | 44 (46) | 23 (15.75) | 3 (4.5) | 75 |
| High school | 15 (17.5) | 100 (92) | 30 (31.50) | 5 (9) | 150 |
| Grade school | 15 (8.75) | 40 (46) | 10 (15.75) | 10 (4.5) | 75 |
| Total | 35 | 184 | 63 | 18 | 300 |

The result indicates a high correlation between education levels and preferences about smoking in public places ($p = 0.001$).

*Note:* An SAS program would include these instructions:

```
DATA;
INPUT EDUCAT POLICY COUNT;
CARDS;
1 1 5
1 2 44
1 3 23
1 4 3
2 1 15
2 2 100
2 3 30
2 4 5
3 1 15
3 2 40
3 3 10
3 4 10;
PROC FREQ;
WEIGHT COUNT;
TABLES EDUCAT*POLICY/CHISQ;
```

Statistical decisions based on Pearson's chi-square statistic make use of the percentiles of the chi-square distribution. Since chi-square is a continuous distribution and categorical data are discrete, some statisticians use a version of Pearson's statistic with a *continuity correction*, called *Yates' corrected chi-square test*, which can be expressed as

$$X_c^2 = \sum_{i,j} \frac{(|x_{ij} - e_{ij}| - 0.5)^2}{e_{ij}}$$

Statisticians still disagree about whether or not a continuity correction is needed. Generally, the corrected version is more conservative and used more widely in the applied literature.

So far, the test for the comparison of two proportions and the test for independence have been presented in very different ways; fortunately, they are not that different. For example, if we apply the test of independence to a $2 \times 2$ table where data came from two binomial samples, say a case–control study, we get same chi-square statistic as if we applied the chi-square test to compare the two proportions against a two-sided alternative. In other words, the chi-square test—presented as a comparison of observed versus expected frequencies—applies regardless of the sampling mechanism. For example, we can use the chi-square test to compare several proportions *simultaneously* using data from several binomial samples (see Example 6.13). [In this type of prob-

**TABLE 6.15**

| Shipbuilding | Cases | Controls | Total |
|---|---|---|---|
| Yes | 11 | 35 | 46 |
| No | 50 | 203 | 253 |
| Total | 61 | 238 | 299 |

lem, we have $k$ independent samples of binary data $(n_1, x_1), (n_2, x_2), \ldots,$ $(n_k, x_k)$, where the $n$'s are sample sizes and the $x$'s are the numbers of positive outcomes in the $k$ samples. For these $k$ independent binomial samples, we consider the null hypothesis

$$H_0: \pi_1 = \pi_2 = \cdots = \pi_k$$

expressing the equality of the $k$ population proportions.]

***Example 6.12***   The case–control study of lung cancer among male residents of coastal Georgia of Example 6.6 was referred to in an attempt to identify reasons for the exceptionally high rate of lung cancer (Table 6.15). We have

$$e_{11} = \frac{(46)(61)}{299}$$

$$= 9.38$$

$$e_{12} = 46 - 9.38$$

$$= 36.62$$

$$e_{21} = 61 - 9.38$$

$$= 51.62$$

$$e_{22} = 253 - 51.62$$

$$= 201.38$$

leading to

$$X^2 = \frac{(11 - 9.38)^2}{9.38} + \frac{(35 - 36.62)^2}{36.62} + \frac{(50 - 51.62)^2}{51.62} + \frac{(203 - 201.38)^2}{201.38}$$

$$= 0.42$$

This result, a chi-square value, is identical to that from Example 6.6, where we obtained a $z$ score of 0.64 [note that $(0.64)^2 = 0.42$].

***Example 6.13***   A study was undertaken to investigate the roles of bloodborne environmental exposures on ovarian cancer from assessment of consumption of

**TABLE 6.16**

| Coffee Drinkers | Cases | Hospital Controls | Population Controls | Total |
|---|---|---|---|---|
| Yes | 177 (170.42) | 249 (253.81) | 233 (234.77) | 659 |
| No | 11 (17.58) | 31 (26.19) | 26 (24.23) | 68 |
| Total | 188 | 280 | 259 | 727 |

coffee, tobacco, and alcohol. Study subjects consist of 188 women in the San Francisco Bay area with epithelial ovarian cancers diagnosed in 1983–1985, and 539 control women. Of the 539 controls, 280 were hospitalized women without ovarian cancer and 259 were chosen from the general population by random telephone dialing. Data for coffee consumption are summarized in Table 6.16 (the numbers in parentheses are expected frequencies). In this example, we want to *compare* the three proportions of coffee drinkers, but we still can apply the same chi-square test:

$$X^2 = \frac{(177 - 170.42)^2}{170.42} + \cdots + \frac{(26 - 24.23)^2}{24.23}$$
$$= 3.83$$

The result indicates that the difference between the three groups is not significant at the 5% level (the cutpoint at the 5% level for chi-square with 2 degrees of freedom is 5.99). In other words, there is enough evidence to implicate coffee consumption in this study of epithelial ovarian cancer. It is important to note that in solving the problem above, a comparison of several proportions, one may be tempted to compare all possible pairs of proportions and do many chi-square tests. What is the matter with this approach to doing many chi-square tests, one for each pair of samples? As the number of groups increases, so does the number of tests to perform; for example, we would have to do 45 tests if we have 10 groups to compare. Obviously, the amount of work is greater, but that is not the critical problem, especially with technological aids such as the use of calculators and computers. So what is the problem? The answer is that performing many tests increases the probability that one or more of the comparisons will result in a type I error (i.e., a significant test result when the null hypothesis is true). This statement should make sense intuitively. For example, suppose that the null hypothesis is true and we perform 100 tests—each has a 0.05 probability of resulting in a type I error; then 5 of these 100 tests would be statistically significant as the results of type I errors. Of course, we usually do not need to do that many tests; however, every time we do more than one, the probability that at least one will result in a type I error exceeds 0.05, indicating a falsely significant difference! What is needed is a method of comparing these proportions *simultaneously*, in one step. The chi-square test for a general two-way table, in this case a 2 × 3 table, achieves just that.

## 6.6 FISHER'S EXACT TEST

Even with a continuity correction, the *goodness-of-fit* test statistic such as Pearson's $X^2$ is not suitable when the sample is small. Generally, statisticians suggest using them only if no expected frequency in the table is less than 5. For studies with small samples, we introduce a method known as *Fisher's exact test*. For tables in which use of the chi-square test $X^2$ is appropriate, the two tests give very similar results.

Our purpose is to find the exact significance level associated with an observed table. The central idea is to enumerate all possible outcomes consistent with a given set of marginal totals and add up the probabilities of those tables more extreme than the one observed. Conditional on the margins, a $2 \times 2$ table is a one-dimensional random variable having a known distribution, so the exact test is relatively easy to implement. The probability of observing a table with cells $a$, $b$, $c$, and $d$ (with total $n$) is

$$\Pr(a, b, c, d) = \frac{(a+b)!\,(c+d)!\,(a+c)!\,(b+d)!}{n!\,a!\,b!\,c!\,d!}$$

The process for doing hand calculations would be as follows:

1. Rearrange the rows and columns of the table observed so the the smaller total is in the first row and the smaller column total is in the first column.
2. Start with the table having 0 in the $(1,1)$ cell (top left cell). The other cells in this table are determined automatically from the fixed row and column margins.
3. Construct the next table by increasing the $(1,1)$ cell from 0 to 1 and decreasing all other cells accordingly.
4. Continue to increase the $(1,1)$ cell by 1 until one of the other cells becomes 0. At that point we have enumerated all possible tables.
5. Calculate and add up the probabilities of those tables with cell $(1,1)$ having values from 0 to the observed frequency (left side for a one-sided test); double the smaller side for a two-sided test.

In practice, the calculations are often tedious and should be left to a computer program to implement.

***Example 6.14*** A study on deaths of men aged over 50 yields the data shown in Table 6.17 (numbers in parentheses are expected frequencies). An application of Fisher's exact test yields a one-sided $p$ value of 0.375 or a two-sided $p$ value of 0.688; we cannot say, on the basis of this limited amount of data, that there is a significant association between salt intake and cause of death even though the proportions of CVD deaths are different (71.4% versus 56.6%). For implementing hand calculations, we would focus on the tables where cell $(1,1)$

**TABLE 6.17**

| | Type of Diet | | |
| Cause of Death | High Salt | Low Salt | Total |
| --- | --- | --- | --- |
| Non-CVD | 2 (2.92) | 23 (22.08) | 25 |
| CVD | 5 (4.08) | 30 (30.92) | 35 |
| Total | 7 | 53 | 60 |

equals 0, 1, and 2 (observed value; the probabilities for these tables are 0.017, 0.105, and 0.252, respectively).

*Note:* An SAS program would include these instructions:

```
DATA;
INPUT CVM DIET COUNT;
CARDS;
1 1 2
1 2 23
2 1 5
2 2 30;
PROC FREQ;
WEIGHT COUNT;
TABLES CVD*DIET/CHISQ;
```

The output also includes Pearson's test ($X^2 = 0.559$; $p = 0.455$) as well.

## 6.7   ORDERED $2 \times k$ CONTINGENCY TABLES

In this section we present an efficient method for use with ordered $2 \times k$ contingency tables, tables with two rows and with $k$ columns having a certain natural ordering. We introduced it, in Chapter 1, to provide a descriptive statistic, the *generalized odds*.

In general, consider an ordered $2 \times k$ table with the frequencies shown in Table 6.18. The number of concordances is calculated by

$$C = a_1(b_2 + \cdots + b_k) + a_2(b_3 + \cdots + b_k) + \cdots + a_{k-1}b_k$$

The number of discordances is

$$D = b_1(a_2 + \cdots + a_k) + b_2(a_3 + \cdots + a_k) + \cdots + b_{k-1}a_k$$

To perform the test, we calculate the statistic

$$S = C - D$$

**TABLE 6.18**

| Row | Column Level | | | | Total |
| | 1 | 2 | $\cdots$ | $k$ | |
| --- | --- | --- | --- | --- | --- |
| 1 | $a_1$ | $a_2$ | $\cdots$ | $a_k$ | $A$ |
| 2 | $b_1$ | $b_2$ | $\cdots$ | $b_k$ | $B$ |
| Total | $n_1$ | $n_2$ | $\cdots$ | $n_k$ | $N$ |

then standardize it to obtain

$$z = \frac{S - \mu_S}{\sigma_D}$$

where $\mu_S = 0$ is the mean of $S$ under the null hypothesis and

$$\sigma_S = \left[ \frac{AB}{3N(N-1)} (N^3 - n_1^3 - n_2^3 - \cdots - n_k^3) \right]^{1/2}$$

The standardized $z$ score is distributed as standard normal if the null hypothesis is true. For a one-sided alternative, which is a natural choice for this type of test, the null hypothesis is rejected at the 5% level if $z > 1.65$ (or $z < -1.65$ if the terms *concordance* and *discordance* are switched).

***Example 6.15***   Consider an example concerning the use of seat belts in automobiles. Each accident in this example is classified according to whether a seat belt was used and the severity of injuries received: none, minor, major, or death (Table 6.19). For this study on the use of seat belts in automobiles, an application of the method above yields

$$C = (75)(175 + 135 + 25) + (160)(135 + 25) + (100)(25)$$
$$= 53{,}225$$
$$D = (65)(160 + 100 + 15) + (175)(100 + 15) + (135)(15)$$
$$= 40{,}025$$

**TABLE 6.19**

| Seat Belt | Extent of Injury Received | | | | Total |
| | None | Minor | Major | Death | |
| --- | --- | --- | --- | --- | --- |
| Yes | 75 | 160 | 100 | 15 | 350 |
| No | 65 | 175 | 135 | 25 | 390 |
| Total | 130 | 335 | 235 | 40 | 740 |

In addition, we have

$$A = 350$$
$$B = 390$$
$$n_1 = 130$$
$$n_2 = 335$$
$$n_3 = 235$$
$$n_4 = 40$$
$$N = 740$$

Substituting these values into the equations of the test statistic, we have

$$S = 53{,}225 - 40{,}025$$
$$= 13{,}200$$

$$\sigma_S = \left[ \frac{(350)(390)}{(3)(740)(739)} (740^3 - 130^3 - 335^3 - 235^3 - 40^3) \right]^{1/2}$$
$$= 5414.76$$

leading to

$$z = \frac{13{,}200}{5414.76}$$
$$= 2.44$$

which shows a high degree of significance (one-sided $p$ value $= 0.0073$). It is interesting to not that in order to compare the extent of injury from those who used seat belts and those who did not, we can perform a chi-square test as presented in Section 6.6. Such an application of the chi-square test yields

$$X^2 = 9.26$$

with 3 df is 7.81 (for $0.01 \leq p \leq 0.05$). Therefore, the difference between the two groups is significant at the 5% level but not at the 1% level, a lower degree of significance (compared to the $p$ value of 0.0073 above). This is because the usual chi-square calculation takes no account of the fact that the extent of injury has a natural ordering: none < minor < major < death. In addition, the percent of seat belt users in each injury group decreases from level "none" to level "death":

$$\text{None:} \qquad \frac{75}{75 + 65} = 54\%$$

$$\text{Minor:} \qquad \frac{160}{160 + 175} = 48\%$$

$$\text{Major:} \qquad \frac{100}{100 + 135} = 43\%$$

$$\text{Death:} \qquad \frac{15}{15 + 25} = 38\%$$

The method introduced in this section seems particularly ideal for the evaluation of ordinal risk factors in case–control studies.

***Example 6.16***  Prematurity, which ranks as the major cause of neonatal morbidity and mortality, has traditionally been defined on the basis of a birth weight under 2500 g. But this definition encompasses two distinct types of infants: infants who are small because they are born early, and infants who are born at or near term but are small because their growth was retarded. *Prematurity* has now been replaced by *low birth weight* to describe the second type and *preterm* to characterize the first type (babies born before 37 weeks of gestation).

A case–control study of the epidemiology of preterm delivery was undertaken at Yale–New Haven Hospital in Connecticut during 1977 (Berkowitz, 1981). The study population consisted of 175 mothers of singleton preterm infants and 303 mothers of singleton full-term infants. Table 6.20 gives the distribution of age of the mother. We have

$$C = (15)(25 + 62 + 122 + 78) + (22)(62 + 122 + 78)$$
$$\quad + (47)(122 + 78) + (56)(78)$$
$$= 20{,}911$$
$$D = (16)(22 + 47 + 56 + 35) + (25)(47 + 56 + 35)$$
$$\quad + (62)(56 + 35) + (122)(35)$$
$$= 15{,}922$$

**TABLE 6.20**

| Age | Cases | Controls | Total |
|---|---|---|---|
| 14–17 | 15 | 16 | 31 |
| 18–19 | 22 | 25 | 47 |
| 20–24 | 47 | 62 | 109 |
| 25–29 | 56 | 122 | 178 |
| ≥30 | 35 | 78 | 113 |
| Total | 175 | 303 | |

$$S = 20{,}911 - 15{,}922$$
$$= 4989$$

$$\sigma_S = \left[ \frac{(175)(303)}{(3)(478)(477)} (478^3 - 31^3 - 47^3 - 109^3 - 178^3 - 113^3) \right]^{1/2}$$
$$= 2794.02$$

leading to

$$z = \frac{4989}{27.94.02}$$
$$= 1.79$$

which shows a significant association between the mother's age and preterm delivery (one-sided $p$ value $= 0.0367$); the younger the mother, the more likely the preterm delivery.

## 6.8  NOTES ON COMPUTATIONS

Samples of SAS program instructions were provided at the end of Examples 6.8 and 6.14 for complicated procedures. Other computations can be implemented easily using a calculator *provided* that data have been summarized in the form of a two-way table. Read Section 1.4 on how to use Excel's *PivotTable* procedure to form a $2 \times 2$ table from a raw data file. After the value of a statistic has been obtain, you can use NORMDIST (Section 3.6) to obtain an exact $p$ value associated with a $z$ score and CHIDIST to obtain an exact $p$ value associated with a chi-square statistic; the CHIDIST procedure can be used similar to the case of a one-tailed TDIST (Section 3.6). For the method of Section 6.5, and instead of writing a SAS program, one can calculate expected frequencies (using *formula* and *drag and fill*, Section 1.4), then input them into the CHITEST procedure of Excel.

## EXERCISES

**6.1**  Consider a sample of $n = 110$ women drawn randomly from the membership list of the National Organization for Women (N.O.W.), $x = 25$ of whom were found to smoke. Use the result of this sample to test whether the rate found is significantly *different* from the U.S. proportion of 0.30 for women.

**6.2**  In a case–control study, 317 patients suffering from endometrial carcinoma were matched individually with 317 other cancer patients in a hospital and the use of estrogen in the six months prior to diagnosis was determined (Table E6.2). Use McNemar's chi-square test to investigate the significance of the association between estrogen use and endometrial carcinoma; state your null and alternative hypotheses.

**TABLE E6.2**

| | Controls | |
|---|---|---|
| Cases | Estrogen | No Estrogen |
| Estrogen | 39 | 113 |
| No estrogen | 15 | 150 |

**6.3** A study in Maryland identified 4032 white persons, enumerated in a nonofficial 1963 census, who became widowed between 1963 and 1974. These people were matched, one to one, to married persons on the basis of race, gender, year of birth, and geography of residence. The matched pairs were followed in a second census in 1975, and the data for men have been analyzed so as to compare the mortality of widowed men versus married men (see Example 6.3). The data for 2828 matched pairs of women are shown in Table E6.3. Test to compare the mortality of widowed women versus married women; state your null and alternative hypotheses.

**TABLE E6.3**

| Widowed | Married Women | |
|---|---|---|
| Women | Dead | Alive |
| Dead | 1 | 264 |
| Alive | 249 | 2314 |

**6.4** It has been noted that metal workers have an increased risk for cancer of the internal nose and paranasal sinuses, perhaps as a result of exposure to cutting oils. Therefore, a study was conducted to see whether this particular exposure also increases the risk for squamous cell carcinoma of the scrotum (see Example 6.2). Cases included all 45 squamous cell carcinomas of the scrotum diagnosed in Connecticut residents from 1955 to 1973, as obtained from the Connecticut Tumor Registry. Matched controls were selected for each case based on the age at death (within eight years), year of death (within three years), and number of jobs as obtained from combined death certificate and directory sources. An occupational indicator of metal worker (yes/no) was evaluated as the possible risk factor in this study. The results are shown in Table 6.2. Test to compare the cases versus the controls using the McNemar's chi square test; state clearly your null and alternative hypotheses and choice of test size.

**6.5**   A matched case–control study on endometrial cancer, where the exposure was "ever having taken any estrogen," yields the data shown in
Table E6.5. Test to compare the cases versus the controls; state clearly
your null and alternative hypotheses and choice of test size (alpha level).

**TABLE E6.5**

|  | Matching Controls | |
| --- | --- | --- |
| Cases | Exposed | Nonexposed |
| Exposed | 27 | 29 |
| Nonexposed | 3 | 4 |

**6.6**   Ninety-eight heterosexual couples, at least one of whom was HIV-
infected, were enrolled in an HIV transmission study and interviewed
about sexual behavior. Table E6.6 provides a summary of condom use
reported by heterosexual partners. Test to compare the men versus the
women; state clearly your null and alternative hypotheses and choice of
test size (alpha level).

**TABLE E6.6**

|  | Man | |
| --- | --- | --- |
| Woman | Ever | Never |
| Ever | 45 | 6 |
| Never | 7 | 40 |

**6.7**   A matched case–control study was conducted by Schwarts et al. (1989)
to evaluate the cumulative effects of acrylate and methacrylate vapors
on olfactory function (Table E6.7). Cases were defined as scoring at or
below the 10th percentile on the UPSIT (University of Pennsylvania
Smell Identification Test). Test to compare the cases versus the controls;
state clearly your null and alternative hypotheses and your choice of test
size.

**TABLE E6.7**

|  | Cases | |
| --- | --- | --- |
| Controls | Exposed | Unexposed |
| Exposed | 25 | 22 |
| Unexposed | 9 | 21 |

**6.8**   Self-reported injuries among left- and right-handed people were compared in a survey of 1896 college students in British Columbia, Canada.

Of the 180 left-handed students, 93 reported at least one injury. In the same period, 619 of the 1716 right-handed students reported at least one injury. Test to compare the proportions of injured students, left-handed versus right-handed; state clearly your null and alternative hypotheses and choice of test size.

**6.9**   In a study conducted to evaluate the hypothesis that tea consumption and premenstrual syndrome are associated, 188 nursing students and 64 tea factory workers were given questionnaires. The prevalence of premenstrual syndrome was 39% among the nursing students and 77% among the tea factory workers. Test to compare the prevalences of premenstrual syndrome, tea factory workers versus nursing students; state clearly your null and alternative hypotheses and choice of test size.

**6.10**  A study was conducted to investigate drinking problems among college students. In 1983, a group of students were asked whether they had ever driven an automobile while drinking. In 1987, after the legal drinking age was raised, a different group of college students were asked the same question. The results are as shown in Table E6.10. Test to compare the proportions of students with a drinking problem, 1987 versus 1983; state clearly your null and alternative hypotheses and choice of test size.

**TABLE E6.10**

| Drove While | Year | |
| Drinking | 1983 | 1987 |
| --- | --- | --- |
| Yes | 1250 | 991 |
| No | 1387 | 1666 |
| Total | 2637 | 2657 |

**6.11**  In August 1976, tuberculosis was diagnosed in a high school student (index case) in Corinth, Mississippi. Subsequently, laboratory studies revealed that the student's disease was caused by drug-resistant tubercule bacilli. An epidemiologic investigation was conducted at the high school. Table E6.11 gives the rates of positive tuberculin reaction determined for various groups of students according to the degree of exposure to the index case. Test to compare the proportions of students with infection, high exposure versus low exposure; state clearly your null and alternative hypotheses and choice of test size.

**TABLE E6.11**

| Exposure Level | Number Tested | Number Positive |
| --- | --- | --- |
| High | 129 | 63 |
| Low | 325 | 36 |

**6.12**    Epidemic keratoconjunctivitis (EKC) or "shipyard eye" is an acute infectious disease of the eye. A case of EKC is defined as an illness consisting of redness, tearing, and pain in one or both eyes for more than three days' duration, diagnosed as EKC by an ophthalmologist. In late October 1977, one (physician A) of the two ophthalmologists providing the majority of specialized eye care to the residents of a central Georgia county (population 45,000) saw a 27-year-old nurse who had returned from a vacation in Korea with severe EKC. She received symptomatic therapy and was warned that her eye infection could spread to others; nevertheless, numerous cases of an illness similar to hers soon occurred in the patients and staff of the nursing home (nursing home A) where she worked (these people came to physician A for diagnosis and treatment). Table E6.12 provides exposure history of 22 persons with EKC between October 27, 1977 and January 13, 1978 (when the outbreak stopped, after proper control techniques were initiated). Nursing home B, included in this table, is the only other area chronic-care facility. Using an appropriate test, compare the proportions of cases from the two nursing homes.

**TABLE E6.12**

| Exposure Cohort | Number Exposed | Number of Cases |
|---|---|---|
| Nursing home A | 64 | 16 |
| Nursing home B | 238 | 6 |

**6.13**    Consider the data taken from a study that attempts to determine whether the use of electronic fetal monitoring (EFM) during labor affects the frequency of cesarean section deliveries. Of the 5824 infants included in the study, 2850 were monitored electronically and 2974 were not. The outcomes are shown in Table E6.13. Test to compare the rates of cesarean section delivery, EFM-exposed versus nonexposed; state clearly your null and alternative hypotheses and choice of test size.

**TABLE E6.13**

| Cesarean Delivery | EFM Exposure | |
|---|---|---|
| | Yes | No |
| Yes | 358 | 229 |
| No | 2492 | 2745 |
| Total | 2850 | 2974 |

**6.14**    A study was conducted to investigate the effectiveness of bicycle safety helmets in preventing head injury. The data consist of a random sample of 793 people involved in bicycle accidents during a one-year period (Table E6.14). Test to compare the proportions with head injury, those with helmets versus those without; state clearly your null and alternative hypotheses and choice of test size.

**TABLE E6.14**

|  | Wearing Helmet | |
| --- | --- | --- |
| Head Injury | Yes | No |
| Yes | 17 | 218 |
| No | 130 | 428 |
| Total | 147 | 646 |

**6.15**  A case–control study was conducted relating to the epidemiology of breast cancer and the possible involvement of dietary fats, along with other vitamins and nutrients. It included 2024 breast cancer cases admitted to Roswell Park Memorial Institute, Erie County, New York, from 1958 to 1965. A control group of 1463 was chosen from patients having no neoplasms and no pathology of gastrointestinal or reproductive systems. The primary factors being investigated were vitamins A and E (measured in international units per month). The data shown in Table E6.15 are for 1500 women over 54 years of age. Test to compare the proportions of subjects who consumed less vitamin A ($\leq 150,500$ IU/ month), cases versus controls; state clearly your null and alternative hypotheses and choice of test size.

**TABLE E6.15**

| Vitamin A (IU/month) | Cases | Controls |
| --- | --- | --- |
| $\leq 150,500$ | 893 | 392 |
| $> 150,500$ | 132 | 83 |
| Total | 1025 | 475 |

**6.16**  In a randomize trial, 111 pregnant women had elective induction of labor between 39 and 40 weeks, and 117 controls were managed expectantly until 41 weeks. The results are shown in Table E6.16. Use Fisher's exact test to verify the alternative that patients with elective induction have less meconium staining in labor than do control patients.

**TABLE E6.16**

|  | Induction Group | Control Group |
| --- | --- | --- |
| Number of patients | 111 | 117 |
| Number with meconium staining | 1 | 13 |

**6.17**   A research report states that "a comparison of 90 normal patients with 10 patients with hypotension shows that only 3.3% of the former group died as compared to 30% of the latter." Put the data into a $2 \times 2$ contingency table and test the significance of blood pressure as a prognostic sign using:

(a) Pearson's chi-square test.

(b) Pearson's chi-square test with Yates's continuity correction.

(c) Fisher's exact test.

**6.18**   In a trial of diabetic therapy, patients were either treated with phenformin or a placebo. The numbers of patients and deaths from cardiovascular causes are listed in Table E6.18. User Fisher's exact test to investigate the difference in cardiovascular mortality between the phenformin and placebo groups; state your null and alternative hypotheses.

**TABLE E6.18**

| Result | Phenformin | Placebo | Total |
|---|---|---|---|
| Cardiovascular deaths | 26 | 2 | 28 |
| Not deaths | 178 | 62 | 240 |
| Total | 204 | 64 | 268 |

**6.19**   In a seroepidemiologic survey of health workers representing a spectrum of exposure to blood and patients with hepatitis B virus (HBV), it was found that infection increased as a function of contact. Table E6.19 provides data for hospital workers with uniform socioeconomic status at an urban teaching hospital in Boston, Massachusetts. Test to compare the proportions of HBV infection between the four groups: physicians with frequent exposure, physicians with infrequent exposure, nurses with frequent exposure, and nurses with infrequent exposure. State clearly your null and alternative hypotheses and choice of test size.

**TABLE E6.19**

| Personnel | Exposure | $n$ | HBV Positive |
|---|---|---|---|
| Physicians | Frequent | 81 | 17 |
|  | Infrequent | 89 | 7 |
| Nurses | Frequent | 104 | 22 |
|  | Infrequent | 126 | 11 |

**6.20**   It has been hypothesized that dietary fiber decreases the risk of colon cancer, while meats and fats are thought to increase this risk. A large study was undertaken to confirm these hypotheses (Graham et al., 1988). Fiber and fat consumptions are classified as low or high and data are tabulated separately for males and females in Table E6.20 ("low" means below the median). Test to investigate the relationship between the disease (two categories: cases and controls) and the diet (four categories: low fat and high fiber, low fat and low fiber, high fat and high fiber, high fat and low fiber). State clearly your null and alternative hypotheses and choice of test size.

**TABLE E6.20**

| Diet | Males | | Females | |
|------|-------|----------|-------|----------|
|      | Cases | Controls | Cases | Controls |
| Low fat, high fiber | 27 | 38 | 23 | 39 |
| Low fat, low fiber  | 64 | 78 | 82 | 81 |
| High fat, high fiber | 78 | 61 | 83 | 76 |
| High fat, low fiber | 36 | 28 | 35 | 27 |

**6.21** A case–control study was conducted in Auckland, New Zealand to investigate the effects of alcohol consumption on both nonfatal myocardial infarction and coronary death in the 24 hours after drinking, among regular drinkers. Data are tabulated separately for men and women in Table E6.21. For each group, men and women, and for each type of event, myocardial infarction and coronary death, test to compare cases versus controls. State, in each analysis, your null and alternative hypotheses and choice of test size.

**TABLE E6.21**

|       | Drink in the Last 24 Hours | Myocardial Infarction | | Coronary Death | |
|-------|----------------------------|----------|-------|----------|-------|
|       |                            | Controls | Cases | Controls | Cases |
| Men   | No  | 197 | 142 | 135 | 103 |
|       | Yes | 201 | 136 | 159 | 69  |
| Women | No  | 144 | 41  | 89  | 12  |
|       | Yes | 122 | 19  | 76  | 4   |

**6.22** Refer to the data in Exercise 6.21, but assume that gender (male/female) may be a confounder but not an effect modifier. For each type of event, myocardial infarction and coronary death, use the Mantel–Haenszel method to investigate the effects of alcohol consumption. State, in each analysis, your null and alternative hypotheses and choice of test size.

**6.23** Since incidence rates of most cancers rise with age, this must always be considered a confounder. Stratified data for an unmatched case–control study are shown in Table E6.23. The disease was esophageal cancer among men, and the risk factor was alcohol consumption. Use the Mantel–Haenszel procedure to compare the cases versus the controls. State your null hypothesis and choice of test size.

**TABLE E6.23**

| Age | | Daily Alcohol Consumption | |
|-----|-----|-----|-----|
| | | 80+ g | 0–79 g |
| 25–44 | Cases | 5 | 5 |
| | Controls | 35 | 270 |
| 45–64 | Cases | 67 | 55 |
| | Controls | 56 | 277 |
| 65+ | Cases | 24 | 44 |
| | Controls | 18 | 129 |

**6.24** Postmenopausal women who develop endometrial cancer are on the whole heavier than women who do not develop the disease. One possible explanation is that heavy women are more exposed to endogenous estrogens, which are produced in postmenopausal women by conversion of steroid precursors to active estrogens in peripheral fat. In the face of varying levels of endogenous estrogen production, one might ask whether the carcinogenic potential of exogenous estrogens would be the peripheral fat. In the face of varying levels of endogenous estrogen production, one might ask whether the carcinogenic potential of exogenous estrogens would be the same in all women. A study has been conducted to examine the relation between weight, replacement estrogen therapy, and endometrial cancer in a case–control study (Table E6.24). Use the Mantel–Haenszel procedure to compare the cases versus the controls. State your null hypothesis and choice of test size.

**TABLE E6.24**

| Weight (kg) | | Estrogen Replacement | |
|-----|-----|-----|-----|
| | | Yes | No |
| <57 | Cases | 20 | 12 |
| | Controls | 61 | 183 |
| 57–75 | Cases | 37 | 45 |
| | Controls | 113 | 378 |
| >75 | Cases | 9 | 42 |
| | Controls | 23 | 140 |

**6.25** Risk factors of gallstone disease were investigated in male self-defense officials who received, between October 1986 and December 1990, a retirement health examination at the Self-Defense Forces Fukuoka

Hospital, Fukuoka, Japan (Table E6.25). For each of the three criteria (smoking, alcohol, and body mass index), test, at the 5% level, for the trend of men with gallstones.

**TABLE E6.25**

| Factor | Number of Men Surveyed | |
|---|---|---|
| | Total | with Number Gallstones |
| Smoking | | |
| Never | 621 | 11 |
| Past | 776 | 17 |
| Current | 1342 | 33 |
| Alcohol | | |
| Never | 447 | 11 |
| Past | 113 | 3 |
| Current | 2179 | 47 |
| Body mass index $(kg/m^2)$ | | |
| <22.5 | 719 | 13 |
| 22.5–24.9 | 1301 | 30 |
| ≥25.0 | 719 | 18 |

**6.26** Prematurity, which ranks as the major cause of neonatal morbidity and mortality, has traditionally been defined on the basis of a birth weight under 2500 g. But this definition encompasses two distinct types of infants: infants who are small because they are born early, and infants who are born at or near term but are small because their growth was retarded. *Prematurity* has now been replaced by *low birth weight* to describe the second type and *preterm* to characterize the first type (babies born before 37 weeks of gestation). A case–control study of the epidemiology of preterm delivery was undertaken at Yale–New Haven Hospital in Connecticut during 1977. The study population consisted of 175 mothers of singleton preterm infants and 303 mothers of singleton full-term infants. In Example 6.16 we have analyzed and found a significant association between the mother's age and preterm delivery; that the younger the mother, the more likely the preterm delivery. Table E6.26

**TABLE E6.26**

| Socioeconomic Level | Cases | Controls |
|---|---|---|
| Upper | 11 | 40 |
| Upper middle | 14 | 45 |
| Middle | 33 | 64 |
| Lower middle | 59 | 91 |
| Lower | 53 | 58 |
| Unknown | 5 | 5 |

gives the distribution of socioeconomic status. Test to prove that there is a significant association between the mother's age and preterm delivery; that the poorer the mother, the more likely the preterm delivery. State your null hypothesis and choice of test size.

**6.27** Postneonatal mortality due to respiratory illnesses is known to be inversely related to maternal age, but the role of young motherhood as a risk factor for respiratory morbidity in infants has not been explored thoroughly. A study was conducted in Tucson, Arizona aimed at the incidence of lower respiratory tract illnesses during the first year of life. In this study, over 1200 infants were enrolled at birth between 1980 and 1984. The data shown in Table E6.27 are concerned with wheezing lower respiratory tract illnesses (wheezing LRIs). For each of the two groups, boys and girls, test to investigate the relationship between maternal age and respiratory illness. State clearly your null and alternative hypotheses and choice of test size.

**TABLE E6.27**

|                      | Boys |     | Girls |     |
| -------------------- | ---- | --- | ----- | --- |
| Maternal Age (years) | No   | Yes | No    | Yes |
| <21                  | 19   | 8   | 20    | 7   |
| 21–25                | 98   | 40  | 128   | 36  |
| 26–30                | 160  | 45  | 148   | 42  |
| >30                  | 110  | 20  | 116   | 25  |

**6.28** Data were collected from 2197 white ovarian cancer patients and 8893 white controls in 12 different U.S. case–control studies conducted by various investigators in the period 1956–1986. These were used to evaluate the relationship of invasive epithelial ovarian cancer to reproductive

**TABLE E6.28a**

|                                          | Cases | Controls |
| ---------------------------------------- | ----- | -------- |
| Duration of unprotected intercourse (years) |       |          |
| <2                                       | 237   | 477      |
| 2–9                                      | 166   | 354      |
| 10–14                                    | 47    | 91       |
| ≥15                                      | 133   | 174      |
| History of infertility                   |       |          |
| No                                       | 526   | 966      |
| Yes                                      |       |          |
| No drug use                              | 76    | 124      |
| Drug use                                 | 20    | 11       |

and menstrual characteristics, exogenous estrogen use, and prior pelvic surgeries (Table E6.28). For each of the two criteria, duration of unprotected intercourse and history of infertility, test to investigate the relationship between that criterion and ovarian cancer. State clearly your null and alternative hypotheses and choice of test size.

**6.29** Table E6.29 lists data from different studies designed to investigate the accuracy of death certificates. The results of 5373 autopsies were compared to the causes of death listed on the certificates. Test to confirm the downward trend of accuracy over time. State clearly your null and alternative hypotheses and choice of test size.

**TABLE E6.29**

|  | Accurate Certificate | | |
| --- | --- | --- | --- |
| Date of Study | Yes | No | Total |
| 1955–1965 | 2040 | 694 | 2734 |
| 1970–1971 | 437 | 203 | 640 |
| 1975–1978 | 1128 | 599 | 1727 |
| 1980 | 121 | 151 | 272 |

**6.30** A study was conducted to ascertain factors that influence a physician's decision to transfuse a patient. A sample of 49 attending physicians was selected. Each physician was asked a question concerning the frequency with which an unnecessary transfusion was give because another physician suggested it. The same question was asked of a sample of 71 residents. The data are shown in Table E6.30. Test the null hypothesis of no association, with attention to the natural ordering of the columns. State clearly your alternative hypothesis and choice of test size.

**TABLE E6.30**

| | Frequency of Unnecessary Transfusion | | | | |
| --- | --- | --- | --- | --- | --- |
| Type of Physician | Very Frequently (1/week) | Frequently (1/two weeks) | Occasionally (1/month) | Rarely (1/two months) | Never |
| Attending | 1 | 1 | 3 | 31 | 13 |
| Resident | 2 | 13 | 28 | 23 | 5 |

# 7

# COMPARISON OF POPULATION MEANS

If each element of a data set may lie at only a few isolated points, we have a *discrete* or *categorical data set*; examples include gender, race, and some sort of artificial grading used as outcomes. If each element of a data set may lie anywhere on the numerical scale, we have a *continuous data set*; examples include blood pressure, cholesterol level, and time to a certain event. In Chapter 6 we dealt with the analysis of categorical data, such as comparing two proportions. In this chapter we focus on continuous measurements, especially comparisons of population means. We follow the same layout, starting with the simplest case, the one-sample problem.

## 7.1  ONE-SAMPLE PROBLEM WITH CONTINUOUS DATA

In this type of problem we have a sample of continuous measurements of size $n$ and we consider the null hypothesis

$$H_0: \mu = \mu_0$$

where $\mu_0$ is a fixed and known number. It is often a standardized or referenced figure; for example, the average blood pressure of men in a certain age group (this figure may come from a sample itself, but the referenced sample is often large enough so as to produce negligible sampling error in $\mu_0$). Or, we could be concerned with a question such as: Is the average birth weight for boys for this particular sub-population below normal average $\mu_0$: say, 7.5 lb? In Exercise 7.1, we try to decide whether air quality on a certain given day in a particular city exceeds regulated limit $\mu_0$ set by a federal agency.

In a typical situation, the null hypothesis of a statistical test is concerned with a parameter; the parameter in this case, with continuous data, is the mean $\mu$. Sample data are summarized into a statistic that is used to estimate the parameter under investigation. Since the parameter under investigation is the population mean $\mu$, our focus in this case is the sample mean $\bar{x}$. In general, a statistic is itself a variable with a specific sampling distribution (in the context of repeated sampling). Our statistic in this case is the sample mean $\bar{x}$; the corresponding sampling distribution is obtained easily by invoking the *central limit theorem*. With large sample size and assuming that the null hypothesis $H_0$ is true, it is the normal distribution with mean and variance given by

$$\mu_{\bar{x}} = \mu_0$$

$$\sigma_{\bar{x}}^2 = \frac{\sigma^2}{n}$$

respectively. The extra needed parameter, the population variance $\sigma^2$, has to be estimated from our data by the sample variance $s^2$. From this sampling distribution, the observed value of the sample mean can be converted to standard units: the number of standard errors away from the hypothesized value of $\mu_0$. In other words, to perform a test of significance for $H_0$, we proceed with the following steps:

1. Decide whether a one- or a two-sided test is appropriate; this decision depends on the research question.
2. Choose a level of significance; a common choice is 0.05.
3. Calculate the $t$ statistic:

$$t = \frac{\bar{x} - \mu_0}{\mathrm{SE}(\bar{x})}$$

$$= \frac{\bar{x} - \mu_0}{s/\sqrt{n}}$$

4. From the table for $t$ distribution (Appendix C) with $(n-1)$ degrees of freedom and the choice of $\alpha$ (e.g., $\alpha = 0.05$), the rejection region is determined by:
   (a) For a one-sided test, use the column corresponding to an upper tail area of 0.05:

   $$t \leq -\text{tabulated value} \quad \text{for } H_A: \mu < \mu_0$$

   $$t \geq \text{tabulated value} \quad \text{for } H_A: \mu > \mu_0$$

   (b) For a two-sided test or $H_A: \mu \neq \mu_0$, use the column corresponding to an upper tail area of 0.025:

$$z \leq -\text{tabulated value} \quad \text{or} \quad z \geq \text{tabulated value}$$

This test is referred to as the *one-sample t test*.

***Example 7.1***    Boys of a certain age have a mean weight of 85 lb. An observation was made that in a city neighborhood, children were underfed. As evidence, all 25 boys in the neighborhood of that age were weighed and found to have a mean $\bar{x}$ of 80.94 lb and a standard deviation $s$ of 11.60 lb. An application of the procedure above yields

$$\text{SE}(\bar{x}) = \frac{s}{\sqrt{n}}$$

$$= \frac{11.60}{\sqrt{25}}$$

$$= 2.32$$

leading to

$$t = \frac{80.94 - 85}{2.32}$$

$$= -1.75$$

The underfeeding complaint corresponds to the one-sided alternative

$$H_A: \mu < 85$$

so that we would reject the null hypothesis if

$$t \leq -\text{tabulated value}$$

From Appendix C and with 24 degrees of freedom $(n - 1)$, we find that

$$\text{tabulated value} = 1.71$$

under the column corresponding to a 0.05 upper tail area; the null hypothesis is rejected at the 0.05 level. In other words, there is enough evidence to support the underfeeding complaint.

## 7.2   ANALYSIS OF PAIR-MATCHED DATA

The method presented in this section applies to cases where each subject or member of a group is observed twice (e.g., before and after certain interventions), or matched pairs are measured for the same continuous characteristic.

In the study reported in Example 7.3, blood pressure was measured from a group of women before and after each took an oral contraceptive. In Exercise 7.2, the insulin level in the blood was measured from dogs before and after some kind of nerve stimulation. In another exercise, we compared self-reported versus measured height. A popular application is an epidemiological design called a *pair-matched case–control study*. In case–control studies, cases of a specific disease are ascertained as they arise from population-based registers or lists of hospital admissions, and controls are sampled either as disease-free individuals from the population at risk or as hospitalized patients having a diagnosis other than the one under investigation. As a technique to control confounding factors, individual cases are matched, often one to one, to controls chosen to have similar values for confounding variables such as age, gender, or race.

Data from matched or before-and-after experiments should never be considered as coming from two independent samples. The procedure is to reduce the data to a one-sample problem by computing before-and-after (or case-and-control) difference for each subject or pairs of matched subjects. By doing this with paired observations, we get a set of differences, each of which is independent of the characteristics of the person on whom measurements were made. The analysis of pair-matched data with a continuous measurement can be seen as follows. What we really want to do is to compare the means, before versus after or cases versus controls, and use of the sample of differences $\{d_i\}$, one for each subject, helps to achieve that. With large sample size and assuming that the null hypothesis $H_0$ of *no difference* is true, the mean $\bar{d}$ of theses differences is distributed as *normal* with mean and variance given by

$$\mu_{\bar{d}} = 0$$

$$\sigma_{\bar{d}}^2 = \frac{\sigma_d^2}{n}$$

respectively. The extra needed parameter, the variance $\sigma_d^2$, has to be estimated from our data by the sample variance $s_d^2$. In other words, the analysis of pair-matched data with a continuous measurement can be seen as a special case of the one-sample problem of Section 7.1 with $\mu_0 = 0$. Recalling the form of the test statistic of Section 7.1, we have

$$t = \frac{\bar{d} - 0}{s_d / \sqrt{n}}$$

and the rejection region is determined using the $t$ distribution at $n - 1$ degrees of freedom. This test is referred to as the *one-sample t test*, the same one-sample $t$ test as in Section 7.1.

***Example 7.2***   Trace metals in drinking water affect the flavor of the water, and unusually high concentrations can pose a health hazard. Table 7.1 shows trace-

**TABLE 7.1**

| Location | Bottom | Surface | Difference, $d_i$ | $d_i^2$ |
|----------|--------|---------|-------------------|---------|
| 1 | 0.430 | 0.415 | 0.015 | 0.000225 |
| 2 | 0.266 | 0.238 | 0.028 | 0.000784 |
| 3 | 0.567 | 0.390 | 0.177 | 0.030276 |
| 4 | 0.531 | 0.410 | 0.121 | 0.014641 |
| 5 | 0.707 | 0.605 | 0.102 | 0.010404 |
| 6 | 0.716 | 0.609 | 0.107 | 0.011449 |
| Total | | | 0.550 | 0.068832 |

metal concentrations (zinc, in mg/L) for both surface water and bottom water at six different river locations (the difference is bottom − surface). The necessary summarized figures are

$$\bar{d} = \text{average difference}$$

$$= \frac{0.550}{6}$$

$$= 0.0917 \text{ mg/L}$$

$$s_d^2 = \frac{0.068832 - (0.550)^2/6}{5}$$

$$= 0.00368$$

$$s_d = 0.061$$

$$\text{SE}(\bar{d}) = \frac{0.061}{\sqrt{6}}$$

$$= 0.0249$$

$$t = \frac{0.0917}{0.0249}$$

$$= 3.68$$

Using the column corresponding to the upper tail area of 0.025 in Appendix C, we have a tabulated value of 2.571 for 5 df. Since

$$t = 3.68 > 2.571$$

we conclude that the null hypothesis of *no difference* should be rejected at the

0.05 level; there is enough evidence to support the hypothesis of *different* mean zinc concentrations (two-sided alternative).

*Note:* An SAS program would include these instructions:

```
DATA;
INPUT BZINC SZINC;
DIFF = BZINC - SZINC;
DATALINES;
.430 .415
.266 .238
.567 .390
.531 .410
.707 .605
.716 .609;
PROC MEANS N MEAN STDERR T PRT;
```

for which we'll get sample size (N), sample mean (MEAN), standard error (STDERR), $t$ statistic (T), and $p$ value (PRT).

***Example 7.3***   The systolic blood pressures of $n = 12$ women between the ages of 20 and 35 were measured before and after administration of a newly developed oral contraceptive. Data are shown in Table 7.2 (the difference is after–before). The necessary summarized figures are

**TABLE 7.2**

| Subject | Systolic Blood Pressure (mmHg) | | Difference, $d_i$ | $d_i^2$ |
|---------|---------|--------|------------------|---------|
|         | Before  | After  |                  |         |
| 1       | 122     | 127    | 5                | 25      |
| 2       | 126     | 128    | 2                | 4       |
| 3       | 132     | 140    | 8                | 64      |
| 4       | 120     | 119    | −1               | 1       |
| 5       | 142     | 145    | 3                | 9       |
| 6       | 130     | 130    | 0                | 0       |
| 7       | 142     | 148    | 6                | 36      |
| 8       | 137     | 135    | −2               | 4       |
| 9       | 128     | 129    | 1                | 1       |
| 10      | 132     | 137    | 5                | 25      |
| 11      | 128     | 128    | 0                | 0       |
| 12      | 129     | 133    | 4                | 16      |
|         |         |        | 31               | 185     |

$$\bar{d} = \text{average difference}$$

$$= \frac{31}{12}$$

$$= 2.58 \text{ mmHg}$$

$$s_d^2 = \frac{185 - (31)^2/12}{11}$$

$$= 9.54$$

$$s_d = 3.09$$

$$\text{SE}(\bar{d}) = \frac{3.09}{\sqrt{12}}$$

$$= 0.89$$

$$t = \frac{2.58}{0.89}$$

$$= 2.90$$

Using the column corresponding to the upper tail area of 0.05 in Appendix C, we have a tabulated value of 1.796 for 11 df. Since

$$t = 2.90 > 2.201$$

we conclude that the null hypothesis of no blood pressure change should be rejected at the 0.05 level; there is enough evidence to support the hypothesis of *increased* systolic blood pressure (one-sided alternative).

***Example 7.4*** Data in epidemiologic studies are sometimes self-reported. Screening data from the hypertension detection and follow-up program in Minneapolis, Minnesota (1973–1974) provided an opportunity to evaluate the accuracy of self-reported height and weight. Table 7.3 gives the percent discrepancy between self-reported and measured height:

$$x = \frac{\text{self-reported height} - \text{measured height}}{\text{measured height}} \times 100\%$$

**TABLE 7.3**

| Education | Men | | | Women | | |
|---|---|---|---|---|---|---|
| | $n$ | Mean | SD | $n$ | Mean | SD |
| ≤High school | 476 | 1.38 | 1.53 | 323 | 0.66 | 1.53 |
| ≥College | 192 | 1.04 | 1.31 | 62 | 0.41 | 1.46 |

Let's focus on the sample of men with a high school education; investigations of other groups and the differences between them are given in the exercises at the end of this chapter. An application of the one-sample $t$ test yields

$$\bar{x} = 1.38$$

$$s_x = 1.53$$

$$\text{SE}(\bar{x}) = \frac{1.53}{\sqrt{476}}$$

$$= 0.07$$

$$t = \frac{1.38}{0.07}$$

$$= 19.71$$

It can be easily seen that the difference between self-reported height and measured height is highly statistically significant ($p < 0.01$: comparing 19.71 versus the cutpoint of 2.58 for a large sample).

## 7.3  COMPARISON OF TWO MEANS

Perhaps one of the most common problems in statistical inference is a comparison of two population means using data from two independent samples; the sample sizes may or may not be equal. In this type of problem, we have two sets of continuous measurents, one of size $n_1$ and one of size $n_2$, and we consider the null hypothesis

$$H_0: \mu_1 = \mu_2$$

expressing the equality of the two population means.

To perform a test of significance for $H_0$, we proceed with the following steps:

1. Decide whether a one-sided test, say

$$H_A: \mu_2 > \mu_1$$

or a two-sided test,

$$H_A: \mu_1 \neq \mu_2$$

is appropriate.
2. Choose a significance level $\alpha$, a common choice being 0.05.
3. Calculate the $t$ statistic,

$$t = \frac{\bar{x}_1 - \bar{x}_2}{\mathrm{SE}(\bar{x}_1 - \bar{x}_2)}$$

where

$$\mathrm{SE}(\bar{x}_1 - \bar{x}_2) = s_p \sqrt{\frac{1}{n_1} + \frac{1}{n_2}}$$

$$s_p^2 = \frac{(n_1 - 1)s_1^2 + (n_2 - 1)s_2^2}{n_1 + n_2 - 2}$$

This test is referred to as a *two-sample t test* and its rejection region is determined using the *t* distribution at $(n_1 + n_2 - 2)$ degrees of freedom:

- For a one-tailed test, use the column corresponding to an upper tail area of 0.05 and $H_0$ is rejected if

$$t \le -\text{tabulated value} \quad \text{for } H_A: \mu_1 < \mu_2$$

or

$$t \ge \text{tabulated value} \quad \text{for } H_A: \mu_1 > \mu_2$$

- For a two-tailed test or $H_A: \mu_1 \ne \mu_2$, use the column corresponding to an upper tail area of 0.025 and $H_0$ is rejected if

$$t \le -\text{tabulated value} \quad \text{or} \quad t \ge \text{tabulated value}$$

**Example 7.5**   In an attempt to assess the physical condition of joggers, a sample of $n_1 = 25$ joggers was selected and their maximum volume of oxygen uptake ($VO_2$) was measured with the following results:

$$\bar{x}_1 = 47.5 \text{ mL/kg} \qquad s_1 = 4.8 \text{ mL/kg}$$

Results for a sample of $n_2 = 26$ nonjoggers were

$$\bar{x}_2 = 37.5 \text{ mL/kg} \qquad s_2 = 5.1 \text{ mL/kg}$$

To proceed with the two-tailed, two-sample *t* test, we have

$$s_p^2 = \frac{(24)(4.8)^2 + (25)(5.1)^2}{49}$$

$$= 24.56$$

$$s_p = 4.96$$

$$\text{SE}(\bar{x}_1 - \bar{x}_2) = 4.96\sqrt{\frac{1}{25} + \frac{1}{26}}$$

$$= 1.39$$

It follows that

$$t = \frac{47.5 - 37.5}{1.39}$$

$$= 7.19$$

indicating a significant difference between joggers and nonjoggers (at 49 degrees of freedom and $\alpha = 0.01$, the tabulated $t$ value, with an upper tail area of 0.025, is about 2.0).

***Example 7.6*** Vision, or more especially visual acuity, depends on a number of factors. A study was undertaken in Australia to determine the effect of one of these factors: racial variation. Visual acuity of recognition as assessed in clinical practice has a defined normal value of 20/20 (or zero in log scale). The following summarized data on monocular visual acuity (expressed in log scale) were obtained from two groups:

1. Australian males of European origin

$$n_1 = 89$$

$$\bar{x}_1 = -0.20$$

$$s_1 = 0.18$$

2. Australian males of Aboriginal origin

$$n_2 = 107$$

$$\bar{x}_2 = -0.26$$

$$s_2 = 0.13$$

To proceed with a two-sample $t$ test, we have

$$s_p^2 = \frac{(88)(0.18)^2 + (106)(0.13)^2}{194}$$

$$= (0.155)^2$$

$$\text{SE}(\bar{x}_1 - \bar{x}_2) = (0.155)\sqrt{\frac{1}{89} + \frac{1}{107}}$$

$$= 0.022$$

$$t = \frac{(-0.20) - (-0.26)}{0.022}$$

$$= 2.73$$

The result indicates that the difference is statistically significant beyond the 0.01 level (at $\alpha = 0.01$, and for a two-sided test the cut point is 2.58 for high df values).

*Example 7.7*    The extend to which an infant's health is affected by parental smoking is an important public health concern. The following data are the urinary concentrations of cotinine (a metabolite of nicotine); measurements were taken both from a sample of infants who had been exposed to household smoke and from a sample of unexposed infants.

| Unexposed ($n_1 = 7$) | 8 | 11 | 12 | 14 | 20 | 43 | 111 | |
|---|---|---|---|---|---|---|---|---|
| Exposed ($n_2 = 8$) | 35 | 56 | 83 | 92 | 128 | 150 | 176 | 208 |

The statistics needed for our two-sample $t$ test are:

1. For unexposed infants:

$$n_1 = 7$$
$$\bar{x}_1 = 31.29$$
$$s_1 = 37.07$$

2. For exposed infants:

$$n_2 = 8$$
$$\bar{x}_2 = 116.00$$
$$s_2 = 59.99$$

To proceed with a two-sample $t$ test, we have

$$s_p^2 = \frac{(6)(37.07)^2 + (7)(59.99)^2}{13}$$

$$= (50.72)^2$$

$$\text{SE}(\bar{x}_1 - \bar{x}_2) = (50.72)\sqrt{\frac{1}{7} + \frac{1}{8}}$$

$$= 26.25$$

$$t = \frac{116.00 - 31.29}{26.25}$$

$$= 3.23$$

The result indicates that the difference is statistically significant beyond the 0.01 level (at $\alpha = 0.01$, and for a two-sided test the cut point is 3.012 for 13 df).

*Note:* An SAS program would include these instructions:

```
DATA;
INPUT GROUP $ COTININE;
DATALINES;
U 8
...
U 111
E 35
...
E 208;
PROC TTEST;
CLASS GROUP;
VAR COTININE;
```

in which independent variable name or group follows CLASS and dependent variable name follows VAR.


## 7.4  NONPARAMETRIC METHODS

Earlier we saw how to apply one- and two-sample $t$ tests to compare population means. However, these methods depend on certain assumptions about distributions in the population; for example, its derivation assumes that the population(s) is (are) normally distributed. It has even been proven that these procedures are robust; that is, they are relatively insensitive to departures from the assumptions made. In other words, departures from those assumptions have very little effect on the results, provided that samples are large enough. But the procedures are all sensitive to extreme observations, a few very small or very large—perhaps erroneous—data values. In this section we learn some nonparametric procedures, or distribution-free methods, where no assumptions about population distributions are made. The results of these nonparametric tests are much less affected by extreme observations.


### 7.4.1  Wilcoxon Rank-Sum Test

The Wilcoxon rank-sum test is perhaps the most popular nonparametric procedure. The Wilcoxon test is a nonparametric counterpart of the two-sample $t$ test; it is used to compare two samples that have been drawn from independent populations. But unlike the $t$ test, the Wilcoxon test does not assume that the underlying populations are normally distributed and is less affected by extreme observations. The Wilcoxon rank-sum test evaluates the null hypothesis that the medians of the two populations are identical (for a normally distributed population, the population median is also the population mean).

For example, a study was designed to test the question of whether cigarette smoking is associated with reduced serum-testosterone levels. To carry out this research objective, two samples, each of size 10, are selected independently. The first sample consists of 10 nonsmokers who have never smoked, and the second sample consists of 10 heavy smokers, defined as those who smoke 30 or more cigarettes a day. To perform the Wilcoxon rank-sum test, we combine the two samples into one large sample (of size 20), arrange the observations from smallest to largest, and assign a rank, from 1 to 20, to each. If there are tied observations, we assign an average rank to all measurements with the same value. For example, if the two observations next to the third smallest are equal, we assign an average rank of $(4 + 5)/2 = 4.5$ to each one. The next step is to find the sum of the ranks corresponding to each of the original samples. Let $n_1$ and $n_2$ be the two sample sizes and $R$ be the sum of the ranks from the sample with size $n_1$.

Under the null hypothesis that the two underlying populations have identical medians, we would expect the averages of ranks to be approximately equal. We test this hypothesis by calculating the statistic

$$z = \frac{R - \mu_R}{\sigma_R}$$

where

$$\mu_R = \frac{n_1(n_1 + n_2 + 1)}{2}$$

is the mean and

$$\sigma_R = \sqrt{\frac{n_1 n_2 (n_1 + n_2 + 1)}{12}}$$

is the standard deviation of $R$. It does not make any difference which rank sum we use. For relatively large values of $n_1$ and $n_2$ (say, both greater than or equal to 10), the sampling distribution of this statistic is approximately standard normal. The null hypothesis is rejected at the 5% level, against a two-sided alternative, if

$$z < -1.96 \quad \text{or} \quad z > 1.96$$

***Example 7.8*** For the study on cigarette smoking above, Table 7.4 shows the raw data, where testosterone levels were measured in μg/dL and the ranks were determined. The sum of the ranks for group 1 (nonsmokers) is

$$R = 143$$

**TABLE 7.4**

| Nonsmokers | | Heavy Smokers | |
|---|---|---|---|
| Measurement | Rank | Measurement | Rank |
| 0.44 | 8.5 | 0.45 | 10 |
| 0.44 | 8.5 | 0.25 | 1 |
| 0.43 | 7 | 0.40 | 6 |
| 0.56 | 14 | 0.27 | 2 |
| 0.85 | 17 | 0.34 | 4 |
| 0.68 | 15 | 0.62 | 13 |
| 0.96 | 20 | 0.47 | 11 |
| 0.72 | 16 | 0.30 | 3 |
| 0.92 | 19 | 0.35 | 5 |
| 0.87 | 18 | 0.54 | 12 |

In addition,

$$\mu_R = \frac{10(10 + 10 + 1)}{2}$$

$$= 105$$

and

$$\sigma_R = \sqrt{\frac{(10)(10)(10 + 10 + 1)}{12}}$$

$$= 13.23$$

Substituting these values into the equation for the test statistic, we have

$$z = \frac{R - \mu_R}{\sigma_R}$$

$$= \frac{143 - 105}{13.23}$$

$$= 2.87$$

Since $z > 1.96$, we reject the null hypothesis at the 5% level. (In fact, since $z > 2.58$, we reject the null hypothesis at the 1% level.) Note that if we use the sum of the ranks for the other group (heavy smokers), the sum of the ranks is 67, leading to a $z$ score of

$$\frac{67 - 105}{13.23} = -2.87$$

and we would come to the same decision.

*Example 7.9*    Refer to the nicotine data of Example 7.6, where measurements were taken both from a sample of infants who had been exposed to household smoke and from a sample of unexposed infants. We have:

| Unexposed ($n_1 = 7$) | 8 | 11 | 12 | 14 | 20 | 43 | 111 |
|---|---|---|---|---|---|---|---|
| Rank | 1 | 2 | 3 | 4 | 5 | 7 | 11 |

| Exposed ($n_2 = 8$) | 35 | 56 | 83 | 92 | 128 | 150 | 176 | 208 |
|---|---|---|---|---|---|---|---|---|
| Rank | 6 | 8 | 9 | 10 | 12 | 13 | 14 | 15 |

The sum of the ranks for the group of exposed infants is

$$R = 87$$

In addition,

$$\mu_R = \frac{(8)(8 + 7 + 1)}{2}$$

$$= 64$$

and

$$\sigma_R = \sqrt{\frac{(8)(7)(8 + 7 + 1)}{12}}$$

$$= 8.64$$

Substituting these values into the equation for the Wilcoxon test, we have

$$z = \frac{R - \mu_R}{\sigma_R}$$

$$= \frac{87 - 64}{8.64}$$

$$= 2.66$$

Since $z > 1.96$, we reject the null hypothesis at the 5% level. In fact, since $z > 2.58$, we reject the null hypothesis at the 1% level; $p$ value $< 0.01$. (It should be noted that the sample sizes of 7 and 8 in this example may be not large enough.)

*Note:* An SAS program would include these instructions:

```
DATA;
INPUT GROUP $ COTININE;
DATALINES;
U 8
...
U 111
E 35
...
E 208;
PROC NPAR1WAY WILCOXON; CLASS GROUP;
VAR COTININE;
```

### 7.4.2  Wilcoxon Signed-Rank Test

The idea of using *ranks*, instead of measured values, to form statistical tests to compare population means applies to the analysis of pair-matched data as well. As with the one-sample *t* test for pair-matched data, we begin by forming differences. Then the absolute values of the differences are assigned ranks; if there are ties in the differences, the average of the appropriate ranks is assigned. Next, we attach a + or a − sign back to each rank, depending on whether the corresponding difference is positive or negative. This is achieved by multiplying each rank by +1, −1, or 0 as the corresponding difference is positive, negative, or zero. The results are *n signed ranks*, one for each pair of observations; for example, if the difference is zero, its signed rank is zero. The basic idea is that if the *mean difference* is positive, there would be more and larger *positive signed ranks*; since if this were the case, most differences would be positive and larger in magnitude than the few negative differences, most of the ranks, especially the larger ones, would then be positively signed. In other words, we can base the test on the *sum R* of the *positive signed ranks*. We test the null hypothesis of no difference by calculating the *standardized test statistic*:

$$z = \frac{R - \mu_R}{\sigma_R}$$

where

$$\mu_R = \frac{(n)(n+1)}{4}$$

is the mean and

$$\sigma_R = \sqrt{\frac{(n)(n+1)(2n+1)}{24}}$$

is the standard deviation of $R$ under the null hypothesis. This normal approximation applies for relatively large samples, $n \geq 20$; the null hypothesis is rejected at the 5% level, against a two-sided alternative, if

$$z < -1.96 \quad \text{or} \quad z > 1.96$$

This test is referred to as *Wilcoxon's signed-rank test*.

*Example 7.10*    Ultrasounds were taken at the time of liver transplant and again 5 to 10 years later to determine the systolic pressure of the hepatic artery. Results for 21 transplants for 21 children are shown in Table 7.5. The sum of the positive signed ranks is

$$13 + 9 + 15.5 + 20 + 5.5 + 4 + 5.5 + 17.5 = 90$$

Its mean and standard deviation under the null hypothesis are

$$\mu_R = \frac{(21)(22)}{4}$$

$$= 115.5$$

**TABLE 7.5**

| Child | Later | At Transplant | Difference | Absolute Value of Difference | Rank | Signed Rank |
|-------|-------|---------------|------------|------------------------------|------|-------------|
| 1 | 46 | 35 | 11 | 11 | 13 | 13 |
| 2 | 40 | 40 | 0 | 0 | 2 | 0 |
| 3 | 50 | 58 | −8 | 8 | 9 | −9 |
| 4 | 50 | 71 | −19 | 19 | 17.5 | −17.5 |
| 5 | 41 | 33 | 8 | 8 | 9 | 9 |
| 6 | 70 | 79 | −9 | 9 | 11 | −11 |
| 7 | 35 | 20 | 15 | 15 | 15.5 | 15.5 |
| 8 | 40 | 19 | 21 | 21 | 20 | 20 |
| 9 | 56 | 56 | 0 | 0 | 2 | 0 |
| 10 | 30 | 26 | 4 | 4 | 5.5 | 5.5 |
| 11 | 30 | 44 | −14 | 14 | 14 | −14 |
| 12 | 60 | 90 | −30 | 30 | 21 | −21 |
| 13 | 43 | 43 | 0 | 0 | 2 | 0 |
| 14 | 45 | 42 | 3 | 3 | 4 | 4 |
| 15 | 40 | 55 | −15 | 15 | 15.5 | −15.5 |
| 16 | 50 | 60 | −10 | 10 | 12 | −12 |
| 17 | 66 | 62 | 4 | 4 | 5.5 | 5.5 |
| 18 | 45 | 26 | 19 | 19 | 17.5 | 17.5 |
| 19 | 40 | 60 | −20 | 20 | 19 | −19 |
| 20 | 35 | 27 | −8 | 8 | 9 | −9 |
| 21 | 25 | 31 | −6 | 6 | 7 | −7 |

$$\sigma_R = \sqrt{\frac{(21)(22)(43)}{24}}$$

$$= 28.77$$

leading to a standardized $z$ score of

$$z = \frac{90 - 115.5}{28.77}$$

$$= -0.89$$

The result indicates that the systolic pressure of the hepatic artery measured five years after the liver transplant, compared to the measurement at transplant, is lower on the average; however, the difference is not statistically significant at the 5% level $(-0.89 > -1.96)$.

*Note:* An SAS program would include these instructions:

```
DATA;
INPUT POST PRE;
DIFF = POST - PRE;
DATALINES;
46 35
...
35 27
25 31;
PROC UNIVARIATE;
```

for which we'll get, among many other things, the test statistic (SGN RANK) and the $p$ value (Prob $> |S|$).


## 7.5  ONE-WAY ANALYSIS OF VARIANCE

Suppose that the goal of a research project is to discover whether there are differences in the means of several independent groups. The problem is how we will measure the extent of differences among the means. If we had two groups, we would measure the difference by the distance between sample means $(\bar{x} - \bar{y})$ and use the two-sample $t$ test. Here we have more than two groups; we could take all possible pairs of means and do many two-sample $t$ tests. What is the matter with this approach of doing many two-sample $t$ tests, one for each pair of samples? As the number of groups increases, so does the number of tests to perform; for example, we would have to do 45 tests if we have 10 groups to compare. Obviously, the amount of work is greater, but that should not be the critical problem, especially with technological aids such as the use of calculators and computers. So what is the problem? The answer is that performing many tests increases the probability that one or more of the comparisons will result in a type I error (i.e., a significant test result when the null hypothesis is true). This statement should make sense intuitively. For example, suppose that the null hypothesis is true and we perform 100 tests—each has a 0.05 probability of resulting in a type I error; then 5 of these 100 tests would be statistically significant as the result of type I errors. Of course, we usually do not need to do that many tests; however, every time we do more than one, the probability that at least one will result in a type I error exceeds 0.05, indicating a falsely significant difference! What is needed is a different way to summarize the differences between several means and a method of *simultaneously* comparing these means in one step. This method is called ANOVA or one-way ANOVA, an abbreviation of *analysis of variance*.

We have continuous measurements $X$'s from $k$ independent samples; the sample sizes may or may not be equal. We assume that these are samples from $k$ normal distributions with a common variance $\sigma^2$, but the means, $\mu_i$'s, may or may not be the same. The case where we apply the two-sample $t$ test is a special case of this one-way ANOVA model with $k = 2$. Data from the $i$th sample can be summarized into sample size $n_i$, sample mean $\bar{x}_i$, and sample variance $s_i^2$. If we pool data together, the (grand) mean of this combined sample can be calculated from

$$\bar{x} = \frac{\sum(n_i)(\bar{x}_i)}{\sum(n_i)}$$

In that combined sample of size $n = \sum n_i$, the variation in $X$ is measured conventionally in terms of the deviations $(x_{ij} - \bar{x})$ (where $x_{ij}$ is the $j$th measurement from the $i$th sample); the total variation, denoted by SST, is the sum of squared deviations:

$$\text{SST} = \sum_{i,j}(x_{ij} - \bar{x})^2$$

For example, SST = 0 when all observation $x_{ij}$ values are the same; SST is the numerator of the sample variance of the combined sample: The higher the SST value, the greater the variation among all $X$ values. The total variation in the combined sample can be decomposed into two components:

$$x_{ij} - \bar{x} = (x_{ij} - \bar{x}_i) + (\bar{x}_i - \bar{x})$$

1. The first term reflects the variation *within* the $i$th sample; the sum

$$\text{SSW} = \sum_{i,j}(x_{ij} - \bar{x}_i)^2$$

$$= \sum_i(n_i - 1)s_i^2$$

is called the *within sum of squares*.

2. The difference between the two sums of squares above,

$$SSB = SST - SSW$$

$$= \sum_{i,j} (\bar{x}_i - \bar{x})^2$$

$$= \sum_i n_i (\bar{x}_i - \bar{x})^2$$

is called the *between sum of squares*. SSB represents the variation or differences between the sample means, a measure very similar to the numerator of a sample variance; the $n_i$ values serve as *weights*.

Corresponding to the partitioning of the total sum of squares SST, there is partitioning of the associated degrees of freedom (df). We have $(n - 1)$ degrees of freedom associated with SST, the denominator of the variance of the combined sample. SSB has $(k - 1)$ degrees of freedom, representing the differences between $k$ groups; the remaining $[n - k = \sum(n_i - 1)]$ degrees of freedom are associated with SSW. These results lead to the usual presentation of the ANOVA process:

1. The *within mean square*

$$MSW = \frac{SSW}{n - k}$$

$$= \frac{\sum_i (n_i - 1) s_i^2}{\sum (n_i - 1)}$$

serves as an estimate of the common variance $\sigma^2$ as stipulated by the one-way ANOVA *model*. In fact, it can be seen that MSW is a natural extension of the pooled estimate $s_p^2$ as used in the two-sample $t$ test; It is a measure of the average variation within the $k$ samples.

2. The *between mean square*

$$MSB = \frac{SSB}{k - 1}$$

represents the *average* variation (or differences) between the $k$ sample means.

3. The breakdowns of the total sum of squares and its associated degrees of freedom are *displayed* in the form of an *analysis of variance table* (Table 7.6). The test statistic $F$ for the one-way analysis of variance above compares MSB (the *average* variation—or differences—between the $k$ sample means) and MSE (the average variation within the $k$ samples). A value

**TABLE 7.6**

| Source of Variation | SS | df | MS | F Statistic | p Value |
|---|---|---|---|---|---|
| Between samples | SSB | $k - 1$ | MSB | MSB/MSW | $p$ |
| Within samples | SSW | $n - k$ | MSW | | |
| Total | SST | $n - 1$ | | | |

near 1 supports the null hypothesis of *no differences between the k population means*. Decisions are made by referring the observed value of the test statistic $F$ to to the $F$ table in Appendix E with $(k - 1, n - k)$ degress of freedom. In fact, when $k = 2$, we have

$$F = t^2$$

where $t$ is the test statistic for comparing the two population means. In other words, when $k = 2$, the $F$ test is equivalent to the two-sided two-sample $t$ test.

***Example 7.11***   Vision, especially visual acuity, depends on a number of factors. A study was undertaken in Australia to determine the effect of one of these factors: racial variation. Visual acuity of recognition as assessed in clinical practice has a defined normal value of 20/20 (or zero on the log scale). The following summarize the data on monocular visual acuity (expressed on a log scale); part of this data set was given in Example 7.6.

1. Australian males of European origin

$$n_1 = 89$$
$$\bar{x}_1 = -0.20$$
$$s_1 = 0.18$$

2. Australian males of Aboriginal origin

$$n_2 = 107$$
$$\bar{x}_2 = -0.26$$
$$s_2 = 0.13$$

3. Australian females of European origin

$$n_3 = 63$$
$$\bar{x}_3 = -0.13$$
$$s_3 = 0.17$$

4. Australian females of Aboriginal origin

$$n_4 = 54$$

$$\bar{x}_4 = -0.24$$

$$s_4 = 0.18$$

To proceed with a one-way analysis of variance, we calculate the mean of the combined sample:

$$\bar{x} = \frac{(89)(-0.20) + (107)(-0.26) + (63)(-0.13) + (54)(-0.24)}{89 + 107 + 63 + 54}$$

$$= -0.213$$

and

$$SSB = (89)(-0.20 + 0.213)^2 + (107)(-0.26 + 0.213)^2$$

$$+ (63)(-0.13 + 0.213)^2 + (54)(-0.24 + 0.213)^2$$

$$= 0.7248$$

$$MSB = \frac{0.7248}{3}$$

$$= 0.2416$$

$$SSW = (88)(0.18)^2 + (106)(0.13)^2 + (62)(0.17)^2 + (53)(0.18)^2$$

$$= 8.1516$$

$$MSW = \frac{8.1516}{309}$$

$$= 0.0264$$

$$F = \frac{0.2416}{0.0264}$$

$$= 9.152$$

The results are summarized in an ANOVA table (Table 7.7). The resulting $F$ test inducates that the overall differences between the four population means is highly significant ($p < 0.00001$).

*Example 7.12*   A study was conducted to test the question as to whether cigarette smoking is associated with reduced serum-testosterone levels in men aged 35 to 45. The study involved the following four groups:

**TABLE 7.7**

| Source of Variation | SS | df | MS | $F$ Statistic | $p$ Value |
|---|---|---|---|---|---|
| Between samples | 0.7248 | 3 | 0.2416 | 9.152 | <0.0001 |
| Within samples | 8.1516 | 309 | 0.0264 | | |
| Total | 8.8764 | 312 | | | |

1. Nonsmokers who had never smoked
2. Former smokers who had quit for at least six months prior to the study
3. Light smokers, defined as those who smoked 10 or fewer cigarettes per day
4. Heavy smokers, defined as those who smoked 30 or more cigarettes per day

Each group consisted of 10 men and Table 7.8 shows raw data, where serum-testosterone levels were measured in µg/dL.

An application of the one-way ANOVA yields Table 7.9. The resulting $F$ test inducates that the overall differences between the four population means is statistically significant at the 5% level but not at the 1% level ($p = 0.0179$).

**TABLE 7.8**

| Nonsmokers | Former Smokers | Light Smokers | Heavy Smokers |
|---|---|---|---|
| 0.44 | 0.46 | 0.37 | 0.44 |
| 0.44 | 0.50 | 0.42 | 0.25 |
| 0.43 | 0.51 | 0.43 | 0.40 |
| 0.56 | 0.58 | 0.48 | 0.27 |
| 0.85 | 0.85 | 0.76 | 0.34 |
| 0.68 | 0.72 | 0.60 | 0.62 |
| 0.96 | 0.93 | 0.82 | 0.47 |
| 0.72 | 0.86 | 0.72 | 0.70 |
| 0.92 | 0.76 | 0.60 | 0.60 |
| 0.87 | 0.65 | 0.51 | 0.54 |

**TABLE 7.9**

| Source of Variation | SS | df | MS | $F$ Statistic | $p$ Value |
|---|---|---|---|---|---|
| Between samples | 0.3406 | 3 | 0.1135 | 3.82 | 0.0179 |
| Within samples | 1.0703 | 36 | 0.0297 | | |
| Total | 1.4109 | 39 | | | |

*Note:* An SAS program would include these instructions:

```
DATA;
INPUT GROUP $ SERUMT;
DATALINES;
N .44
...
N .87
F .46
...
F .65
L .37
...
L .51
H .44
...
H .54;
PROC ANOVA;
CLASS GROUP;
MODEL SERUMT = GROUP;
MEANS GROUP;
```

The last row is an option to provide the sample mean of each group.

## 7.6   BRIEF NOTES ON THE FUNDAMENTALS

There are two simple, different, but equivalent ways to view Wilcoxon's rank-sum test. In the first approach, one can view the ranks $R_1, R_2, \ldots, R_n$ from a combined sample, $n = n_1 + n_2$, as forming finite population. The mean and variance of this finite population are

$$\mu = \frac{\sum_{i=1}^{n} R_i}{n}$$

$$= \frac{n+1}{2}$$

$$\sigma^2 = \frac{\sum_{i=1}^{n} (R_i - \mu)^2}{n}$$

$$= \frac{n^2 - 1}{12}$$

Under the null hypothesis, the rank sum $R$, from the sample with size $n_1$, is the total from a sample of size $n_1$ from the finite population above, so that

$$\mu_R = n_1 \mu$$

$$\sigma_R^2 = n_1^2 \frac{\sigma^2}{n_1} \left( \frac{n - n_1}{n - 1} \right)$$

The last term of the last equation is called the *correction factor for finite population*.

Equivalently, the ranks $R_1, R_2, \ldots, R_n$ from a combined sample can be thought of as a standardized representation of the numbers $x_1, x_2, \ldots, x_{n_1}$ from sample 1 and $y_1, y_2, \ldots, y_{n_2}$ from sample 2, which preserves all order relationships. Therefore, it is plausible to contruct a test statistic by substituting $R_1, R_2, \ldots, R_{n_1}$ for $x_1, x_2, \ldots, x_{n_1}$ and $R_1, R_2, \ldots, R_{n_2}$ for $y_1, y_2, \ldots, y_{n_2}$. The result is equivalent to Wilcoxon's rank-sum test.

## 7.7  NOTES ON COMPUTATIONS

Samples of SAS program instructions were provided for all procedures, at the end of Examples 7.2, 7.7, 7.9, 7.10, and 7.12. Since all $p$ values provided by SAS are for two-sided alternatives. The one-sample and two-sample $t$ tests can also be implemented easily using Microsoft's Excel. The first two steps are the same as in obtaining descriptive statistics: (1) click the *paste function icon*, f*, and (2) click *Statistical*. Among the functions available, choose TTEST. A box appears with four rows to be filled. The first two are for data in the two groups to be compared; in each you identify the range of cells, say B2:B23. The third box asking for "tails," enter "1" ("2") for one-sided (two-sided) alternative. Enter the "type" of test on the last row, "1" ("2") for a one-sample (two-sample) $t$ test.

## EXERCISES

**7.1**  The criterion for issuing a smog alert is established at greater than 7 ppm of a particular pollutant. Samples collected from 16 stations in a certain city give a $\bar{x}$ value of 7.84 ppm with a standard deviation of $s = 2.01$ ppm. Do these findings indicate that the smog alert criterion has been exceeded? State clearly your null and alternative hypotheses and choice of test size (alpha level).

**7.2**  The purpose of an experiment is to investigate the effect of vagal nerve stimulation on insulin secretion. The subjects are mongrel dogs with varying body weights. Table E7.2 gives the amount of immunoreactive insulin in pancreatic venous plasma just before stimulation of the left vagus and the amount measured 5 minutes after stimulation for seven dogs. Test the null hypothesis that the stimulation of the vagus nerve has no effect on the blood level of immunoreactive insulin; that is,

$$H_0: \mu_{before} = \mu_{after}$$

State your alternative hypothesis and choice of test size, and draw the appropriate conclusion.

**TABLE E7.2**

| Dog | Before | After |
|-----|--------|-------|
|     | Blood Levels of Immunoreactive Insulin ($\mu$U/mL) | |
| 1 | 350 | 480 |
| 2 | 200 | 130 |
| 3 | 240 | 250 |
| 4 | 290 | 310 |
| 5 | 90 | 280 |
| 6 | 370 | 1450 |
| 7 | 240 | 280 |

**7.3**  In a study of saliva cotinine, seven subjects, all of whom had abstained from smoking for a week, were asked to smoke a single cigareete. The cotinine levels at 12 and 24 hours after smoking are given in Table E7.3. Test to compare the mean cotinine levels at 12 and 24 hours after smoking. State clearly your null and alternative hypotheses and choice of test size.

**TABLE E7.3**

| Subject | After 12 Hours | After 24 Hours |
|---------|----------------|----------------|
|         | Cotinine Levels (mmol/L) | |
| 1 | 73 | 24 |
| 2 | 58 | 27 |
| 3 | 67 | 49 |
| 4 | 93 | 59 |
| 5 | 33 | 0 |
| 6 | 18 | 11 |
| 7 | 147 | 43 |

**7.4**  Dentists often make many people nervous. To see if such nervousness elevates blood pressure, the systolic blood pressures of 60 subjects were measured in a dental setting, then again in a medical setting. Data for 60 matched pairs (dental–medical) are summarized as follows:

$$\text{mean} = 4.47$$

$$\text{standard deviation} = 8.77$$

Test to compare the means blood pressure under two different settings. Name the test and state clearly your null and alternative hypotheses and choice of test size.

7.5    In Example 7.10 a study with 21 transplants for 21 children was reported where ultrasounds were taken at the time of liver transplant and again 5 to 10 years later to determine the systolic pressure of the hepatic artery. In that example, Wilcoxon's signed-rank test was applied to compare the hepatic systolic pressures measured at two different times. Table E7.5 gives the diastolic presures obtained from the study. Test to compare the mean diastolic hepatic pressures. Name the test and state clearly your null and alternative hypotheses and choice of test size.

**TABLE E7.5**

| Subject | Diastolic Hepatic Pressure | |
|---|---|---|
| | 5–10 Years Later | At Transplant |
| 1 | 14 | 25 |
| 2 | 10 | 10 |
| 3 | 20 | 23 |
| 4 | 10 | 14 |
| 5 | 4 | 19 |
| 6 | 20 | 12 |
| 7 | 10 | 5 |
| 8 | 18 | 4 |
| 9 | 12 | 23 |
| 10 | 18 | 8 |
| 11 | 10 | 10 |
| 12 | 10 | 20 |
| 13 | 15 | 12 |
| 14 | 10 | 10 |
| 15 | 10 | 19 |
| 16 | 15 | 20 |
| 17 | 26 | 26 |
| 18 | 20 | 8 |
| 19 | 10 | 10 |
| 20 | 10 | 11 |
| 21 | 10 | 16 |

7.6    A study was conducted to investigate whether oat bran cereal helps to lower serum cholesterol in men with high cholesterol levels. Fourteen men were randomly placed on a diet which included either oat bran or cornflakes; after two weeks their low-density-lipoprotein cholesterol levels were recorded. Each man was then switched to the alternative diet. After a second two-week period, the LDL cholesterol level of each person was recorded again (Table E7.6). Test to compare the means LDL cholesterol level. Name the test and state clearly your null and alternative hypotheses and choice of test size.

**TABLE E7.6**

| Subject | LDL (mmol/L) | |
|---|---|---|
| | Cornflakes | Oat Bran |
| 1 | 4.61 | 3.84 |
| 2 | 6.42 | 5.57 |
| 3 | 5.40 | 5.85 |
| 4 | 4.54 | 4.80 |
| 5 | 3.98 | 3.68 |
| 6 | 3.82 | 2.96 |
| 7 | 5.01 | 4.41 |
| 8 | 4.34 | 3.72 |
| 9 | 3.80 | 3.49 |
| 10 | 4.56 | 3.84 |
| 11 | 5.35 | 5.26 |
| 12 | 3.89 | 3.73 |
| 13 | 2.25 | 1.84 |
| 14 | 4.24 | 4.14 |

7.7 Data in epidemiologic studies are sometimes self-reported. Screening data from the hypertension detection and follow-up program in Minneapolis, Minnesota (1973–1974) provided an opportunity to evaluate the accuracy of self-reported height and weight (see Example 7.4). Table 7.3 gives the percent discrepancy between self-reported and measured height:

$$x = \frac{\text{self-reported height} - \text{measured height}}{\text{measured height}} \times 100\%$$

Example 7.4 was focused on the sample of men with a high school education. Using the same procedure, investigate the difference between self-reported height and measured height among:

(a) Men with a college education.

(b) Women with a high school education.

(c) Women with a college education.

In each case, name the test and state clearly your null and alternative hypotheses and choice of test size. Also, compare the mean difference in percent discrepancy between:

**(d)** Men with different education levels.

**(e)** Women with different education levels.

**(f)** Men versus women at each educational level.

In each case, name the test and state clearly your null and alternative hypotheses and choice of test size.

**7.8** A case–control study was undertaken to study the relationship between hypertension and obesity. Persons aged 30 to 49 years who were clearly nonhypertensive at their first multiphasic health checkup and became hypertensive by age 55 were sought and identified as cases. Controls were selected from among participants in a health plan, those who had the first checkup and no sign of hypertension in subsequent checkups. One control was matched to each case based on gender, race, year of birth, and year of entrance into the health plan. Data for 609 matched pairs are summarized in Table E7.8. Compare the cases versus the controls using each measured characteristic. In each case, name the test and state clearly your null and alternative hypotheses and choice of test size.

**TABLE E7.8**

| Variable | Paired Difference | |
| --- | --- | --- |
| | Mean | Standard Deviation |
| Systolic blood pressure (mmHg) | 6.8 | 13.86 |
| Diastolic blood pressure (mmHg) | 5.4 | 12.17 |
| Body mass index ($kg/m^2$) | 1.3 | 4.78 |

**7.9** The Australian study of Example 7.6 also provided these data on monocular acuity (expressed in log scale) for two female groups of subjects:

**(1)** Australian females of European origin

$$n_1 = 63$$
$$\bar{x}_1 = -0.13$$
$$s_1 = 0.17$$

**(2)** Australian females of Aboriginal origin

$$n_2 = 54$$
$$\bar{x}_2 = -0.24$$
$$s_2 = 0.18$$

Do these indicate a racial variation among women? Name your test and state clearly your null and alternative hypotheses and choice of test size.

**7.10**  The ages (in days) at time of death for samples of 11 girls and 16 boys who died of sudden infant death syndrome are given in Table E7.10. Do these indicate a gender difference? Name your test and state clearly your null and alternative hypotheses and choice of test size.

**TABLE E7.10**

| Females | Males | |
|---------|-------|-----|
| 53 | 46 | 115 |
| 56 | 52 | 133 |
| 60 | 58 | 134 |
| 60 | 59 | 175 |
| 78 | 77 | 175 |
| 87 | 78 | |
| 102 | 80 | |
| 117 | 81 | |
| 134 | 84 | |
| 160 | 103 | |
| 277 | 114 | |

**7.11**  An experimental study was conducted with 136 five-year-old children in four Quebec schools to investigate the impact of simulation games designed to teach children to obey certain traffic safety rules. The transfer of learning was measured by observing children's reactions to a quasi-real-life model of traffic risks. The scores on the transfer of learning for the control and attitude/behavior simulation game groups are summarized in Table E7.11. Test to investigate the impact of simulation games. Name your test and state clearly your null and alternative hypotheses and choice of test size.

**TABLE E7.11**

| Summarized Data | Control | Simulation Game |
|-----------------|---------|-----------------|
| $n$ | 30 | 33 |
| $\bar{x}$ | 7.9 | 10.1 |
| $s$ | 3.7 | 2.3 |

**7.12**  In a trial to compare a stannous fluoride dentifrice (A) with a commercially available fluoride-free dentifrice (D), 270 children received A and 250 received D for a period of three years. The number $x$ of DMFS increments (i.e., the number of new decayed, missing, and filled tooth surfaces) was obtained for each child. Results were:

$$\text{Dentifrice A:} \quad \bar{x}_A = 9.78$$
$$s_A = 7.51$$
$$\text{Dentifrice D:} \quad \bar{x}_D = 12.83$$
$$s_D = 8.31$$

Do the results provide strong enough evidence to suggest a real effect of fluoride in *reducing* the mean DMFS?

**7.13**   An experiment was conducted at the University of California–Berkeley to study the psychological environment effect on the anatomy of the brain. A group of 19 rats was randomly divided into two groups. Twelve animals in the treatment group lived together in a large cage, furnished with playthings that were changed daily, while animals in the control group lived in isolation with no toys. After a month the experimental animals were killed and dissected. Table E7.13 gives the cortex weights (the thinking part of the brain) in milligrams. Use the two-sample $t$ test to compare the means of the two groups and draw appropriate conclusions.

**TABLE E7.13**

| Treatment | | Control |
|---|---|---|
| 707 | 696 | 669 |
| 740 | 712 | 650 |
| 745 | 708 | 651 |
| 652 | 749 | 627 |
| 649 | 690 | 656 |
| 676 | | 642 |
| 699 | | 698 |

**7.14**   Depression is one of the most commonly diagnosed conditions among hospitalized patients in mental institutions. The occurrence of depression was determined during the summer of 1979 in a multiethnic probability sample of 1000 adults in Los Angeles County, as part of a community survey of the epidemiology of depression and help-seeking behavior. The primary measure of depression was the CES-D scale developed by the

**TABLE E7.14**

| | | CES-D Score | |
|---|---|---|---|
| | Cases | $\bar{x}$ | $s$ |
| Male | 412 | 7.6 | 7.5 |
| Female | 588 | 10.4 | 10.3 |

Center for Epidemiologic Studies. On a scale of 0 to 60, a score of 16 or higher was classified as depression. Table E7.14 gives the average CES-D score for the two genders. Use a $t$ test to compare the males versus the females and draw appropriate conclusions.

**7.15**  A study was undertaken to study the relationship between exposure to polychlorinated biphenyls (PCBs) and reproduction among women occupationally exposed to PCBs during the manufacture of capacitors in upstate New York. To ascertain information on reproductive outcomes, interviews were conducted in 1982 with women who had held jobs with direct exposure and women who had never held a direct-exposure job. Data are summarized in Table E7.15. Test to evaluate the effect of direct exposure (compared to indirect exposure) using each measured characteristic. In each case, name the test and state clearly your null and alternative hypotheses and choice of test size.

**TABLE E7.15**

|  | Exposure | | | |
|---|---|---|---|---|
|  | Direct ($n = 172$) | | Indirect ($n - 184$) | |
| Variable | Mean | SD | Mean | SD |
| Weight gain during pregnancy (lb) | 25.5 | 14.0 | 29.0 | 14.7 |
| Birth weight (g) | 3313 | 456 | 3417 | 486 |
| Gestational age (days) | 279.0 | 17.0 | 279.3 | 13.5 |

**7.16**  The following data are taken from a study that compares adolescents who have bulimia to healthy adolescents with similar body compositions and levels of physical activity. Table E7.16 provides measures of daily caloric intake (kcal/kg) for random samples of 23 bulimic adolescents and 15 healthy ones. Use the Wilcoxon test to compare the two populations.

**TABLE E7.16**

| Bulimic Adolescents | | | Healthy Adolescents | |
|---|---|---|---|---|
| 15.9 | 17.0 | 18.9 | 30.6 | 40.8 |
| 16.0 | 17.6 | 19.6 | 25.7 | 37.4 |
| 16.5 | 28.7 | 21.5 | 25.3 | 37.1 |
| 18.9 | 28.0 | 24.1 | 24.5 | 30.6 |
| 18.4 | 25.6 | 23.6 | 20.7 | 33.2 |
| 18.1 | 25.2 | 22.9 | 22.4 | 33.7 |
| 30.9 | 25.1 | 21.6 | 23.1 | 36.6 |
| 29.2 | 24.5 |  | 23.8 |  |

**7.17** A group of 19 rats was divided randomly into two groups. The 12 animals in the experimental group lived together in a large cage furnished with playthings that were changed daily, while the seven animals in the control group lived in isolation without toys. Table E7.17 provides the cortex weights (the thinking part of the brain) in milligrams. Use the Wilcoxon test to compare the two populations.

**TABLE E7.17**

| Experimental Group | Control Group |
|---|---|
| 707, 740, 745, 652, 649, 676, 699, 696, 712, 708, 749, 690 | 669, 650, 651, 627, 656, 642, 698 |

**7.18** College students were assigned to three study methods in an experiment to determine the effect of study technique on learning. The three methods are: read only, read and underline, and read and take notes. The test scores are recorded in Table E7.18. Test to compare the three groups simultaneously. Name your test and state clearly your null and alternative hypotheses and choice of test size.

**TABLE E7.18**

| Technique | Test Score | | | | | |
|---|---|---|---|---|---|---|
| Read only | 15 | 14 | 16 | 13 | 11 | 14 |
| Read and underline | 15 | 14 | 25 | 10 | 12 | 14 |
| Read and take notes | 18 | 18 | 18 | 16 | 18 | 20 |

**7.19** Four different brands of margarine were analyzed to determine the level of some unsaturated fatty acids (as a percentage of fats; Table E7.19). Test to compare the four groups simultaneously. Name your test and state clearly your null and alternative hypotheses and choice of test size.

**TABLE E7.19**

| Brand | Fatty Acids (%) | | | | |
|---|---|---|---|---|---|
| A | 13.5 | 13.4 | 14.1 | 14.2 | |
| B | 13.2 | 12.7 | 12.6 | 13.9 | |
| C | 16.8 | 17.2 | 16.4 | 17.3 | 18.0 |
| D | 18.1 | 17.2 | 18.7 | 18.4 | |

**7.20** A study was done to determine if simplification of smoking literature improved patient comprehension. All subjects were administered a pretest. Subjects were then randomized into three groups. One group received no booklet, one group received one written at the fifth-grade reading level, and the third received one written at the tenth-grade reading level. After booklets were received, all subjects were administered a second test. The mean score differences (postscore − prescore) are given in Table E7.20 along with their standard deviations and sample sizes. Test to compare the three groups simultaneously. Name your test and state clearly your null and alternative hypotheses and choice of test size.

**TABLE E7.20**

|  | No Booklet | 5th-Grade Level | 10th-Grade Level |
|---|---|---|---|
| $\bar{x}$ | 0.25 | 1.57 | 0.63 |
| $s$ | 2.28 | 2.54 | 2.38 |
| $n$ | 44 | 44 | 41 |

**7.21** A study was conducted to investigate the risk factors for peripheral arterial disease among persons 55 to 74 years of age. Table E7.21 provides data on LDL cholesterol levels (mmol/L) from four different subgroups of subjects. Test to compare the three groups simultaneously. Name your test and state clearly your null and alternative hypotheses and choice of test size.

**TABLE E7.21**

| Group | $n$ | $\bar{x}$ | $s$ |
|---|---|---|---|
| 1. Patients with intermittent claudication | 73 | 6.22 | 1.62 |
| 2. Major asymptotic disease cases | 105 | 5.81 | 1.43 |
| 3. Minor asymptotic disease cases | 240 | 5.77 | 1.24 |
| 4. Those with no disease | 1080 | 5.47 | 1.31 |

**7.22** A study was undertaken to clarify the relationship between heart disease and occupational carbon disulfide exposure along with another important factor, elevated diastolic blood pressure (DBP), in a data set obtained from a 10-year prospective follow-up of two cohorts of over 340 male industrial workers in Finland. Carbon disulfide is an industrial solvent that is used all over the world in the production of viscose rayon fibers. Table E7.22 gives the mean and standard deviation (SD) of serum cholesterol (mg/100 mL) among exposed and nonexposed cohorts, by diastolic blood pressure (DBP). Test to compare simultaneously, separately for the exposed and nonexposed groups, the mean serum cholesterol levels at the three DBP levels using one-way ANOVA. Also, compare serum cholesterol levels between exposed and nonexposed cohorts at each level of DBP by using two-sample $t$ tests. Draw your conclusions.

**TABLE E7.22**

| DBP (mmHg) | Exposed | | | Nonexposed | | |
|---|---|---|---|---|---|---|
| | $n$ | Mean | SD | $n$ | Mean | SD |
| <95 | 205 | 220 | 50 | 271 | 221 | 42 |
| 95–100 | 92 | 227 | 57 | 53 | 236 | 46 |
| ≥100 | 20 | 233 | 41 | 10 | 216 | 48 |

**7.23**   When a patient is diagnosed as having cancer of the prostate, an important question in deciding on treatment strategy is whether or not the cancer has spread to the neighboring lymph nodes. The question is so critical in prognosis and treatment that it is customary to operate on the patient (i.e., perform a laparotomy) for the sole purpose of examining

**TABLE E7.23**

| X-ray | Grade | Stage | Age | Acid | Nodes | X-ray | Grade | Stage | Age | Acid | Nodes |
|---|---|---|---|---|---|---|---|---|---|---|---|
| 0 | 1 | 1 | 64 | 40 | 0 | 0 | 0 | 0 | 60 | 78 | 0 |
| 0 | 0 | 1 | 63 | 40 | 0 | 0 | 0 | 0 | 52 | 83 | 0 |
| 1 | 0 | 0 | 65 | 46 | 0 | 0 | 0 | 1 | 67 | 95 | 0 |
| 0 | 1 | 0 | 67 | 47 | 0 | 0 | 0 | 0 | 56 | 98 | 0 |
| 0 | 0 | 0 | 66 | 48 | 0 | 0 | 0 | 1 | 61 | 102 | 0 |
| 0 | 1 | 1 | 65 | 48 | 0 | 0 | 0 | 0 | 64 | 187 | 0 |
| 0 | 0 | 0 | 60 | 49 | 0 | 1 | 0 | 1 | 58 | 48 | 1 |
| 0 | 0 | 0 | 51 | 49 | 0 | 0 | 0 | 1 | 65 | 49 | 1 |
| 0 | 0 | 0 | 66 | 50 | 0 | 1 | 1 | 1 | 57 | 51 | 1 |
| 0 | 0 | 0 | 58 | 50 | 0 | 0 | 1 | 0 | 50 | 56 | 1 |
| 0 | 1 | 0 | 56 | 50 | 0 | 1 | 1 | 0 | 67 | 67 | 1 |
| 0 | 0 | 1 | 61 | 50 | 0 | 0 | 0 | 1 | 67 | 67 | 1 |
| 0 | 1 | 1 | 64 | 50 | 0 | 0 | 1 | 1 | 57 | 67 | 1 |
| 0 | 0 | 0 | 56 | 52 | 0 | 0 | 1 | 1 | 45 | 70 | 1 |
| 0 | 0 | 0 | 67 | 52 | 0 | 0 | 0 | 1 | 46 | 70 | 1 |
| 1 | 0 | 0 | 49 | 55 | 0 | 1 | 0 | 1 | 51 | 72 | 1 |
| 0 | 1 | 1 | 52 | 55 | 0 | 1 | 1 | 1 | 60 | 76 | 1 |
| 0 | 0 | 0 | 68 | 56 | 0 | 1 | 1 | 1 | 56 | 78 | 1 |
| 0 | 1 | 1 | 66 | 59 | 0 | 1 | 1 | 1 | 50 | 81 | 1 |
| 1 | 0 | 0 | 60 | 62 | 0 | 0 | 0 | 0 | 56 | 82 | 1 |
| 0 | 0 | 0 | 61 | 62 | 0 | 0 | 0 | 1 | 63 | 82 | 1 |
| 1 | 1 | 1 | 59 | 63 | 0 | 1 | 1 | 1 | 65 | 84 | 1 |
| 0 | 0 | 0 | 51 | 65 | 0 | 1 | 0 | 1 | 64 | 89 | 1 |
| 0 | 1 | 1 | 53 | 66 | 0 | 0 | 1 | 0 | 59 | 99 | 1 |
| 0 | 0 | 0 | 58 | 71 | 0 | 1 | 1 | 1 | 68 | 126 | 1 |
| 0 | 0 | 0 | 63 | 75 | 0 | 1 | 0 | 0 | 61 | 136 | 1 |
| 0 | 0 | 1 | 53 | 76 | 0 | | | | | | |

*Note:* This is a very long data file; its electronic copy, in a Web-based form, is available from the author upon request.

the nodes and removing tissue samples to examine under the microscope for evidence of cancer. However, certain variables that can be measured without surgery are predictive of the nodal involvement. The purpose of the study presented here was to examine the data for 53 prostate cancer patients receiving surgery, to determine which of five preoperative variables are predictive of nodal involvement. Table E7.23 presents the complete data set. For each of the 53 patients, there are two continuous independent variables: age at diagnosis and level of serum acid phosphatase ($\times 100$; called "acid"); and three binary variables: x-ray reading, pathology reading (grade) of a biopsy of the tumor obtained by needle before surgery, and a rough measure of the size and location of the tumor (stage) obtained by palpation with the fingers via the rectum. In addition, the sixth column presents the findings at surgery—the primary outcome of interest, which is binary, a value of 1 denoting nodal involvement, and a value of 0 denoting no nodal involvement found at surgery. The three binary factors have been investigated previously; this exercise is focused on the effects of the two continuous factors (age and acid phosphatase). Test to compare the group with nodal involvement and the group without, using:

**(a)** The two-sample $t$ test.

**(b)** Wilcoxon's rank-sum test.

# 8

# CORRELATION AND REGRESSION

Methods discussed in Chapters 6 and 7 are tests of significance; they provide analyses of data where a single measurement was made on each element of a sample, and the study may involve one, two, or several samples. If the measurement made is binary or categorical, we are often concerned with a comparison of proportions, the topics of Chapter 6. If the measurement made is continuous, we are often concerned with a comparison of means, the topics of Chapter 7. The main focus of both chapters was the *difference* between populations or subpopulations. In many other studies, however, the purpose of the research is to assess relationships among a set of variables. For example, the sample consists of pairs of values, say a mother's weight and her newborn's weight measured from each of 50 sets of mother and baby, and the research objective is concerned with the association between these weights. Regression analysis is a technique for investigating relationships between variables; it can be used both for assessment of association and for prediction. Consider, for example, an analysis of whether or not a woman's age is predictive of her systolic blood pressure. As another example, the research question could be whether or not a leukemia patient's white blood count is predictive of his survival time. Research designs may be classified as experimental or observational. Regression analyses are applicable to both types; yet the confidence one has in the results of a study can vary with the research type. In most cases, one variable is usually taken to be the response or dependent variable, that is, a variable to be predicted from or explained by other variables. The other variables are called *predictors*, or *explanatory variables* or *independent variables*. The examples above, and others, show a wide range of applications in which the dependent variable is a continuous measurement. Such a variable is often assumed to be normally distributed and a *model* is formulated to express the *mean* of this normal distribution as a function of potential independent variables under investigation. The dependent variable is denoted by $Y$, and the study often involves a number of *risk factors* or *predictor variables*: $X_1, X_2, \ldots, X_k$.

## 8.1  SIMPLE REGRESSION ANALYSIS

In this section we discuss the basic ideas of simple regression analysis when only one predictor or independent variable is available for predicting the response of interest. In the interpretation of the primary parameter of the model, we discuss both scales of measurement, discrete and continuous, even though in practical applications, the independent variable under investigation is often on a continuous scale.

### 8.1.1  Simple Linear Regression Model

Choosing an appropriate model and analytical technique depends on the type of variable under investigation. In a variety of applications, the dependent variable of interest is a continuous variable that we can assume may, after an approropriate transformation, be normally distributed. The *regression model* describes the *mean* of that normally distributed dependent variable $Y$ as a function of the predictor or independent variable $X$:

$$Y_i = \beta_0 + \beta_1 x_i + \varepsilon_i$$

where $Y_i$ is the value of the response or dependent variable from the $i$th pair, $\beta_0$ and $\beta_1$ are the two unknown parameters, $x_i$ is the value of the independent variable from the $i$th pair, and $\varepsilon_i$ is a random error term which is distributed as normal with mean zero and variance $\sigma^2$, so that $(\beta_0 + \beta_1 x_i)$ is the mean $\mu_i$ of $Y_i$. The model above is referred to as the *simple linear regression model*. It is *simple* because it contains only one independent variable. It is *linear* because the independent variable appears only in the first power; if we graph the mean of $Y$ versus $X$, the graph is a *straight line* with *intercept* $\beta_0$ and *slope* $\beta_1$.

### 8.1.2  Scatter Diagram

As mentioned above, and stipulated by the *simple linear regression model*, if we graph the mean of $Y$ versus $X$, the graph is a *straight line*. But that is the line for the means of $Y$; at each level of $X$, the *observed value* of $Y$ may exceed or fall short of its mean. Therefore, when we graph the *observed* value of $Y$ versus $X$, the points do not fall perfectly on any line. This is an important characteristic for *statistical* relationships, as pointed out in Chapter 2. If we let each pair of numbers $(x, y)$ be represented by a dot in a diagram with the $x$'s on the horizontal axis, we have the figure called a *scatter diagram*, as seen in Chapter 2 and again in the next few examples. The scatter diagram is a useful diagnostic tool for checking out the validity of features of the simple linear regression model. For example, if dots fall around a curve, not a straight line, the *linearity* assumption may be violated. In addition, the model stipulates that for each level of $X$, the normal distribution for $Y$ has constant variance not depending on the value of $X$. That would lead to a scatter diagram with dots spreading out around the line evenly accross levels of $X$. In most cases an appropriate transformation, such as taking the logarithm of $Y$ or $X$, would improve and bring the data closer to fitting the model.

### 8.1.3   Meaning of Regression Parameters

The parameters $\beta_0$ and $\beta_1$ are called *regression coefficients*. The parameter $\beta_0$ is the intercept of the regression line. If the scope of the model includes $X = 0$, $\beta_0$ gives the mean of $Y$ when $X = 0$; when the scope of the model does not cover $X = 0$, $\beta_0$ does not have any particular meaning as a separate term in the regression model. As for the meaning of $\beta_1$, our more important parameter, it can be seen as follows. We first consider the case of a binary dependent variable with the conventional coding

$$X_i = \begin{cases} 0 & \text{if the patient is not exposed} \\ 1 & \text{if the patient is exposed} \end{cases}$$

Here the term *exposed* may refer to a risk factor such as smoking or a patient's characteristic such as race (white/nonwhite) or gender (male/female). It can be seen that:

1. For a nonexposed subject (i.e., $X = 0$)

$$\mu_y = \beta_0$$

2. For an exposed subject (i.e., $X = 1$)

$$\mu_y = \beta_0 + \beta_1$$

Hence, $\beta_1$ represents the *increase* (or *decrease*, if $\beta_1$ is negative) in the mean of $Y$ associated with the exposure. Similarly, we have for a continuous covariate $X$ and any value $x$ of $X$:

1. When $X = x$,

$$\mu_y = \beta_0 + \beta_1 x$$

2. Whereas if $X = x + 1$,

$$\mu_y = \beta_0 + \beta_1(x + 1)$$

It can be seen that by taking the difference, $\beta_1$ represents the *increase* (or *decrease*, if $\beta_1$ is negative) in the mean of $Y$ associated with a 1-unit increase in the value of $X$, $X = x + 1$ versus $X = x$. For an *m*-unit increase in the value of $X$, say $X = x + m$ versus $X = x$, the corresponding increase (or decrease) in the mean of $Y$ is $m\beta_1$.

### 8.1.4   Estimation of Parameters

To find *good* estimates of the unknown parameters $\beta_0$ and $\beta_1$, statisticians use a method called *least squares*, which is described as follows. For each subject or pair of values $(X_i, Y_i)$, we consider the deviation from the observed value $Y_i$ to its *expected value*, the mean $\beta_0 + \beta_1 X_i$,

$$Y_i - (\beta_0 + \beta_1 X_i) = \varepsilon_i$$

In particular, the method of least squares requires that we consider the *sum of squared deviations*:

$$S = \sum_{i=1}^{n}(Y_i - \beta_0 - \beta_1 X_i)^2$$

According to the method of least squares, the good estimates of $\beta_0$ and $\beta_1$ are values $b_0$ and $b_1$, respectively, which *minimize* the sum $S$. The results are

$$b_1 = \frac{\sum xy - (\sum x)(\sum y)/n}{\sum x^2 - (\sum x)^2/n}$$

$$b_0 = \bar{y} - b_1 \bar{x}$$

Given the estimates $b_0$ and $b_1$ obtained from the sample, we estimate the mean response by

$$\hat{Y} = b_0 + b_1 X$$

This is our *predicted value* for (the mean of) $Y$ at a given level or value of $X$.

*Example 8.1*   In Table 8.1, the first two columns give the values for the birth weight ($x$, in ounces) and the increase in weight between days 70 and 100 of life, expressed as a percentage of the birth weight ($y$) for 12 infants. We first let

**TABLE 8.1**

| $x$ | $y$ | $x^2$ | $y^2$ | $xy$ |
|---|---|---|---|---|
| 112 | 63 | 12,544 | 3,969 | 7,056 |
| 111 | 66 | 12,321 | 4,356 | 7,326 |
| 107 | 72 | 11,449 | 5,184 | 7,704 |
| 119 | 52 | 14,161 | 2,704 | 6,188 |
| 92 | 75 | 8,464 | 5,625 | 6,900 |
| 80 | 118 | 6,400 | 13,924 | 9,440 |
| 81 | 120 | 6,561 | 14,400 | 9,720 |
| 84 | 114 | 7,056 | 12,996 | 9,576 |
| 118 | 42 | 13,924 | 1,764 | 4,956 |
| 106 | 72 | 11,236 | 5,184 | 7,632 |
| 103 | 90 | 10,609 | 8,100 | 9,270 |
| 94 | 91 | 8,836 | 8,281 | 8,554 |
| 1,207 | 975 | 123,561 | 86,487 | 94,322 |

each pair of numbers $(x, y)$ be represented by a dot in a diagram with the $x$'s on the horizontal axis, and we have the scatter diagram shown in Figure 8.1. The dots do not fall perfectly on a straight line, but scatter around a line, very typical for statistical relationships. However, a straight line seems to fit very well. Generally, the 12 dots go from upper left to lower right, and we have a negative association. As shown in Example 2.8, we obtained a Pearson's correlation coefficient of $r = -0.946$, indicating a very strong negative association.

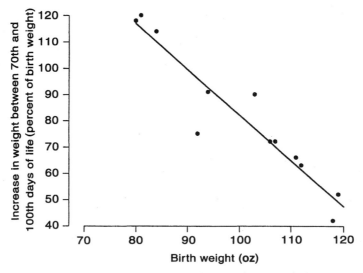

**Figure 8.1**  Graphical display for data in Example 8.1.

Applying the formulas, we obtain estimates for the slope and intercept as follows:

$$b_1 = \frac{94{,}322 - (1207)(975)/12}{123{,}561 - (1207)^2/12}$$

$$= -1.74$$

$$\bar{x} = \frac{1207}{12}$$

$$= 100.6$$

$$\bar{y} = \frac{975}{12}$$

$$= 81.3$$

$$b_0 = 81.3 - (-1.74)(100.6)$$

$$= 256.3$$

For example, if the birth weight is 95 oz, it is predicted that the increase between days 70 and 100 of life would be

$$\hat{y} = 256.3 + (-1.74)(95)$$

$$= 90.1\% \text{ of birth weight}$$

*Note:* An SAS program would include these instructions:

```
DATA;
INPUT WEIGHT GAIN;
DATALINES;
112 63
111 66
. . .
103 90
94 91
;
PROC REG;
MODEL = WEIGHT;
PLOT GAIN*WEIGHT;
```

for which we will get the analysis as well as the scatter diagram.

**Example 8.2**  In Table 8.2 the first two columns give the values for age ($x$, in years) and systolic blood pressure ($y$, in mmHg) for 15 women. We first let

**TABLE 8.2**

| $x$ | $y$ | $x^2$ | $y^2$ | $xy$ |
|---|---|---|---|---|
| 42 | 130 | 1,764 | 16,900 | 5,460 |
| 46 | 115 | 2,116 | 13,225 | 5,290 |
| 42 | 148 | 1,764 | 21,904 | 6,216 |
| 71 | 100 | 5,041 | 10,000 | 7,100 |
| 80 | 156 | 6,400 | 24,336 | 12,480 |
| 74 | 162 | 5,476 | 26,224 | 11,988 |
| 70 | 151 | 4,900 | 22,801 | 10,570 |
| 80 | 156 | 6,400 | 24,336 | 12,480 |
| 85 | 162 | 7,225 | 26,224 | 13,770 |
| 72 | 158 | 5,184 | 24,964 | 11,376 |
| 64 | 155 | 4,096 | 24,025 | 9,920 |
| 81 | 160 | 6,561 | 25,600 | 12,960 |
| 41 | 125 | 1,681 | 15,625 | 5,125 |
| 61 | 150 | 3,721 | 22,500 | 9,150 |
| 75 | 165 | 5,625 | 27,225 | 12,375 |
| 984 | 2,193 | 67,954 | 325,889 | 146,260 |

each pair of numbers $(x, y)$ be represented by a dot in a diagram with the $x$'s on the horizontal axis; we have the scatter diagram shown in Figure 8.2. Again the dots do not fall perfectly on a straight line, but scatter around a line, very typical for statistical relationships. In this example, a straight line still seems to fit, too; however, the dots spread more and cluster less around the line, indicating a weaker association. Generally, the 15 dots go from lower left to upper right, and we have a positive association. As shown in Example 2.9, we obtained a Pearson's correlation coefficient of $r = 0.566$, indicating a moderately positive association, confirming the observation from the graph.

Applying the formulas, we obtain estimates for the slope and intercept as follows:

$$b_1 = \frac{146,260 - (984)(2193)/15}{67,954 - (984)^2/15}$$

$$= 0.71$$

$$\bar{x} = \frac{984}{15}$$

$$= 65.6$$

$$\bar{y} = \frac{2193}{15}$$

$$= 146.2$$

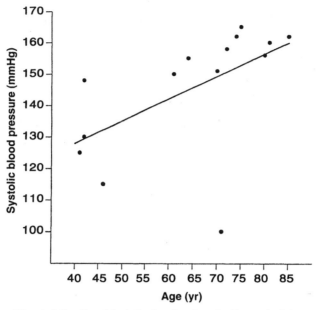

**Figure 8.2**  Graphical display for data in Example 8.2.

$$b_0 = 146.2 - (0.71)(65.6)$$
$$= 99.6$$

For example, for a 50-year-old woman, it is predicted that her systolic blood pressure would be about

$$\hat{y} = 99.6 + (0.71)(50)$$
$$= 135 \text{ mmHg}$$

### 8.1.5  Testing for Independence

In addition to being able to *predict* the (mean) response at various levels of the independent variable, regression data can also be used to test for the independence between the two variables under investigation. Such a statistical test can be viewed or approached in two ways: through the coefficient of correlation or through the slope.

1. The *correlation coefficient r* measures the strength of the relationship between two variables. It is an estimate of an unknown population correlation coefficient $\rho$ (rho), the same way the sample mean $\bar{x}$ is used as an estimate of an unknown population mean $\mu$. We are usually interested in

knowing if we may conclude that $\rho \neq 0$, that is, that the two variables under investigation are really correlated. The test statistic is

$$t = r\sqrt{\frac{n-2}{1-r^2}}$$

The procedure is often performed as two-sided, that is,

$$H_A: \rho \neq 0$$

and it is a $t$ test with $n - 2$ degrees of freedom, the same $t$ test as used in the comparisons of population means in Chapter 7.

2. The role of the slope $\beta_1$ can be seen as follows. Since the regression model describes the mean of the dependent variable $Y$ as a function of the predictor or independent variable $X$,

$$\mu_y = \beta_0 + \beta_1 x$$

$Y$ and $X$ would be independent if $\beta_1 = 0$. The test for

$$H_0: \beta_1 = 0$$

can be performed similar to the method for one-sample problems in Sections 6.1 and 7.1. In that process the observed/estimated value $b_1$ is converted to a standard unit: the number of standard errors away from the hypothesized value of zero. The formula for a standard error of $b_1$ is rather complicated; fortunately, the resulting test is *identical* to the $t$ test above. Whenever needed, for example in the computation of confidence intervals for the slope, we can always obtain the numerical value of its standard error from computer output.

When the $t$ test for independence above is significant, the value of $X$ has a real effect on the distribution of $Y$. To be more precise, the square of the correlation coefficient, $r^2$, represents the proportion of the variability of $Y$ accounted for by $X$. For example, $r^2 = 0.25$ indicates that the total variation in $Y$ is reduced by 25% by the use of information about $X$. In other words, if we have a sample of the same size, with all $n$ subjects having the same $X$ value, the variation in $Y$ (say, measured by its variance) is 25% less than the variation of $Y$ in the current sample. It is interesting to note that $r = 0.5$ would give an impression of greater association between $X$ and $Y$, but a 25% reduction in variation would not. The parameter $r^2$ is called the *coefficient of determination*, an index with a clearer operational interpretation than the coefficient of correlation $r$.

***Example 8.3*** For the birth-weight problem of Examples 2.8 and 8.1, we have

$$n = 12$$
$$r = -0.946$$

leading to

$$t = (-0.946)\sqrt{\frac{10}{1 - (-0.946)^2}}$$
$$= -9.23$$

At $\alpha = 0.05$ and df $= 10$, the tabulated $t$ coefficient is 2.228, indicating that the null hypothesis of independence should be rejected ($t = -9.23 < -2.228$). In this case, the weight on day 70 ($X$) would account for

$$r^2 = 0.895$$

or 89.5% of the variation in growth rates.

***Example 8.4*** For the blood pressure problem of Examples 2.9 and 8.2, we have

$$n = 15$$
$$r = 0.566$$

leading to

$$t = (0.566)\sqrt{\frac{13}{1 - (0.566)^2}}$$
$$= 2.475$$

At $\alpha = 0.05$ and df $= 13$, the tabulated $t$ value is 2.16. Since

$$t > 2.16$$

we have to conclude that the null hypothesis of independence should be rejected; that is, the relationship between age and systolic blood pressure is real. However, a woman's age ($X$) would account for only

$$r^2 = 0.32$$

or 32% of the variation among systolic blood pressures.

### 8.1.6   Analysis-of-Variance Approach

The variation in $Y$ is conventionally measured in terms of the deviations $(Y_i - \bar{Y})$; the total variation, denoted by SST, is the sum of squared deviations:

$$\text{SST} = \sum (Y_i - \bar{Y})^2$$

For example, SST $= 0$ when all observations are the same. SST is the numerator of the sample variance of $Y$. The larger the SST, the greater the variation among $Y$ values.

When we use the regression approach, the variation in $Y$ is decomposed into two components:

$$Y_i - \bar{Y} = (Y_i - \hat{Y}_i) + (\hat{Y}_i - \bar{Y})$$

1. The first term reflects the *variation around the regression line*; the part that cannot be explained by the regression itself with the sum of squared deviations:

$$\text{SSE} = \sum (Y_i - \hat{Y}_i)^2$$

called the *error sum of squares*.

2. The difference between the two sums of squares,

$$\text{SSR} = \text{SST} - \text{SSE}$$

$$= \sum (\hat{Y}_i - \bar{Y})^2$$

is called the *regression sum of squares*. SSR may be considered a measure of the variation in $Y$ associated with the regression line. In fact, we can express the coefficient of determination as

$$r^2 = \frac{\text{SSR}}{\text{SST}}$$

Corresponding to the partitioning of the total sum of squares SST, there is partitioning of the associated degrees of freedom (df). We have $n - 1$ degrees of freedom associated with SST, the denominator of the sample variance of $Y$; SSR has 1 degree of freedom representing the slope, the remaining $n - 2$ are associated with SSE. These results lead to the usual presentation of regression analysis by most computer programs:

1. The *error mean square*,

$$\text{MSE} = \frac{\text{SSE}}{n - 2}$$

**TABLE 8.3**

| Source of Variation | SS | df | MS | F Statistic | p Value |
|---|---|---|---|---|---|
| Regression | SSR | 1 | MSR $=$ SSR$/1$ | $F =$ MSR$/$MSE | $p$ |
| Error | SSE | $n - 2$ | MSE $=$ SSE$/(n - 2)$ | | |
| Total | SST | $n - 1$ | | | |

serves as an estimate of the constant variance $\sigma^2$ as stipulated by the regression *model*.

2. The breakdowns of the total sum of squares and its associated degree of freedom are *displayed* in the form of an *analysis-of-variance* (ANOVA) *table* (Table 8.3). The test statistic $F$ for the analysis-of-variance approach above compares MSR and MSE. A value near 1 supports the null hypothesis of independence. In fact, we have

$$F = t^2$$

where $t$ is the test statistic for testing whether or not $\beta_1 = 0$; the $F$ test is equivalent to the two-sided $t$ test when refered to the $F$ table in Appendix E with $(1, n - 2)$ degrees of freedom.

***Example 8.5*** For the birth-weight problem of Examples 8.1 and 8.3, we have the analysis of variance shown in Table 8.4.

***Example 8.6*** For the blood pressure problem of Examples 8.2 and 8.4, we have the analysis of variance shown in Table 8.5.

**TABLE 8.4**

| Source of Variation | SS | df | MS | F Statistic | p Value |
|---|---|---|---|---|---|
| Regression | 6508.43 | 1 | 6508.43 | 85.657 | 0.0001 |
| Error | 759.82 | 10 | 75.98 | | |
| Total | 7268.25 | 11 | | | |

**TABLE 8.5**

| Source of Variation | SS | df | MS | F Statistic | p Value |
|---|---|---|---|---|---|
| Regression | 1691.20 | 1 | 1691.20 | 6.071 | 0.0285 |
| Error | 3621.20 | 13 | 278.55 | | |
| Total | 5312.40 | 14 | | | |

## 8.2  MULTIPLE REGRESSION ANALYSIS

The effect of some factor on a dependent or response variable may be influenced by the presence of other factors because of redundancies or effect modifications (i.e., interactions). Therefore, to provide a more comprehensive analysis, it may be desirable to consider a large number of factors and sort out which ones are most closely related to the dependent variable. In this section we discuss a multivariate method for this type of risk determination. This method, which is multiple regression analysis, involves a linear combination of the explanatory or independent variables, also called *covariates*; the variables must be quantitative with particular numerical values for each subject in the sample. A covariate or independent variable, such as a patient characteristic, may be dichotomous, polytomous, or continuous (categorical factors will be represented by dummy variables). Examples of *dichotomous covariates* are gender and presence/absence of certain comorbidity. Polytomous covariates include race and different grades of symptoms; these can be covered by the use of *dummy variables*. *Continuous covariates* include patient age and blood pressure. In many cases, data transformations (e.g., taking the logarithm) may be needed to satisfy the linearity assumption.

### 8.2.1  Regression Model with Several Independent Variables

Suppose that we want to consider $k$ independent variables simultaneously. The simple linear model of Section 8.1 can easily be generalized and expressed as

$$Y_i = \beta_0 + \sum_{j=1}^{k} \beta_j x_{ji} + \varepsilon_i$$

where $Y_i$ is the value of the response or dependent variable from the $i$th subject; $\beta_0, \beta_1, \ldots, \beta_k$ are the $k + 1$ unknown parameters ($\beta_0$ is the intercept and the $\beta_i$'s are the slopes, one for each independent variables); $X_{ij}$ is the value of the $j$th independent variable ($j = 1$ to $k$) from the $i$th subject ($i = 1$ to $n$); and $\varepsilon_i$ is a random error term which is distributed as normal with mean zero and variance $\sigma^2$, so that the mean of $Y_i$ is

$$\mu_i = \beta_0 + \sum_{j=1}^{k} \beta_j x_{ji}$$

The model above is referred to as the *multiple linear regression model*. It is *multiple* because it contains several independent variables. It is still *linear* because the independent variables appear only in the first power; this feature is rather difficult to check because we do not have a scatter diagram to rely on as in the case of simple linear regression. In addition, the model can be modified to include higher powers of independent variables as well as their various products, as we'll see subsequently.

## 8.2.2   Meaning of Regression Parameters

Similar to the univariate case, $\beta_i$ represents one of the following:

1. The *increase* (or *decrease*, if $\beta_i$ is negative) in the mean of $Y$ associated with the exposure if $X_i$ is binary (exposed $X_i = 1$ versus unexposed $X_i = 0$), *assuming* that other independent variables are fixed; or
2. The *increase* (or *decrease*, if $\beta_i$ is negative) in the mean of $Y$ associated with a 1-unit increase in the value of $X_i$, $X_i = x + 1$ versus $X_i = x$. For an *m*-unit increase in the value of $X_i$, say $X_i = x + m$ versus $X_i = x$, the corresponding increase (or decrease) in the mean of $Y$ is $m\beta_i$ *assuming* that other independent variables are fixed. In other words, $\beta_i$ represents the *additional contribution* of $X_i$ in the explanation of variation among $y$ values. Of course, before such analyses are done, the problem and the data have to be examined carefully. If some of the variables are highly correlated, one or fewer of the correlated factors are likely to be as good predictors as all of them; information from similar studies also has to be incorporated so as to drop some of these correlated explanatory variables.

## 8.2.3   Effect Modifications

Consider a multiple regression model involving *two* independent variables:

$$Y_i = \beta_0 + \beta_1 x_{1i} + \beta_2 x_{2i} + \beta_3 x_{1i} x_{2i} + \varepsilon_i$$

It can be seen that the meaning of $\beta_1$ and $\beta_2$ here is not the same as that given earlier because of the cross-product term $\beta_3 x_1 x_2$. Suppose, for simplicity, that both $X_1$ and $X_2$ are binary; then:

1. For $X_2 = 1$ or exposed, we have

$$\mu_y = \begin{cases} \beta_0 + \beta_1 + \beta_3 & \text{if exposed to } X_1 \\ \beta_0 & \text{if not exposed to } X_1 \end{cases}$$

so that the increase (or decrease) in the mean of $Y$ due to an exposure to $X_1$ is $\beta_1 + \beta_3$, whereas

2. For $X_2 = 0$ or not exposed, we have

$$\mu_y = \begin{cases} \beta_0 + \beta_1 & \text{if exposed to } X_1 \\ \beta_0 & \text{if not exposed to } X_1 \end{cases}$$

so that the increase (or decrease) in the mean of $Y$ due to an exposure to $X_1$ is $\beta_1$.

In other words, the effect of $X_1$ depends on the level (presence or absence) of $X_2$, and vice versa. This phenomenon is called *effect modification* (i.e., one factor modifies the effect of the other). The cross-product term $x_1 x_2$ is called an *interaction term*. Use of these products will help in the investigation of possible effect modifications. If $\beta_3 = 0$, the effect of two factors acting together (represented by $\beta_1 + \beta_2$), is equal to the combined effects of two factors acting separately. If $\beta_3 > 0$, we have a synergistic interaction; if $\beta_3 < 0$, we have an antagonistic interaction.

### 8.2.4   Polynomial Regression

Consider the multiple regression model involving *one* independent variable:

$$Y_i = \beta_0 + \beta_1 x_i + \beta_2 x_i^2 + \varepsilon_i$$

or it can be written as a multiple model:

$$Y_i = \beta_0 + \beta_1 x_{1i} + \beta_2 x_{2i}^2 + \varepsilon_i$$

with $X_1 = X$ and $X_2 = X^2$, where $X$ is a continuous independent variable. The meaning of $\beta_1$ here is not the same as that given earlier because of the quadratic term $\beta_2 x_i^2$. We have, for example,

$$\mu_y = \begin{cases} \beta_0 + \beta_1 x + \beta_2 x^2 & \text{when } X = x \\ \beta_0 + \beta_1(x+1) + \beta_2(x+1)^2 & \text{when } X = x + 1 \end{cases}$$

so that the difference is

$$\beta_1 + \beta_2(2x + 1)$$

a function of $x$.

Polynomial models with an independent variable present in higher powers than the second are not often used. The second-order or quadratic model has two basic type of uses: (1) when the true relationship is a second-degree polynomial or when the true relationship is unknown but the second-degree polynomial provides a better fit than a linear one, but (2) more often, a quadratic model is fitted for the purpose of establishing the linearity. The key item to look for is whether $\beta_2 = 0$. The use of polynomial models, however, is not without drawbacks. The most potential drawback is that $X$ and $X^2$ are strongly related, especially if $X$ is restricted to a narrow range; in this case the standard errors are often very large.

### 8.2.5   Estimation of Parameters

To find *good* estimates of the $k + 1$ unknown parameters $\beta_0$ and $\beta_i$'s, statisticians use the same method of *least squares* described earlier. For each subject with data values $(Y_i, X_i$'s), we consider the deviation from the observed value

$Y_i$ to its *expected value*:

$$\mu_y = \beta_0 + \sum_{j=1}^{k} \beta_j x_{ji}$$

In particular, the method of least squares requires that we consider the *sum of squared deviations*:

$$S = \sum_{i=1}^{n} \left( Y_i - \beta_0 - \sum_{j=1}^{k} \beta_j x_{ji} \right)^2$$

According to the method of least squares, the good estimates of $\beta_0$ and $\beta_i$'s are values $b_0$ and $b_i$'s, respectively, which *minimize* the sum $S$. The method is the same, but the results are much more difficult to obtain; fortunately, these results are provided by most standard computer programs, such as Excel and SAS. In addition, computer output also provides standard errors for all estimates of regression coefficients.

### 8.2.6 Analysis-of-Variance Approach

The total sum of squares,

$$\text{SST} = \sum (Y_i - \bar{Y})^2$$

and its associated degree of freedom $(n - 1)$ are defined and partitioned the same as in the case of simple linear regression. The results are *displayed* in the form of an *analysis-of-variance* (ANOVA) *table* (Table 8.6) of the same form, where $k$ is the number of independent variables. In addition:

1. The coefficient of multiple determination is defined as

$$R^2 = \frac{\text{SSR}}{\text{SST}}$$

It measures the proportionate reduction of total variation in $Y$ associated with the use of the set of independent varables. As for $r^2$ of the simple

**TABLE 8.6**

| Source of Variation | SS | df | MS | F Statistic | p Value |
|---|---|---|---|---|---|
| Regression | SSR | $k$ | $\text{MSR} = \text{SSR}/k$ | $F = \text{MSR}/\text{MSE}$ | $p$ |
| Error | SSE | $n - k - 1$ | $\text{MSE} = \text{SSE}/(n - k - 1)$ | | |
| Total | SST | $n - 1$ | | | |

linear regression, we have

$$0 \leq R^2 \leq 1$$

and $R^2$ only assumes the value 0 when all $\beta_i = 0$.

2. The *error mean square*,

$$\text{MSE} = \frac{\text{SSE}}{n-2}$$

serves as an estimate of the constant variance $\sigma^2$ as stipulated by the regression model.

### 8.2.7   Testing Hypotheses in Multiple Linear Regression

Once we have fit a multiple regression model and obtained estimates for the various parameters of interest, we want to answer questions about the contributions of various factors to the prediction of the binary response variable. There are three types of such questions:

1. *Overall test.* Taken collectively, does the entire set of explanatory or independent variables contribute significantly to the prediction of the response (or the explanation of variation among responses)?
2. *Test for the value of a single factor.* Does the addition of one particular variable of interest add significantly to the prediction of response over and above that achieved by other independent variables?
3. *Test for contribution of a group of variables.* Does the addition of a group of variables add significantly to the prediction of response over and above that achieved by other independent variables?

***Overall Regression Tests***   We now consider the first question stated above concerning an overall test for a model containg $k$ factors. The null hypothesis for this test may be stated as: "All $k$ independent variables *considered together* do not explain the variation in the responses." In other words,

$$H_0: \beta_1 = \beta_2 = \cdots = \beta_k = 0$$

This *global* null hypothesis can be tested using the $F$ statistic in Table 8.6:

$$F = \frac{\text{MSR}}{\text{MSE}}$$

an $F$ test at $(k, n - k - 1)$ degrees of freedom.

***Tests for a Single Variable***   Let us assume that we now wish to test whether the addition of one particular independent variable of interest adds significantly to the prediction of the response over and above that achieved by other factors already present in the model. The null hypothesis for this test may be stated as: "Factor $X_i$ does not have any value added to the prediction of the response *given that other factors are already included in the model*." In other words,

$$H_0: \beta_i = 0$$

To test such a null hypothesis, one can use

$$t_i = \frac{\hat{\beta}_i}{SE(\hat{\beta}_i)}$$

in a $t$ test with $n - k - 1$ degrees of freedom, where $\hat{\beta}_i$ is the corresponding estimated regression coefficient and $SE(\hat{\beta}_i)$ is the estimate of the standard error of $\hat{\beta}_i$, both of which are printed by standard computer packaged programs.

***Example 8.7***   Ultrasounds were taken at the time of liver transplant and again five to ten years later to determine the systolic pressure of the hepatic artery. Results for 21 transplants for 21 children are shown in Table 8.7; also available are gender (1 = male, 2 = female) and age at the second measurement.

**TABLE 8.7**

| Child | 5–10 Years Later | At Transplant | Gender | Age |
|-------|------------------|---------------|--------|-----|
| 1 | 46 | 35 | 2 | 16 |
| 2 | 40 | 40 | 2 | 19 |
| 3 | 50 | 58 | 2 | 19 |
| 4 | 50 | 71 | 1 | 23 |
| 5 | 41 | 33 | 1 | 16 |
| 6 | 70 | 79 | 1 | 23 |
| 7 | 35 | 20 | 1 | 13 |
| 8 | 40 | 19 | 1 | 19 |
| 9 | 56 | 56 | 1 | 11 |
| 10 | 30 | 26 | 2 | 14 |
| 11 | 30 | 44 | 1 | 15 |
| 12 | 60 | 90 | 2 | 12 |
| 13 | 43 | 43 | 2 | 15 |
| 14 | 45 | 42 | 1 | 14 |
| 15 | 40 | 55 | 1 | 14 |
| 16 | 50 | 60 | 2 | 17 |
| 17 | 66 | 62 | 2 | 21 |
| 18 | 45 | 26 | 2 | 21 |
| 19 | 40 | 60 | 1 | 11 |
| 20 | 35 | 27 | 1 | 9 |
| 21 | 25 | 31 | 1 | 9 |

**TABLE 8.8**

| Source of Variation | SS | df | MS | F Statistic | p Value |
|---|---|---|---|---|---|
| Regression | 1810.93 | 3 | 603.64 | 12.158 | 0.0002 |
| Error | 844.02 | 17 | 49.65 | | |
| Total | 2654.95 | 20 | | | |

Using the second measurement of the systolic pressure of the hepatic artery as our dependent variable, the resulting ANOVA table is shown in Table 8.8. The result of the overall $F$ test ($p = 0.0002$) indicates that taken collectively, the three independent variables (systolic pressure at transplant, gender, and age) contribute significantly to the prediction of the dependent variable. In addition, we have the results shown in Table 8.9. The effects of pressure at transplant and age are significant at the 5% level, whereas the effect of gender is not ($p = 0.5982$).

*Note:* An SAS program would include these instructions:

```
DATA;
INPUT POST PRE SEX AGE;
DATALINES;
46 35 2 16
40 40 2 19
. . .
35 27 1 9
25 31 1 9
;
PROC REG;
MODEL POST = PRE SEX AGE;
```

which gives us all of the results above.

***Example 8.8*** There have been times the city of London experienced periods of dense fog. Table 8.10 shows such data for a very severe 15-day period which included the number of deaths in each day ($y$), the mean atmospheric smoke ($x_1$, in mg/m$^3$), and the mean atmospheric sulfur dioxide content ($x_2$, in ppm).

**TABLE 8.9**

| Variable | Coefficient | Standard Error | t Statistic | p Value |
|---|---|---|---|---|
| Pressure at transplant | 0.381 | 0.082 | 4.631 | 0.0002 |
| Gender | 1.740 | 3.241 | 0.537 | 0.5982 |
| Age | 0.935 | 0.395 | 2.366 | 0.0301 |

**TABLE 8.10**

| Number of Deaths | Smoke | Sulfur Dioxide |
|---|---|---|
| 112 | 0.30 | 0.09 |
| 140 | 0.49 | 0.16 |
| 143 | 0.61 | 0.22 |
| 120 | 0.49 | 0.14 |
| 196 | 2.64 | 0.75 |
| 294 | 3.45 | 0.86 |
| 513 | 4.46 | 1.34 |
| 518 | 4.46 | 1.34 |
| 430 | 1.22 | 0.47 |
| 274 | 1.22 | 0.47 |
| 255 | 0.32 | 0.22 |
| 236 | 0.29 | 0.23 |
| 256 | 0.50 | 0.26 |
| 222 | 0.32 | 0.16 |
| 213 | 0.32 | 0.16 |

**TABLE 8.11**

| Source of Variation | SS | df | MS | F Statistic | p Value |
|---|---|---|---|---|---|
| Regression | 205,097.52 | 2 | 102,548.76 | 36.566 | 0.0001 |
| Error | 33,654.20 | 12 | 2,804.52 | | |
| Total | 238,751.73 | 14 | | | |

Using the number of deaths in each day as our dependent variable, Table 8.11 is the resulting ANOVA table. The result of the overall $F$ test ($p = 0.0001$) indicates that taken collectively, the two independent variables contribute significantly to the prediction of the dependent variable. In addition, we have the results shown in Table 8.12. The effects of both factors, the mean atmospheric smoke and the mean atmospheric sulfur dioxide content, are significant even at the 1% level (both $p < 0.001$).

***Contribution of a Group of Variables*** This testing procedure addresses the more general problem of assessing the additional contribution of two or more factors to prediction of the response over and above that made by other variables already in the regression model. In other words, the null hypothesis is of the form

$$H_0: \beta_1 = \beta_2 = \cdots = \beta_m = 0$$

To test such a null hypothesis, one can fit two regression models, one with all $X$'s included to obtain the regression sum of squares ($SSR_1$) and one with all

**TABLE 8.12**

| Variable | Coefficient | Standard Error | $t$ Statistic | $p$ Value |
|----------|-------------|----------------|---------------|-----------|
| Smoke    | −220.324    | 58.143         | −3.789        | 0.0026    |
| Sulfur   | 1051.816    | 212.596        | 4.947         | 0.0003    |

other $X$'s with the $X$'s under investigation deleted to obtain the regression sum of squares ($SSR_2$). Define the mean square due to $H_0$ as

$$\text{MSR} = \frac{\text{SSR}_1 - \text{SSR}_2}{m}$$

Then $H_0$ can be tested using

$$F = \frac{\text{MSR}}{\text{MSE}}$$

an $F$ test at $(m, n - k - 1)$ degrees of freedom. This *multiple contribution procedure* is very useful for assessing the importance of potential explanatory variables. In particular, it is often used to test whether a similar group of variables, such as *demographic characteristics*, is important for prediction of the response; these variables have some trait in common. Another application would be a collection of powers *and/or* product terms (referred to as *interaction variables*). It is often of interest to assess the interaction effects collectively before trying to consider individual interaction terms in a model as suggested previously. In fact, such use reduces the total number of tests to be performed, and this, in turn, helps to provide better control of overall type I error rates, which may be inflated due to multiple testing.

*Example 8.9*    Refer to the data on liver transplants of Example 8.7 consisting of three independent variables: hepatic systolic pressure at transplant time (called pressure$_1$), age (at the second measurement time), and gender of the child. Let us consider all five quadratic terms and products of these three original factors ($x_1 = \text{pressure}_1^2$, $x_2 = \text{age}^2$, $x_3 = \text{pressure}_1 \times \text{age}$, $x_4 = \text{pressure}_1 \times \text{gender}$, and $x_5 = \text{age} \times \text{gender}$). Using the second measurement of the systolic pressure of the hepatic artery as our dependent variable and fitting the multiple regression model with all eight independent variables (three original plus five newly defined terms), we have

$$\text{SSR} = 1944.70 \text{ with 8 df}$$

$$\text{SSE} = 710.25 \text{ with 12 df}, \quad \text{or}$$

$$\text{MSE} = 19.19$$

as compared to the results from Example 8.7:

$$SSR = 1810.93 \text{ with 3 df}$$

The significance of the additional contribution of the five new factors, considered together, is judged using the $F$ statistic:

$$F = \frac{(1944.70 - 1810.93)/5}{19.19}$$
$$= 0.80 \text{ at } (5, 12) \text{ df}$$

In other words, all five quadratic and product terms considered together do not contribute significantly to the prediction/explanation of the second measurement of the systolic pressure of the hepatic artery; the model with three original factors is adequate.

***Stepwise Regression***   In many applications, our major interest is to identify important risk factors. In other words, we wish to identify from many available factors a small subset of factors that relate significantly to the outcome (e.g., the disease under investigation). In that identification process, of course, we wish to avoid a large type I (false positive) error. In a regression analysis, a type I error corresponds to including a predictor that has no real relationship to the outcome; such an inclusion can greatly confuse the interpretation of the regression results. In a standard multiple regression analysis, this goal can be achieved by using a strategy that adds into or removes from a regression model one factor at a time according to a certain order of relative importance. Therefore, the two important steps are as follows:

1. Specify a criterion or criteria for selecting a model.
2. Specify a strategy for applying the criterion or criteria chosen.

*Strategies*   This is concerned with specifying the strategy for selecting variables. Traditionally, such a strategy is concerned with whether and which particular variable should be added to a model or whether any variable should be deleted from a model at a particular stage of the process. As computers became more accessible and more powerfull, these practices became more popular.

- **Forward selection procedure**
  1. Fit a simple linear regression model to each factor, one at a time.
  2. Select the most important factor according to certain predetermined criterion.
  3. Test for the significance of the factor selected in step 2 and determine, according to a certain predetermined criterion, whether or not to add this factor to the model.

4. Repeat steps 2 and 3 for those variables not yet in the model. At any subsequent step, if none meets the criterion in step 3, no more variables are included in the model and the process is terminated.

• **Backward elimination procedure**

1. Fit the multiple regression model containing all available independent variables:

2. Select the least important factor according to a certain predetermined criterion; this is done by considering one factor at a time and treat it as though it were the last variable to enter.

3. Test for the significance of the factor selected in step 2 and determine, according to a certain predetermined criterion, whether or not to delete this factor from the model.

4. Repeat steps 2 and 3 for those variables still in the model. At any subsequent step, if none meets the criterion in step 3, no more variables are removed in the model and the process is terminated.

• **Stepwise regression procedure.** Stepwise regression is a modified version of forward regression that permits reexamination, at every step, of the variables incorporated in the model in previous steps. A variable entered at an early stage may become superfluous at a later stage because of its relationship with other variables now in the model; the information it provides becomes redundant. That variable may be removed, if meeting the elimination criterion, and the model is refitted with the remaining variables, and the forward process goes on. The entire process, one step forward followed by one step backward, continues until no more variables can be added or removed.

*Criteria* For the first step of the forward selection procedure, decisions are based on individual score test results [$t$ test, $(n-2)$ df]. In subsequent steps, both forward and backward, the decision is made as follows. Suppose that there are $r$ independent variables already in the model and a decision is needed in the forward selection process. Two regression models are now fitted, one with all $r$ current $X$'s included to obtain the regression sum of squares ($SSR_1$) and one with all $r$ $X$'s plus the $X$ under investigation to obtain the regression sum of squares ($SSR_2$). Define the mean square due to addition (or elimination) as

$$MSR = \frac{SSR_2 - SSR_1}{1}$$

Then the decision concerning the candidate variable is based on

$$F = \frac{MSR}{MSE}$$

an $F$ test at $(1, n - r - 1)$ degrees of freedom.

The following example is used to illustrate the process; however, the process is most useful when we have a large number of independent variables.

***Example 8.10***   Refer to the data on liver transplants of Example 8.7 consisting of three independent variables: hepatic systolic pressure at transplant time (called pressure$_1$), age (at the second measurement time), and gender of the child. The results for individual terms were shown in Example 8.7; these indicate that the pressure at transplant time (pressure$_1$) is the most significant variable.

*Step 1:* Variable PRESSURE1 is entered. The model with only pressure at transplant time yields

$$\text{SSR} = 1460.47 \text{ with 1 df}$$

Analysis of variables not in the model: With the addition of age, we have

$$\text{SSR} = 1796.61 \text{ with 2 df}$$

$$\text{MSE} = 47.69 \ (\text{df} = 18)$$

leading to an *F* statistic of 7.05 ($p = 0.0161$). Variable AGE is entered next; the remaining variable (gender) does not meet the criterion of 0.1 (or 10%) level to enter the model.

*Step 2:* Variable AGE is entered. The final model consists of two independent variables with the results shown in Table 8.13.

*Note:* The SAS program of Example 8.7 should be changed to

```
PROC REG;
MODEL POST = PRE SEX AGE/SELECTION = STEPWISE;
```

to specify the stepwise process.

## 8.3   NOTES ON COMPUTATIONS

Samples of SAS program instructions were provided for all procedures at the end of Examples 8.1, 8.7, and 8.10. Regression analyses can also be implemented easily using Microsoft's Excel; however, you need *Data Analysis,* an

**TABLE 8.13**

| Factor | Coefficient | Standard Error | *F* Statistic | *p* Value |
|---|---|---|---|---|
| Pressure$_1$ | 0.3833 | 0.0806 | 22.60 | 0.0002 |
| Age | 0.9918 | 0.3735 | 7.05 | 0.0161 |

Excel add-in option that is available from the Excel installation CD. After installation, it is listed in your *Tools* menu. The process is rather simple: Click (1) the *Tools*, then (2) *Data Analysis.* Among the functions available, choose REGRESSION. A box appears; use the cursor to fill in the ranges of $Y$ and $X$'s. The results include all items mentioned in this chapter, plus confidence intervals for regression coefficients.

## EXERCISES

**8.1** Trace metals in drinking water affect the flavor of the water, and unusually high concentration can pose a health hazard. Table E8.1 shows trace-metal concentrations (zinc, in mg/L) for both surface water and bottom water at six different river locations. Our aim is to see if surface water concentration $(x)$ is predictive of bottom water concentration $(y)$.

**TABLE E8.1**

| Location | Bottom | Surface |
|---|---|---|
| 1 | 0.430 | 0.415 |
| 2 | 0.266 | 0.238 |
| 3 | 0.567 | 0.390 |
| 4 | 0.531 | 0.410 |
| 5 | 0.707 | 0.605 |
| 6 | 0.716 | 0.609 |

(a) Draw a scatter diagram to show a possible association between the concentrations and check to see if a linear model is justified.

(b) Estimate the regression parameters, the bottom water concentration for location with a surface water concentration of 0.5 mg/L, and draw the regression line on the same graph with the scatter diagram.

(c) Test to see if the two concentrations are independent; state your hypotheses and choice of test size.

(d) Calculate the coefficient of determination and provide your interpretation.

**8.2** In a study of saliva cotinine, seven subjects, all of whom had abstained from smoking for a week, were asked to smoke a single cigarette. The cotinine levels at 12 and 24 hours after smoking are provided in Table E8.2.

(a) Draw a scatter diagram to show a possible association between the cotinine levels (24-hour measurement as the dependent variable) and check to see if a linear model is justified.

**TABLE E8.2**

| | Cotinine Level (mmol/L) | |
|---|---|---|
| Subject | After 12 hours | After 24 hours |
| 1 | 73 | 24 |
| 2 | 58 | 27 |
| 3 | 67 | 49 |
| 4 | 93 | 59 |
| 5 | 33 | 0 |
| 6 | 18 | 11 |
| 7 | 147 | 43 |

**(b)** Estimate the regression parameters, the 24-hour measurement for a subject with a 12-hour cotinine level of 60 mmol/L, and draw the regression line on the same graph with the scatter diagram.

**(c)** Test to see if the two cotinine levels are independent; state your hypotheses and choice of test size.

**(d)** Calculate the coefficient of determination and provide your interpretation.

**8.3** Table E8.3 gives the net food supply ($x$, number of calories per person per day) and the infant mortality rate ($y$, number of infant deaths per 1000 live births) for certain selected countries before World War II.

**TABLE E8.3**

| Country | $x$ | $y$ | Country | $x$ | $y$ |
|---|---|---|---|---|---|
| Argentina | 2730 | 98.8 | Iceland | 3160 | 42.4 |
| Australia | 3300 | 39.1 | India | 1970 | 161.6 |
| Austria | 2990 | 87.4 | Ireland | 3390 | 69.6 |
| Belgium | 3000 | 83.1 | Italy | 2510 | 102.7 |
| Burma | 1080 | 202.1 | Japan | 2180 | 60.6 |
| Canada | 3070 | 67.4 | Netherlands | 3010 | 37.4 |
| Chile | 2240 | 240.8 | New Zealand | 3260 | 32.2 |
| Cuba | 2610 | 116.8 | Sweden | 3210 | 43.3 |
| Egypt | 2450 | 162.9 | U.K. | 3100 | 55.3 |
| France | 2880 | 66.1 | U.S. | 3150 | 53.2 |
| Germany | 2960 | 63.3 | Uruguay | 2380 | 94.1 |

**(a)** Draw a scatter diagram to show a possible association between the infant mortality rate (used as the dependent variable) and the net food supply and check to see if a linear model is justified.

**(b)** Estimate the regression parameters, the infant mortality rate for a country with a net food supply of 2900 calories per person per day, and draw the regression line on the same graph with the scatter diagram.

(c) Test to see if the two factors are independent; state your hypotheses and choice of test size.

(d) Calculate the coefficient of determination and provide your interpretation.

**8.4**  Refer to the data in Exercise 8.3, but in the context of a multiple regression problem with two independent variables: the net food supply $(x_1 = x)$ and its square $(x_2 = x^2)$.

(a) Taken collectively, do the two independent variables contribute significantly to the variation in the number of infant deaths?

(b) Calculate the coefficient of multiple determination and provide your interpretation.

(c) Fit the multiple regression model to obtain estimates of individual regression coefficients and their standard errors, and draw your conclusions: especially, the conditional contribution of the quadratic term.

**8.5**  The following are the heights (measured to the nearest 2 cm) and the weights (measured to the nearest kilogram) of 10 men:

| Height | 162 | 168 | 174 | 176 | 180 | 180 | 182 | 184 | 186 | 186 |
|--------|-----|-----|-----|-----|-----|-----|-----|-----|-----|-----|
| Weight | 65  | 65  | 84  | 63  | 75  | 76  | 82  | 65  | 80  | 81  |

and 10 women:

| Height | 152 | 156 | 158 | 160 | 162 | 162 | 164 | 164 | 166 | 166 |
|--------|-----|-----|-----|-----|-----|-----|-----|-----|-----|-----|
| Weight | 52  | 50  | 47  | 48  | 52  | 55  | 55  | 56  | 60  | 60  |

Separately for each group, men and women:

(a) Draw a scatter diagram to show a possible association between the weight (used as the dependent variable) and the height and check to see if a linear model is justified.

(b) Estimate the regression parameters, the weight for a subject who is 160 cm (does gender have an effect on this estimate?), and draw the regression line on the same graph with the scatter diagram.

(c) Test to see if the two factors are independent; state your hypotheses and choice of test size.

(d) Calculate the coefficient of determination and provide your interpretation.

(e) Is there evidence of an effect modification? (Compare the two coefficients of determination/correlation informally.)

**8.6**  Refer to the data in Exercise 8.5, but in the context of a multiple regression problem with three independent variables: height, gender, and product height by gender.

**(a)** Fit the multiple regression model to obtain estimates of individual regression coefficients and their standard errors. Draw your conclusion concerning the conditional contribution of each factor.

**(b)** Within the context of the multiple regression model in part (a), does gender alter the effect of height on weight?

**(c)** Taken collectively, do the three independent variables contribute significantly to the variation in weights?

**(d)** Calculate the coefficient of multiple determination and provide your interpretation.

**8.7** In an assay of heparin, a standard preparation is compared with a test preparation by observing the log clotting times ($y$, in seconds) of blood containing different doses of heparin ($x$ is the log dose) (Table E8.7). Replicate readings are made at each dose level). Separately for each preparation, standard and test:

**TABLE E8.7**

| Log Clotting Times | | | | Log Dose |
|---|---|---|---|---|
| Standard | | Test | | |
| 1.806 | 1.756 | 1.799 | 1.763 | 0.72 |
| 1.851 | 1.785 | 1.826 | 1.832 | 0.87 |
| 1.954 | 1.929 | 1.898 | 1.875 | 1.02 |
| 2.124 | 1.996 | 1.973 | 1.982 | 1.17 |
| 2.262 | 2.161 | 2.140 | 2.100 | 1.32 |

**(a)** Draw a scatter diagram to show a possible association between the log clotting time (used as the dependent variable) and the log dose and check to see if a linear model is justified.

**(b)** Estimate the regression parameters, the log clotting time for a log dose of 1.0 (are estimates for different preparations different?), and draw the regression line on the same graph with the scatter diagram.

**(c)** Test to see if the two factors are independent; state your hypotheses and choice of test size.

**(d)** Calculate the coefficient of determination and provide your interpretation.

**(e)** Is there evidence of an effect modification? (Compare the two coefficients of determination/correlation informally.)

**8.8** Refer to the data in Exercise 8.7, but in the context of a multiple regression problem with three independent variables: log dose, preparation, and product log dose by preparation.

**(a)** Fit the multiple regression model to obtain estimates of individual regression coefficients and their standard errors. Draw your conclusion concerning the conditional contribution of each factor.

**(b)** Within the context of the multiple regression model in part (a), does preparation alter the effect of log dose on the log clotting time?

**(c)** Taken collectively, do the three independent variables contribute significantly to the variation in log clotting times?

**(d)** Calculate the coefficient of multiple determination and provide your interpretation.

**8.9**    Data are shown in Table E8.9 for two groups of patients who died of acute myelogenous leukemia. Patients were classified into the two groups according to the presence or absence of a morphologic characteristic of white cells. Patients termed AG positive were identified by the presence of Auer rods and/or significant granulature of the leukemic cells in the bone marrow at diagnosis. For AG-negative patients, these factors were absent. Leukemia is a cancer characterized by an overproliferation of white blood cells; the higher the white blood count (WBC), the more severe the disease. Separately for each morphologic group, AG positive and AG negative:

**(a)** Draw a scatter diagram to show a possible association between the log survival time (take the log yourself and use as the dependent variable) and the log WBC (take the log yourself) and check to see if a linear model is justified.

**TABLE E8.9**

| AG Positive, $N = 17$ | | AG Negative, $N = 16$ | |
|---|---|---|---|
| WBC | Survival Time (weeks) | WBC | Survival Time (weeks) |
| 2,300 | 65 | 4,400 | 56 |
| 750 | 156 | 3,000 | 65 |
| 4,300 | 100 | 4,000 | 17 |
| 2,600 | 134 | 1,500 | 7 |
| 6,000 | 16 | 9,000 | 16 |
| 10,500 | 108 | 5,300 | 22 |
| 10,000 | 121 | 10,000 | 3 |
| 17,000 | 4 | 19,000 | 4 |
| 5,400 | 39 | 27,000 | 2 |
| 7,000 | 143 | 28,000 | 3 |
| 9,400 | 56 | 31,000 | 8 |
| 32,000 | 26 | 26,000 | 4 |
| 35,000 | 22 | 21,000 | 3 |
| 100,000 | 1 | 79,000 | 30 |
| 100,000 | 1 | 100,000 | 4 |
| 52,000 | 5 | 100,000 | 43 |
| 100,000 | 65 | | |

**(b)** Estimate the regression parameters, the survival time for a patient with a WBC of 20,000 (are estimates for different groups different?), and draw the regression line on the same graph with the scatter diagram.

**(c)** Test to see if the two factors are independent; state your hypotheses and choice of test size.

**(d)** Calculate the coefficient of determination and provide your interpretation.

**(e)** Is there evidence of an effect modification? (Compare the two coefficients of determination/correlation informally.)

**8.10** Refer to the data in Exercise 8.9, but in the context of a multiple regression problem with three independent variables: log WBC, the morphologic characteristic (AG, represented by a binary indicator: 0 if AG negative and 1 if AG positive), and product log WBC by morphologic characteristic (AG).

**(a)** Fit the multiple regression model to obtain estimates of individual regression coefficients and their standard errors. Draw your conclusion concerning the conditional contribution of each factor.

**(b)** Within the context of the multiple regression model in part (a), does the morphologic characteristic (AG) alter the effect of log WBC on log survival time?

**(c)** Taken collectively, do the three independent variables contribute significantly to the variation in log survival times?

**(d)** Calculate the coefficient of multiple determination and provide your interpretation.

**8.11** The purpose of this study was to examine the data for 44 physicians working for an emergency at a major hospital so as to determine which of a number of factors are related to the number of complaints received during the preceding year. In addition to the number of complaints, data available consist of the number of visits (which serves as the *size* for the observation unit), the physician, and four other factors under investigation. Table E8.11 presents the complete data set. For each of the 44 physicians there are two continuous explanatory factors: the revenue (dollars per hour) and the workload at the emergency service (hours), and two binary variables: gender (female/male) and residency training in emergency services (no/yes). Divide the number of complaints by the number of visits and use this ratio (number of complaints per visit) as the primary *outcome* or dependent variable *Y*. Individually for each of the two continuous explanatory factors, revenue (dollars per hour) and workload at the emergency service (hours):

**(a)** Draw a scatter diagram to show a possible association with the number of complaints per visit, and check to see if a linear model is justified.

**TABLE E8.11**

| No. of Visits | Complaint | Residency | Gender | Revenue | Hours |
|---|---|---|---|---|---|
| 2014 | 2 | Y | F | 263.03 | 1287.25 |
| 3091 | 3 | N | M | 334.94 | 1588.00 |
| 879 | 1 | Y | M | 206.42 | 705.25 |
| 1780 | 1 | N | M | 226.32 | 1005.50 |
| 3646 | 11 | N | M | 288.91 | 1667.25 |
| 2690 | 1 | N | M | 275.94 | 1517.75 |
| 1864 | 2 | Y | M | 295.71 | 967.00 |
| 2782 | 6 | N | M | 224.91 | 1609.25 |
| 3071 | 9 | N | F | 249.32 | 1747.75 |
| 1502 | 3 | Y | M | 269.00 | 906.25 |
| 2438 | 2 | N | F | 225.61 | 1787.75 |
| 2278 | 2 | N | M | 212.43 | 1480.50 |
| 2458 | 5 | N | M | 211.05 | 1733.50 |
| 2269 | 2 | N | F | 213.23 | 1847.25 |
| 2431 | 7 | N | M | 257.30 | 1433.00 |
| 3010 | 2 | Y | M | 326.49 | 1520.00 |
| 2234 | 5 | Y | M | 290.53 | 1404.75 |
| 2906 | 4 | N | M | 268.73 | 1608.50 |
| 2043 | 2 | Y | M | 231.61 | 1220.00 |
| 3022 | 7 | N | M | 241.04 | 1917.25 |
| 2123 | 5 | N | F | 238.65 | 1506.25 |
| 1029 | 1 | Y | F | 287.76 | 589.00 |
| 3003 | 3 | Y | F | 280.52 | 1552.75 |
| 2178 | 2 | N | M | 237.31 | 1518.00 |
| 2504 | 1 | Y | F | 218.70 | 1793.75 |
| 2211 | 1 | N | F | 250.01 | 1548.00 |
| 2338 | 6 | Y | M | 251.54 | 1446.00 |
| 3060 | 2 | Y | M | 270.52 | 1858.25 |
| 2302 | 1 | N | M | 247.31 | 1486.25 |
| 1486 | 1 | Y | F | 277.78 | 933.75 |
| 1863 | 1 | Y | M | 259.68 | 1168.25 |
| 1661 | 0 | N | M | 260.92 | 877.25 |
| 2008 | 2 | N | M | 240.22 | 1387.25 |
| 2138 | 2 | N | M | 217.49 | 1312.00 |
| 2556 | 5 | N | M | 250.31 | 1551.50 |
| 1451 | 3 | Y | F | 229.43 | 973.75 |
| 3328 | 3 | Y | M | 313.48 | 1638.25 |
| 2927 | 8 | N | M | 293.47 | 1668.25 |
| 2701 | 8 | N | M | 275.40 | 1652.75 |
| 2046 | 1 | Y | M | 289.56 | 1029.75 |
| 2548 | 2 | Y | M | 305.67 | 1127.00 |
| 2592 | 1 | N | M | 252.35 | 1547.25 |
| 2741 | 1 | Y | F | 276.86 | 1499.25 |
| 3763 | 10 | Y | M | 308.84 | 1747.50 |

*Note:* This is a very long data file; its electronic copy, in a Web-based form, is available from the author upon request.

    **(b)** Estimate the regression parameters, the number of complaints per visit for a physician having the (sample) *mean* level of the explanatory factor, and draw the regression line on the same graph with the scatter diagram.

    **(c)** Test to see if the factor and the number of complaints per visit are independent; state your hypotheses and choice of test size.

    **(d)** Calculate the coefficient of determination and provide your interpretation.

**8.12** Refer to the data in Exercise 8.11, but consider all four explanatory factors and the product residency training by workload simultaneously.

    **(a)** Fit the multiple regression model to obtain estimates of individual regression coefficients and their standard errors. Draw conclusions concerning the conditional contribution of each factor.

    **(b)** Within the context of the multiple regression model in part (a), does the residency training alter the effect of workload on the number of complaints per visit?

    **(c)** Taken collectively, do the five independent variables contribute significantly to the variation in log survival times?

    **(d)** Calculate the coefficient of multiple determination and provide your interpretation.

# 9

# LOGISTIC REGRESSION

The purpose of many research projects is to assess relationships among a set of variables and regression techniques often used as statistical analysis tools in the study of such relationships. Research designs may be classified as experimental or observational. Regression analyses are applicable to both types; yet the confidence one has in the results of a study can vary with the research type. In most cases, one variable is usually taken to be the response or dependent variable, that is, a variable to be predicted from or explained by other variables. The other variables are called *predictors, explanatory variables,* or *independent variables.* Choosing an appropriate model and analytical technique depends on the type of variable under investigation. Methods and regression models of Chapter 8 deal with cases where the dependent variable of interest is a continous variable which we assume, perhaps after an appropriate transformation, to be normally distributed. In Chapter 8 we considered cases with one independent variable (simple regression) and cases with several independent variables or covariates (multiple regression). However, in a variety of other applications, the dependent variable of interest is not on a continuous scale; it may have only two possible outcomes and therefore can be represented by an indicator variable taking on values 0 and 1. Consider, for example, an analysis of whether or not business firms have a day-care facility; the corresponding independent variable, for example, is the number of female employees. The dependent variable in this study was defined to have two possible outcomes: (1) the firm has a day-care facility, and (2) the firm does not have a day-care facility, which may be coded as 1 and 0, respectively. As another example, consider a study of drug use among middle school students as a function of gender, age, family structure (e.g., who is the head of household), and family income. In this

314

study, the dependent variable $Y$ was defined to have two possible outcomes: (1) the child uses drugs and (2) the child does not use drugs. Again, these two outcomes may be coded 1 and 0, respectively.

The examples above, and others, show a wide range of applications in which the dependent variable is dichotomous and hence may be represented by a variable taking the value 1 with probability $\pi$ and the value 0 with probability $1 - \pi$. Such a variable is a *point binomial variable*, that is, a binomial variable with $n = 1$ trial, and the model often used to express the probability $\pi$ as a function of potential independent variables under investigation is the logistic regression model. It should be noted that the regression models of Chapter 8 do not apply here because linear combinations of independent variables are not bounded between 0 and 1 as required in the applications above. Instead of regression models imposed to describe the mean of a normal variate, the logistic model has been used extensively and successfully in the health sciences to describe the probability (or risk) of developing a condition—say, a disease—over a specified time period as a function of certain risk factors $X_1, X_2, \ldots, X_k$. The following is such a typical example.

***Example 9.1***    When a patient is diagnosed as having cancer of the prostate, an important question in deciding on treatment strategy for the patient is whether or not the cancer has spread to neighboring lymph nodes. The question is so critical in prognosis and treatment that it is customary to operate on the patient (i.e., perform a laparotomy) for the sole purpose of examining the nodes and removing tissue samples to examine under the microscope for evidence of cancer. However, certain variables that can be measured without surgery are predictive of the nodal involvement; and the purpose of the study presented in Brown (1980) was to examine the data for 53 prostate cancer patients receiving surgery, to determine which of five preoperative variables are predictive of nodal involvement. In particular, the principal investigator was interested in the predictive value of the level of acid phosphatase in blood serum. Table 9.1 presents the complete data set. For each of the 53 patients, there are two continuous independent variables: age at diagnosis and level of serum acid phosphatase ($\times 100$; called "acid"), and three binary variables: x-ray reading, pathology reading (grade) of a biopsy of the tumor obtained by needle before surgery, and a rough measure of the size and location of the tumor (stage) obtained by palpation with the fingers via the rectum. For these three binary independent variables a value of 1 signifies a positive or more serious state and a 0 denotes a negative or less serious finding. In addition, the sixth column presents the finding at surgery—the primary binary response or dependent variable $Y$, a value of 1 denoting nodal involvement, and a value of 0 denoting no nodal involvement found at surgery.

A careful reading of the data reveals, for example, that a positive x-ray or an elevated acid phosphatase level, in general, seems likely being associated with nodal involvement found at surgery. However, predictive values of other variables are not clear, and to answer the question, for example, concerning

**TABLE 9.1**

| X-ray | Grade | Stage | Age | Acid | Nodes | X-ray | Grade | Stage | Age | Acid | Nodes |
|---|---|---|---|---|---|---|---|---|---|---|---|
| 0 | 1 | 1 | 64 | 40 | 0 | 0 | 0 | 0 | 60 | 78 | 0 |
| 0 | 0 | 1 | 63 | 40 | 0 | 0 | 0 | 0 | 52 | 83 | 0 |
| 1 | 0 | 0 | 65 | 46 | 0 | 0 | 0 | 1 | 67 | 95 | 0 |
| 0 | 1 | 0 | 67 | 47 | 0 | 0 | 0 | 0 | 56 | 98 | 0 |
| 0 | 0 | 0 | 66 | 48 | 0 | 0 | 0 | 1 | 61 | 102 | 0 |
| 0 | 1 | 1 | 65 | 48 | 0 | 0 | 0 | 0 | 64 | 187 | 0 |
| 0 | 0 | 0 | 60 | 49 | 0 | 1 | 0 | 1 | 58 | 48 | 1 |
| 0 | 0 | 0 | 51 | 49 | 0 | 0 | 0 | 1 | 65 | 49 | 1 |
| 0 | 0 | 0 | 66 | 50 | 0 | 1 | 1 | 1 | 57 | 51 | 1 |
| 0 | 0 | 0 | 58 | 50 | 0 | 0 | 1 | 0 | 50 | 56 | 1 |
| 0 | 1 | 0 | 56 | 50 | 0 | 1 | 1 | 0 | 67 | 67 | 1 |
| 0 | 0 | 1 | 61 | 50 | 0 | 0 | 0 | 1 | 67 | 67 | 1 |
| 0 | 1 | 1 | 64 | 50 | 0 | 0 | 1 | 1 | 57 | 67 | 1 |
| 0 | 0 | 0 | 56 | 52 | 0 | 0 | 1 | 1 | 45 | 70 | 1 |
| 0 | 0 | 0 | 67 | 52 | 0 | 0 | 0 | 1 | 46 | 70 | 1 |
| 1 | 0 | 0 | 49 | 55 | 0 | 1 | 0 | 1 | 51 | 72 | 1 |
| 0 | 1 | 1 | 52 | 55 | 0 | 1 | 1 | 1 | 60 | 76 | 1 |
| 0 | 0 | 0 | 68 | 56 | 0 | 1 | 1 | 1 | 56 | 78 | 1 |
| 0 | 1 | 1 | 66 | 59 | 0 | 1 | 1 | 1 | 50 | 81 | 1 |
| 1 | 0 | 0 | 60 | 62 | 0 | 0 | 0 | 0 | 56 | 82 | 1 |
| 0 | 0 | 0 | 61 | 62 | 0 | 0 | 0 | 1 | 63 | 82 | 1 |
| 1 | 1 | 1 | 59 | 63 | 0 | 1 | 1 | 1 | 65 | 84 | 1 |
| 0 | 0 | 0 | 51 | 65 | 0 | 1 | 0 | 1 | 64 | 89 | 1 |
| 0 | 1 | 1 | 53 | 66 | 0 | 0 | 1 | 0 | 59 | 99 | 1 |
| 0 | 0 | 0 | 58 | 71 | 0 | 1 | 1 | 1 | 68 | 126 | 1 |
| 0 | 0 | 0 | 63 | 75 | 0 | 1 | 0 | 0 | 61 | 136 | 1 |
| 0 | 0 | 1 | 53 | 76 | 0 | | | | | | |

*Note:* This is a very long data file; its electronic copy, in a Web-based form, is available from the author upon request.

the usefulness of acid phosphatase as a prognostic variable, we need a more detailed analysis before a conclusion can be made.

## 9.1    SIMPLE REGRESSION ANALYSIS

Following the outline of Chapter 8, in this section we discuss the basic ideas of simple regression analysis when only one predictor or independent variable is available for predicting the response of interest. Again, the response of interest is binary taking the value 1 with probability $\pi$ and the value 0 with probability $1 - \pi$. In the interpretation of the primary parameter of the model, we discuss both scales of measurement, discrete and continuous, even though in most practical applications, the independent variable under investigation is often on a continuous scale.

### 9.1.1  Simple Logistic Regression Model

The usual regression analysis goal, as seen in various sections of Chapter 8, is to describe the mean of a dependent variable $Y$ as a function of a set of predictor variables. The logistic regression, however, deals with the case where the basic random variable $Y$ of interest is a dichotomous variable taking the value 1 with probability $\pi$ and the value 0 with probability $(1 - \pi)$. Such a random variable is called a *point-binomial* or *Bernouilli variable*, and it has the simple discrete probability distribution

$$\Pr(Y = y) = \pi^y (1 - \pi)^{1-y} \qquad y = 0, 1$$

Suppose that for the $i$th individual of a sample $(i = 1, 2, \ldots, n)$, $Y_i$ is a Bernouilli variable with

$$\Pr(Y_i = y_i) = \pi_i^{y_i} (1 - \pi_i)^{1-y_i} \qquad y_i = 0, 1$$

The logistic regression analysis assumes that the relationship between $\pi_i$ and the covariate value $x_i$ of the same person is described by the logistic function

$$\pi_i = \frac{1}{1 + \exp[-(\beta_0 + \beta_1 x_i)]} \qquad i = 1, 2, \ldots, n,$$

The basic logistic function is given by

$$f(z) = \frac{1}{1 + e^{-z}}$$

where, as in this simple regression model,

$$z_i = \beta_0 + \beta_1 x_i$$

or, in the multiple regression model of subsequent sections,

$$z_i = \beta_0 + \sum_{j=1}^{k} \beta_j x_{ji}$$

representing an index of combined risk factors. There are two important reasons that make logistic regression popular:

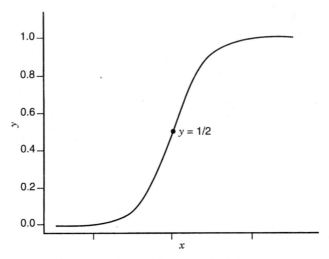

**Figure 9.1** General form of a logistic curve.

1. The range of the logistic function is between 0 and 1; that makes it suitable for use as a probability model, representing individual risk.
2. The logistic curve has an increasing S-shape with a threshold (Figure 9.1); that makes it suitable for use as a biological model, representing risk due to exposure.

Under the simple logistic regression model, the likelihood function is given by

$$L = \prod_{i=1}^{n} \Pr(Y_i = y_i)$$

$$= \prod_{i=1}^{n} \frac{[\exp(\beta_0 + \beta_1 x_i)]^{y_i}}{1 + \exp(\beta_0 + \beta_1 x_i)} \qquad y_i = 0, 1$$

from which we can obtain maximum likelihood estimates of the parameters $\beta_0$ and $\beta_1$. As mentioned previously, the logistic model has been used both extensively and successfully to describe the probability of developing $(Y = 1)$ some disease over a specified time period as a function of a risk factor $X$.

### 9.1.2  Measure of Association

Regression analysis serves two major purposes: (1) control or intervention, and (2) prediction. In many studies, such as the one in Example 9.1, one important objective is measuring the strength of a statistical relationship between the binary dependent variable and each independent variable or covariate measured from patients; findings may lead to important decisions in patient management (or public health interventions in other examples). In epidemiological studies, such effects are usually measured by the *relative risk* or *odds ratio*; when the logistic model is used, the measure is the odds ratio.

For the case of the logistic regression model, the logistic function for the probability $\pi_i$ can also be expressed as a linear model in the log scale (of the odds):

$$\ln \frac{\pi_i}{1 - \pi_i} = \beta_0 + \beta_1 x_i$$

We first consider the case of a binary covariate with the conventional coding:

$$X_i = \begin{cases} 0 & \text{if the patient is not exposed} \\ 1 & \text{if the patient is exposed} \end{cases}$$

Here, the term *exposed* may refer to a risk factor such as smoking, or a patient's characteristic such as race (white/nonwhite) or gender (male/female). It can be seen that from the log-linear form of the logistic regression model,

$$\ln(\text{odds; nonexposed}) = \beta_0$$

$$\ln(\text{odds; exposed}) = \beta_0 + \beta_1$$

So that after exponentiating, the difference leads to

$$e^{\beta_1} = \frac{(\text{odds; exposed})}{(\text{odds; nonexposed})}$$

represents the odds ratio (OR) associated with the exposure, exposed versus nonexposed. In other words, the primary regression coefficient $\beta_1$ is the value of the odds ratio on the log scale.

Similarly, we have for a continuous covariate $X$ and any value $x$ of $X$,

$$\ln(\text{odds; } X = x) = \beta_0 + \beta_1(x)$$

$$\ln(\text{odds; } X = x + 1) = \beta_0 + \beta_1(x + 1)$$

So that after exponentiating, the difference leads to

$$e^{\beta_1} = \frac{(\text{odds; } X = x + 1)}{(\text{odds; } X = x)}$$

represents the odds ratio (OR) associated with a 1-unit increase in the value of $X$, $X = x + 1$ versus $X = x$. For example, a systolic blood pressure of 114 mmHg versus 113 mmHg. For an $m$-unit increase in the value of $X$, say $X = x + m$ versus $X = x$, the corresponding odds ratio is $e^{m\beta_1}$.

The primary regression coefficient $\beta_1$ (and $\beta_0$, which is often not needed) can be estimated iteratively using a computer-packaged program such as SAS. From the results, we can obtained a point estimate

$$\widehat{OR} = e^{\hat{\beta}_1}$$

and its 95% confidence interval

$$\exp[\hat{\beta}_1 \pm 1.96\ SE(\hat{\beta}_1)]$$

### 9.1.3   Effect of Measurement Scale

It should be noted that the odds ratio, used as a measure of association between the binary dependent variable and a covariate, depends on the coding scheme for a binary covariate and for a continuous covariate $X$, the scale with which to measure $X$. For example, if we use the following coding for a factor,

$$X_i = \begin{cases} -1 & \text{if the subject is not exposed} \\ 1 & \text{if the subject is exposed} \end{cases}$$

then

$$\ln(\text{odds; nonexposed}) = \beta_0 - \beta_1$$
$$\ln(\text{odds; exposed}) = \beta_0 + \beta_1$$

so that

$$OR = \exp[\ln(\text{odds; exposed}) - \ln(\text{odds; nonexposed})]$$
$$= e^{2\beta_1}$$

and its 95% confidence interval,

$$\exp[2(\hat{\beta}_1 \pm 1.96\ SE(\hat{\beta}_1))]$$

Of course, the estimate of $\beta_1$ under the new coding scheme is only half of that under the former scheme; therefore, the (numerical) estimate of the OR remains unchanged. The following example, however, will show the clear effect of a measurement scale in the case of a continuous independent variable.

*Example 9.2*   Refer to the data for patients diagnosed as having cancer of the prostate in Example 9.1 (Table 9.1) and suppose that we want to investigate the relationship between nodal involvement found at surgery and the level of acid phosphatase in blood serum in two different ways using either (a) $X = \text{acid}$ or (b) $X = \log_{10}(\text{acid})$.

(a) For $X =$ acid, we find that

$$\hat{\beta}_1 = 0.0204$$

from which the odds ratio for (acid $= 100$) versus (acid $= 50$) would be

$$OR = \exp[(100 - 50)(0.0204)]$$
$$= 2.77$$

(b) For $X = \log_{10}(\text{acid})$, we find that

$$\hat{\beta}_1 = 5.1683$$

from which the odds ratio for (acid $= 100$) versus (acid $= 50$) would be

$$OR = \exp\{[\log_{10}(100) - \log_{10}(50)](5.1683)\}$$
$$= 4.74$$

*Note:* If $X =$ acid is used, an SAS program would include these instructions:

```
PROC LOGISTIC DESCENDING
DATA = CANCER;
MODEL NODES = ACID;
```

where CANCER is the name assigned to the data set, NODES is the variable name for nodal involvement, and ACID is the variable name for our covariate, the level of acid phosphatase in blood serum. The option DESCENDING is needed because PROC LOGISTIC models $\Pr(Y = 0)$ instead of $\Pr(Y = 1)$.

The results above are different for two different choices of $X$ and this seems to cause an obvious problem of choosing an appropriate measurement scale. Of course, we assume a *linear model* and one choice of scale for $X$ would fit better than the other. However, it is very difficult to compare different scales unless there were replicated data at each level of $X$; if such replications are available, one can simply graph a scatter diagram of log(odds) versus the $X$ value and check for linearity of each choice of scale of measurement for $X$.

### 9.1.4 Tests of Association

Sections 9.1.2 and 9.1.3 deal with inferences concerning the primary regression coefficient $\beta_1$, including both point and interval estimation of this parameter and the odds ratio. Another aspect of statistical inference concerns the test of significance; the null hypothesis to be considered is

$$H_0: \beta_1 = 0$$

The reason for interest in testing whether or not $\beta_1 = 0$ is that $\beta_1 = 0$ implies that there is no relation between the binary dependent variable and the covariate $X$ under investigation. Since the likelihood function is rather simple, one can easily derive, say, the score test for the null hypothesis above; however, nothing would be gained by going through this exercise. We can simply apply a chi-square test (if the covariate is binary or categorical) or $t$ test or Wilcoxon test (if the covariate under investigation is on a continuous scale). Of course, the application of the logistic model is still desirable, at least in the case of a continuous covariate, because it would provide a measure of association.

### 9.1.5  Use of the Logistic Model for Different Designs

Data for risk determination may come from different sources, with the two fundamental designs being retrospective and prospective. Prospective studies enroll group or groups of subjects and follow them over certain periods of time—examples include occupational mortality studies and clinical trials—and observe the occurrence of a certain event of interest such as a disease or death. Retrospective studies gather past data from selected cases and controls to determine differences, if any, in the exposure to a suspected risk factor. They are commonly referred to as case–control studies. It can be seen that the logistic model fits in very well with the prospective or follow-up studies and has been used successfully to model the "risk" of developing a condition—say, a disease—over a specified time period as a function of a certain risk factor. In such applications, after a logistic model has been fitted, one can estimate the individual risks $\pi(x)$'s—given the covariate value $x$—as well as any risks ratio or relative risk,

$$\text{RR} = \frac{\pi(x_i)}{\pi(x_j)}$$

As for case–control studies, it can be shown, using the Bayes' theorem, that if we have for the population

$$\Pr(Y = 1; \text{ given } x) = \frac{1}{1 + \exp[-(\beta_0 + \beta_1 x)]}$$

then

$$\Pr(Y = 1; \text{ given the samples and given } x) = \frac{1}{1 + \exp[-(\beta_0^* + \beta_1 x)]}$$

with

$$\beta_0^* = \beta_0 + \frac{\theta_1}{\theta_0}$$

where $\theta_1$ is the probability that a case was sampled and $\theta_0$ is the probability that a control was sampled. This result indicates the following points for a case–control study.

1. We *cannot* estimate individual risks, or relative risk, unless $\theta_0$ and $\theta_1$ are known, which are unlikely. The value of the intercept provided by the computer output is meaningless.
2. However, since we have the same $\beta_1$ as with a prospective model, we still can estimate the odds ratio and if the rare disease assumption applies, can interpret the numerical result as an approximate relative risk.

### 9.1.6  Overdispersion

This section introduces a new issue, the issue of *overdispersion*, which is of practical importance. However, the presentation also involves somewhat more advanced statistical concept, such as invoking the variance of the binomial distribution, which was introduced very briefly in Chapter 3. Because of that, student readers, especially beginners, may decide to skip without having any discontinuity. Logistic regression is based on the *point binomial* or *Bernouilli distribution*; its mean is $\pi$ and the variance is $(\pi)(1 - \pi)$. If we use the variance/mean ratio as a dispersion parameter, it is 1 in a standard logistic model, less than 1 in an underdispersed model, and greater than 1 in an overdispersed model. Overdispersion is a common phenomenon in practice and it causes concerns because the implication is serious; the analysis, which assumes the logistic model, often underestimates standard error(s) and thus wrongly inflates the level of significance.

*Measuring and Monitoring Dispersion*    After a logistic regression model is fitted, dispersion is measured by the scaled deviance or scaled Pearson chi-square; it is the deviance or Pearson chi-square divided by the degrees of freedom. The deviance is defined as twice the difference between the maximum achievable log likelihood and the log likelihood at the maximum likelihood estimates of the regression parameters. Suppose that data are with replications consisting of $m$ subgroups (with identical covariate values); then the Pearson chi-square and deviance are given by

$$X_P^2 = \sum_i \sum_j \frac{(r_{ij} - n_i p_{ij})^2}{n_i p_{ij}}$$

$$X_D^2 = \sum_i \sum_j r_{ij} \log \frac{r_{ij}}{n_i p_{ij}}$$

Each of these goodness-of-fit statistics devided by the appropriate degrees of freedom, called the *scaled Pearson chi-square* and *scaled deviance*, respectively, can be used as a measure for overdispersion (underdispersion, with those measures less than 1, occurs much less often in practice). When their values are much larger than 1, the assumption of binomial variability may not be valid and the data are said to exhibit overdispersion. Several factors can cause overdispersion; among these are such problems as outliers in the data, omitting important covariates in the model, and the need to transform some explanatory factors. PROC LOGISTIC of SAS, has an option, called AGGREGATE, that can be used to form subgroups. Without such a grouping, data may be too sparse, the Pearson chi-square and deviance do not have a chi-square distribution, and the scaled Pearson chi-square and scaled deviance cannot be used as indicators of overdispersion. A large difference between the scaled Pearson chi-square and scaled deviance provides evidence of this situation.

***Fitting an Overdispersed Logistic Model***    One way of correcting overdispersion is to multiply the covariance matrix by the value of the overdispersion parameter $\phi$, scaled Pearson chi-square, or scaled deviance (as used in weighted least-squares fitting):

$$E(p_i) = \pi_i$$
$$\text{Var}(p_i) = \phi \pi_i (1 - pi_i)$$

In this correction process, the parameter estimates are not changed. However, their standard errors are adjusted (increased), affecting their significant levels (reduced).

***Example 9.3***    In a study of the toxicity of certain chemical compound, five groups of 20 rats each were fed for four weeks by a diet mixed with that compound at five different doses. At the end of the study, their lungs were harvested and subjected to histopathological examinations to observe for sign(s) of toxicity (yes $= 1$, no $= 0$). The results are shown in Table 9.2. A routine fit of the simple logistic regression model yields Table 9.3. In addition, we obtained the results in Table 9.4 for the monitoring of overdispersion.

**TABLE 9.2**

| Group | Dose (mg) | Number of Rats | Number of Rats with Toxicity |
|-------|-----------|----------------|------------------------------|
| 1 | 5 | 20 | 1 |
| 2 | 10 | 20 | 3 |
| 3 | 15 | 20 | 7 |
| 4 | 20 | 20 | 14 |
| 5 | 30 | 20 | 10 |

**TABLE 9.3**

| Variable | Coefficient | Standard Error | $z$ Statistic | $p$ Value |
|---|---|---|---|---|
| Intercept | −2.3407 | 0.5380 | −4.3507 | 0.0001 |
| Dose | 0.1017 | 0.0277 | 3.6715 | 0.0002 |

**TABLE 9.4**

| Parameter | Chi-Square | Degrees of Freedom | Scaled Parameter |
|---|---|---|---|
| Pearson | 10.9919 | 3 | 3.664 |
| Deviance | 10.7863 | 3 | 3.595 |

**TABLE 9.5**

| Variable | Coefficient | Standard Error | $z$ Statistic | $p$ Value |
|---|---|---|---|---|
| Intercept | −2.3407 | 1.0297 | −2.2732 | 0.0230 |
| Dose | 0.1017 | 0.0530 | 1.9189 | 0.0548 |

*Note:* An SAS program would include these instructions:

```
INPUT DOSE N TOXIC;
PROC LOGISTIC DESCENDING;
MODEL NODES TOXIC/N = DOSE/SCALE = NONE;
```

The results above indicate an obvious sign of overdispersion. By fitting an overdispersed model, controlling for the scaled deviance, we have Table 9.5. Compared to the previous results, the point estimates remain the same but the standard errors are larger. The effect of dose is no longer significant at the 5% level.

*Note:* An SAS program would include these instructions:

```
INPUT DOSE N TOXIC;
PROC LOGISTIC DESCENDING;
MODEL NODES TOXIC/N = DOSE/SCALE = D;
```

## 9.2  MULTIPLE REGRESSION ANALYSIS

The effect of some factor on a dependent or response variable may be influenced by the presence of other factors through effect modifications (i.e., interactions). Therefore, to provide a more comprehensive analysis, it is very desirable to consider a large number of factors and sort out which ones are most closely related to the dependent variable. In this section we discuss a multivariate method for risk determination. This method, which is multiple logistic regression analysis, involves a linear combination of the explanatory or independent variables; the variables must be quantitative with particular numerical values for each patient. A covariate or independent variable, such as a patient characteristic, may be dichotomous, polytomous, or continuous (categorical factors will be represented by dummy variables). Examples of dichotomous covariates are gender and presence/absence of certain comorbidity. Polytomous covariates include race and different grades of symptoms; these can be covered by the use of *dummy variables*. Continuous covariates include patient age and blood pressure. In many cases, data transformations (e.g., taking the logarithm) may be desirable to satisfy the linearity assumption.

### 9.2.1   Logistic Regression Model with Several Covariates

Suppose that we want to consider $k$ covariates simultaneously; the simple logistic model of Section 9.1 can easily be generalized and expressed as

$$\pi_i = \frac{1}{1 + \exp[-(\beta_0 + \sum_{j=1}^{k} \beta_j x_{ji})]} \qquad i = 1, 2, \ldots, n$$

or, equivalently,

$$\ln \frac{\pi_i}{1 - \pi_i} = \beta_0 + \sum_{j=1}^{k} \beta_j x_{ji}$$

This leads to the likelihood function

$$L = \prod_{i=1}^{n} \frac{[\exp(\beta_0 + \sum_{i=1}^{k} \beta_j x_{ji})]^{y_i}}{1 + \exp(\beta_0 + \sum_{j=1}^{k} \beta_j x_{ji})} \qquad y_i = 0, 1$$

from which parameters can be estimated iteratively using a computer-packaged program such as SAS.

Also similar to the univariate case, $\exp(\beta_i)$ represents one of the following:

1. The odds ratio associated with an exposure if $X_i$ is binary (exposed $X_i = 1$ versus unexposed $X_i = 0$); or
2. The odds ratio due to a 1-unit increase if $X_i$ is continuous ($X_i = x + 1$ versus $X_i = x$).

After $\hat{\beta}_i$ and its standard error have been obtained, a 95% confidence interval for the odds ratio above is given by

$$\exp[\hat{\beta}_i \pm 1.96\ \text{SE}(\hat{\beta}_i)]$$

These results are necessary in the effort to identify important risk factors for the binary outcome. Of course, before such analyses are done, the problem and the data have to be examined carefully. If some of the variables are highly correlated, one or fewer of the correlated factors are likely to be as good predictors as all of them; information from similar studies also has to be incorporated so as to drop some of these correlated explanatory variables. The use of products such as $X_1 X_2$ and higher power terms such as $X_1^2$ may be necessary and can improve the goodness of fit. It is important to note that we are assuming a *(log)linear* regression model, in which, for example, the odds ratio due to a 1-unit increase in the value of a continuous $X_i$ ($X_i = x + 1$ versus $X_i = x$) is independent of $x$. Therefore, if this *linearity* seems to be violated, the incorporation of powers of $X_i$ should be seriously considered. The use of products will help in the investigation of possible effect modifications. Finally, there is the messy problem of missing data; most packaged programs would delete a subject if one or more covariate values are missing.

## 9.2.2  Effect Modifications

Consider the model

$$\pi_i = \frac{1}{1 + \exp[-(\beta_0 + \beta_1 x_{1i} + \beta_2 x_{2i} + \beta_3 x_{1i} x_{2i})]} \qquad i = 1, 2, \ldots, n$$

The meaning of $\beta_1$ and $\beta_2$ here is not the same as that given earlier because of the cross-product term $\beta_3 x_1 x_2$. Suppose that both $X_1$ and $X_2$ are binary.

1. For $X_2 = 1$, or exposed, we have

   $$(\text{odds ratio; not exposed to } X_1) = e^{\beta_0 + \beta_2}$$

   $$(\text{odds ratio; exposed to } X_1) = e^{\beta_0 + \beta_1 + \beta_2 + \beta_3}$$

   so that the ratio of these ratios, $e^{\beta_1 + \beta_3}$, represents the odds ratio associated with $X_1$, exposed versus nonexposed, in the presence of $X_2$, whereas

2. For $X_2 = 0$, or not exposed, we have

   $$(\text{odds ratio; not exposed to } X_1) = e^{\beta_0} \quad (\text{i.e., baseline})$$

   $$(\text{odds ratio; exposed to } X_1) = e^{\beta_0 + \beta_1}$$

   so that the ratio of these ratios, $e^{\beta_1}$, represents the odds ratio associated with $X_1$, exposed versus nonexposed, in the absence of $X_2$. In other words, the effect of $X_1$ depends on the level (presence or absence) of $X_2$, and vice versa.

This phenomenon is called *effect modification* (i.e., one factor modifies the effect of the other. The cross-product term $x_1 x_2$ is called an *interaction term*; The use of these products will help in the investigation of possible effect modifications. If $\beta_3 = 0$, the effect of two factors acting together, as measured by the odds ratio, is equal to the combined effects of two factors acting separately, as measured by the product of two odds ratios:

$$e^{\beta_1 + \beta_2} = e^{\beta_1} e^{\beta_2}$$

This fits the classic definition of *no interaction* on a multiplicative scale.

### 9.2.3  Polynomial Regression

Consider the model

$$\pi_i = \frac{1}{1 + \exp[-(\beta_0 + \beta_1 x_i + \beta_2 x_i^2)]} \qquad i = 1, 2, \ldots, n$$

where $X$ is a continuous covariate. The meaning of $\beta_1$ here is not the same as that given earlier because of the quadratic term $\beta_2 x_i^2$. We have, for example,

$$\ln(\text{odds}; X = x) = \beta_0 + \beta_1 x + \beta_2 x^2$$

$$\ln(\text{odds}; X = x + 1) = \beta_0 + \beta_1(x + 1) + \beta_2(x + 1)^2$$

so that after exponentiating, the difference leads to

$$\text{OR} = \frac{(\text{odds}; X = x + 1)}{(\text{odds}; X = x)}$$

$$= \exp[\beta_1 + \beta_2(2x + 1)]$$

a function of $x$.

Polynomial models with an independent variable present in higher powers than the second are not often used. The second-order or quadratic model has two basic type of uses: (1) when the true relationship is a second-degree polynomial or when the true relationship is unknown but the second-degree polynomial provides a better fit than a linear one, but (2) more often, a quadratic model is fitted for the purpose of establishing the linearity. The key item to look for is whether $\beta_2 = 0$.

The use of polynomial models is not without drawbacks. The most potential drawback is that multicollinearity is unavoidable. Especially if the covariate is restricted to a narrow range, the degree of multicollinearity can be quite high; in this case the standard errors are often very large. Another problem arises when one wants to use the stepwise regression search method. In addition, finding a satisfactory interpretation for the *curvature effect coefficient* $\beta_2$ is not easy. Perhaps a rather interesting application would be finding a value $x$ of the covariate $X$ so as to maximize or minimize the

$$\ln(\text{odds}; X = x) = \beta_0 + \beta_1 x + \beta_2 x^2$$

### 9.2.4 Testing Hypotheses in Multiple Logistic Regression

Once we have fit a multiple logistic regression model and obtained estimates for the various parameters of interest, we want to answer questions about the contributions of various factors to the prediction of the binary response variable. There are three types of such questions:

1. *Overall test.* Taken collectively, does the entire set of explatory or independent variables contribute significantly to the prediction of response?
2. *Test for the value of a single factor.* Does the addition of one particular variable of interest add significantly to the prediction of response over and above that achieved by other independent variables?
3. *Test for contribution of a group of variables.* Does the addition of a group of variables add significantly to the prediction of response over and above that achieved by other independent variables?

***Overall Regression Tests***  We now consider the first question stated above concerning an overall test for a model containg $k$ factors, say,

$$\pi_i = \frac{1}{1 + \exp[-(\beta_0 + \sum_{j=1}^{k} \beta_j x_{ji})]} \qquad i = 1, 2, \ldots, n$$

The null hypothesis for this test may be stated as: "All $k$ independent variables *considered together* do not explain the variation in the responses." In other words,

$$H_0: \beta_1 = \beta_2 = \cdots = \beta_k = 0$$

Two likelihood-based statistics can be used to test this *global* null hypothesis; each has a symptotic chi-square distribution with $k$ degrees of freedom under $H_0$.

1. Likelihood ratio test:

$$\chi_{\text{LR}}^2 = 2[\ln L(\hat{\beta}) - \ln L(0)]$$

2. Score test:

$$\chi_S^2 = \left[\frac{\delta \ln L(0)}{\delta \beta}\right]^T \left[-\frac{\delta^2 \ln L(0)}{\delta \beta^2}\right]^{-1} \left[\frac{\delta \ln L(0)}{\delta \beta}\right]$$

Both statistics are provided by most standard computer programs, such as SAS, and they are asymptotically equivalent, yielding identical statistical decisions most of the times.

***Example 9.4***    Refer to the data set on prostate cancer of Example 9.1 (Table 9.1). With all five covariates, we have the following test statistics for the global null hypothesis:

1. Likelihood test:

$$\chi^2_{LR} = 22.126 \text{ with 5 df;} \quad p = 0.0005$$

2. Score test:

$$\chi^2_S = 19.451 \text{ with 5 df;} \quad p = 0.0016$$

*Note:* An SAS program would include these instructions:

```
PROC LOGISTIC DESCENDING
DATA = CANCER;
MODEL NODES = X-RAY, GRADE, STAGE, AGE, ACID;
```

where CANCER is the name assigned to the data set, NODES is the variable name for nodal involvement, and X-RAY, GRADE, STAGE, AGE, and ACID are the variable names assigned to the five covariates.

***Tests for a Single Variable***    Let us assume that we now wish to test whether the addition of one particular independent variable of interest adds significantly to the prediction of the response over and above that achieved by other factors already present in the model. The null hypothesis for this test may stated as: "Factor $X_i$ does not have any value added to the prediction of the response *given that other factors are already included in the model.*" In other words,

$$H_0: \beta_i = 0$$

To test such a null hypothesis, one can perform a likelihood ratio chi-squared test, with 1 df, similar to that for the global hypothesis above:

$$\chi^2_{LR} = 2[\ln L(\hat{\beta}; \text{ all } X\text{'s}) - \ln L(\hat{\beta}; \text{ all other } X\text{'s with } X_i \text{ deleted})]$$

A much easier alternative method is using

$$z_i = \frac{\hat{\beta}_i}{\text{SE}(\hat{\beta}_i)}$$

**TABLE 9.6**

| Variable | Coefficient | Standard Error | $z$ Statistic | $p$ Value |
|---|---|---|---|---|
| Intercept | 0.0618 | 3.4599 | 0.018 | 0.9857 |
| X-ray | 2.0453 | 0.8072 | 2.534 | 0.0113 |
| Stage | 1.5641 | 0.7740 | 2.021 | 0.0433 |
| Grade | 0.7614 | 0.7708 | 0.988 | 0.3232 |
| Age | −0.0693 | 0.0579 | −1.197 | 0.2314 |
| Acid | 0.0243 | 0.0132 | 1.850 | 0.0643 |

where $\hat{\beta}_i$ is the corresponding estimated regression coefficient and $SE(\hat{\beta}_i)$ is the estimate of the standard error of $\hat{\beta}_i$, both of which are printed by standard computer-packaged programs. In performing this test, we refer the value of the $z$ statistic to percentiles of the standard normal distribution.

***Example 9.5***   Refer to the data set on prostate cancer of Example 9.1 (Table 9.1). With all five covariates, we have the results shown in Table 9.6. The effects of x-ray and stage are significant at the 5% level, whereas the effect of acid is marginally significant ($p = 0.0643$).
   *Note:* Use the same SAS program as in Example 9.4.

Given a continuous variable of interest, one can fit a polynomial model and use this type of test to check for linearity. It can also be used to check for a single product representing an effect modification.

***Example 9.6***   Refer to the data set on prostate cancer of Example 9.1 (Table 9.1), but this time we investigate only one covariate, the level of acid phosphatase (acid). After fitting the second-degree polinomial model,

$$\pi_i = \frac{1}{1 + \exp[-(\beta_0 + \beta_1(\text{acid}) + \beta_2(\text{acid})^2)]} \qquad i = 1, 2, \ldots, n$$

we obtained the results shown in Table 9.7, indicating that the *curvature effect* should not be ignored ($p = 0.0437$).

**TABLE 9.7**

| Factor | Coefficient | Standard Error | $z$ Statistic | $p$ Value |
|---|---|---|---|---|
| Intercept | −7.3200 | 2.6229 | −2.791 | 0.0053 |
| Acid | 0.1489 | 0.0609 | 2.445 | 0.0145 |
| Acid$^2$ | −0.0007 | 0.0003 | −2.017 | 0.0437 |

*Contribution of a Group of Variables* This testing procedure addresses the more general problem of assessing the additional contribution of two or more factors to the prediction of the response over and above that made by other variables already in the regression model. In other words, the null hypothesis is of the form

$$H_0: \beta_1 = \beta_2 = \cdots = \beta_m = 0$$

To test such a null hypothesis, one can perform a likelihood ratio chi-square test, with $m$ df,

$$\chi^2_{LR} = 2[\ln L(\hat{\beta}; \text{ all } X\text{'s}) - \ln L(\hat{\beta}; \text{ all other } X\text{'s with } X\text{'s under}$$
$$\text{investigation deleted})]$$

As with the $z$ test above, this *multiple contribution procedure* is very useful for assessing the importance of potential explanatory variables. In particular it is often used to test whether a similar group of variables, such as *demographic characteristics*, is important for the prediction of the response; these variables have some trait in common. Another application would be a collection of powers *and/or* product terms (referred to as *interaction variables*). It is often of interest to assess the interaction effects collectively before trying to consider individual interaction terms in a model as suggested previously. In fact, such use reduces the total number of tests to be performed, and this, in turn, helps to provide better control of overall type I error rate, which may be inflated due to multiple testing.

*Example 9.7* Refer to the data set on prostate cancer of Example 9.1 (Table 9.1) with all five covariates. We consider, collectively, these four interaction terms: acid × x-ray, acid × stage, acid × grade, and acid × age. The basic idea is to see if *any* of the other variable would modify the effect of the level of acid phosphatase on the response.

1. With the original five variables, we obtained $\ln L = -24.063$.
2. With all nine variables, five original plus four products, we obtained $\ln L = -20.378$.

Therefore,

$$\chi^2_{LR} = 2[\ln L(\hat{\beta}; \text{ nine variables}) - \ln L(\hat{\beta}; \text{ five original variables})]$$
$$= 7.371; 4 \text{ df}, 0.05 \leq p\text{-value} \leq 0.10$$

In other words, all four interaction terms, *considered together*, are marginally significant ($0.05 \leq p$-value $\leq 0.10$); there may be some weak effect modification and that the effect of acid phosphatase on the response may be somewhat stronger for a certain combination of levels of the other four variables.

***Stepwise Regression***   In many applications (e.g., a case–control study on a specific disease) our major interest is to identify important risk factors. In other words, we wish to identify from many available factors a small subset of factors that relate significantly to the outcome (e.g., the disease under investigation). In that identification process, of course, we wish to avoid a large type I (false positive) error. In a regression analysis, a type I error corresponds to including a predictor that has no real relationship to the outcome; such an inclusion can greatly confuse the interpretation of the regression results. In a standard multiple regression analysis, this goal can be achieved by using a strategy that adds into or removes from a regression model one factor at a time according to a certain order of relative importance. Therefore, the two important steps are as follows:

1. Specify a criterion or criteria for selecting a model.
2. Specify a strategy for applying the criterion or criteria chosen.

*Strategies*   This is concerned with specifying the strategy for selecting variables. Traditionally, such a strategy is concerned with whether and which a particular variable should be added to a model or whether any variable should be deleted from a model at a particular stage of the process. As computers became more accessible and more powerful, these practices became more popular.

- **Forward selection procedure**
  1. Fit a simple logistic linear regression model to each factor, one at a time.
  2. Select the most important factor according to a certain predetermined criterion.
  3. Test for the significance of the factor selected in step 2 and determine, according to a certain predetermined criterion, whether or not to add this factor to the model.
  4. Repeat steps 2 and 3 for those variables not yet in the model. At any subsequent step, if none meets the criterion in step 3, no more variables are included in the model and the process is terminated.
- **Backward elimination procedure**
  1. Fit the multiple logistic regression model containing all available independent variables:
  2. Select the least important factor according to a certain predetermined criterion; this is done by considering one factor at a time and treating it as though it were the last variable to enter.

3. Test for the significance of the factor selected in step 2 and determine, according to a certain predetermined criterion, whether or not to delete this factor from the model.

4. Repeat steps 2 and 3 for those variables still in the model. At any subsequent step, if none meets the criterion in step 3, no more variables are removed in the model and the process is terminated.

- **Stepwise regression procedure.**   Stepwise regression is a modified version of forward regression that permits reexamination, at every step, of the variables incorporated in the model in previous steps. A variable entered at an early stage may become superfluous at a later stage because of its relationship with other variables now in the model; the information it provides becomes redundant. That variable may be removed if meeting the elimination criterion, and the model is re-fitted with the remaining variables, and the forward process goes on. The entire process, one step forward followed by one step backward, continues until no more variables can be added or removed.

*Criteria*   For the first step of the forward selection procedure, decisions are based on individual score test results (chi-square, 1 df). In subsequent steps, both forward and backward, the ordering of levels of importance (step 2) and the selection (test in step 3) are based on the likelihood ratio chi-square statistic:

$$\chi^2_{\text{LR}} = 2[\ln L(\hat{\beta}; \text{ all other } X\text{'s}) - \ln L(\hat{\beta}; \text{ all other } X\text{'s with one } X \text{ deleted})]$$

***Example 9.8***   Refer to the data set on prostate cancer of Example 9.1 (Table 9.1) with all five covariates: x-ray, stage, grade, age, and acid. This time we perform a stepwise regression analysis in which we specify that a variable has to be significant at the 0.10 level before it can enter into the model and that a variable in the model has to be significant at the 0.15 for it to remain in the model (most standard computer programs allow users to make these selections; default values are available). First, we get these individual score test results for all variables (Table 9.8). These indicate that x-ray is the most significant variable.

**TABLE 9.8**

| Variable | Score $\chi^2$ | $p$ Value |
|----------|----------|---------|
| X-ray | 11.2831 | 0.0008 |
| Stage | 7.4383 | 0.0064 |
| Grade | 4.0746 | 0.0435 |
| Age | 1.0936 | 0.2957 |
| Acid | 3.1172 | 0.0775 |

**TABLE 9.9**

| Variable | Score $\chi^2$ | $p$ Value |
|---|---|---|
| Stage | 5.6394 | 0.0176 |
| Grade | 2.3710 | 0.1236 |
| Age | 1.3523 | 0.2449 |
| Acid | 2.0733 | 0.1499 |

**TABLE 9.10**

| Factor | Coefficient | Standard Error | $z$ Statistic | $p$ Value |
|---|---|---|---|---|
| Intercept | −2.0446 | 0.6100 | −3.352 | 0.0008 |
| X-ray | 2.1194 | 0.7468 | 2.838 | 0.0045 |
| Stage | 1.5883 | 0.7000 | 2.269 | 0.0233 |

**TABLE 9.11**

| Variable | Score $\chi^2$ | $p$ value |
|---|---|---|
| Grade | 0.5839 | 0.4448 |
| Age | 1.2678 | 0.2602 |
| Acid | 3.0917 | 0.0787 |

**TABLE 9.12**

| Factor | Coefficient | Standard Error | $z$ Statistic | $p$ Value |
|---|---|---|---|---|
| Intercept | −3.5756 | 1.1812 | −3.027 | 0.0025 |
| X-ray | 2.0618 | 0.7777 | 2.651 | 0.0080 |
| Stage | 1.7556 | 0.7391 | 2.375 | 0.0175 |
| Acid | 0.0206 | 0.0126 | 1.631 | 0.1029 |

**TABLE 9.13**

| Variable | Score $\chi^2$ | $p$ Value |
|---|---|---|
| Grade | 1.065 | 0.3020 |
| Age | 1.5549 | 0.2124 |

- *Step 1:* Variable "x-ray" is entered. Analysis of variables not in the model is shown in Table 9.9.
- *Step 2:* Variable "stage" is entered. Analysis of variables in the model (Table 9.10) shows that neither variable is removed. Analysis of variables not in the model is shown in Table 9.11.

- *Step 3:* Variable "acid" is entered. Analysis of variables in the model is shown in Table 9.12. None of the variables are removed. Analysis of variables not in the model is shown in Table 9.13. No (additional) variables meet the 0.1 level for entry into the model.

*Note:* An SAS program would include these instructions:

```
PROC LOGISTIC DESCENDING
DATA = CANCER;
MODEL NODES = X-RAY, GRADE, STAGE, AGE, ACID
/SELECTION = STEPWISE SLE = .10 SLS = .15 DETAILS;
```

where CANCER is the name assigned to the data set, NODES is the variable name for nodal involvement, and X-RAY, GRADE, STAGE, AGE, and ACID are the variable names assigned to the five covariates. The option DETAILS provides step-by-step detailed results; without specifying it, we would have only the final fitted model (which is just fine in practical applications). The default values for SLE (entry) and SLS (stay) probabilities are 0.05 and 0.10, respectively.

### 9.2.5   Receiver Operating Characteristic Curve

Screening tests, as presented in Chapter 1, were focused on binary test outcome. However, it is often true that the result of the test, although dichotomous, is based on the dichotomization of a continuous variable—say, $X$—herein referred to as the *separator variable*. Let us assume without loss of generality that smaller values of $X$ are associated with the diseased population, often called the *population of the cases*. Conversely, larger values of the separator are assumed to be associated with the control or nondiseased population.

A test result is classified by choosing a cutoff $X = x$ against which the observation of the separator is compared. A test result is positive if the value of the separator does not exceed the cutoff; otherwise, the result is classified as negative. Most diagnostic tests are imperfect instruments, in the sense that healthy persons will occasionally be classified wrongly as being ill, while some people who are really ill may fail to be detected as such. Therefore, there is the ensuing conditional probability of the correct classification of a randomly selected case, or the sensitivity of the test as defined in Section 1.1.2, which is estimated by the proportion of cases with $X \leq x$. Similarly, the conditional probability of the correct classification of a randomly selected control, or the specificity of the test as defined in Section 1.1.2, which can be estimated by the proportion of controls with $X \geq x$. A *receiver operating characteristic* (ROC) curve, the trace of the sensitivity versus $(1 - \text{specificity})$ of the test, is generated as the cutoff $x$ moves through its range of possible values. The ROC curve goes from left bottom corner $(0,0)$ to right top corner $(1,1)$ as shown in Figure 9.2.

Being able to estimate the ROC curve, you would be able to do a number of things.

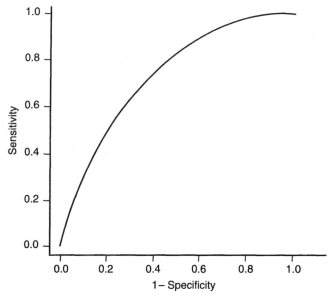

**Figure 9.2**  Receiving operating characteristic (ROC) curve.

1. We can determine the *optimal* cut point, which is nearest to the upper left corner $(0, 1)$. This corner corresponds to 100% sensitivity and 100% specificity.

2. We can estimate the *separation power* of the separator $X$, which is estimated by the area under the ROC curve estimated above. Given two available separators, the better separator is the one with the higher separation power.

Given two independent samples, $\{x_{1i}\}_{i=1}^{m}$ and $\{x_{2j}\}_{j=1}^{n}$, from $m$ controls and $n$ cases, respectively, the *estimated ROC curve*, often called the *nonparametric ROC curve*, is defined as the *random walk* from the left bottom corner $(0, 0)$ to the right top corner $(1, 1)$ whose next step is $1/m$ to the right or $1/n$ up, according to whether the next observation in the ordered combined sample is a control $(x_1)$ or a case $(x_2)$. For example, suppose that we have the samples

$$x_{21} < x_{22} < x_{11} < x_{23} < x_{12} \qquad (n = 3, m = 2)$$

Then the nonparametric ROC curve is as shown in Figure 9.3.

### 9.2.6  ROC Curve and Logistic Regression

In the usual (also referred to as Gaussian) regression analyses of Chapter 8, $R^2$ gives the proportional reduction in variation in comparing the conditional variation of the response to the marginal variation. It describes the strength of

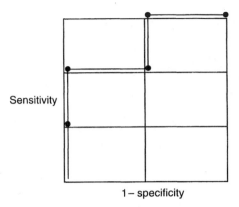

Sensitivity

1 – specificity

**Figure 9.3**   Nonparametric ROC curve.

the association between the response and the set of independent variables considered together; for example, with $R^2 = 1$, we can predict the response perfectly.

For logistic regression analyses, after fitting a logistic regression model, each subject's fitted response probability, $\hat{\pi}_i$, can be calculated. Using these probabilities *as values of a separator*, we can construct a nonparametric ROC curve-tracing sensitivities against the estimated false positivities for various cut points. Such an ROC curve not only makes it easy to determine an optimal cut point [the point on the curve nearest the top left corner (0,1) which corresponds to 1.0 sensitivity and 1.0 specificity] but also shows the overall performance of the fitted logistic regression model; the better the performance, the farther away the curve is from the diagonal. The area $C$ under this ROC curve can be used as a measure of goodness of fit. The measure $C$ represents the *separation power* of the logistic model under consideration; for example, with $C = 1$, the fitted response probabilities for subjects with $y = 1$ and the fitted response probabilities for subjects with $y = 0$ are *separated* completely.

*Example 9.9*   Refer to the data set on prostate cancer of Example 9.1 (Table 9.1) with all five covariates and fitted results shown in Example 9.4. Using the estimated regression parameters obtained from Example 9.4, we have $C = 0.845$.

*Note:* The area under the ROC curve, measure $C$, is provided by SAS's PROC LOGISTIC.

Since the measure of goodness of fit $C$ has a meaningful interpretation and increases when we add an explanatory variable to the model, it can be used as a criterion in performing stepwise logistic regression instead of the $p$ value, which is easily influenced by the sample size. For example, in the forward selection procedure, we proceed as follows:

1. Fit a simple logistic linear regression model to each factor, one at a time.
2. Select the most important factor defined as the one with the largest value of the measure of goodness of fit $C$.
3. Compare this value of $C$ for the factor selected in step 2 and determine, according to a predetermined criterion, whether or not to add this factor to the model—say, to see if $C \geq 0.53$, an increase of 0.03 or 3% over 0.5 when no factor is considered.
4. Repeat steps 2 and 3 for those variables not yet in the model. At any subsequent step, if none meets the criterion in step 3—say, increase the separation power by 0.03, no more variables are included in the model and the process is terminated.

## 9.3   BRIEF NOTES ON THE FUNDAMENTALS

Here are a few more remarks on the use of the logistic regression model as well as a new approach to forming one. The usual approach to regression modeling is (1) to assume that independent variable $X$ is fixed, not random, and (2) to assume a functional relationship between a parameter characterizing the distribution of the dependent variable $Y$ and the measured value of the independent variable. For example:

1. In the simple (Gaussian) regression model of Chapter 8, the *model* describes the *mean* of that normally distributed dependent variable $Y$ as a function of the predictor or independent variable $X$,

$$\mu_i = \beta_0 + \beta_1 x_i$$

2. In the simple logistic regression of this chapter, the *model* assumes that the dependent variable $Y$ of interest is a dichotomous variable taking the value 1 with probability $\pi$ and the value 0 with probability $(1 - \pi)$, and that the relationship between $\pi_i$ and the covariate value $x_i$ of the same person is described by the logistic function

$$\pi_i = \frac{1}{1 + \exp[-(\beta_0 + \beta_1 x_i)]}$$

In both cases, the $x_i$'s are treated as fixed values.

A different approach to the same logistic regression model can be described as follows. Assume that the independent variable $X$ is also a random variable following, say, a normal distribution. Then using the Bayes' theorem of Chapter 3, we can express the ratio of posterior probabilities (after data on $X$ were obtained) as the ratio of prior probabilities (before data on $X$ were obtained) times the likelihood ratio:

$$\frac{\Pr(Y=1; X=x)}{\Pr(Y=0; X=x)} = \frac{\Pr(Y=1)}{\Pr(Y=0)} \frac{\Pr(X=x; Y=1)}{\Pr(X=x; Y=0)}$$

On the right-hand side, the ratio of prior probabilities is a constant with respect to $x$, and with our assumption that $X$ has a normal distribution, the likelihood ratio is the ratio of two normal densities. Let

$$\mu_i = E(X; Y = i) \quad \text{for } i = 0, 1)$$

$$\sigma_i^2 = \text{Var}(X; Y = i) \quad \text{for } i = 0, 1)$$

denote the means and variances of the subjects with events (e.g., cases, $Y = 1$) and the subjects without events (e.g., controls, $Y = 0$), respectively, we can write

$$\text{logit} = \ln \frac{p_x}{1 - p_x}$$

$$= \text{constant} + \left( \frac{\mu_1}{\sigma_1^2} - \frac{\mu_0}{\sigma_0^2} \right) x + \frac{1}{2} \left( \frac{1}{\sigma_0^2} - \frac{1}{\sigma_1^2} \right) x^2$$

This result indicates that if $\sigma_1^2$ and $\sigma_0^2$ are not equal, we should have a quadratic model; the model is linear if and only if $\sigma_0^2 = \sigma_1^2$. We often drop the quadratic term, but the robustness has not been investigated fully.

Let us assume that

$$\sigma_0^2 = \sigma_1^2 = \sigma^2$$

so that we have the linear model

$$\text{logit} = \ln \frac{p_x}{1 - p_x}$$

$$= \text{constant} + \frac{\mu_1 - \mu_0}{\sigma^2} x$$

the very same linear logistic model as Section 9.1. It can be seen that with this approach,

$$\beta_1 = \frac{\mu_1 - \mu_0}{\sigma^2}$$

which can easily be estimated using sample means of Chapter 2 and pooled sample variance as used in the $t$ tests of Chapter 7. That is,

$$\hat{\beta}_1 = \frac{\bar{x}_1 - \bar{x}_0}{s_p^2}$$

and it has been shown that this estimate works quite well even if the distribution of $X$ is not normal.

## EXERCISE

**9.1** Radioactive radon is an inert gas that can migrate from soil and rock and accumulate in enclosed areas such as underground mines and homes. The radioactive decay of trace amounts of uranium in Earth's crust through radium is the source of radon, or more precisely, the isotope radon-222. Radon-222 emits alpha particles; when inhaled, alpha particles rapidly diffuse across the alveolar membrane of the lung and are transported by the blood to all parts of the body. Due to the relatively high flow rate of blood in bone marrow, this may be a biologically plausible mechanism for the development of leukemia. Table E9.1 provides some data from a case–control study to investigate the association between indoor residential radon exposure and risk of childhood acute myeloid leukemia. The variables are:

- Disease (1, case; 2, control)
- Radon (radon concentration in $Bq/m^3$)
- Some characteristics of the child: gender (1, male; 2, female), race (1, white; 2, black; 3, Hispanic; 4, Asian; 5, others), Down's syndrome (a known risk factor for leukemia; 1, no; 2, yes)
- Risk factors from the parents: Msmoke (1, mother a current smoker; 2, no; 0, unknown), Mdrink (1, mother a current alcohol drinker; 2, no; 0, unknown), Fsmoke (1, father a current smoker; 2, no; 0, unknown), Fdrink (1, father a current alcohol drinker; 2, no; 0, unknown).

(*Note*: An electronic copy of this file is available upon request.)

**(a)** Taken collectively, do the covariates contribute significantly to separation of the cases and controls? Give your interpretation for the measure of goodness of fit $C$.

**(b)** Fit the multiple regression model to obtain estimates of individual regression coefficients and their standard errors. Draw your conclusion concerning the conditional contribution of each factor.

**(c)** Within the context of the multiple regression model in part (b), does gender alter the effect of Down's syndrome?

**(d)** Within the context of the multiple regression model in part (b), does Down's syndrome alter the effect of radon exposure?

**(e)** Within the context of the multiple regression model in part (b), taken collectively, do the smoking–drinking variables (by the father or mother) relate significanty to the disease of the child?

**(f)** Within the context of the multiple regression model in part (b), is the effect of radon concentration linear?

**(g)** Focus on radon exposure as the primary factor, taken collectively, was this main effect altered by any other covariates?

TABLE E9.1

| DISEASE | GENDER | RACE | RADON | MSMOKE | MDRINK | FSMOKE | FDRINK | DOWNS |
|---|---|---|---|---|---|---|---|---|
| 1 | 2 | 1 | 17 | 1 | 1 | 1 | 2 | 1 |
| 1 | 2 | 1 | 8 | 1 | 2 | 2 | 0 | 2 |
| 2 | 2 | 1 | 8 | 1 | 2 | 1 | 2 | 2 |
| 2 | 1 | 1 | 1 | 2 | 0 | 2 | 0 | 2 |
| 1 | 1 | 1 | 4 | 2 | 0 | 2 | 0 | 2 |
| 2 | 1 | 1 | 4 | 1 | 1 | 1 | 1 | 2 |
| 1 | 2 | 1 | 5 | 2 | 0 | 2 | 0 | 2 |
| 2 | 1 | 1 | 4 | 2 | 0 | 2 | 0 | 2 |
| 1 | 2 | 1 | 7 | 1 | 1 | 2 | 0 | 1 |
| 1 | 1 | 1 | 15 | 1 | 1 | 1 | 2 | 1 |
| 2 | 1 | 1 | 16 | 2 | 0 | 1 | 1 | 1 |
| 1 | 1 | 1 | 12 | 2 | 0 | 1 | 2 | 1 |
| 2 | 2 | 1 | 14 | 1 | 1 | 1 | 1 | 1 |
| 2 | 1 | 1 | 12 | 2 | 0 | 2 | 0 | 1 |
| 2 | 2 | 1 | 14 | 1 | 2 | 1 | 2 | 2 |
| 2 | 1 | 1 | 9 | 1 | 2 | 1 | 1 | 2 |
| 1 | 1 | 1 | 4 | 2 | 2 | 1 | 1 | 1 |
| 2 | 2 | 1 | 2 | 2 | 0 | 1 | 1 | 2 |
| 1 | 2 | 1 | 12 | 2 | 0 | 1 | 1 | 1 |
| 1 | 2 | 1 | 13 | 2 | 0 | 2 | 0 | 2 |
| 1 | 1 | 1 | 13 | 2 | 0 | 2 | 0 | 1 |
| 2 | 1 | 1 | 18 | 2 | 0 | 2 | 0 | 1 |
| 1 | 1 | 1 | 13 | 1 | 2 | 1 | 1 | 1 |
| 1 | 2 | 1 | 16 | 2 | 0 | 2 | 0 | 2 |
| 2 | 2 | 1 | 10 | 1 | 1 | 2 | 0 | 1 |
| 2 | 1 | 1 | 11 | 2 | 0 | 2 | 0 | 1 |
| 2 | 2 | 1 | 4 | 1 | 1 | 1 | 1 | 2 |
| 1 | 2 | 1 | 1 | 1 | 2 | 1 | 1 | 2 |

| | | | | | | | | |
|---|---|---|---|---|---|---|---|---|
| 1 | 0 | 2 | 0 | 2 | 9 | 2 | 1 | 1 |
| 1 | 2 | 1 | 1 | 1 | 15 | 1 | 2 | 1 |
| 1 | 2 | 1 | 0 | 2 | 17 | 1 | 2 | 2 |
| 1 | 0 | 2 | 0 | 2 | 9 | 1 | 1 | 1 |
| 1 | 1 | 1 | 0 | 2 | 15 | 1 | 1 | 2 |
| 1 | 0 | 2 | 1 | 1 | 10 | 1 | 1 | 1 |
| 1 | 0 | 2 | 2 | 1 | 11 | 1 | 1 | 2 |
| 2 | 0 | 2 | 0 | 2 | 8 | 1 | 2 | 1 |
| 2 | 2 | 1 | 2 | 1 | 14 | 1 | 1 | 1 |
| 2 | 1 | 1 | 2 | 1 | 14 | 1 | 2 | 2 |
| 2 | 1 | 1 | 0 | 2 | 1 | 1 | 2 | 1 |
| 2 | 2 | 2 | 0 | 2 | 1 | 1 | 1 | 2 |
| 1 | 0 | 2 | 2 | 1 | 6 | 1 | 2 | 1 |
| 2 | 0 | 1 | 0 | 2 | 16 | 1 | 2 | 2 |
| 2 | 2 | 1 | 0 | 2 | 3 | 1 | 1 | 1 |
| 1 | 9 | 1 | 0 | 2 | 5 | 2 | 2 | 2 |
| 2 | 2 | 1 | 0 | 2 | 15 | 1 | 2 | 1 |
| 2 | 2 | 2 | 2 | 1 | 17 | 1 | 1 | 2 |
| 2 | 0 | 2 | 0 | 2 | 17 | 1 | 1 | 1 |
| 2 | 0 | 2 | 0 | 2 | 3 | 1 | 1 | 2 |
| 2 | 1 | 1 | 2 | 1 | 11 | 1 | 2 | 1 |
| 1 | 1 | 2 | 1 | 0 | 14 | 1 | 1 | 2 |
| 2 | 0 | 2 | 2 | 1 | 17 | 1 | 2 | 1 |
| 2 | 0 | 2 | 0 | 2 | 1 | 1 | 1 | 2 |
| 1 | 1 | 1 | 1 | 1 | 10 | 1 | 1 | 2 |
| 2 | 2 | 1 | 2 | 1 | 14 | 3 | 1 | 1 |
| 2 | 0 | 2 | 0 | 2 | 4 | 1 | 1 | 2 |
| 2 | 0 | 2 | 1 | 1 | 12 | 1 | 2 | 1 |
| 2 | 2 | 1 | 0 | 2 | 9 | 1 | 2 | 1 |
| 2 | 2 | 1 | 2 | 1 | 7 | 1 | 2 | 2 |
| | | | | | 5 | | | 1 |

(Continued)

| DISEASE | GENDER | RACE | RADON | MSMOKE | MDRINK | FSMOKE | FDRINK | DOWNS |
|---|---|---|---|---|---|---|---|---|
| 1 | 1 | 1 | 8 | 2 | 0 | 2 | 0 | 2 |
| 2 | 1 | 1 | 9 | 2 | 0 | 2 | 0 | 2 |
| 2 | 1 | 1 | 15 | 1 | 2 | 1 | 2 | 2 |
| 1 | 2 | 1 | 10 | 2 | 0 | 2 | 0 | 1 |
| 2 | 2 | 1 | 10 | 2 | 0 | 2 | 0 | 1 |
| 2 | 1 | 1 | 1 | 2 | 0 | 2 | 0 | 2 |
| 2 | 2 | 1 | 1 | 2 | 1 | 1 | 1 | 2 |
| 1 | 2 | 1 | 9 | 1 | 1 | 2 | 0 | 1 |
| 1 | 2 | 5 | 14 | 1 | 0 | 2 | 0 | 2 |
| 1 | 2 | 1 | 8 | 2 | 0 | 1 | 1 | 2 |
| 2 | 2 | 1 | 7 | 2 | 0 | 2 | 0 | 2 |
| 2 | 2 | 1 | 13 | 2 | 0 | 1 | 1 | 2 |
| 1 | 2 | 1 | 1 | 2 | 0 | 2 | 0 | 2 |
| 2 | 2 | 1 | 12 | 2 | 0 | 1 | 2 | 2 |
| 1 | 1 | 5 | 11 | 1 | 2 | 1 | 2 | 1 |
| 2 | 1 | 1 | 2 | 2 | 0 | 2 | 0 | 1 |
| 2 | 2 | 1 | 3 | 1 | 1 | 1 | 2 | 2 |
| 2 | 2 | 1 | 6 | 1 | 2 | 1 | 2 | 2 |
| 2 | 2 | 1 | 3 | 2 | 0 | 1 | 0 | 2 |
| 1 | 1 | 3 | 1 | 2 | 0 | 2 | 2 | 2 |
| 2 | 2 | 5 | 2 | 1 | 2 | 1 | 2 | 2 |
| 1 | 1 | 5 | 14 | 1 | 1 | 1 | 1 | 1 |
| 1 | 1 | 1 | 1 | 1 | 1 | 1 | 2 | 2 |
| 2 | 1 | 1 | 12 | 2 | 0 | 1 | 1 | 1 |
| 2 | 2 | 1 | 13 | 1 | 1 | 1 | 2 | 2 |
| 1 | 2 | 1 | 11 | 2 | 0 | 2 | 0 | 1 |
| 2 | 1 | 1 | 11 | 2 | 0 | 2 | 0 | 1 |

(Continued)

| | | | | | | | | |
|---|---|---|---|---|---|---|---|---|
| 1 | 0 | 2 | 2 | 1 | 16 | 1 | 2 | 1 |
| 2 | 2 | 1 | 0 | 2 | 3 | 1 | 1 | 2 |
| 1 | 0 | 2 | 0 | 2 | 13 | 1 | 1 | 1 |
| 1 | 0 | 2 | 0 | 2 | 12 | 1 | 2 | 2 |
| 1 | 0 | 2 | 0 | 2 | 12 | 1 | 1 | 2 |
| 2 | 2 | 1 | 2 | 1 | 3 | 1 | 1 | 1 |
| 2 | 1 | 1 | 0 | 2 | 5 | 1 | 2 | 2 |
| 1 | 1 | 1 | 1 | 1 | 7 | 1 | 2 | 1 |
| 1 | 2 | 1 | 0 | 2 | 7 | 1 | 2 | 2 |
| 2 | 2 | 1 | 0 | 2 | 2 | 1 | 1 | 1 |
| 2 | 1 | 1 | 1 | 1 | 2 | 1 | 1 | 2 |
| 2 | 0 | 2 | 0 | 2 | 2 | 1 | 1 | 1 |
| 2 | 1 | 1 | 1 | 2 | 3 | 1 | 2 | 2 |
| 2 | 0 | 2 | 2 | 1 | 3 | 1 | 2 | 1 |
| 2 | 0 | 2 | 0 | 2 | 14 | 1 | 2 | 2 |
| 2 | 0 | 2 | 1 | 2 | 15 | 1 | 2 | 2 |
| 1 | 2 | 1 | 0 | 2 | 1 | 1 | 1 | 1 |
| 1 | 1 | 1 | 1 | 1 | 1 | 3 | 1 | 2 |
| 2 | 2 | 1 | 0 | 2 | 1 | 3 | 1 | 2 |
| 2 | 2 | 1 | 0 | 2 | 10 | 1 | 2 | 2 |
| 2 | 2 | 1 | 0 | 2 | 9 | 1 | 2 | 1 |
| 1 | 1 | 1 | 0 | 2 | 14 | 1 | 2 | 2 |
| 1 | 0 | 1 | 1 | 1 | 9 | 1 | 1 | 2 |
| 2 | 1 | 2 | 1 | 1 | 9 | 1 | 2 | 2 |
| 1 | 2 | 1 | 0 | 2 | 17 | 1 | 1 | 2 |
| 2 | 1 | 1 | 1 | 1 | 3 | 1 | 1 | 2 |
| 1 | 1 | 1 | 0 | 1 | 5 | 1 | 1 | 1 |
| 2 | 2 | 1 | 2 | 1 | 15 | 1 | 2 | 2 |
| 1 | 0 | 1 | 1 | 1 | 14 | 1 | 1 | 2 |
| 2 | | 2 | | | | | | 2 |

**TABLE E9.1** (*Continued*)

| DISEASE | GENDER | RACE | RADON | MSMOKE | MDRINK | FSMOKE | FDRINK | DOWNS |
|---|---|---|---|---|---|---|---|---|
| 2 | 2 | 1 | 5 | 1 | 2 | 1 | 2 | 2 |
| 2 | 2 | 3 | 13 | 2 | 0 | 1 | 1 | 1 |
| 2 | 1 | 1 | 15 | 2 | 0 | 1 | 1 | 2 |
| 2 | 1 | 1 | 12 | 2 | 0 | 1 | 1 | 1 |
| 1 | 1 | 1 | 1 | 2 | 0 | 1 | 2 | 2 |
| 2 | 2 | 1 | 2 | 2 | 0 | 2 | 0 | 2 |
| 2 | 2 | 3 | 1 | 2 | 0 | 1 | 2 | 2 |
| 2 | 2 | 5 | 3 | 2 | 0 | 2 | 0 | 2 |
| 2 | 2 | 1 | 15 | 2 | 0 | 2 | 0 | 1 |
| 1 | 1 | 2 | 6 | 1 | 2 | 2 | 0 | 1 |
| 2 | 1 | 1 | 5 | 2 | 0 | 2 | 1 | 2 |
| 1 | 1 | 1 | 2 | 1 | 1 | 1 | 0 | 2 |
| 1 | 1 | 1 | 15 | 2 | 0 | 2 | 0 | 1 |
| 1 | 1 | 2 | 5 | 1 | 1 | 2 | 1 | 2 |
| 1 | 1 | 1 | 2 | 2 | 2 | 1 | 1 | 2 |
| 2 | 2 | 1 | 2 | 1 | 0 | 1 | 2 | 2 |
| 1 | 1 | 1 | 17 | 1 | 0 | 1 | 2 | 2 |
| 2 | 2 | 1 | 8 | 2 | 0 | 2 | 0 | 2 |
| 2 | 2 | 1 | 6 | 2 | 0 | 1 | 1 | 2 |
| 1 | 1 | 1 | 1 | 1 | 2 | 1 | 1 | 2 |
| 1 | 2 | 1 | 13 | 2 | 0 | 2 | 2 | 1 |
| 2 | 1 | 1 | 12 | 2 | 0 | 2 | 1 | 1 |
| 2 | 1 | 1 | 9 | 2 | 0 | 2 | 0 | 1 |
| 1 | 1 | 1 | 10 | 1 | 1 | 1 | 0 | 1 |
| 2 | 2 | 1 | 15 | 1 | 1 | 2 | 0 | 1 |
| 1 | 2 | 1 | 3 | 2 | 0 | 1 | 2 | 2 |
| 2 | 1 | 1 | 2 | 1 | 2 | 1 | 1 | 2 |
| 1 | 2 | 1 | 11 | 1 | 2 | 1 | 2 | 2 |

( Continued )

2 2 2 1 2 2 2 2 2 2 2 2 2 1 1 2 2 2 1 2 2 2 2 1 2 1 2 2 2 2 2 2

2 0 0 1 2 2 0 1 1 1 2 1 0 1 1 0 0 0 0 2 0 2 2 1 2 2 0 0 0 0

1 2 2 1 1 2 1 1 1 1 2 1 1 2 2 2 2 1 2 1 1 1 1 1 2 2 2 2

0 0 0 0 0 1 0 2 1 2 0 1 0 2 0 0 0 0 0 0 0 2 2 1 2 2 0 1 0 0

2 2 2 2 2 1 2 1 1 1 2 1 2 1 2 2 2 2 2 1 1 1 1 1 2 1 2 2

11 4 2 12 1 15 8 1 2 8 2 1 12 12 1 12 16 9 7 2 1 11 13 7 13 14 2 1 2 2

1 1 1 1 1 1 1 1 1 1 1 1 1 1 1 1 1 1 1 1 3 4 1 1 1 1 1 1 1 1 1 1 1

2 1 2 2 2 1 2 2 2 1 1 1 2 1 2 1 2 2 2 2 1 2 2 2 1 2 1 2 2 2

2 2 1 1 1 1 2 2 2 1 2 1 2 1 2 1 2 2 1 2 1 2 1 1 2 2 2 2 1 2 1 2 1

**347**

**TABLE E9.1** (*Continued*)

| DISEASE | GENDER | RACE | RADON | MSMOKE | MDRINK | FSMOKE | FDRINK | DOWNS |
|---|---|---|---|---|---|---|---|---|
| 2 | 2 | 1 | 1 | 1 | 2 | 2 | 0 | 2 |
| 2 | 1 | 5 | 1 | 2 | 0 | 1 | 1 | 2 |
| 2 | 1 | 1 | 9 | 1 | 1 | 1 | 2 | 1 |
| 1 | 1 | 1 | 13 | 1 | 2 | 2 | 0 | 2 |
| 2 | 1 | 1 | 12 | 2 | 0 | 2 | 0 | 1 |
| 1 | 2 | 1 | 17 | 2 | 0 | 2 | 0 | 1 |
| 2 | 2 | 1 | 10 | 2 | 2 | 2 | 2 | 1 |
| 2 | 1 | 1 | 11 | 1 | 0 | 1 | 0 | 1 |
| 2 | 2 | 1 | 5 | 2 | 0 | 2 | 0 | 2 |
| 2 | 2 | 1 | 15 | 2 | 0 | 2 | 1 | 1 |
| 1 | 2 | 1 | 2 | 2 | 1 | 1 | 0 | 2 |
| 2 | 2 | 1 | 2 | 1 | 1 | 2 | 1 | 2 |
| 2 | 2 | 1 | 11 | 1 | 0 | 1 | 9 | 2 |
| 1 | 1 | 1 | 10 | 2 | 1 | 1 | 2 | 1 |
| 1 | 1 | 1 | 5 | 1 | 1 | 1 | 0 | 2 |
| 2 | 1 | 1 | 4 | 1 | 2 | 2 | 1 | 2 |
| 2 | 2 | 1 | 1 | 1 | 1 | 1 | 0 | 2 |
| 2 | 2 | 1 | 1 | 2 | 0 | 2 | 0 | 2 |
| 1 | 2 | 1 | 2 | 2 | 0 | 2 | 0 | 2 |
| 2 | 2 | 1 | 10 | 1 | 2 | 2 | 0 | 2 |
| 1 | 2 | 1 | 9 | 2 | 0 | 2 | 0 | 1 |
| 1 | 1 | 1 | 13 | 2 | 0 | 2 | 0 | 2 |
| 2 | 1 | 1 | 12 | 2 | 0 | 2 | 0 | 1 |
| 2 | 2 | 1 | 14 | 2 | 0 | 2 | 1 | 2 |
| 1 | 1 | 1 | 3 | 1 | 2 | 1 | 0 | 2 |
| 1 | 1 | 5 | 12 | 2 | 0 | 2 | 0 | 2 |
| 1 | 1 | 1 | 2 | 2 | 0 | 1 | 2 | 2 |
| 2 | 1 | 1 | 3 | 1 | 1 | 2 | 0 | 2 |

# 10

# METHODS FOR COUNT DATA

Chapter 1 was devoted to descriptive methods for categorical data, but most topics were centered on binary data and the binomial distribution. Chapter 9 continues on that direction with logistic regression methods for binomial-distributed responses. This chapter is devoted to a different type of categorical data, count data; the eventual focus is the Poisson regression model; the Poisson distribution was introduced very briefly in Chapter 3. As usual, the purpose of the research is to assess relationships among a set of variables, one of which is taken to be the response or dependent variable, that is, a variable to be predicted from or explained by other variables; other variables are called predictors, explanatory variables, or independent variables. Choosing an appropriate model and analytical technique depends on the type of response variable or dependent variable under investigation. The Poisson regression model applies when the dependent variable follows a Poisson distribution.

## 10.1 POISSON DISTRIBUTION

The binomial distribution is used to characterize an experiment when each trial of the experiment has two possible outcomes (often referred to as *failure* and *success*. Let the probabilities of failure and success be, respectively, $1 - \pi$ and $\pi$; the target for the binomial distribution is the total number $X$ of successes in $n$ trials. The Poisson model, on the other hand, is used when the random variable $X$ is supposed to represent the number of occurrences of some random event in an interval of time or space, or some volume of matter, so that it is not bounded by $n$ as in the binomial distribution; numerous applications in health sciences have been documented. For example, the number of viruses in a solution, the number of defective teeth per person, the number of focal lesions in virology, the number of victims of specific diseases, the number of cancer deaths per household, and the number of infant deaths in a certain locality during a given year, among others.

The probability density function of a Poisson distribution is given by

$$\Pr(X = x) = \frac{\theta^x e^{-\theta}}{x!} \qquad \text{for } x = 0, 1, 2, \ldots$$

$$= p(x; \theta)$$

The mean and variance of the Poisson distribution $P(\theta)$ are

$$\mu = \theta$$

$$\sigma^2 = \theta$$

However, the binomial distribution and the Poisson distribution are not totally unrelated. In fact, it can be shown that a binomial distribution with large $n$ and small $\pi$ can be approximated by the Poisson distribution with $\theta = n\pi$.

Given a sample of counts from the Poisson distribution $P(\theta)$, $\{x_i\}_{i=1}^{n}$, the sample mean $\bar{x}$ is an unbiased estimator for $\theta$; its standard error is given by

$$\text{SE}(\bar{x}) = \sqrt{\frac{\bar{x}}{n}}$$

***Example 10.1***  In estimating the infection rates in populations of organisms, sometimes it is impossible to assay each organism individually. Instead, the organisms are randomly divided into a number of pools and each pool is tested as a unit. Let

$N$ = number of insects in the sample

$n$ = number of pools used in the experiment

$m$ = number of insects per pool, $N = nm$ (for simplicity, assume that $m$ is the same for every pool)

The random variable $X$ concerned is the number of pools that show negative test results (i.e., none of the insects are infected).

Let $\lambda$ be the population infection rate; the probability that all $m$ insects in a pool are negative (in order to have a negative pool) is given by

$$\pi = (1 - \lambda)^m$$

Designating a negative pool as "success," we have a binomial distribution for

$X$, which is $B(n, \pi)$. In situations where the infection rate $\lambda$ is very small, the Poisson distribution could be used as an approximation, with $\theta = m\lambda$ being the expected number of infected insects in a pool. The Poisson probability of this number being zero is

$$\pi = \exp(-\theta)$$

and we have the same binomial distribution $B(n, \pi)$. It is interesting to note that testing for syphilis (as well as other very rare diseases) in the U.S. Army used to be done this way.

**Example 10.2**    For the year of 1981, the infant mortality rate (IMR) for the United States was 11.9 deaths per 1000 live births. For the same period, the New England states (Connecticut, Maine, Massachusetts, New Hampshire, Rhode Island, and Vermont) had 164,200 live births and 1585 infant deaths. If the national IMR applies, the mean and vaiance of the number of infant deaths in the New England states would be

$$(164.2)(11.9) = 1954$$

From the $z$ score,

$$z = \frac{1585 - 1954}{\sqrt{1954}}$$
$$= -8.35$$

it is clear that the IMR in the New England states is below the national average.

**Example 10.3**    Cohort studies are designs in which one enrolls a group of healthy persons and follows them over certain periods of time; examples include occupational mortality studies. The cohort study design focuses attention on a particular exposure rather than a particular disease as in case–control studies. Advantages of a longitudinal approach include the opportunity for more accurate measurement of exposure history and a careful examination of the time relationships between exposure and disease.

The observed mortality of the cohort under investigation often needs to be compared with that expected from the death rates of the national population (served as standard), with allowance made for age, gender, race, and time period. Rates may be calculated either for total deaths or for separate causes of interest. The statistical method is often referred to as the *person-years method*. The basis of this method is the comparison of the observed number of deaths, $d$, from the cohort with the mortality that would have been expected if the group had experienced death rates similar to those of the standard population of which the cohort is a part. The expected number of deaths is usually calculated using published national life tables and the method is similar to that of indirect standardization of death rates.

Each member of the cohort contributes to the calculation of the expected deaths for those years in which he or she was at risk of dying during the study. There are three types of subjects:

1. Subjects still alive on the analysis date.
2. Subjects who died on a known date within the study period.
3. Subjects who are lost to follow-up after a certain date. These cases are a potential source of bias; effort should be expended in reducing the number of subjects in this category.

Figure 10.1 shows the sitution illustrated by one subject of each type, from enrollment to the study termination. Each subject is represented by a diagonal line that starts at the age and year at which the subject entered the study and continues as long as the subject is at risk of dying in the study. In the figure, each subject is represented by a line starting from the year and age at entry and continuing until the study date, the date of death, or the date the subject was lost to follow-up. Period and age are divided into five-year intervals corresponding to the usual availability of referenced death rates. Then a quantity, $r$,

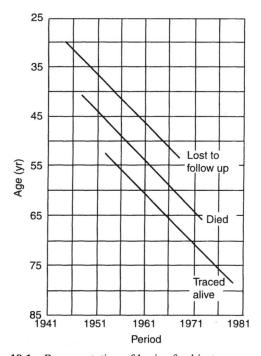

**Figure 10.1**    Representation of basis of subject-years method.

is defined for each person as the cumulative risk over the follow-up period:

$$r = \sum x\omega$$

where summation is over each square in Figure 10.1 entered by the follow-up line, $x$ the time spent in a square, and $\omega$ the corresponding death rate for the given age–period combination. For the cohort, the individual values of $r$ are added to give the expected number of deaths:

$$m = \sum r$$

For various statistical analyses, the observed number of deaths $d$ may be treated as a Poisson variable with mean $\theta = m\rho$, where $\rho$ is the relative risk of being a member of the cohort as compared to the standard population.

## 10.2  TESTING GOODNESS OF FIT

A goodness-of-fit test is used when one wishes to decide if an observed distribution of frequencies is incompatible with some hypothesized distribution. The Poisson is a very special distribution; its mean and its variance are equal. Therefore, given a sample of count data $\{x_i\}_{i=1}^{n}$, we often wish to know whether these data provide sufficient evidence to indicate that the sample did not come from a Poisson-distributed population. The hypotheses are as follows:

$H_0$: The sampled population is distributed as Poisson.

$H_A$: The sampled population is not distributed as Poisson.

The most frequent violation is an overdispersion; the variance is larger than the mean. The implication is serious; the analysis assumes that the Poisson model often underestimates standard error(s) and thus wrongly inflates the level of significance.

The test statistic is the familiar Pearson chi-square:

$$X^2 = \sum_{i}^{k} \frac{(O_i - E_i)^2}{E_i}$$

where $O_i$ and $E_i$ refer to the $i$th observed and expected frequencies, respectively (we used the notations $x_{ij}$ and $\widehat{m}_{ij}$ in Chapter 6). In this formula, $k$ is the number of groups for which observed and expected frequencies are available. When the null hypothesis is true, the test statistic is distributed as chi-square with $(k - 2)$ degrees of freedom; 1 degree of freedom was lost because the mean needs to be estimated and 1 was lost because of the constraint $\sum O_i = \sum E_i$. It is also recommended that adjacent groups at the bottom of the table be combined to avoid having any expected frequencies less than 1.

***Example 10.4***  The purpose of this study was to examine the data for 44 physicians working for an emergency department at a major hospital. The response variable is the number of complaints received during the preceding year; other details of the study are given in Example 10.5. For the purpose of testing the goodness of fit, the data are summarized in Table 10.1.

To obtain the frequencies expected, we first obtain relative frequencies by evaluating the Poisson probability for each value of $X = x$,

$$\Pr(X = x) = \frac{\theta^x e^{-\theta}}{x!} \qquad \text{for } x = 0, 1, 2, \ldots$$

Since

$$\hat{\theta} = \frac{\sum x_i}{44}$$

$$= 3.34$$

we have, for example,

$$\Pr(X = 2) = \frac{3.34^2 e^{-3.34}}{2!}$$

$$= 0.198$$

Each of the expected relative frequencies is multiplied by the sample size, 44, to

**TABLE 10.1**

| Number of Complaints | Observed $O_i$ | Expected $E_i$ |
|---|---|---|
| 0 | 1 | 1.54 |
| 1 | 12 | 5.19 |
| 2 | 12 | 8.71 |
| 3 | 5 | 9.68 |
| 4 | 1 | 8.10 |
| 5 | 4 | 5.46 |
| 6 | 2 | 2.99 |
| 7 | 2 | 1.45 |
| 8 | 2 | 0.62 |
| 9 | 1 | 0.22 |
| 10 | 1 | 0.09 |
| 11 | 1 | 0.04 |

obtain the expected frequencies; for example,

$$E_2 = (44)(0.198)$$
$$= 8.71$$

To avoid having any expected frequencies less than 1, we combine the last five groups together, resulting in eight groups available for testing goodness of fit with

$$O_8 = 2 + 2 + 1 + 1 + 1$$
$$= 7$$
$$E_8 = 1.45 + 0.62 + 0.22 + 0.09 + 0.04$$
$$= 2.42$$

The result is

$$X^2 = \frac{(1 - 1.59)^2}{1.19} + \frac{(12 - 5.19)^2}{5.19} + \cdots + \frac{(7 - 2.42)^2}{2.42}$$
$$= 28.24$$

with $8 - 2 = 6$ degrees of freedom, indicating a significant deviation from the Poisson distribution ($p < 0.005$). A simple inspection of Table 10.1 reveals an obvious overdispersion.

## 10.3   POISSON REGRESSION MODEL

As mentioned previously, the Poisson model is often used when the random variable $X$ is supposed to represent the number of occurrences of some random event in an interval of time or space, or some volume of matter, and numerous applications in health sciences have been documented. In some of these applications, one may be interested to see if the Poisson-distributed dependent variable can be predicted from or explained by other variables. The other variables are called predictors, explanatory variables, or independent variables. For example, we may be interested in the number of defective teeth per person as a function of gender and age of a child, brand of toothpaste, and whether or not the family has dental insurance. In this and other examples, the dependent variable $Y$ is assumed to follow a Poisson distribution with mean $\theta$.

The Poisson regression model expresses this mean as a function of certain independent variables $X_1, X_2, \ldots, X_k$, in addition to the *size* of the observation unit from which one obtained the count of interest. For example, if $Y$ is the number of virus in a solution, the *size* is the volume of the solution; or if $Y$ is the number of defective teeth, the *size* is the total number of teeth for that same person. The following is the continuation of Example 10.4 on the emergency service data; but data in Table 10.2 also include information on four covariates.

***Example 10.5***  The purpose of this study was to examine the data for 44 physicians working for an emergency at a major hospital so as to determine which of four variables are related to the number of complaints received during the preceding year. In addition to the number of complaints, served as the dependend variable, data available consist of the number of visits (which serves as the *size* for the observation unit, the physician) and four covariates. Table 10.2 presents the complete data set. For each of the 44 physicians there are two continuous independent variables, the revenue (dollars per hour) and the workload at the emergency service (hours) and two binary variables, gender (female/male) and residency traning in emergency services (no/yes).

### 10.3.1  Simple Regression Analysis

In this section we discuss the basic ideas of simple regression analysis when only one predictor or independent variable is available for predicting the response of interest.

***Poisson Regression Model***  In our framework, the dependent variable $Y$ is assumed to follow a Poisson distribution; its values $y_i$'s are available from $n$ *observation units*, which is also characterized by an independent variable $X$. For the observation unit $i$ $(1 \leq n)$, let $s_i$ be the size and $x_i$ be the covariate value.

The Poisson regression model assumes that the relationship between the mean of $Y$ and the covariate $X$ is described by

$$E(Y_i) = s_i \lambda(x_i)$$
$$= s_i \exp(\beta_0 + \beta_1 x_i)$$

where $\lambda(x_i)$ is called the *risk* of observation unit $i$ $(1 \leq n)$. Under the assumption that $Y_i$ is Poisson, the likelihood function is given by

$$L(y; \beta) = \prod_{i=1}^{n} \frac{[s_i \lambda(x_i)]^{y_i} \exp[-s_i \lambda(x_i)]}{y_i!}$$

$$\ln L = \sum_{i=1}^{n} [y_i \ln s_i - \ln y_i! + y_i(\beta_0 + \beta_1 x_i) - s_i \exp(\theta_0 + \beta_1 x_i)]$$

from which estimates for $\beta_0$ and $\beta_1$ can be obtained by the maximum likelihood procedure.

***Measure of Association***  Consider the case of a binary covariate $X$: say, representing an exposure $(1 = \text{exposed}, 0 = \text{not exposed})$. We have:

**TABLE 10.2**

| No. of Visits | Complaint | Residency | Gender | Revenue | Hours |
|---|---|---|---|---|---|
| 2014 | 2 | Y | F | 263.03 | 1287.25 |
| 3091 | 3 | N | M | 334.94 | 1588.00 |
| 879 | 1 | Y | M | 206.42 | 705.25 |
| 1780 | 1 | N | M | 226.32 | 1005.50 |
| 3646 | 11 | N | M | 288.91 | 1667.25 |
| 2690 | 1 | N | M | 275.94 | 1517.75 |
| 1864 | 2 | Y | M | 295.71 | 967.00 |
| 2782 | 6 | N | M | 224.91 | 1609.25 |
| 3071 | 9 | N | F | 249.32 | 1747.75 |
| 1502 | 3 | Y | M | 269.00 | 906.25 |
| 2438 | 2 | N | F | 225.61 | 1787.75 |
| 2278 | 2 | N | M | 212.43 | 1480.50 |
| 2458 | 5 | N | M | 211.05 | 1733.50 |
| 2269 | 2 | N | F | 213.23 | 1847.25 |
| 2431 | 7 | N | M | 257.30 | 1433.00 |
| 3010 | 2 | Y | M | 326.49 | 1520.00 |
| 2234 | 5 | Y | M | 290.53 | 1404.75 |
| 2906 | 4 | N | M | 268.73 | 1608.50 |
| 2043 | 2 | Y | M | 231.61 | 1220.00 |
| 3022 | 7 | N | M | 241.04 | 1917.25 |
| 2123 | 5 | N | F | 238.65 | 1506.25 |
| 1029 | 1 | Y | F | 287.76 | 589.00 |
| 3003 | 3 | Y | F | 280.52 | 1552.75 |
| 2178 | 2 | N | M | 237.31 | 1518.00 |
| 2504 | 1 | Y | F | 218.70 | 1793.75 |
| 2211 | 1 | N | F | 250.01 | 1548.00 |
| 2338 | 6 | Y | M | 251.54 | 1446.00 |
| 3060 | 2 | Y | M | 270.52 | 1858.25 |
| 2302 | 1 | N | M | 247.31 | 1486.25 |
| 1486 | 1 | Y | F | 277.78 | 933.75 |
| 1863 | 1 | Y | M | 259.68 | 1168.25 |
| 1661 | 0 | N | M | 260.92 | 877.25 |
| 2008 | 2 | N | M | 240.22 | 1387.25 |
| 2138 | 2 | N | M | 217.49 | 1312.00 |
| 2556 | 5 | N | M | 250.31 | 1551.50 |
| 1451 | 3 | Y | F | 229.43 | 973.75 |
| 3328 | 3 | Y | M | 313.48 | 1638.25 |
| 2927 | 8 | N | M | 293.47 | 1668.25 |
| 2701 | 8 | N | M | 275.40 | 1652.75 |
| 2046 | 1 | Y | M | 289.56 | 1029.75 |
| 2548 | 2 | Y | M | 305.67 | 1127.00 |
| 2592 | 1 | N | M | 252.35 | 1547.25 |
| 2741 | 1 | Y | F | 276.86 | 1499.25 |
| 3763 | 10 | Y | M | 308.84 | 1747.50 |

*Note:* This is a very long data file; its electronic copy, in a Web-based form, is available from the author upon request.

1. If the observation unit is exposed,

$$\ln \lambda_i(\text{exposed}) = \beta_0 + \beta_1$$

whereas

2. If the observation unit is not exposed,

$$\ln \lambda_i(\text{not exposed}) = \beta_0$$

or, in other words,

$$\frac{\lambda_i(\text{exposed})}{\lambda_i(\text{not exposed})} = e^{\beta_1}$$

This quantity is called the *relative risk associated with the exposure.*

Similarly, we have for a continuous covariate $X$ and any value $x$ of $X$,

$$\ln \lambda_i(X = x) = \beta_0 + \beta_1 x$$
$$\ln \lambda_i(X = x + 1) = \beta_0 + \beta_1(x + 1)$$

so that

$$\frac{\lambda_i(X = x + 1)}{\lambda_i(X = x)} = e^{\beta_1}$$

representing the relative risk associated with a 1-unit increase in the value of $X$.

The basic rationale for using the terms *risk* and *relative risk* is the approximation of the binomial distribution by the Poisson distribution. Recall from Section 10.2 that when $n \to \infty$, $\pi \to 0$ while $\theta = n\pi$ remains constant, the binomial distribution $B(n, \pi)$ can be approximated by the Poisson distribution $P(\theta)$. The number $n$ is the size of the observation unit; so the ratio between the mean and the size represents the $\pi$ [or $\lambda(x)$ in the new model], the probability or *risk* and the ratio of risks is the risks ratio or *relative risk*.

***Example 10.6***  Refer to the emergency service data in Example 10.5 (Table 10.2) and suppose that we want to investigate the relationship between the number of complaints (adjusted for number of visits) and residency training. It may be perceived that by having training in the specialty a physician would perform better and therefore would be less likely to provoke complaints. An application of the simple Poisson regression analysis yields the results shown in Table 10.3.

The result indicates that the common perception is almost true, that the relationship between the number of complaints and no residency training in emergency service is marginally significant ($p = 0.0779$); the relative risk asso-

**TABLE 10.3**

| Variable | Coefficient | Standard Error | z Statistic | p Value |
|----------|-------------|----------------|-------------|---------|
| Intercept | −6.7566 | 0.1387 | −48.714 | <0.0001 |
| No residency | 0.3041 | 0.1725 | 1.763 | 0.0779 |

ciated with no residency training is

$$\exp(0.3041) = 1.36$$

Those without previous training is 36% more likely to receive the same number of complaints as those who were trained in the specialty.

*Note:* An SAS program would include these instructions:

```
DATA EMERGENCY;
INPUT VISITS CASES RESIDENCY;
LN = LOG(VISITS);
CARDS;
(Data);
PROC GENMOD DATA = EMERGENCY;
CLASS RESIDENCY;
MODEL CASES = RESIDENCY/DIST = POISSON LINK = LOG
OFFSET = LN;
```

where EMERGENCY is the name assigned to the data set, VISITS is the number of visits, CASES is the number of complaints, and RESIDENCY is the binary covariate, indicating whether the physician received residency training in the specialty. The option CLASS is used to declare that the covariate is categorical.

### 10.3.2 Multiple Regression Analysis

The effect of some factor on a dependent or response variable may be influenced by the presence of other factors through effect modifications (i.e., interactions). Therefore, to provide a more comprehensive analysis, it is very desirable to consider a large number of factors and sort out which are most closely related to the dependent variable. This method, which is multiple Poisson regression analysis, involves a linear combination of the explanatory or independent variables; the variables must be quantitative with particular numerical values for each observation unit. A covariate or independent variable may be dichotomous, polytomous, or continuous; categorical factors will be represented by dummy variables. In many cases, data transformations of continuous measurements (e.g., taking the logarithm) may be desirable so as to satisfy the linearity assumption.

***Poisson Regression Model with Several Covariates***     Suppose that we want to consider $k$ covariates, $X_1, X_2, \ldots, X_k$, simultaneously. The simple Poisson regression model of Section 10.3.1 can easily be generalized and expressed as

$$E(Y_i) = s_i \lambda(x_{ji}\text{'s})$$

$$= s_i \exp\left( \beta_0 + \sum_{j=1}^{k} \beta_j x_{ji} \right)$$

where $Y$ is the Poisson-distributed dependent variable and $\lambda(x_{ji}\text{'s})$ is the *risk* of observation unit $i$ $(1 \leq n)$.

Under the assumption that $Y_i$ is Poisson, the likelihood function is given by

$$L(y; \beta) = \prod_{i=1}^{n} \frac{[s_i \lambda(x_i)]^{y_i} \exp[-s_i \lambda(x_{ji}\text{'s})]}{y_i!}$$

$$= \sum_{i=1}^{n} \left[ y_i \ln s_i - \ln y_i! + y_i \left( \beta_0 + \sum_{j=1}^{k} \beta_j x_{ji} \right) - s_i \exp\left( \theta_0 + \sum_{j=1}^{k} \beta_j x_{ji} \right) \right]$$

from which estimates for $\beta_0, \beta_1, \ldots, \beta_k$ can be obtained by the maximum likelihood procedure.

Also similar to the simple regression case, $\exp(\beta_i)$ represents:

1. The relative risk associated with an exposure if $X_i$ is binary (exposed $X_i = 1$ versus unexposed $X_i = 0$), or
2. The relative risk due to a 1-unit increase if $X_i$ is continuous ($X_i = x + 1$ versus $X_i = x$).

After $\hat{\beta}_i$ and its standard error have been obtained, a 95% confidence interval for the relative risk above is given by

$$\exp[\hat{\beta}_i \pm 1.96\text{SE}(\hat{\beta}_i)]$$

These results are necessary in the effort to identify important risk factors for the Poisson outcome, the *count*. Of course, before such analyses are done, the problem and the data have to be examined carefully. If some of the variables are highly correlated, one or fewer of the correlated factors are likely to be as good predictors as all of them; information from other similar studies also has to be incorporated so as to drop some of these correlated explanatory variables. The use of products such as $X_1 X_2$ and higher power terms such as $X_1^2$ may be necessary and can improve the goodness of fit. It is important to note that we are assuming a *(log)linear regression model* in which, for example, the relative risk due to a 1-unit increase in the value of a continuous $X_i$ ($X_i = x + 1$ versus $X_i = x$) is independent of $x$. Therefore, if this *linearity* seems to be violated, the incorporation of powers of $X_i$ should be considered seriously. The use of products will help in the investigation of possible effect modifications. Finally, there is the messy problem of missing data; most packaged programs would delete a subject if one or more covariate values are missing.

***Testing Hypotheses in Multiple Poisson Regression***   Once we have fit a multiple Poisson regression model and obtained estimates for the various parameters of interest, we want to answer questions about the contributions of various factors to the prediction of the Poisson-distributed response variable. There are three types of such questions:

1. *Overall test.* Taken collectively, does the entire set of explanatory or independent variables contribute significantly to the prediction of response?
2. *Test for the value of a single factor.* Does the addition of one particular variable of interest add significantly to the prediction of response over and above that achieved by other independent variables?
3. *Test for contribution of a group of variables.* Does the addition of a group of variables add significantly to the prediction of response over and above that achieved by other independent variables?

*Overall Regression Test*   We now consider the first question stated above concerning an overall test for a model containing $k$ factors. The null hypothesis for this test may stated as: "All $k$ independent variables *considered together* do not explain the variation in the response any more than the size alone." In other words,

$$H_0: \beta_1 = \beta_2 = \cdots = \beta_k = 0$$

This can be tested using the likelihood ratio chi-square test at $k$ degrees of freedom:

$$\chi^2 = 2(\ln L_k - \ln L_0)$$

where $\ln L_k$ is the log likelihood value for the model containing all $k$ covariates and $\ln L_0$ is the log likelihood value for the model containing only the intercept. A computer-packaged program such as **SAS PROC GENMOD** provides these log likelihood values but in separate runs.

***Example 10.7***   Refer to the data set on emergency service of Example 10.5 (Table 10.2) with four covariates: gender, residency, revenue, and workload (hours). We have:

1. With all four covariates included, $\ln L_4 = 47.783$, whereas
2. With no covariates included, $\ln L_0 = 43.324$

leading to $\chi^2 = 8.918$ with 5 df ($p = .0636$), indicating that at least one covariate must be moderately related significantly to the number of complaints.

*Note:* For model 1, the SAS program would include this instruction:

```
MODEL CASES = GENDER RESIDENCY REVENUE HOURS/
DIST = POISSON LINK = LOG OFFSET = LN;
```

and for model 2,

```
MODEL CASES = /DIST = POISSON LINK = LOG
OFFSET = LN;
```

(See the note after Example 10.6 for other details of the program.)

*Test for a Single Variable*    Let us assume that we now wish to test whether the addition of one particular independent variable of interest adds significantly to the prediction of the response over and above that achieved by other factors already present in the model. The null hypothesis for this test may be stated as: "Factor $X_i$ does not have any value added to the prediction of the response *given that other factors are already included in the model.*" In other words,

$$H_0: \beta_i = 0$$

To test such a null hypothesis, one can use

$$z_i = \frac{\hat{\beta}_i}{\text{SE}(\hat{\beta}_i)}$$

where $\hat{\beta}_i$ is the corresponding estimated regression coefficient and $\text{SE}(\hat{\beta}_i)$ is the estimate of the standard error of $\hat{\beta}_i$, both of which are printed by standard computer-packaged programs such as SAS. In performing this test, we refer the value of the $z$ score to percentiles of the standard normal distribution; for example, we compare the absolute value of $z$ to 1.96 for a two-sided test at the 5% level.

***Example 10.8***    Refer to the data set on emergency service of Example 10.5 (Table 10.2) with all four covariates. We have the results shown in Table 10.4. Only the effect of workload (hours) is significant at the 5% level.

*Note:* Use the same SAS program as in Examples 10.6 and 10.7.

Given a continuous variable of interest, one can fit a polynomial model and use this type of test to check for linearity (see *type 1 analysis* in the next section). It can also be used to check for a single product representing an effect modification.

**TABLE 10.4**

| Variable | Coefficient | Standard Error | $z$ Statistic | $p$ Value |
|----------|-------------|----------------|---------------|-----------|
| Intercept | −8.1338 | 0.9220 | −8.822 | <0.0001 |
| No residency | 0.2090 | 0.2012 | 1.039 | 0.2988 |
| Female | −0.1954 | 0.2182 | −0.896 | 0.3703 |
| Revenue | 0.0016 | 0.0028 | 0.571 | 0.5775 |
| Hours | 0.0007 | 0.0004 | 1.750 | 0.0452 |

**Example 10.9** Refer to the data set on mergency service of Example 10.5 (Table 10.2), but this time we investigate only one covariate, the workload (hours). After fitting the second-degree polinomial model,

$$E(Y_i) = s_i \exp(\beta_0 + \beta_1 \cdot hour_i + \beta_2 \cdot hour_i^2)$$

we obtained a result which indicates that the *curvature effect* is negligible ($p = 0.8797$).

*Note:* An SAS program would include the instruction

```
MODEL CASES = HOURS HOURS*HOURS/ DIST = POISSON
LINK = LOG OFFSET = LN;
```

The following is another interesting example comparing the incidences of nonmelanoma skin cancer among women from two major metropolitan areas, one in the south and one in the north.

**Example 10.10** In this example, the dependent variable is the number of cases of skin cancer. Data were obtained from two metropolitan areas: Minneapolis–St. Paul and Dallas–Ft. Worth. The population of each area is divided into eight age groups and the data are shown in Table 10.5.

**TABLE 10.5**

| Age Group | Minneapolis–St. Paul | | Dallas–Ft. Worth | |
|-----------|-------|------------|-------|------------|
| | Cases | Population | Cases | Population |
| 15–24 | 1 | 172,675 | 4 | 181,343 |
| 25–34 | 16 | 123,065 | 38 | 146,207 |
| 35–44 | 30 | 96,216 | 119 | 121,374 |
| 45–54 | 71 | 92,051 | 221 | 111,353 |
| 55–64 | 102 | 72,159 | 259 | 83,004 |
| 65–74 | 130 | 54,722 | 310 | 55,932 |
| 75–84 | 133 | 32,185 | 226 | 29,007 |
| 85+ | 40 | 8,328 | 65 | 7,538 |

**TABLE 10.6**

| Variable | Coefficient | Standard Error | $z$ Statistic | $p$ Value |
|----------|-------------|----------------|---------------|-----------|
| Intercept | −5.4797 | 0.1037 | 52.842 | <0.0001 |
| Age 15–24 | −6.1782 | 0.4577 | −13.498 | <0.0001 |
| Age 25–34 | −3.5480 | 0.1675 | −21.182 | <0.0001 |
| Age 35–44 | −2.3308 | 0.1275 | −18.281 | <0.0001 |
| Age 45–54 | −1.5830 | 0.1138 | −13.910 | <0.0001 |
| Age 55–64 | −1.0909 | 0.1109 | −9.837 | <0.0001 |
| Age 65–74 | −0.5328 | 0.1086 | −4.906 | <0.0001 |
| Age 75–84 | −0.1196 | 0.1109 | −1.078 | 0.2809 |
| Dallas–Ft. Worth | 0.8043 | 0.0522 | 15.408 | <0.0001 |

This problem involves two covariates: age and location; both are categorical. Using seven dummy variables to represent the eight age groups (with $85^+$ being the baseline) and one for location (with Minneapolis–St. Paul as the baseline, we obtain the results in Table 10.6. These results indicate a clear upward trend of skin cancer incidence with age, and with Minneapolis–St. Paul as the baseline, the relative risk associated with Dallas–Ft. Worth is

$$RR = \exp(0.8043)$$
$$= 2.235$$

an increase of more than twofold for this southern metropolitan area.
   *Note:* An SAS program would include the instruction

```
INPUT AGEGROUP CITY $ POP CASES;
LN = LOG(POP);
MODEL CASES = AGEGROUP CITY/DIST = POISSON
LINK = LOG OFFSET = LN;
```

*Contribution of a Group of Variables*   This testing procedure addresses the more general problem of assessing the additional contribution of two or more factors to the prediction of the response over and above that made by other variables already in the regression model. In other words, the null hypothesis is of the form

$$H_0: \beta_1 = \beta_2 = \cdots = \beta_m = 0$$

To test such a null hypothesis, one can perform a likelihood ratio chi-square test, with $m$ df,

$$\chi^2_{LR} = 2[\ln L(\hat{\beta}; \text{ all } X\text{'s}) - \ln L(\hat{\beta}; \text{ all other } X\text{'s with } X\text{'s under}$$
$$\text{investigation deleted})]$$

As with the *z test* above, this *multiple contribution* procedure is very useful for assessing the importance of potential explanatory variables. In particular, it is often used to test whether a similar group of variables, such as *demographic characteristics*, is important for the prediction of the response; these variables have some trait in common. Another application would be a collection of powers *and/or* product terms (referred to as *interaction variables*). It is often of interest to assess the interaction effects collectively before trying to consider individual interaction terms in a model as suggested previously. In fact, such use reduces the total number of tests to be performed, and this, in turn, helps to provide better control of overall type I error rates, which may be inflated due to multiple testing.

***Example 10.11***   Refer to the data set on skin cancer of Example 10.10 (Table 10.5) with all eight covariates, and we consider collectively the seven dummy variables representing the age. The basic idea is to see if there are any differences without drawing seven separate conclusion comparing each age group versus the baseline.

1. With all eight variables included, we obtained $\ln L = 7201.864$.
2. When the seven age variables were deleted, we obtained $\ln L = 5921.076$.

Therefore,

$$\chi^2_{LR} = 2[\ln L(\hat{\beta}; \text{ eight variables}) - \ln L(\hat{\beta}; \text{ only location variable})]$$
$$= 2561.568; 7 \text{ df}, p\text{-value} < 0.0001$$

In other words, the difference between the age group is highly significant; in fact, it is more so than the difference between the cities.

*Main Effects*   The *z* tests for single variables are sufficient for investigating the effects of continuous and binary covariates. For categorical factors with several categories, such as the age group in the skin cancer data of Example 10.10, this process in PROC GENMOD would choose a *baseline* category and compare each other category with the baseline category chosen. However, the importance of the *main effects* is usually of interest (i.e., one statistical test for each covariate, not each category of a covariate). This can be achieved using PROC GENMOD by two different ways: (1) treating the several category-specific effects as a *group* as seen in Example 10.11 (this would requires two sepate computer runs), or (2) requesting the *type 3 analysis* option as shown in the following example.

***Example 10.12***   Refer to the skin cancer data of Example 10.10 (Table 10.5). Type 3 analysis yields the results shown in Table 10.7. The result for the *age group* main effect is identical to that of Example 10.11.

**TABLE 10.7**

| Source | df | LR $\chi^2$ | $p$ Value |
|---|---|---|---|
| Age group | 7 | 2561.57 | <0.0001 |
| City | 1 | 258.72 | <0.0001 |

*Note:* An SAS program would include the instruction

```
MODEL CASES = AGEGROUP CITY/DIST = POISSON
LINK = LOG OFFSET = LN TYPE3;
```

*Specific and Sequencial Adjustments*   In type 3 analysis, or any other multiple regression analysis, we test the significance of the effect of each factor *added* to the model containing *all other factors*; that is, to investigate the *additional* contribution of the factor to the explanation of the dependent variable. Sometimes, however, we may be interested in a hierarchical or sequential adjustment. For example, we have Poisson-distributed response $Y$ and three covariates, $X_1$, $X_2$, and $X_3$; we want to investigate the effect of $X_1$ on $Y$ (unadjusted), the effect of $X_2$ added to the model containing $X_1$, and the effect of $X_3$ added to the model containing $X_1$ and $X_2$. This can be achieved using PROC GENMOD by requesting a *type 1 analysis* option.

***Example 10.13***   Refer to the data set on emergency service of Example 10.5 (Table 10.2). Type 3 analysis yields the results shown in Table 10.8 and type 1 analysis yields Table 10.9. The results for physician hours are identical because it is adjusted for all other three covariates in both types of analysis. However, the results for other covariates are very different. The effect of residency is marginally significant in type 1 analysis ($p = 0.0741$, unadjusted) and is not significant in type 3 analysis after adjusting for the other three covariates. Similarly, the results for revenue are also different; in type 1 analysis it is adjusted only for residency and gender ($p = 0.3997$; the ordering of variables is specified in the INPUT statement of the computer program), whereas in type 3 analysis it is adjusted for all three other covariates ($p = 0.5781$).

*Note:* An SAS program would include the instruction

```
MODEL CASES = RESIDENCY GENDER REVENUE HOURS/
DIST = POISSON LINK = LOG OFFSET = LN TYPE1 TYPE3;
```

**TABLE 10.8**

| Source | df | LR $\chi^2$ | $p$ Value |
|---|---|---|---|
| Residency | 1 | 1.09 | 0.2959 |
| Gender | 1 | 0.82 | 0.3641 |
| Revenue | 1 | 0.31 | 0.5781 |
| Hours | 1 | 4.18 | 0.0409 |

**TABLE 10.9**

| Source | df | LR $\chi^2$ | p Value |
|--------|-----|------|---------|
| Residency | 1 | 3.199 | 0.0741 |
| Gender | 1 | 0.84 | 0.3599 |
| Revenue | 1 | 0.71 | 0.3997 |
| Hours | 1 | 4.18 | 0.0409 |

### 10.3.3  Overdispersion

The Poisson is a very special distribution; its mean $\mu$ and variance $\sigma^2$ are equal. If we use the variance/mean ratio as a dispersion parameter, it is 1 in a standard Poisson model, less than 1 in an underdispersed model, and greater than 1 in an overdispersed model. Overdispersion is a common phenomenon in practice and it causes concerns because the implication is serious; analysis that assumes the Poisson model often underestimates standard error(s) and, thus wrongly inflates the level of significance.

***Measuring and Monitoring Dispersion***  After a Poisson regression model is fitted, dispersion is measured by the scaled deviance or scaled Peason chi-square; it is the deviance or Pearson chi-square divided by the degrees of freedom (the number of observations minus the number of parameters). The *deviance* is defined as twice the difference between the maximum achievable log likelihood and the log likelihood at the maximum likelihood estimates of the regression parameters. The contribution to the Pearson chi-square from the $i$th observation is

$$\frac{(y_i - \hat{\mu}_i)^2}{\hat{\mu}_i}$$

***Example 10.14***  Refer to the data set on emergency service of Example 10.5 (Table 10.2). With all four covariates, we have the results shown in Table 10.10. Both indices are greater than 1, indicating an overdispersion. In this example we have a sample size of 44 but 5 df lost, due to the estimation of the five regression parameters, including the intercept.

***Fitting an Overdispersed Poisson Model***  PROC GENMOD allows the specification of a scale parameter to fit overdispersed Poisson regression models. The GENMOD procedure does not use the Poisson density function; it fits gener-

**TABLE 10.10**

| Criterion | df | Value | Scaled Value |
|-----------|-----|-------|-------|
| Deviance | 39 | 54.518 | 1.398 |
| Pearson chi-square | 39 | 54.417 | 1.370 |

alized linear models of which Poisson model is a special case. Instead of

$$\text{Var}(Y) = \mu$$

it allows the variance function to have a multiplicative overdispersion factor $\phi$:

$$\text{Var}(Y) = \phi\mu$$

The models are fit in the usual way, and the point estimates of regression coefficients are not affected. The covariance matrix, however, is multiplied by $\phi$. There are two options available for fitting overdispersed models; the users can control either the scaled deviance (by specifying DSCALE in the model statement) or the scaled Pearson chi-square (by specifying PSCALE in the model statement). The value of the controlled index becomes 1; the value of the other index is close to, but may not be equal, 1.

***Example 10.15*** Refer to the data set on emergency service of Example 10.5 (Table 10.2) with all four covariates. By fitting an overdispersed model controlling the scaled deviance, we have the results shown in Table 10.11. As compared to the results in Example 10.8, the point estimates remain the same but the standard errors are larger; the effect of workload (hours) is no longer significant at the 5% level.

*Note:* An SAS program would include the instruction

```
MODEL CASES = GENDER RESIDENCY REVENUE HOURS/
DIST = POISSON LINK = LOG OFFSET = LN DSCALE;
```

and the measures of dispersion become those shown in Table 10.12. We would obtain similar results by controlling the scaled Pearson chi-square.

**TABLE 10.11**

| Variable | Coefficient | Standard Error | $z$ Statistic | $p$ Value |
|---|---|---|---|---|
| Intercept | −8.1338 | 1.0901 | −7.462 | <0.0001 |
| No residency | 0.2090 | 0.2378 | 0.879 | 0.3795 |
| Female | −0.1954 | 0.2579 | −0.758 | 0.4486 |
| Revenue | 0.0016 | 0.0033 | 0.485 | 0.6375 |
| Hours | 0.0007 | 0.0004 | 1.694 | 0.0903 |

**TABLE 10.12**

| Criterion | df | Value | Scaled Value |
|---|---|---|---|
| Deviance | 39 | 39.000 | 1.000 |
| Pearson chi-square | 39 | 38.223 | 0.980 |

### 10.3.4  Stepwise Regression

In many applications, our major interest is to identify important risk factors. In other words, we wish to identify from many available factors a small subset of factors that relate significantly to the outcome (e.g., the disease under investigation). In that identification process, of course, we wish to avoid a large type I (false positive) error. In a regression analysis, a type I error corresponds to including a predictor that has no real relationship to the outcome; such an inclusion can greatly confuse the interpretation of the regression results. In a standard multiple regression analysis, this goal can be achieved by using a strategy that adds to, or removes from, a regression model one factor at a time according to a certain order of relative importance. Therefore, the two important steps are as follows:

1. Specify a criterion or criteria for selecting a model.
2. Specify a strategy for applying the criterion or criteria chosen.

*Strategies*  This is concerned with specifying the strategy for selecting variables. Traditionally, such a strategy is concerned with whether and which particular variable should be added to a model or whether any variable should be deleted from a model at a particular stage of the process. As computers became more accessible and more powerfull, these practices became more popular.

- **Forward selection procedure**
  1. Fit a simple logistic linear regression model to each factor, one at a time.
  2. Select the most important factor according to a certain predetermined criterion.
  3. Test for the significance of the factor selected in step 2 and determine, according to a certain predetermined criterion, whether or not to add this factor to the model.
  4. Repeat steps 2 and 3 for those variables not yet in the model. At any subsequent step, if none meets the criterion in step 3, no more variables are included in the model and the process is terminated.
- **Backward elimination procedure**
  1. Fit the multiple logistic regression model containing all available independent variables.
  2. Select the least important factor according to a certain predetermined criterion; this is done by considering one factor at a time and treating it as though it were the last variable to enter.
  3. Test for the significance of the factor selected in step 2 and determine, according to a certain predetermined criterion, whether or not to delete this factor from the model.
  4. Repeat steps 2 and 3 for those variables still in the model. At any subsequent step, if none meets the criterion in step 3, no more variables are removed from the model and the process is terminated.

- **Stepwise regression procedure.** Stepwise regression is a modified version of forward regression that permits reexamination, at every step, of the variables incorporated in the model in previous steps. A variable entered at an early stage may become superfluous at a later stage because of its relationship with other variables now in the model; the information it provides becomes redundant. That variable may be removed, if meeting the elimination criterion, and the model is refitted with the remaining variables, and the forward process goes on. The entire process, one step forward followed by one step backward, continues until no more variables can be added or removed. Without an automatic computer algorithm, this comprehensive strategy may be too tedious to implement.

*Criteria* For the first step of the forward selection procedure, decisions are based on individual score test results (chi-square, 1 df). In subsequent steps, both forward and backward, the ordering of levels of importance (step 2) and the selection (test in step 3) are based on the likelihood ratio chi-square statistic:

$$\chi^2_{LR} = 2[\ln L(\hat{\beta}; \text{ all other } X\text{'s}) - \ln L(\hat{\beta}; \text{ all other } X\text{'s with one } X \text{ deleted})]$$

In the case of Poisson regression, a computer-packaged program such as SAS's PROC GENMOD does not have an automatic stepwise option. Therefore, the implementation is much more tedious and time consuming. In selecting the first variable (step 1), we have to fit simple regression models to every factor separately, then decide, based on the computer output, on the first selection before coming back for computer runs in step 2. At subsequent steps we can tave advantage of *type 1 analysis* results.

***Example 10.16*** Refer to the data set on emergency service of Example 10.5 (Table 10.2) with all four covariates: workload (hours), residency, gender, and revenue. This time we perform a regression analysis using forward selection in which we specify that a variable has to be significant at the 0.10 level before it can enter into the model. In addition, we fit all overdispersed models using DSCALE option in PROC GENMOD.

The results of the four simple regression analyses are shown in Table 10.13. Workload (hours) meets the entrance criterion and is selected. In the next step, we fit three models each with two covariates: hours and residency, hours and

**TABLE 10.13**

| Variable | LR $\chi^2$ | $p$ Value |
|---|---|---|
| Hours | 4.136 | 0.0420 |
| Residency | 2.166 | 0.1411 |
| Gender | 0.845 | 0.3581 |
| Revenue | 0.071 | 0.7897 |

**TABLE 10.14**

| Variable | LR $\chi^2$ | $p$ Value |
|----------|-------------|-----------|
| Residency | 0.817 | 0.3662 |
| Gender | 1.273 | 0.2593 |
| Revenue | 0.155 | 0.6938 |

gender, and hours and revenue. Table 10.14 shows the significance of each added variable to the model containg hours using type 1 analysis. None of these three variables meets the 0.10 level for entry into the model.

**EXERCISE**

**10.1** Inflammation of the middle ear, *otitis media* (OM), is one of the most common childhood illnesses and accounts for one-third of the practice of pediatrics during the first five years of life. Understanding the natural history of otitis media is of considerable importance, due to the morbidity for children as well as concern about long-term effects on behavior, speech, and language development. In an attempt to understand that natural history, a large group of pregnant women were enrolled and their newborns were followed from birth. The response variable is the number of episodes of otitis media in the first six months (NBER), and potential factors under investigation are upper respiratory infection (URI), sibling history of otitis media (SIBHX; 1 for yes), day care, number of cigarettes consumed a day by parents (CIGS), cotinin level (CNIN) measured from the urine of the baby (a marker for exposure to cigarette smoke), and whether the baby was born in the fall season (FALL). Table E10.1 provides about half of our data set.

(a) Taken collectively, do the covariates contribute significantly to the prediction of the number of otitis media cases in the first six months?

(b) Fit the multiple regression model to obtain estimates of individual regression coefficients and their standard errors. Draw your conclusions concerning the conditional contribution of each factor.

(c) Is there any indication of overdispersion? If so, fit an appropriate overdispersed model and compare the results to those in part (b).

(d) Refit the model in part (b) to implement this sequential adjustment:

$$\text{URI} \rightarrow \text{SIBHX} \rightarrow \text{DAYCARE} \rightarrow \text{CIGS} \rightarrow \text{CNIN} \rightarrow \text{FALL}$$

(e) Within the context of the multiple regression model in part (b), does day care alter the effect of sibling history?

(f) Within the context of the multiple regression model in part (b), is the effect of the cotinin level linear?

(g) Focus on the sibling history of otitis media (SIBHX) as the primary factor. Taken collectively, was this main effect altered by any other covariates?

(*Note*: An electronic copy, in a Web-based form, is available upon request.)

**TABLE E10.1**

| URI | SIBHX | DAYCARE | CIGS | CNIN | FALL | NBER |
|---|---|---|---|---|---|---|
| 1 | 0 | 0 | 0 | 0.00 | 0 | 2 |
| 1 | 0 | 0 | 0 | 27.52 | 0 | 3 |
| 1 | 0 | 1 | 0 | 0.00 | 0 | 0 |
| 1 | 0 | 0 | 0 | 0.00 | 0 | 0 |
| 0 | 1 | 1 | 0 | 0.00 | 0 | 0 |
| 1 | 0 | 0 | 0 | 0.00 | 0 | 0 |
| 1 | 0 | 1 | 0 | 0.00 | 0 | 0 |
| 0 | 0 | 1 | 0 | 0.00 | 0 | 0 |
| 0 | 1 | 0 | 8 | 83.33 | 0 | 0 |
| 1 | 0 | 1 | 0 | 89.29 | 0 | 0 |
| 0 | 0 | 1 | 0 | 0.00 | 0 | 1 |
| 0 | 1 | 1 | 0 | 32.05 | 0 | 0 |
| 1 | 0 | 0 | 0 | 471.40 | 0 | 0 |
| 1 | 0 | 0 | 0 | 0.00 | 0 | 1 |
| 1 | 0 | 1 | 0 | 12.10 | 0 | 0 |
| 0 | 1 | 0 | 5 | 26.64 | 0 | 0 |
| 0 | 0 | 1 | 0 | 40.00 | 0 | 0 |
| 1 | 0 | 1 | 0 | 512.05 | 0 | 0 |
| 0 | 0 | 1 | 0 | 77.59 | 0 | 0 |
| 1 | 0 | 1 | 0 | 0.00 | 0 | 0 |
| 1 | 0 | 1 | 0 | 0.00 | 0 | 0 |
| 0 | 0 | 1 | 0 | 0.00 | 0 | 0 |
| 1 | 0 | 1 | 0 | 0.00 | 0 | 3 |
| 1 | 0 | 0 | 0 | 0.00 | 0 | 1 |
| 0 | 0 | 1 | 0 | 21.13 | 0 | 0 |
| 1 | 0 | 0 | 0 | 15.96 | 0 | 1 |
| 1 | 0 | 0 | 0 | 0.00 | 0 | 1 |
| 0 | 0 | 0 | 0 | 0.00 | 0 | 0 |
| 1 | 0 | 0 | 0 | 9.26 | 0 | 0 |
| 1 | 0 | 1 | 0 | 0.00 | 0 | 0 |
| 1 | 0 | 0 | 0 | 0.00 | 0 | 1 |
| 0 | 0 | 0 | 0 | 0.00 | 0 | 0 |
| 0 | 1 | 0 | 0 | 0.00 | 0 | 0 |
| 0 | 0 | 1 | 0 | 0.00 | 0 | 0 |
| 1 | 0 | 1 | 0 | 525.00 | 0 | 2 |

(*Continued*)

**TABLE E10.1** *(Continued)*

| URI | SIBHX | DAYCARE | CIGS | CNIN | FALL | NBER |
|-----|-------|---------|------|------|------|------|
| 0 | 0 | 0 | 0 | 0.00 | 0 | 0 |
| 1 | 0 | 1 | 0 | 0.00 | 0 | 0 |
| 1 | 0 | 1 | 0 | 0.00 | 0 | 1 |
| 1 | 0 | 0 | 0 | 0.00 | 0 | 0 |
| 0 | 0 | 0 | 0 | 57.14 | 0 | 0 |
| 1 | 1 | 1 | 0 | 125.00 | 0 | 0 |
| 1 | 0 | 0 | 0 | 0.00 | 0 | 0 |
| 0 | 1 | 0 | 0 | 0.00 | 0 | 0 |
| 1 | 0 | 1 | 0 | 0.00 | 0 | 3 |
| 1 | 0 | 1 | 0 | 0.00 | 0 | 0 |
| 0 | 0 | 0 | 0 | 0.00 | 0 | 1 |
| 1 | 0 | 1 | 0 | 0.00 | 0 | 1 |
| 1 | 0 | 1 | 0 | 80.25 | 0 | 2 |
| 0 | 0 | 0 | 0 | 0.00 | 0 | 0 |
| 1 | 0 | 1 | 0 | 219.51 | 0 | 1 |
| 1 | 0 | 0 | 0 | 0.00 | 0 | 0 |
| 1 | 0 | 0 | 0 | 0.00 | 0 | 1 |
| 0 | 0 | 0 | 0 | 0.00 | 0 | 0 |
| 1 | 0 | 0 | 0 | 0.00 | 0 | 0 |
| 1 | 0 | 1 | 0 | 0.00 | 0 | 2 |
| 1 | 0 | 1 | 0 | 8.33 | 0 | 2 |
| 1 | 0 | 0 | 0 | 12.02 | 0 | 5 |
| 0 | 1 | 1 | 40 | 297.98 | 0 | 0 |
| 1 | 0 | 1 | 0 | 13.33 | 0 | 0 |
| 1 | 0 | 0 | 0 | 0.00 | 0 | 3 |
| 1 | 1 | 1 | 25 | 110.31 | 0 | 3 |
| 1 | 0 | 1 | 0 | 0.00 | 0 | 1 |
| 1 | 0 | 1 | 0 | 0.00 | 0 | 1 |
| 1 | 0 | 0 | 0 | 0.00 | 0 | 1 |
| 0 | 0 | 0 | 0 | 0.00 | 0 | 0 |
| 0 | 0 | 0 | 0 | 0.00 | 0 | 0 |
| 1 | 0 | 1 | 0 | 0.00 | 0 | 1 |
| 1 | 0 | 1 | 0 | 0.00 | 0 | 1 |
| 1 | 0 | 0 | 0 | 285.28 | 0 | 1 |
| 1 | 0 | 0 | 0 | 0.00 | 0 | 1 |
| 1 | 0 | 0 | 0 | 15.00 | 0 | 2 |
| 1 | 0 | 0 | 0 | 0.00 | 0 | 0 |
| 1 | 0 | 0 | 0 | 13.40 | 0 | 1 |
| 1 | 0 | 1 | 0 | 0.00 | 0 | 2 |
| 1 | 0 | 1 | 0 | 46.30 | 0 | 0 |
| 1 | 0 | 1 | 0 | 0.00 | 0 | 0 |
| 1 | 0 | 1 | 0 | 0.00 | 0 | 1 |
| 0 | 0 | 0 | 0 | 0.00 | 1 | 0 |
| 1 | 0 | 1 | 0 | 0.00 | 1 | 0 |
| 1 | 0 | 1 | 0 | 0.00 | 1 | 2 |
| 1 | 0 | 1 | 0 | 0.00 | 1 | 0 |

**TABLE E10.1** *(Continued)*

| URI | SIBHX | DAYCARE | CIGS | CNIN | FALL | NBER |
|-----|-------|---------|------|----------|------|------|
| 1 | 0 | 1 | 0 | 0.00 | 1 | 1 |
| 1 | 0 | 0 | 0 | 0.00 | 1 | 2 |
| 1 | 1 | 0 | 0 | 0.00 | 1 | 6 |
| 1 | 0 | 0 | 0 | 53.46 | 1 | 0 |
| 1 | 0 | 1 | 0 | 0.00 | 1 | 0 |
| 1 | 0 | 0 | 0 | 0.00 | 1 | 0 |
| 1 | 0 | 1 | 0 | 3.46 | 1 | 3 |
| 1 | 0 | 1 | 0 | 125.00 | 1 | 1 |
| 0 | 0 | 1 | 0 | 0.00 | 1 | 0 |
| 1 | 0 | 1 | 0 | 0.00 | 1 | 2 |
| 1 | 0 | 1 | 0 | 0.00 | 1 | 3 |
| 1 | 0 | 1 | 0 | 0.00 | 1 | 0 |
| 1 | 0 | 0 | 0 | 0.00 | 1 | 2 |
| 1 | 0 | 0 | 0 | 0.00 | 1 | 0 |
| 1 | 1 | 0 | 0 | 0.00 | 1 | 1 |
| 0 | 0 | 0 | 0 | 0.00 | 1 | 0 |
| 0 | 0 | 1 | 0 | 0.00 | 1 | 0 |
| 1 | 0 | 0 | 0 | 0.00 | 1 | 0 |
| 1 | 0 | 0 | 0 | 2.80 | 1 | 0 |
| 0 | 0 | 1 | 0 | 1,950.00 | 1 | 0 |
| 1 | 0 | 1 | 0 | 69.44 | 1 | 2 |
| 1 | 0 | 1 | 0 | 0.00 | 1 | 3 |
| 1 | 0 | 0 | 0 | 0.00 | 1 | 0 |
| 1 | 0 | 0 | 0 | 0.00 | 1 | 2 |
| 1 | 1 | 0 | 0 | 0.00 | 1 | 0 |
| 1 | 1 | 1 | 0 | 0.00 | 1 | 4 |
| 1 | 0 | 0 | 0 | 0.00 | 0 | 0 |
| 1 | 0 | 1 | 0 | 0.00 | 0 | 0 |
| 1 | 0 | 1 | 0 | 0.00 | 0 | 1 |
| 1 | 0 | 0 | 0 | 0.00 | 0 | 1 |
| 1 | 0 | 1 | 0 | 0.00 | 0 | 1 |
| 0 | 0 | 1 | 0 | 31.53 | 0 | 0 |
| 0 | 0 | 1 | 0 | 0.00 | 0 | 0 |
| 1 | 0 | 1 | 0 | 11.40 | 0 | 3 |
| 0 | 0 | 0 | 0 | 0.00 | 0 | 0 |
| 0 | 1 | 0 | 0 | 750.00 | 0 | 0 |
| 1 | 0 | 0 | 0 | 0.00 | 0 | 0 |
| 0 | 0 | 0 | 0 | 0.00 | 0 | 1 |
| 1 | 1 | 0 | 0 | 0.00 | 0 | 0 |
| 1 | 1 | 0 | 0 | 0.00 | 0 | 0 |
| 0 | 0 | 0 | 0 | 0.00 | 1 | 0 |
| 1 | 1 | 0 | 0 | 0.00 | 1 | 2 |
| 0 | 0 | 0 | 0 | 0.00 | 1 | 1 |
| 1 | 0 | 1 | 0 | 0.00 | 1 | 1 |
| 0 | 1 | 1 | 22 | 824.22 | 1 | 1 |
| 1 | 0 | 0 | 0 | 0.00 | 1 | 0 |

*(Continued)*

**TABLE E10.1** (*Continued*)

| URI | SIBHX | DAYCARE | CIGS | CNIN | FALL | NBER |
|---|---|---|---|---|---|---|
| 0 | 0 | 0 | 0 | 0.00 | 1 | 2 |
| 1 | 1 | 0 | 0 | 0.00 | 1 | 1 |
| 1 | 1 | 1 | 25 | 384.98 | 1 | 2 |
| 1 | 0 | 1 | 0 | 0.00 | 1 | 2 |
| 0 | 0 | 0 | 0 | 0.00 | 1 | 0 |
| 1 | 0 | 1 | 0 | 29.41 | 1 | 0 |
| 1 | 0 | 0 | 0 | 0.00 | 1 | 0 |
| 1 | 0 | 0 | 0 | 0.00 | 1 | 0 |
| 0 | 0 | 1 | 0 | 0.00 | 1 | 0 |
| 0 | 0 | 0 | 0 | 35.59 | 1 | 0 |
| 0 | 0 | 0 | 0 | 0.00 | 1 | 0 |
| 1 | 0 | 1 | 0 | 0.00 | 1 | 3 |
| 1 | 0 | 1 | 0 | 0.00 | 1 | 4 |
| 1 | 0 | 0 | 0 | 0.00 | 1 | 1 |
| 1 | 0 | 0 | 0 | 0.00 | 1 | 0 |
| 1 | 0 | 0 | 0 | 0.00 | 1 | 1 |
| 1 | 1 | 1 | 35 | 390.80 | 1 | 2 |
| 1 | 0 | 1 | 0 | 0.00 | 1 | 0 |
| 1 | 0 | 1 | 0 | 0.00 | 1 | 2 |
| 0 | 0 | 1 | 0 | 0.00 | 1 | 0 |
| 0 | 0 | 1 | 0 | 0.00 | 1 | 0 |
| 1 | 1 | 1 | 0 | 0.00 | 1 | 3 |
| 1 | 1 | 0 | 22 | 1,101.45 | 1 | 3 |
| 1 | 0 | 0 | 0 | 0.00 | 1 | 2 |
| 0 | 0 | 1 | 0 | 0.00 | 1 | 0 |
| 1 | 0 | 1 | 0 | 57.14 | 1 | 0 |
| 1 | 1 | 1 | 40 | 306.23 | 1 | 2 |
| 1 | 0 | 1 | 0 | 300.00 | 1 | 6 |
| 1 | 0 | 1 | 0 | 0.00 | 1 | 2 |
| 0 | 1 | 1 | 0 | 0.00 | 1 | 0 |
| 0 | 0 | 0 | 0 | 43.86 | 1 | 0 |
| 0 | 0 | 0 | 0 | 0.00 | 1 | 3 |
| 1 | 1 | 0 | 0 | 0.00 | 1 | 2 |
| 1 | 1 | 0 | 0 | 0.00 | 1 | 3 |
| 0 | 0 | 0 | 0 | 0.00 | 1 | 0 |
| 0 | 0 | 0 | 0 | 0.00 | 1 | 0 |
| 1 | 0 | 0 | 0 | 0.00 | 1 | 2 |
| 1 | 0 | 1 | 0 | 0.00 | 1 | 0 |
| 1 | 1 | 1 | 0 | 0.00 | 1 | 2 |
| 0 | 1 | 1 | 10 | 1,000.00 | 1 | 1 |
| 0 | 1 | 0 | 10 | 0.00 | 1 | 0 |
| 0 | 1 | 1 | 1 | 0.00 | 0 | 0 |
| 1 | 0 | 0 | 0 | 0.00 | 0 | 1 |
| 1 | 0 | 0 | 0 | 0.00 | 0 | 3 |
| 1 | 0 | 0 | 0 | 0.00 | 0 | 0 |

**TABLE E10.1** *(Continued)*

| URI | SIBHX | DAYCARE | CIGS | CNIN | FALL | NBER |
|-----|-------|---------|------|------|------|------|
| 0 | 0 | 0 | 0 | 0.00 | 0 | 0 |
| 0 | 0 | 1 | 0 | 0.00 | 0 | 0 |
| 1 | 0 | 0 | 0 | 0.00 | 0 | 1 |
| 0 | 0 | 0 | 0 | 0.00 | 0 | 3 |
| 1 | 0 | 0 | 0 | 0.00 | 0 | 1 |
| 0 | 1 | 1 | 23 | 400.00 | 0 | 1 |
| 1 | 1 | 1 | 0 | 0.00 | 0 | 1 |
| 0 | 1 | 0 | 10 | 0.00 | 0 | 0 |
| 1 | 0 | 1 | 0 | 0.00 | 0 | 3 |
| 0 | 0 | 1 | 0 | 0.00 | 0 | 1 |
| 1 | 0 | 1 | 0 | 0.00 | 0 | 3 |
| 0 | 0 | 1 | 0 | 0.00 | 0 | 1 |
| 1 | 1 | 1 | 0 | 0.00 | 0 | 0 |
| 0 | 0 | 0 | 0 | 0.00 | 0 | 0 |
| 0 | 0 | 0 | 0 | 0.00 | 0 | 1 |
| 0 | 0 | 1 | 10 | 1,067.57 | 0 | 1 |
| 1 | 1 | 1 | 3 | 1,492.31 | 0 | 0 |
| 0 | 0 | 1 | 0 | 0.00 | 0 | 2 |
| 1 | 0 | 0 | 0 | 0.00 | 0 | 0 |
| 1 | 0 | 0 | 0 | 0.00 | 0 | 0 |
| 1 | 0 | 1 | 0 | 9.41 | 0 | 1 |
| 1 | 0 | 0 | 0 | 0.00 | 0 | 0 |
| 1 | 0 | 1 | 0 | 9.84 | 0 | 2 |
| 1 | 0 | 1 | 10 | 723.58 | 0 | 2 |
| 1 | 0 | 0 | 0 | 0.00 | 0 | 2 |
| 0 | 0 | 0 | 0 | 15.63 | 0 | 0 |
| 1 | 0 | 0 | 0 | 0.00 | 0 | 0 |
| 1 | 1 | 1 | 30 | 106.60 | 0 | 0 |
| 0 | 0 | 0 | 0 | 0.00 | 0 | 0 |
| 0 | 0 | 1 | 0 | 0.00 | 0 | 0 |
| 1 | 0 | 1 | 0 | 0.00 | 0 | 1 |
| 1 | 0 | 0 | 0 | 0.00 | 0 | 0 |
| 1 | 0 | 1 | 0 | 0.00 | 0 | 0 |
| 0 | 0 | 1 | 0 | 0.00 | 0 | 0 |
| 0 | 0 | 1 | 0 | 0.00 | 0 | 0 |
| 1 | 1 | 0 | 0 | 0.00 | 0 | 0 |
| 1 | 0 | 1 | 0 | 0.00 | 0 | 1 |
| 1 | 0 | 1 | 0 | 0.00 | 0 | 1 |
| 1 | 0 | 0 | 0 | 0.00 | 0 | 0 |
| 1 | 0 | 1 | 0 | 0.00 | 0 | 2 |
| 0 | 1 | 1 | 30 | 15,375.00 | 0 | 0 |
| 0 | 1 | 0 | 75 | 11,000.00 | 0 | 0 |
| 0 | 1 | 1 | 0 | 0.00 | 0 | 0 |
| 1 | 0 | 1 | 0 | 0.00 | 0 | 1 |
| 1 | 0 | 0 | 0 | 0.00 | 0 | 1 |

*(Continued)*

**TABLE E10.1** *(Continued)*

| URI | SIBHX | DAYCARE | CIGS | CNIN | FALL | NBER |
|-----|-------|---------|------|------|------|------|
| 0 | 0 | 0 | 0 | 0.00 | 0 | 0 |
| 0 | 0 | 1 | 0 | 17.39 | 0 | 0 |
| 0 | 0 | 0 | 0 | 0.00 | 0 | 0 |
| 0 | 1 | 1 | 0 | 0.00 | 0 | 0 |
| 0 | 0 | 0 | 0 | 0.00 | 0 | 0 |
| 1 | 0 | 0 | 0 | 0.00 | 0 | 0 |
| 0 | 0 | 1 | 0 | 0.00 | 0 | 0 |
| 0 | 1 | 1 | 6 | 44.19 | 0 | 0 |
| 1 | 1 | 0 | 1 | 0.00 | 0 | 0 |
| 0 | 0 | 1 | 0 | 0.00 | 0 | 1 |
| 1 | 1 | 1 | 30 | 447.15 | 0 | 5 |
| 0 | 0 | 0 | 0 | 0.00 | 0 | 0 |
| 0 | 1 | 0 | 20 | 230.43 | 0 | 1 |
| 1 | 1 | 1 | 0 | 0.00 | 0 | 1 |
| 0 | 0 | 1 | 0 | 0.00 | 0 | 0 |
| 0 | 0 | 1 | 0 | 0.00 | 0 | 0 |
| 0 | 0 | 1 | 0 | 217.82 | 0 | 0 |
| 0 | 0 | 1 | 0 | 0.00 | 0 | 0 |
| 1 | 0 | 0 | 0 | 0.00 | 0 | 0 |
| 1 | 0 | 1 | 0 | 32.41 | 0 | 0 |
| 1 | 1 | 0 | 0 | 0.00 | 0 | 0 |
| 1 | 1 | 1 | 8 | 43.22 | 0 | 0 |
| 1 | 1 | 1 | 28 | 664.77 | 0 | 2 |
| 1 | 0 | 1 | 0 | 0.00 | 0 | 0 |

# 11

# ANALYSIS OF SURVIVAL DATA AND DATA FROM MATCHED STUDIES

Study data may be collected in many different ways. In addition to surveys, which are cross-sectional, as shown in many examples in previous chapters, biomedical research data may come from different sources, the two fundamental designs being retrospective and prospective. *Retrospective studies* gather past data from selected cases and controls to determine differences, if any, in exposure to a suspected risk factor. They are commonly referred to as *case–control studies*; each case–control study is focused on a particular disease. In a typical case–control study, cases of a specific disease are ascertained as they arise from population-based registers or lists of hospital admissions, and controls are sampled either as disease-free persons from the population at risk or as hospitalized patients having a diagnosis other than the one under study. The advantages of a retrospective study are that it is economical and provides answers to research questions relatively quickly because the cases are already available. Major limitations are due to the inaccuracy of the exposure histories and uncertainty about the appropriateness of the control sample; these problems sometimes hinder retrospective studies and make them less preferred than prospective studies. *Prospective studies*, also called *cohort studies*, are epidemiological designs in which one enrolls a group of persons and follows them over certain periods of time; examples include occupational mortality studies and clinical trials. The cohort study design focuses on a particular exposure rather than a particular disease as in case–control studies. Advantages of a longitudinal approach include the opportunity for more accurate measurement of exposure history and a careful examination of the time relationships between exposure and any disease under investigation. An important subset of cohort studies consists of randomized clinical trials where follow-up starts from the date of enrollment and randomization of each subject.

Methodology discussed in this book has mostly been directed to the analysis of cross-sectional and retrospective studies; this chapter is an exception. The topics covered here in the first few sections—basic survival analysis and Cox's proportional hazards regression—were developed to deal with survival data resulting from prospective or cohort studies. Readers should focus on the nature of the various designs because the borderline between categorical and survival data may be somewhat vague, especially for beginning students. Survival analysis, which was developed to deal with data resulting from prospective studies, is also focused on the occurrence of an *event*, such as death or relapse of a disease, after some initial treatment—a binary outcome. Therefore, for beginners, it may be confused with the type of data that require the logistic regression analysis discussed in Chapter 9. The basic difference is that for survival data, studies have staggered entry, and subjects are followed for varying lengths of time; they do not have the same probability for the event to occur even if they have identical characteristics, a basic assumption of the logistic regression model. Second, each member of the cohort belongs to one of three types of termination:

1. Subjects still alive on the analysis date
2. Subjects who died on a known date within the study period
3. Subjects who are lost to follow-up after a certain date

That is, for many study subjects, the observation may be terminated before the occurrence of the main event under investigation: for example, subjects in types 1 and 3.

In the last few sections of this chapter we return to topics of retrospective studies, the analysis of data from matched case–control studies. We put together the analyses of two very different types of data, which come from two very different designs, for a good reason. First, they are not totally unrelated; statistical tests for a comparison of survival distributions are special forms of the Mantel–Haenszel method of Chapter 6. Most methodologies used in survival analysis are generalizations of those for categorical data. In addition, the conditional logistic regression model needed for an analysis of data from matched case–control studies and Cox's regression model for the analysis of some type of survival data correspond to the same likelihood function and are analyzed using the same computer program. For students in applied fields such as epidemiology, access to the methods in this chapter would be beneficial because most may not be adequately prepared for the level of sophistication of a full course in survival analysis. This makes it more difficult to learn the analysis of matched case–control studies. In Sections 11.1 and 11.2 we introduce some basic concepts and techniques of survival analysis; Cox's regression models are covered in Sections 11.3 and 11.4. Methods for matched data begin in Section 11.5.

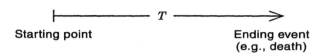

Starting point                          Ending event
                                        (e.g., death)

**Figure 11.1**    Survival time.


## 11.1  SURVIVAL DATA

In prospective studies, the important feature is not only the outcome event, such as death, but the time to that event, the *survival time*. To determine the survival time $T$, three basic elements are needed:

1. A time origin or starting point
2. An ending event of interest
3. A measurement scale for the passage of time

These may be, for example, the life span $T$ from birth (starting point) to death (ending event) in years (measurement scale) (Figure 11.1). The time origin or starting point should be defined precisely, but it need not be birth; it could be the start of a new treatment (randomization date in a clinical trial) or admission to a hospital or nursing home. The ending event should also be defined precisely, but it need not be death; a nonfatal event such as the relapse of a disease (e.g., leukemia) or a relapse from smoking cessation or discharge to the community from a hospital or nursing home satisfy the definition and are acceptable choices. The use of calendar time in health studies is common and meaningful; however, other choices for a time scale are justified—for example, hospital cost (in dollars) from admission (starting point) to discharge (ending event).

Distribution of the survival time $T$ from enrollment or starting point to the event of interest, considered as a random variable, is characterized by either one of two equivalent functions: the survival function and the hazard function. The *survival function*, denoted $S(t)$, is defined as the probability that a person survives longer than $t$ units of time:

$$S(t) = \Pr(T > t)$$

$S(t)$ is also known as the *survival rate*; for example, if times are in years, $S(2)$ is the two-year survival rate, $S(5)$ is the five-year survival rate, and so on. A graph of $S(t)$ versus $t$ is called a *survival curve* (Figure 11.2).

The *hazard* or *risk function* $\lambda(t)$ gives the *instantaneous* failure rate and is defined as follows. Assuming that a typical patient has survived to time $t$, and for a small time increment $\delta$, the probability of an event occurring during time interval $(t, t + \delta)$ to that person is given approximately by

$$\lambda(t)\delta \simeq \Pr(t \leq T \leq t + \delta \,|\, t \leq T)$$

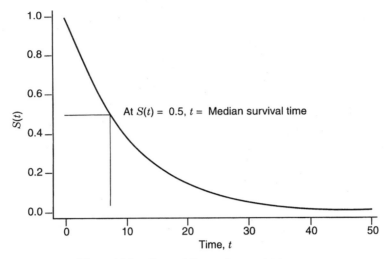

**Figure 11.2**    General form of a survival curve.

In other words, the hazard or risk function $\lambda(t)$ approximates the proportion of subjects dying or having events per unit time around time $t$. Note that this differs from the density function represented by the usual histogram; in the case of $\lambda(t)$, the numerator is a *conditional* probability. $\lambda(t)$ is also known as the *force of mortality* and is a measure of the proneness to failure as a function of the person's age. When a population is subdivided into two subpopulations, $E$ (exposed) and $E'$ (nonexposed), by the presence or absence of a certain characteristic (an exposure such as smoking), each subpopulation corresponds to a hazard or risk function, and the ratio of two such functions,

$$\mathrm{RR}(t) = \frac{\lambda(t; E)}{\lambda(t; E')}$$

is called the *relative risk* associated with exposure to factor $E$, risk of the exposured subjects *relative* to the risk of nonexposed subjects. In general, the relative risk $\mathrm{RR}(t)$ is a function of time and measures the magnitude of an effect; when it remains constant, $\mathrm{RR}(t) = p$, we have a *proportional hazards model* (PHM):

$$\lambda(t; E) = p\lambda(t; E')$$

with the risk of the nonexposed subpopulation served as the baseline. This is a multiplicative model; that is, exposure raises the risk by a multiplicative constant. Another way to express this model is

$$\lambda(t) = \lambda_0(t)e^{\beta x}$$

where $\lambda_0(t)$ is $\lambda(t; E')$—the hazard function of the unexposed subpopulation—and the indicator (or *covariate*) $x$ is defined as

$$x = \begin{cases} 0 & \text{if unexposed} \\ 1 & \text{if exposed} \end{cases}$$

The regression coefficient $\beta$ represents the relative risk on the log scale. This model works with any covariate $X$, continuous or categorical; the binary covariate above is only a very special case. Of course, the model can be extended to include several covariates; it is usually referred to as *Cox's regression model*.

A special source of difficulty in the analysis of survival data is the possibility that some subjects may not be observed for the full time to failure or event. Such *random censoring* arises in medical applications with animal studies, epidemiological applications with human studies, or in clinical trials. In these cases, observation is terminated before the occurrence of the event. In a clinical trial, for example, patients may enter the study at different times; then each is treated with one of several possible therapies after a randomization.

Figure 11.3 shows a description of a typical clinical trial. Of course, in conducting this trial, we want to observe their lifetimes of all subjects from enrollment, but censoring may occur in one of the following forms:

- Loss to follow-up (the patient may decide to move elsewhere)
- Dropout (a therapy may have such bad effects that it is necessary to discontinue treatment)
- Termination of the study
- Death due to a cause not under investigation (e.g., suicide)

To make it possible for statistical analysis we make the crucial assumption that conditionally on the values of any explanatory variables (or covariates), the prognosis for any person who has survived to a certain time $t$ should not be affected if the person is censored at $t$. That is, a person who is censored at $t$ should be representative of all those subjects with the same values of explana-

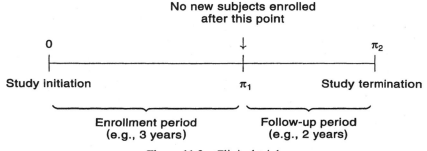

**Figure 11.3**    Clinical trial.

tory variables who survive to $t$. In other words, survival condition and reason of loss are independent; under this assumption, there is no need to distinguish the four forms of censoring described above.

We assume that observations available on the failure time of $n$ subjects are usually taken to be independent. At the end of the study, our sample consists of $n$ pairs of numbers $(t_i, \delta_i)$. Here $\delta_i$ is an indicator variable for survival status ($\delta_i = 0$ if the $i$th individual is censored; $\delta_i = 1$ if the $i$th individual failed) and $t_i$ is the time to failure/event (if $\delta_i = 1$) or the censoring time (if $\delta_i = 0$); $t_i$ is also called the *duration time*. We may also consider, in addition to $t_i$ and $\delta_i$, $(x_{1i}, x_{2i}, \ldots, x_{ki})$, a set of $k$ covariates associated with the $i$th individual representing such cofactors as age, gender, and treatment.

## 11.2  INTRODUCTORY SURVIVAL ANALYSES

In this section we introduce a popular method for estimation of the survival function and a family of statistical tests for comparison of survival distributions.

### 11.2.1  Kaplan–Meier Curve

In this section we introduce the *product-limit* (PL) *method* of estimating survival rates, also called the *Kaplan–Meier method*. Let

$$t_1 < t_2 < \cdots < t_k$$

be the distinct observed death times in a sample of size $n$ from a homogeneous population with survival function $S(t)$ to be estimated ($k \leq n$; $k$ could be less than $n$ because some subjects may be censored and some subjects may have events at the same time). Let $n_i$ be the number of subjects at risk at a time just prior to $t_i$ ($1 \leq i \leq k$; these are cases whose duration time is at least $t_i$), and $d_i$ the number of deaths at $t_i$. The survival function $S(t)$ is estimated by

$$\hat{S}(t) = \prod_{t_i \leq t}\left(1 - \frac{d_i}{n_i}\right)$$

which is called the *product-limit estimator* or *Kaplan–Meier estimator* with a 95% confidence given by

$$\hat{S}(t) \exp[\pm 1.96\hat{s}(t)]$$

where

$$\hat{s}^2(t) = \sum_{t_i \leq t} \frac{d_i}{n_i(n_i - d_i)}$$

The explanation could be simple, as follows: (1) $d_i/n_i$ is the proportion (or estimated probability of having an event in the interval from $t_{i-1}$ to $t_i$, (2) $1 - d_i/n_i$ represents the proportion (or estimated probability of surviving that same interval), and (3) the product in the formula for $\hat{S}(t)$ follows from the product rule for probabilities of Chapter 3.

***Example 11.1***   The remission times of 42 patients with acute leukemia were reported from a clinical trial undertaken to assess the ability of the drug 6-mercaptopurine (6-MP) to maintain remission (Figure 11.4). Each patient was randomized to receive either 6-MP or placebo. The study was terminated after one year; patients have different follow-up times because they were enrolled sequentially at different times. Times in weeks were:

- *6-MP group:* 6, 6, 6, 7, 10, 13, 16, 22, 23, 6+, 9+, 10+, 11+, 17+, 19+, 20+, 25+, 32+, 32+, 34+, 35+
- *Placebo group:* 1, 1, 2, 2, 3, 4, 4, 5, 5, 8, 8, 8, 8, 11, 11, 12, 12, 15, 17, 22, 23

in which $t+$ denotes a censored observation (i.e., the case was censored after $t$ weeks without a relapse). For example, "10+" is a case enrolled 10 weeks before study termination and still remission-free at termination.

According to the product-limit method, survival rates for the 6-MP group are calculated by constructing a table such as Table 11.1 with five columns;

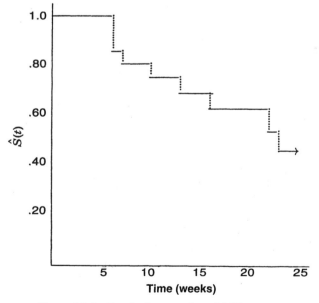

**Figure 11.4**   Survival curve: drug 6-MP group.

**TABLE 11.1**

| (1) | (2) | (3) | (4) | (5) |
|-----|-----|-----|-----|-----|
| $t_i$ | $n_i$ | $d_i$ | $1 - \dfrac{d_i}{n_i}$ | $\hat{S}(t_i)$ |
| 6 | 21 | 3 | 0.8571 | 0.8571 |
| 7 | 17 | 1 | 0.9412 | 0.8067 |
| 10 | 15 | 1 | 0.9333 | 0.7529 |
| 13 | 12 | 1 | 0.9167 | 0.6902 |
| 16 | 11 | 1 | 0.9091 | 0.6275 |
| 22 | 7 | 1 | 0.8571 | 0.5378 |
| 23 | 6 | 1 | 0.8333 | 0.4482 |

to obtain $\hat{S}(t)$, multiply all values in column (4) up to and including $t$. From Table 11.1, we have, for example:

7-week survival rate is 80.67%

22-week survival rate is 53.78%

and a 95% confidence interval for $S(7)$ is (.6531, .9964).

*Note:* An SAS program would include these instructions:

```
PROC LIFETEST METHOD = KM;
TIME WEEKS*RELAPSE(0);
```

where WEEKS is the variable name for duration time, RELAPSE the variable name for survival status, "0" is the coding for censoring, and KM stands for the Kaplan–Meier method.

### 11.2.2   Comparison of Survival Distributions

Suppose that there are $n_1$ and $n_2$ subjects, corresponding to treatment groups 1 and 2, respectively. The study provides two samples of survival data:

$$\{(t_{1i}, \delta_{1i})\} \qquad i = 1, 2, \ldots, n_1$$

and

$$\{(t_{2j}, \delta_{2j})\} \qquad j = 1, 2, \ldots, n_2$$

In the presence of censored observations, tests of significance can be constructed as follows:

**TABLE 11.2**

| Sample | Status | | Total |
|---|---|---|---|
| | Dead | Alive | |
| 1 | $d_{1i}$ | $a_{1i}$ | $n_{1i}$ |
| 2 | $d_{2i}$ | $a_{2i}$ | $n_{2i}$ |
| Total | $d_i$ | $a_i$ | $n_i$ |

1. Pool data from two samples together and let

$$t_1 < t_2 < \cdots < t_m \qquad m \leq d \leq n_1 + n_2$$

be the distinct times with at least one event at each ($d$ is the total number of deaths).

2. At ordered time $t_i$, $1 \leq i \leq m$, the data may be summarized into a $2 \times 2$ table (Table 11.2) where

$n_{1i}$ = number of subjects from sample 1 who were at risk just before time $t_i$

$n_{2i}$ = number of subjects from sample 2 who were at risk just before time $t_i$

$n_i = n_{1i} + n_{2i}$

$d_i$ = number of deaths at $t_i$, $d_{1i}$ of them from sample 1 and $d_{2i}$ of them from sample 2

$\quad = d_{1i} + d_{2i}$

$a_i = n_i - d_i$

$\quad = a_{1i} + a_{2i}$

$\quad$ = number of survivors

$d = \sum d_i$

In this form, the null hypothesis of equal survival functions implies the independence of "sample" and "status" in Table 11.2. Therefore, under the null hypothesis, the expected value of $d_{1i}$ is

$$E_0(d_{1i}) = \frac{n_{1i}d_i}{n_i}$$

($d_{1i}$ being the observed value) following the formula for expected values used in Chapter 6. The variance is estimated by (hypergeometric model)

$$\text{Var}_0(d_{1i}) = \frac{n_{1i}n_{2i}a_id_i}{n_i^2(n_i - 1)}$$

the formula we used in the Mantel–Haenszel method in Chapter 6.

After constructing a $2 \times 2$ table for each uncensored observation, the evidence against the null hypothesis can be summarized in the following statistic:

$$\theta = \sum_{i=1}^{m} w_i[d_{1i} - E_0(d_{1i})]$$

where $w_i$ is the weight associated with the $2 \times 2$ table at $t_i$. We have under the null hypothesis:

$$E_0(\theta) = 0$$

$$\text{Var}_0(\theta) = \sum_{i=1}^{m} w_i^2 \, \text{Var}_0(d_{1i})$$

$$= \sum_{i=1}^{m} \frac{w_i^2 n_{1i}n_{2i}a_id_i}{n_i^2(n_i - 1)}$$

The evidence against the null hypothesis is summarized in the standardized statistic

$$z = \frac{\theta}{[\text{Var}_0(\theta)]^{1/2}}$$

which is referred to the standard normal percentile $z_{1-\alpha}$ for a specified size $\alpha$ of the test. We may also refer $z^2$ to a chi-square distribution at 1 degree of freedom.

There are two important special cases:

1. The choice $w_i = n_i$ gives the *generalized Wilcoxon test*; it is reduced to the Wilcoxon test in the absence of censoring.
2. The choice $w_i = 1$ gives the *log-rank test* (also called the *Cox–Mantel test*; it is similar to the Mantel–Haenszel procedure of Chapter 6 for the combination of several $2 \times 2$ tables in the analysis of categorical data).

There are a few other interesting issues:

1. Which test should we use? The generalized Wilcoxon statistic puts more weight on the beginning observations, and because of that its use is more powerful in detecting the effects of short-term risks. On the other hand, the log-rank statistic puts equal weight on each observation and there-

fore, by default, is more sensitive to exposures with a constant relative risk (the proportional hazards effect; in fact, we have derived the log-rank test as a score test using the proportional hazards model). Because of these characteristics, applications of both tests may reveal not only whether or not an exposure has any effect, but also the nature of the effect, short term or long term.

2. Because of the way the tests are formulated (terms in the summation are not squared),

$$\sum_{\text{all } i} w_i[d_{1i} - E_0(d_{1i})]$$

they are powerful only when one risk is greater than the other at all times. Otherwise, some terms in this sum are positive, other terms are negative, and they cancel each other out. For example, the tests are virtually powerless for the case of crossing survival curves; in this case the assumption of proportional hazards is severely violated.

3. Some cancer treatments (e.g., bone marrow transplantation) are thought to have cured patients within a short time following initiation. Then, instead of all patients having the same hazard, a biologically more appropriate model, the cure model, assumes that an unknown proportion $(1 - \pi)$ are still at risk, whereas the remaining proportion $(\pi)$ have essentially no risk. If the aim of the study is to compare the cure proportions $\pi$ values, neither the generalized Wilcoxon nor log-rank test is appropriate (low power). One may simply choose a time point $t$ far enough out for the curves to level off, then compare the estimated survival rates by referring to percentiles of the standard normal distribution:

$$z = \frac{\hat{S}_1(t) - \hat{S}_2(t)}{\{\text{Var}[\hat{S}_1(t)] + \text{Var}[\hat{S}_2(t)]\}^{1/2}}.$$

Estimated survival rates, $\hat{S}_i(t)$, and their variances are obtained as discussed in Section 11.2.1 (the Kaplan–Meier procedure).

***Example 11.2*** Refer back to the clinical trial (Example 11.1) to evaluate the effect of 6-mercaptopurine (6-MP) to maintain remission from acute leukemia. The results of the tests indicate a highly significant difference between survival patterns of the two groups (Figure 11.5). The generalized Wilcoxon test shows a slightly larger statistic, indicating that the difference is slightly larger at earlier times; however, the log-rank test is almost equally significant, indicating that the use of 6-MP has a long-term effect (i.e., the effect does not wear off).

$$\text{Generalized Wilcoxon}: \quad \chi^2 = 13.46(1 \text{ df}); p < 0.0002$$

$$\text{Log-rank}: \quad \chi^2 = 16.79(1 \text{ df}); p = 0.0001$$

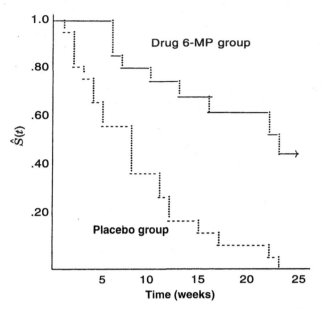

**Figure 11.5**   Two survival curves: drug 6-MP group and placebo group.

*Note:* An SAS program would include these instructions:

```
PROC LIFETEST METHOD = KM;
TIME WEEKS*RELAPSE(0);
STRATA DRUG;
```

where KM stands for the Kaplan–Meier method, WEEKS is the variable name for duration time, RELAPSE is the variable name for survival status, "0" is the coding for censoring, and DRUG is the variable name specifying groups to be compared.

The tests above are applicable to the simultaneous comparison of several samples; when $k$ groups are to be compared, the chi-square tests, both the log-rank and generalized Wilcoxon, have $(k - 1)$ degrees of freedom.

## 11.3   SIMPLE REGRESSION AND CORRELATION

In this section we discuss the basic ideas of simple regression analysis when only one predictor or independent variable is available for predicting the survival of interest. The following example is used for illustration in this and the following sections.

*Example 11.3*   A group of patients who died of acute myelogenous leukemia were classified into two subgroups according to the presence or absence of a

**TABLE 11.3**

| AG Positive, $n_1 = 17$ | | AG Negative, $n_0 = 16$ | |
|---|---|---|---|
| WBC | Survival Time (weeks) | WBC | Survival Time (weeks) |
| 2,300 | 65 | 4,400 | 56 |
| 750 | 156 | 3,000 | 65 |
| 4,300 | 100 | 4,000 | 17 |
| 2,600 | 134 | 1,500 | 7 |
| 6,000 | 16 | 9,000 | 16 |
| 10,500 | 108 | 5,300 | 22 |
| 10,000 | 121 | 10,000 | 3 |
| 17,000 | 4 | 19,000 | 4 |
| 5,400 | 39 | 27,000 | 2 |
| 7,000 | 143 | 28,000 | 3 |
| 9,400 | 56 | 31,000 | 8 |
| 32,000 | 26 | 26,000 | 4 |
| 35,000 | 22 | 21,000 | 3 |
| 100,000 | 1 | 79,000 | 30 |
| 100,000 | 1 | 100,000 | 4 |
| 52,000 | 5 | 100,000 | 43 |
| 100,000 | 65 | | |

morphologic characteristic of white cells (Table 11.3). Patients termed AG positive were identified by the presence of Auer rods and/or significant granulature of leukemic cells in the bone marrow at diagnosis. These factors were absent for AG-negative patients.

Leukemia is a cancer characterized by an overproliferation of white blood cells; the higher the white blood count (WBC), the more severe the disease. Data in Table 11.3 clearly suggest that there is such a relationship, and thus when predicting a leukemia patient's survival time, it is realistic to make a prediction dependent on the WBC (and any other covariates that are indicators of the progression of the disease).

### 11.3.1  Model and Approach

The association between two random variables $X$ and $T$, the second of which, survival time $T$, may be only partially observable, due to right censoring, has been the focus of many investigations. Cox's regression model or proportional hazards model (PHM) expresses a relationship between $X$ and the hazard function of $T$ as follows:

$$\lambda(t \mid X = x) = \lambda_0(t)e^{\beta x}$$

In this model, $\lambda_0(t)$ is an *unspecified* baseline hazard (i.e, hazard at $X = 0$) and

$\beta$ is an *unknown* regression coefficient. The estimation of $\beta$ and subsequent analyses are performed as follows. Denote the ordered distinct death times by

$$t_1 < t_2 < \cdots < t_m$$

and let $R_i$ be the risk set just before time $t_i$, $n_i$ the number of subjects in $R_i$, $D_i$ the death set at time $t_i$, $d_i$ the number of subjects (i.e., deaths) in $D_i$, and $C_i$ the collection of all possible combinations of subjects from $R_i$. Each combination, or subset of $R_i$, has $d_i$ members, and $D_i$ is itself one of these combinations. For example, if three subjects (A, B, and C) are at risk just before time $t_i$ and two of them (A and B) die at $t_i$, then

$$R_i = \{A, B, C\} \qquad n_i = 3$$
$$D_i = \{A, B\} \qquad d_i = 2$$
$$C_i = \{\{A, B\} = D_i, \{A, C\}, \{B, C\}\}$$

The product of the probabilities,

$$L = \prod_{i=1}^{m} \Pr(d_i \mid R_i, d_i)$$
$$= \prod_{i=1}^{m} \frac{\exp(\beta s_i)}{\sum_{C_i} \exp(\beta s_u)}$$

will serve as a likelihood function, called the *partial likelihood function*, in which

$$s_i = \sum_{D_i} x_j$$
$$s_u = \sum_{D_u} x_j \qquad D_u \in C_i$$

### 11.3.2   Measures of Association

We first consider the case of a binary covariate with the conventional coding

$$X_i = \begin{cases} 0 & \text{if the patient is not exposed} \\ 1 & \text{if the patient is exposed} \end{cases}$$

Here, the term *exposed* may refer to a risk factor such as smoking, or a patient's characteristic such as race (white/nonwhite) or gender (male/female). It can be seen that from the proportional hazards model,

$$\lambda(t; \text{ nonexposed}) = \lambda_0(t)$$

$$\lambda(t; \text{ exposed}) = \lambda_0(t)e^\beta$$

so that the ratio

$$e^\beta = \frac{\lambda(t; \text{ exposed})}{\lambda(t; \text{ nonexposed})}$$

represents the relative risk (RR) of the exposure, exposed versus nonexposed. In other words, the regression coefficient $\beta$ is the value of the relative risk on the log scale.

Similarly, we have for a continuous covariate $X$ and any value $x$ of $X$,

$$\lambda(t; X = x) = \lambda_0(t)e^{\beta x}$$

$$\lambda(t; X = x + 1) = \lambda_0(t)e^{\beta(x+1)}$$

so that the ratio

$$e^\beta = \frac{\lambda(t; X = x + 1)}{\lambda(t; X = x)}$$

represents the relative risk (RR) due to a 1-unit increase in the value of $X$, $X = x + 1$ versus $X = x$. For example, a systolic blood pressure of 114 mmHg versus 113 mmHg. For an $m$-unit increase in the value of $X$, say $X = x + m$ versus $X = x$, the corresponding relative risk is $e^{m\beta}$.

The regression coefficient $\beta$ can be estimated iteratively using the first and second derivatives of the partial likelihood function. From the results we can obtain a point estimate

$$\widehat{RR} = e^{\hat\beta}$$

and its 95% confidence interval

$$\exp[\hat\beta \pm 1.96 \, \text{SE}(\hat\beta)]$$

It should be noted that the calculation of the relative risk, used as a measure of association between survival time and a covariate, depends on the coding scheme for a binary factor and for a continuous covariate $X$, the scale with which to measure $X$. For example, if we use the following coding for a factor:

$$X_i = \begin{cases} -1 & \text{if the patient is not exposed} \\ 1 & \text{if the patient is exposed} \end{cases}$$

then

$$\lambda(t; \text{ nonexposed}) = \lambda_0(t)e^{-\beta}$$

$$\lambda(t; \text{ exposed}) = \lambda_0(t)e^{\beta}$$

so that

$$\text{RR} = \frac{\lambda(t; \text{ exposed})}{\lambda(t; \text{ nonexposed})}$$

$$= e^{2\beta}$$

and its 95% confidence interval

$$\exp[2(\hat{\beta} \pm 1.96 \text{ SE}(\hat{\beta}))]$$

Of course, the estimate of $\beta$ under the new coding scheme is only half of that under the former scheme; therefore, the estimate of RR remains unchanged.

The following example, however, will show the clear effect of measurement scale on the value of the relative risk in the case of a continuous measurement.

***Example 11.4***    Refer to the data for patients with acute myelogenous leukemia in Example 11.3 (Table 11.3) and suppose that we want to investigate the relationship between survival time of AG-positive patients and white blood count (WBC) in two different ways using either (a) $X = \text{WBC}$ or (b) $X = \log(\text{WBC})$.

(a) For $X = \text{WBC}$, we find that

$$\hat{\beta} = 0.0000167$$

from which the relative risk for $(\text{WBC} = 100{,}000)$ versus $(\text{WBC} = 50{,}000)$ would be

$$\text{RR} = \exp[(100{,}000 - 50{,}000)(0.0000167)]$$

$$= 2.31$$

(b) For $X = \log(\text{WBC})$, we find that

$$\hat{\beta} = 0.612331$$

from which the relative risk for $(\text{WBC} = 100{,}000)$ versus $(\text{WBC} = 50{,}000)$ would be

$$\text{RR} = \exp\{[\log(100{,}000) - \log(50{,}000)](0.612331)\}$$

$$= 1.53$$

The results above are different for two different choices of $X$, and this causes an obvious problem of choosing an appropriate measurement scale. Of course, we assume a *linear model*, and one choice of $X$ would fit better than the other (there are methods for checking this assumption).

*Note:* An SAS program would include these instructions:

```
PROC PHREG DATA = CANCER;
MODEL WEEKS*DEATH(0) = WBC;
```

where CANCER is the name assigned to the data set, WEEKS is the variable name for duration time, DEATH is the variable name for survival status, and "0" is the coding for censoring.

### 11.3.3  Tests of Association

The null hypothesis to be considered is

$$H_0: \beta = 0$$

The reason for interest in testing whether or not $\beta = 0$ is that $\beta = 0$ implies that there is no relation between survival time $T$ and the covariate $X$ under investigation. For the case of a categorical covariate, the test based on the score statistic of Cox's regression model is identical to the log-rank test of Section 11.2.2.

## 11.4  MULTIPLE REGRESSION AND CORRELATION

The effect of some factor on survival time may be influenced by the presence of other factors through effect modifications (i.e., interactions). Therefore, to provide a more comprehensive prediction of the future of patients with respect to duration, course, and outcome of a disease, it is very desirable to consider a large number of factors and sort out which are most closely related to diagnosis. In this section we discuss a multivariate method for risk determination. This method, multiple regression analysis, involves a linear combination of the explanatory or independent variables; the variables must be quantitative with particular numerical values for each patient. Information concerning possible factors is usually obtained as a subsidiary aspect from clinical trials that were designed to compare treatments. A covariate or prognostic patient characteristic may be dichotomous, polytomous, or continuous (categorical factors will be represented by dummy variables). Examples of dichotomous covariates are gender, and presence/absence of certain comorbidity. Polytomous covariates include race and different grades of symptoms; these can be covered by the use of *dummy variables*. Continuous covariates include patient age and blood pressure. In many cases, data transformations (e.g., taking the logarithm) may be desirable.

### 11.4.1    Proportional Hazards Model with Several Covariates

Suppose that we want to consider $k$ covariates simultaneously. The proportional hazards model of Section 11.3 can easily be generalized and expressed as

$$\lambda[t \mid \mathbf{X} = (x_1, x_2, \ldots, x_k)] = \lambda_0(t)e^{\beta_1 x_1 + \beta_2 x_2 + \cdots + \beta_k x_k}$$

where $\lambda_0(t)$ is an *unspecified* baseline hazard (i.e., hazard at $\mathbf{X} = 0$), $\mathbf{X}$ is the vector representing the $k$ covariates, and $\beta^T = (\beta_1, \beta_2, \ldots, \beta_k)$ are $k$ *unknown* regression coefficients. To have a meaningful baseline hazard, it may be necessary to standardize continuous covariates about their means:

$$\text{new } x = x - \bar{x}$$

so that $\lambda_0(t)$ is the hazard function associated with *a typical patient* (i.e., a hypothetical patient who has all covariates at their average values).

The estimation of $\beta$ and subsequent analyses are performed similar to the univariate case using the partial likelihood function:

$$L = \prod_{i=1}^{m} \Pr(d_i \mid R_i, d_i)$$

$$= \prod_{i=1}^{m} \frac{\exp(\sum_{j=1}^{k} \beta_j s_{ji})}{\sum_{C_i} \exp(\sum_{j=1}^{k} \beta_j s_{ju})}$$

where

$$s_{ji} = \sum_{l \in D_i} x_{jl}$$

$$s_{ju} = \sum_{l \in D_u} x_{jl} \qquad D_u \in C_i$$

Also similar to the univariate case, $\exp(\beta_i)$ represents one of the following:

1. The relative risk associated with an exposure if $X_i$ is binary (exposed $X_i = 1$ versus unexposed $X_i = 0$); or
2. The relative risk due to a 1-unit increase if $X_i$ is continuous ($X_i = x + 1$ versus $X_i = x$)

After $\hat{\beta}_i$ and its standard error have been obtained, a 95% confidence interval for the relative risk above is given by

$$\exp[\hat{\beta}_i \pm 1.96 \; \text{SE}(\hat{\beta}_i)]$$

These results are necessary in an effort to identify important prognostic or risk factors. Of course, before such analyses are done, the problem and the data have to be examined carefully. If some of the variables are highly correlated, one or fewer of the correlated factors are likely to be as good predictors as all of them; information from similar studies also has to be incorporated so as to drop some of the correlated explanatory variables. The uses of products such as $X_1 X_2$, and higher power terms such as $X_1^2$, may be necessary and can improve the goodness of fit. It is important to note that we are assuming a *linear regression model* in which, for example, the relative risk due to a 1-unit increase in the value of a continuous $X_i$ ($X_i = x + 1$ versus $X_i = x$) is independent of $x$. Therefore, if this *linearity* seems to be violated, incorporation of powers of $X_i$ should be considered seriously. The use of products will help in the investigation of possible effect modifications. Finally, there is the messy problem of missing data; most packaged programs would delete the patient if one or more covariate values are missing.

## 11.4.2   Testing Hypotheses in Multiple Regression

Once we have fit a multiple proportional hazards regression model and obtained estimates for the various parameters of interest, we want to answer questions about the contributions of various factors to the prediction of the future of patients. There are three types of such questions:

1. *Overall test.* Taken collectively, does the entire set of explatory or independent variables contribute significantly to the prediction of survivorship?
2. *Test for the value of a single factor.* Does the addition of one particular factor of interest add significantly to the prediction of survivorship over and above that achieved by other factors?
3. *Test for contribution of a group of variables.* Does the addition of a group of factors add significantly to the prediction of survivorship over and above that achieved by other factors?

***Overall Regression Tests***   We now consider the first question stated above concerning an overall test for a model containg $k$ factors: say,

$$\lambda[t \mid \mathbf{X} = (x_1, x_2, \dots, x_k)] = \lambda_0(t) e^{\beta_1 x_1 + \beta_2 x_2 + \dots + \beta_k x_k}$$

The null hypothesis for this test may stated as: "All $k$ independent variables *considered together* do not explain the variation in survival times." In other words,

$$H_0: \beta_1 = \beta_2 = \dots = \beta_k = 0$$

Three likelihood-based statistics can be used to test this *global* null hypothesis; each has a symptotic chi-square distribution with $k$ degrees of freedom under $H_0$: the likelihood ratio test, Wald's test, and the score test. All three chi-square statistics are provided by most standard computer programs.

***Tests for a Single Variable***    Let us assume that we now wish to test whether the addition of one particular factor of interest adds significantly to the prediction of survivorship over and above that achieved by other factors already present in the model. The null hypothesis for this test may be stated as: "Factor $X_i$ does not have any value added to the prediction of survivorship *given that other factors are already included in the model*." In other words,

$$H_0: \beta_i = 0$$

To test such a null hypothesis, one can perform a likelihood ratio chi-square test, with 1 df, similar to that for the global hypothesis above:

$$\chi^2_{LR} = 2[\ln L(\hat{\beta}; \text{ all } X\text{'s}) - \ln L(\hat{\beta}; \text{ all other } X\text{'s with } X_i \text{ deleted})]$$

A much easier alternative method is to use

$$z_i = \frac{\hat{\beta}_i}{\text{SE}(\hat{\beta}_i)}$$

where $\hat{\beta}_i$ is the corresponding estimated regression coefficient and $\text{SE}(\hat{\beta}_i)$ is the estimate of the standard error of $\hat{\beta}_i$, both of which are printed by standard packaged programs. In performing this test, we refer the value of the $z$ statistic to percentiles of the standard normal distribution. This is equivalent to Wald's chi-square test as applied to one parameter.

***Contribution of a Group of Variables***    This testing procedure addresses the more general problem of assessing the additional contribution of two or more factors to the prediction of survivorship over and above that made by other variables already in the regression model. In other words, the null hypothesis is of the form

$$H_0: \beta_1 = \beta_2 = \cdots = \beta_m = 0$$

To test such a null hypothesis, one can perform a likelihood ratio chi-square test with $m$ df:

$$\chi^2_{LR} = 2[\ln L(\hat{\beta}; \text{ all } X\text{'s})$$

$$- \ln L(\hat{\beta}; \text{ all other } X\text{'s with } X\text{'s under investigation deleted})]$$

As with the *z test* above, this *multiple contribution procedure* is very useful for assessing the importance of potential explanatory variables. In particular it is often used to test whether a similar group of variables, such as *demographic characteristics*, is important for the prediction of survivorship; these variables have some trait in common. Another application would be a collection of powers *and/or* product terms (referred to as *interaction variables*). It is often of interest to assess the interaction effects collectively before trying to consider individual interaction terms in a model. In fact, such use reduces the total number of tests to be performed, and this, in turn, helps to provide better control of overall type I error rates, which may be inflated due to multiple testing.

***Stepwise Procedure***    In applications, our major interest is to identify important prognostic factors. In other words, we wish to identify from many available factors a small subset of factors that relate significantly to the length of survival time of patients. In that identification process, of course, we wish to avoid a type I (false positive) error. In a regression analysis, a type I error corresponds to including a predictor that has no real relationship to survivorship; such an inclusion can greatly confuse interpretation of the regression results. In a standard multiple regression analysis, this goal can be achieved by using a strategy that adds into or removes from a regression model one factor at a time according to a certain order of relative importance. The details of this stepwise process for survival data are similar to those for logistic regression given in Chapter 9.

***Stratification***    The proportional hazards model requires that for a covariate $X$—say, an exposure—the hazards functions at different levels, $\lambda(t;$ exposed) and $\lambda(t;$ nonexposed) are proportional. Of course, sometimes there are factors the different levels of which produce hazard functions that deviate markedly from proportionality. These factors may not be under investigation themselves, especially those of no intrinsic interest, those with a large number of levels, and/or those where interventions are not possible. But these factors may act as important confounders which must be included in any meaningful analysis so as to improve predictions concerning other covariates. Common examples include gender, age, and neighborhood. To accommodate such confounders, an extension of the proportional hazards model is desirable. Suppose there is a factor that occurs on $q$ levels for which the proportional hazards model may be violated. If this factor is under investigation as a *covariate*, the model and subsequent analyses are not applicable. However, if this factor is *not* under investigation and is considered only as a confounder so as to improve analyses and/ or predictions concerning other covariates, *we can treat it as a stratification factor*. By doing that we will get no results concerning this factor (which are not wanted), but in return we do not have to assume that the hazard functions corresponding to different levels are proportional (which may be severely violated). Suppose that the stratification factor $Z$ has $q$ levels; this factor is not clinically important itself, but adjustments are still needed in efforts to investigate other covariates. We define the hazard function for a person in the $j$th stratum (or level) of this factor as

$$\lambda[t \mid \mathbf{X} = (x_1, x_2, \ldots, x_k)] = \lambda_{0j}(t)e^{\beta_1 x_1 + \beta_2 x_2 + \cdots + \beta_k x_k}$$

for $j = 1, 2, \ldots, q$, where $\lambda_0(t)$ is an *unspecified* baseline hazard for the $j$th stratum and $\mathbf{X}$ represents other $k$ covariates under investigation (excluding the stratification itself). The baseline hazard functions are allowed to be arbitrary and are completely unrelated (and, of course, *not* proportional). The basic additional assumption here, which is the same as that in *analysis of covariance*, requires the the $\beta$'s are the same across strata (called the *parallel lines assumption*, which is testable).

In the analysis we identify distinct times of events for the $j$th stratum and form the partial likelihood $L_j(\beta)$ as in earlier sections. The *overall* partial likelihood of $\beta$ is then the product of those $q$ stratum-specific likelihoods:

$$L(\beta) = \prod_{j=1}^{q} L_j(\beta)$$

Subsequent analyses, finding maximum likelihood estimates as well as using score statistics, are straightforward. For example, if the null hypothesis $\beta = 0$ for a given covariate is of interest, the score approach would produce a *stratified* log-rank test. An important application of stratification, the analysis of epidemiologic matched studies resulting in the conditional logistic regression model, is presented in Section 11.5.

*Example 11.5* Refer to the myelogenous leukemia data of Example 11.3 (Table 11.3). Patients were classified into the two groups according to the presence or absence of a morphologic characteristic of white cells, and the primary covariate is the white blood count (WBC). Using

$$X_1 = \ln(\text{WBC})$$

$$X_2 = \text{AG group (0 if negative and 1 if positive)}$$

we fit the following model with one interaction term:

$$\lambda[t \mid \mathbf{X} = (x_1, x_2)] = \lambda_0(t)e^{\beta_1 x_1 + \beta_2 x_2 + \beta_3 x_1 x_2}$$

From the results shown in Table 11.4, it can be seen that the interaction effect

**TABLE 11.4**

| Factor | Coefficient | Standard Error | $z$ Statistic | $p$ Value |
|---|---|---|---|---|
| WBC | 0.14654 | 0.17869 | 0.821 | 0.4122 |
| AG group | −5.85637 | 2.75029 | −2.129 | 0.0332 |
| Product | 0.49527 | 0.27648 | 1.791 | 0.0732 |

is almost significant at the 5% level ($p = 0.0732$); that is, the presence of the morphologic characteristic modifies substantially the effect of WBC.

### 11.4.3  Time-Dependent Covariates and Applications

In prospective studies, since subjects are followed over time, values of many independent variables or covariates may be changing; covariates such as patient age, blood pressure, and treatment. In general, covariates are divided into two categories: fixed and time dependent. A covariate is *time dependent* if the *difference* between covariate values from two different subjects may be changing with time. For example, gender and age are *fixed covariates*; a patient's age is increasing by one a year, but the difference in age between two patients remains unchanged. On the other hand, blood pressure is an obvious time-dependent covariate.

*Examples*  The following are three important groups of time-dependent covariates.

1. *Personal characteristics whose measurements are made periodically during the course of a study.* Blood pressure fluctuates; so do cholesterol level and weight. Smoking and alcohol consumption habits may change.

2. *Cumulative exposure.* In many studies, exposures such as smoking are often dichotomized; subjects are classified as exposed or unexposed. But this may be oversimplified, leading to loss of information; the length of exposure may be important. As time goes on, a nonsmoker remains a nonsmoker, but "years of smoking" for a smoker increases.

3. *Switching treatments.* In a clinical trial, a patient may be transferred from one treatment to another due to side effects or even by a patient's request. Organ transplants form another category with switching treatments; when a suitable donor is found, a subject is switched from the non-transplanted group to the transplanted group. The case of intensive care units is even more complicated, as a patient may be moved in and out more than once.

*Implementation*  Recall that in analysis using proportional hazards model, we order the death times and form the partial likelihood function:

$$L = \prod_{i=1}^{m} \Pr(d_i \mid R_i, d_i)$$

$$= \prod_{i=1}^{m} \frac{\exp(\sum_{j=1}^{k} \beta_j s_{ji})}{\sum_{C_i} \exp(\sum_{j=1}^{k} \beta_j s_{ju})}$$

where

$$s_{ji} = \sum_{l \in D_i} x_{jl}$$

$$s_{ju} = \sum_{l \in D_u} x_{jl} \qquad D_u \in C_i$$

and $R_i$ is the risk set just before time $t_i$, $n_i$ the number of subjects in $R_i$, $D_i$ the death set at time $t_i$, $d_i$ the number of subjects (i.e., deaths) in $D_i$, and $C_i$ the collection of all possible combinations of subjects from $R_i$. In this approach we try to explain why *event(s)* occurred to *subject(s)* in $D_i$ whereas all subjects in $R_i$ are equally at risk. *This explanation, through the use of $s_{ji}$ and $s_{ju}$, is based on the covariate values measured at time $t_i$.* Therefore, this needs some modification in the presence of time-dependent covariates because events at time $t_i$ should be explained by *values of covariates measured at that particular moment.* Blood pressure measured years before, for example, may become irrelevant.

First, notations are expanded to handle time-dependent covariates. Let $x_{jil}$ be the value of factor $x_j$ measured from subject l at time $t_i$; then the likelihood function above becomes

$$L = \prod_{i=1}^{m} \Pr(d_i \mid R_i, d_i)$$

$$= \prod_{i=1}^{m} \frac{\exp(\sum_{j=1}^{k} \beta_j s_{jii})}{\sum_{C_i} \exp(\sum_{j=1}^{k} \beta_j s_{jiu})}$$

where

$$s_{jii} = \sum_{l \in D_i} x_{jil}$$

$$s_{jiu} = \sum_{l \in D_u} x_{jil} \qquad D_u \in C_i$$

From this new likelihood function, applications of subsequent steps (estimation of $\beta$'s, formation of test statistics, and estimation of the baseline survival function) are straightforward. In practical implementation, most standard computer programs have somewhat different procedures for two categories of time-dependent covariates: those that can be defined by a mathematical equation (external) and those measured directly from patients (internal); the former categories are implemented much more easily.

***Simple Test of Goodness of Fit***    Treatment of time-dependent covariates leads to a simple test of goodness of fit. Consider the case of a fixed covariate, denoted by $X_1$. Instead of the basic proportional hazards model,

$$\lambda(t \mid X_1 = x_1) = \lim_{\delta \downarrow 0} \frac{\Pr(t \le T \le t + \delta \mid t \le T, X_1 = x_1)}{\delta}$$

$$= \lambda_0(t)e^{\beta x_1}$$

we can define an additional time-dependent covariate $X_2$,

$$X_2 = X_1 t$$

Consider the expanded model,

$$\lambda(t; X_1 = x_1) = \lambda_0(t)e^{\beta_1 x_1 + \beta_2 x_2}$$

$$= \lambda_0(t)e^{\beta_1 x_1 + \beta_2 x_1 t}$$

and examine the significance of

$$H_0: \beta_2 = 0$$

The reason for interest in testing whether or not $\beta_2 = 0$ is that $\beta_2 = 0$ implies a goodness of fit of the proportional hazards model for the factor under investigation, $X_1$. Of course, in defining the new covariate $X_2$, $t$ could be replaced by any function of $t$; a commonly used function is

$$X_2 = X_1 \log(t)$$

This simple approach results in a test of a specific alternative to the proportionality. The computational implementation here is very similar to the case of cumulative exposures; however, $X_1$ may be binary or continuous. We may even investigate the goodness of fit for several variables simultaneously.

***Example 11.6***    Refer to the data set in Example 11.1 (Table 11.1), where the remission times of 42 patients with acute leukemia were reported from a clinical trial undertaken to assess the ability of the drug 6-mercaptopurine (6-MP) to maintain remission. Each patient was randomized to receive either 6-MP or placebo. The study was terminated after one year; patients have different follow-up times because they were enrolled sequentially at different times. Times in weeks were:

- *6-MP group:* 6, 6, 6, 7, 10, 13, 16, 22, 23, 6+, 9+, 10+, 11+, 17+, 19+, 20+, 25+, 32+, 32+, 34+, 35+
- *Placebo group:* 1, 1, 2, 2, 3, 4, 4, 5, 5, 8, 8, 8, 8, 11, 11, 12, 12, 15, 17, 22, 23

in which $t+$ denotes a censored observation (i.e., the case was censored after $t$ weeks without a relapse). For example, "10+" is a case enrolled 10 weeks before study termination and still remission-free at termination.

Since the proportional hazards model is often assumed in the comparison of two survival distributions, such as in this example (see also Example 11.2), it is desirable to check it for validity (if the proportionality is rejected, it would lend support to the conclusion that this drug does have some cumulative effects). Let $X_1$ be the indicator variable defined by

$$X_1 = \begin{cases} 0 & \text{if placebo} \\ 1 & \text{if treated by 6-MP} \end{cases}$$

and

$$X_2 = X_1 t$$

representing treatment weeks (time $t$ is recorded in weeks). To judge the validity of the proportional hazards model with respect to $X_1$, it is the effect of this newly defined covariate, $X_2$, that we want to investigate. We fit the following model;

$$\lambda[t \,|\, \mathbf{X} = (x_1, x_2)] = \lambda_0(t) e^{\beta_1 x_1 + \beta_2 x_2}$$

and from the results shown in Table 11.5, it can be seen that the accumulation effect or lack of fit, represented by $X_2$, is insignificant; in other words, there is not enough evidence to be concerned about the validity of the proportional hazards model.

**TABLE 11.5**

| Factor | Coefficient | Standard Error | $z$ Statistic | $p$ Value |
|--------|-------------|----------------|---------------|-----------|
| $X_1$  | −1.55395    | 0.81078        | −1.917        | 0.0553    |
| $X_2$  | −0.00747    | 0.06933        | −0.108        | 0.9142    |

*Note:* An SAS program would include these instructions:

```
PROC PHREG DATA = CANCER;
MODEL WEEKS*RELAPSE(0) = DRUG TESTEE;
TESTEE = DRUG*WEEKS;
```

where WEEKS is the variable name for duration time, RELAPSE is the variable name for survival status, "0" is the coding for censoring, DRUG is the 0/1 group indicator (i.e., $X_1$), and TESTEE is the newly created variable (i.e., $X_2$).

## 11.5   PAIR-MATCHED CASE–CONTROL STUDIES

Case–control studies have been perhaps the most popular form of research design in epidemiology. They generally can be carried out in a much shorter period of time than cohort studies and are cost-effective. As a technique for controlling the effects of confounders, randomization and stratification are possible solutions at the design stage, and statistical adjustments can be made at the analysis stage. Statistical adjustments are done using regression methods, such as the logistic regression described in Chapter 9. Stratification is more often introduced at the analysis stage, and methods such as the Mantel–Haenszel method (Chapters 1 and 6) are available to complete the task.

Stratification can also be introduced at the design stage; its advantage is that one can avoid inefficiencies resulting from having some strata with a gross imbalance of cases and controls. A popular form of stratified design occurs when each case is *matched individually* with one or more controls chosen to have similar characteristics (i.e., values of confounding or matching variables). Matched designs have several advantages. They make it possible to control for confounding variables that are difficult to measure directly and therefore difficult to adjust at the analysis stage. For example, subjects can be matched using area of residence so as to control for environmental exposure. Matching also provides more adequate control of confounding than can adjustment in analysis using regression because matching does not need specific assumptions as to functional form, which may be needed in regression models. Of course, matching also has disadvantages. Matches for subjects with unusual characteristics are hard to find. In addition, when cases and controls are matched on a certain specific characteristic, the influence of that characterisric on a disease can no longer be studied. Finally, a sample of matched cases and controls is not usually representative of any specific population, which may reduce our ability to generalize analysis results.

One-to-one matching is a cost-effective design and is perhaps the most popular form used in practice. It is conceptually easy and usually leads to a simple analysis.

**TABLE 11.6**

| Factor | Disease + | Disease − | Total |
|---|---|---|---|
| + | $P_1$ | $P_3$ | $P_1 + P_3$ |
| − | $P_2$ | $P_4$ | $P_2 + P_4$ |
| Total | $P_1 + P_2$ | $P_3 + P_4$ | 1 |

### 11.5.1 Model

Consider a case–control design and suppose that each person in a large population has been classified as exposed or not exposed to a certain factor, and as having or not having some disease. The population may then be enumerated in a $2 \times 2$ table (Table 11.6), with entries being the proportions of the total population.

Using these proportions, the association (if any) between the factor and the disease could be measured by the ratio of risks (or relative risk) of being disease positive for those with or without the factor:

$$\text{relative risk} = \frac{P_1/(P_1 + P_3)}{P_2/(P_2 + P_4)}$$

$$= \frac{P_1(P_2 + P_4)}{P_2(P_1 + P_3)}$$

since in many (although not all) situations, the proportions of subjects classified as disease positive will be small. That is, $P_1$ is small in comparison with $P_3$, and $P_2$ will be small in comparison with $P_4$. In such a case, the relative risk is almost equal to

$$\text{OR} = \frac{P_1 P_4}{P_2 P_3}$$

$$= \frac{P_1/P_3}{P_2/P_4}$$

the odds ratio of being disease positive, or

$$= \frac{P_1/P_2}{P_3/P_4}$$

the odds ratio of being exposed. This justifies use of the odds ratio to measure differences, if any, in exposure to a suspected risk factor.

As a technique to control confounding factors in a designed study, individual cases are matched, often one to one, to a set of controls chosen to have

**TABLE 11.7**

| Control | Case + | Case − | Total |
|---|---|---|---|
| + | $n_{11}$ | $n_{01}$ | $n_{11} + n_{01}$ |
| − | $n_{10}$ | $n_{00}$ | $n_{10} + n_{00}$ |
| Total | $n_{11} + n_{10}$ | $n_{01} + n_{00}$ | $n$ |

similar values for the important confounding variables. The simplest example of pair-matched data occurs with a single binary exposure (e.g., smoking versus nonsmoking). The data for outcomes can be represented by a $2 \times 2$ table (Table 11.7) where $(+, -)$ denotes (exposed, unexposed). For example, $n_{10}$ denotes the number of pairs where the case is exposed but the matched control is unexposed. The most suitable statistical model for making inferences about the odds ratio $\theta$ is to use the conditional probability of the number of exposed cases among the discordant pairs. Given $(n_{10} + n_{01})$ as fixed, $n_{10}$ has the binomial distribution $B(n_{10} + n_{01}, \pi)$, that is, the binomial distribution with $n = n_{10} + n_{01}$ trials, each with probability of success

$$\pi = \frac{\text{OR}}{1 + \text{OR}}$$

### 11.5.2  Analysis

Using the binomial model above with the likelihood function

$$\left(\frac{\text{OR}}{1 + \text{OR}}\right)^{n_{10}} \left(\frac{1}{1 + \text{OR}}\right)^{n_{01}}$$

from which one can estimate the odds ratio, the results are

$$\widehat{\text{OR}} = \frac{n_{10}}{n_{01}}$$

$$\widehat{\text{Var}}(\widehat{\text{OR}}) = \frac{n_{10}(n_{10} + n_{01})}{n_{01}^3}$$

For example, with large samples, a 95% confidence interval for the odds ratio is given by

$$\widehat{\text{OR}} \pm (1.96)[\widehat{\text{Var}}(\widehat{\text{OR}})]^{1/2}$$

The null hypothesis of no risk effect (i.e., $H_0$: OR $= 1$) can be tested where the $z$

**TABLE 11.8**

| | Disease Classification | | |
| Exposure | Cases | Controls | Total |
| --- | --- | --- | --- |
| Yes | $a_i$ | $b_i$ | $a_i + b_i$ |
| No | $c_i$ | $d_i$ | $c_i + d_i$ |
| Total | 1 | 1 | 2 |

statistic,

$$z = \frac{n_{10} - n_{01}}{\sqrt{n_{10} + n_{01}}}$$

is compared to percentiles of the standard normal distribution. The corresponding two-tailed procedure based on

$$X^2 = \frac{(n_{10} - n_{01})^2}{n_{10} + n_{01}}$$

is often called McNemar's chi-square test (1 df) (introduced in Section 6.2). It is interesting to note that if we treat a matched pair as a group or a level of a confounder, and present the data in the form of a $2 \times 2$ table (Table 11.8), the Mantel–Haenszel method of Section 1.3.4 would yield the same estimate for the odds ratio:

$$\widehat{OR}_{MH} = \widehat{OR}$$

$$= \frac{n_{10}}{n_{01}}$$

As for the task of forming a 95% confidence interval, an alternative to the preceding formula is first to estimate the odds ratio on a log scale with estimated variance

$$\widehat{Var}(\log(\widehat{OR})) = \frac{1}{n_{10}} + \frac{1}{n_{01}}$$

leading to a 95% confidence interval of

$$\frac{n_{10}}{n_{01}} \exp\left(\pm 1.96 \sqrt{\frac{1}{n_{10}} + \frac{1}{n_{01}}}\right)$$

***Example 11.7*** In a study of endometrial cancer in which the investigators identified 63 cases occurring in a retirement community near Los Angeles,

**TABLE 11.9**

| Control | Case + | − | Total |
|---------|--------|---|-------|
| + | 27 | 3 | 30 |
| − | 29 | 4 | 33 |
| Total | 66 | 7 | 73 |

California from 1971 to 1975, each disease person was matched with $R = 4$ controls who were alive and living in the community at the time the case was diagnosed, who were born within one year of the case, who had the same marital status, and who had entered the community at approximately the same time. The risk factor was previous use of estrogen (yes/no) and the data in Table 11.9 were obtained from the first-found matched control (the complete data set with four matched controls will be given later). An application of the methods above yields

$$\widehat{OR} = \frac{29}{3}$$
$$= 9.67$$

and a 95% confidence interval for OR is $(2.95, 31.74)$.

## 11.6  MULTIPLE MATCHING

One-to-one matching is a cost-effective design. However, an increase in the number of controls may give a study more power. In epidemiologic studies, there are typically a small number of cases and a large number of potential controls to select from. When the controls are more easily available than cases, it is more efficient and effective to select more controls for each case. The efficiency of an $M:1$ control–case ratio for estimating a risk relative to having complete information on the control population (i.e., $M = \infty$) is $M/(M + 1)$. Hence, a 1:1 matching is 50% efficient, 4:1 matching is 80% efficient, 5:1 matching is 83% efficient, and so on. The gain in efficiency diminishes rapidly for designs with $M \geq 5$.

### 11.6.1  Conditional Approach

The analysis of a 1:1 matching design was conditional on the number of pairs showing differences in exposure history: $(-, +)$ and $(+, -)$ pairs. Similarly, considering an $M:1$ matching design, we use a conditional approach, fixing the number $m$ of exposed persons in a matched set, and the sets with $m = 0$ or

$m = M + 1$ [similar to $(-, -)$ and $(+, +)$ cells in the 1:1 matching design] will be ignored.

If we fix the number $m$ of exposed persons in each stratum, $1 \le m \le M$, then

$$\text{Pr(cases exposed}/m \text{ exposed in a stratum)} = \frac{m(\text{OR})}{m(\text{OR}) + M - m + 1}$$

where OR is the odds ratio representing the effect of exposure. The result for pair-matched design in Section 11.5 is a special case where $M = m = 1$.

For the strata, or matched sets with exactly $m$ $(m = 1, 2, \ldots, M)$ exposed persons, let

$$n_{1,m} = \text{number of sets with an exposed case}$$

$$n_{0,m-1} = \text{number of sets with an unexposed case}$$

$$n_m = n_{1,m} + n_{0,m-1}$$

Then given $n_m$ fixed, $n_{1,m}$ has $B(n_m, p_m)$, where

$$p_m = \frac{m(\text{OR})}{m(\text{OR}) + M - m + 1}$$

In the special case of 1:1 matching, $M = 1$, and the result is reduced to the probability $\pi$ of Section 11.5.1.

### 11.6.2 Estimation of the Odds Ratio

From the joint (conditional) likelihood function

$$L(\text{OR}) = \prod_{m=1}^{M} \left[ \frac{m(\text{OR})}{m(\text{OR}) + M - m + 1} \right]^{n_{1,m-1}} \left[ \frac{M - m + 1}{m(\text{OR}) + M - m + 1} \right]^{n_{0,m}}$$

one can obtain the estimate OR, but such a solution requires an iterative procedure and a computer algorithm. We will return to this topic in the next section.

Another simple method for estimating the odds ratio would be to treat a matched set consisting of one case and $M$ matched controls as a stratum. We then present the data from this stratum in the form of a $2 \times 2$ table (Table 11.10) and obtain the Mantel–Haenszel estimate for the odds ratio:

$$\widehat{\text{OR}}_{\text{MH}} = \frac{\sum(ad/(M + 1))}{\sum(bc/(M + 1))}$$

**TABLE 11.10**

| Exposure | Disease Classification | | Total |
| | Cases | Controls | |
|---|---|---|---|
| Yes | $a$ | $b$ | $a+b$ |
| No | $c$ | $d$ | $c+d$ |
| Total | 1 | $M$ | $M+1$ |

The result turns out to be, quite simply,

$$\widehat{\text{OR}_{\text{MH}}} = \frac{\sum(M-m+1)n_{1,m-1}}{\sum mn_{0,m}}$$

and the Mantel–Haenszel estimate has been used widely in the analysis of case–control studies with multiple matching.

### 11.6.3   Testing for Exposure Effect

From the same likelihood function as seen above, a test for

$$H_0: \text{OR} = 1$$

was derived and is given by:

$$\chi^2_{ES} = \frac{\{\sum_{m=1}^{M}[n_{1,m-1} - mn_m/(M+1)]\}^2}{[1/(M+1)^2]\sum_{m=1}^{M} mn_m(M-m+1)}$$

a chi-square test with 1 degree of freedom.

***Example 11.8***   In a study on endometrial cancer, the investigators identified 63 cases of endometrial cancer occurring in a retirement community near Los Angeles, California from 1971 to 1975. Each diseased person was matched with $R = 4$ controls who were alive and living in the community at the time the case was diagnosed, who were born within one year of the case, who had the same marital status, and who had entered the community at approximately the same time. The risk factor was previous use of estrogen (yes/no), and the data in Example 11.7 involve only the first-found matched control. We are now able to analyze the complete data set by 4:1 matching (Table 11.11). Using these 52 sets with four matched controls, we have

$$\chi^2_{ES}$$

$$= \frac{(25)\{[4-(1\times10)/5]+[17-(2\times20)/5]+[11-(3\times12)/5]+[9-(4\times10)/5]\}^2}{(1\times10\times4)+(2\times20\times3)+(3\times12\times2)+(4\times10\times1)}$$

$$= 22.95$$

The Mantel–Haenszel estimate for the odds ratio is

$$\widehat{OR}_{MH} = \frac{(4)(4) + (3)(17) + (2)(11) + (1)(9)}{(1)(6) + (2)(3) + (3)(1) + (4)(1)}$$

$$= 5.16$$

When the number of controls matched to a case, $M$, is variable (due primarily to missing data), the test for exposure effects should incorporate data from all strata:

$$\chi^2_{ES} = \frac{\left\{\sum_M \sum_{m=1}^{M} [n_{1,m-1} - mn_m/(M+1)]\right\}^2}{\sum_M \sum_{m=1}^{M} (mn_m(M-m+1)/(M+1))}$$

The corresponding Mantel–Haenszel estimate for the odds ratio is

$$\widehat{OR}_{MH} = \frac{\sum_M \sum (M-m+1)n_{1,m-1}}{\sum_M \sum mn_{0,m}}$$

***Example 11.9***    Refer to the data on endometrial cancer in Table 11.11; due to missing data, we have some cases matching to four controls and some matching to three controls. In addition to 4:1 matching, we have 3:1 matching (Table 11.12). With the inclusion of the four sets having three matched controls, the result becomes

**TABLE 11.11**

|  | Number of Exposed Persons in Each Matched Set | | | |
|---|---|---|---|---|
| Case | 1 | 2 | 3 | 4 |
| Exposed | 4 | 17 | 11 | 9 |
| Unexposed | 6 | 3 | 1 | 1 |
| Total | 10 | 20 | 12 | 10 |

**TABLE 11.12**

|  | Number of Exposed Persons in Each Matched Set | | |
|---|---|---|---|
| Case | 1 | 2 | 3 |
| Exposed | 1 | 3 | 0 |
| Unexposed | 0 | 0 | 0 |
| Total | 1 | 3 | 0 |

$$\chi_{ES}^2 = \frac{\left\{\left[\left(4 - \frac{1 \times 10}{2}\right) + \left(17 - \frac{2 \times 20}{5}\right) + \left(11 - \frac{3 \times 12}{5}\right) + \left(9 - \frac{4 \times 10}{5}\right)\right] + \left[\left(1 - \frac{1 \times 1}{4}\right) + \left(3 - \frac{2 \times 3}{4}\right)\right]\right\}^2}{\frac{1}{25}[(1 \times 10 \times 4) + (2 \times 20 \times 3) + (3 \times 12 \times 2) + (4 \times 10 \times 1)] + \frac{1}{16}[1 \times 1 \times 3 + 2 \times 3 \times 2]}$$

$$= 27.57$$

Mantel–Haenszel's estimate for the odds ratio is

$$\widehat{OR}_{MH} = \frac{[(4)(4) + (3)(17) + (2)(11) + (1)(9)] + [(3)(1) + (2)(3) + (1)(0)]}{[(1)(6) + (2)(3) + (3)(1) + (4)(1)] + [(1)(0) + (2)(0) + (3)(0)]}$$

$$= 5.63$$

## 11.7   CONDITIONAL LOGISTIC REGRESSION

Recall from Chapter 9 that in a variety of applications using regression analysis, the dependent variable of interest has only two possible outcomes and therefore can be represented by an indicator variable taking on values 0 and 1. An important application is the analysis of case–control studies where the dependent variable represents the disease status, 1 for a case and 0 for a control. The methods that have been used widely and successfully for these applications are based on the logistic model. In this section we also deal with cases where the dependent variable of interest is binary, following a binomial distrubution—the same as those using logistic regression analyses but the data are matched. Again, the term *matching* refers to the pairing of one or more controls to each case on the basis of their similarity with respect to selected variables used as *matching criteria*, as seen earlier. Although the primary objective of matching is the elimination of biased comparison between cases and controls, this objective can be accomplished only if matching is followed by an analysis that corresponds to the design matched. Unless the analysis accounts properly for the matching used in the selection phase of a case–control study, the results can be biased. In other words, matching (which refers to the selection process) is only the first step of a two-step process that can be used effectively to control for confounders: (1) matching design, followed by (2) matched analysis. Suppose that the purpose of the research is to assess relationships between the disease and a set of covariates using a matched case–control design. The regression techniques for the statistical analysis of such relationships is based on the conditional logistic model.

The following are two typical examples; the first one is a case–control study of vaginal carcinoma which involves two binary risk factors, and in the second example, one of the four covariates is on a continuous scale.

*Example 11.10*   The cases were eight women 15 to 22 years of age who were diagnosed with vaginal carcinoma between 1966 and 1969. For each case, four controls were found in the birth records of patients having their babies deliv-

**TABLE 11.13**

| Set | Case | Control Subject Number | | | |
|-----|------|------|------|------|------|
|     |      | 1 | 2 | 3 | 4 |
| 1 | (N, Y) | (N, Y) | (N, N) | (N, N) | (N, N) |
| 2 | (N, Y) | (N, Y) | (N, N) | (N, N) | (N, N) |
| 3 | (Y, N) | (N, Y) | (N, N) | (N, N) | (N, N) |
| 4 | (Y, Y) | (N, N) | (N, N) | (N, N) | (N, N) |
| 5 | (N, N) | (Y, Y) | (N, N) | (N, N) | (N, N) |
| 6 | (Y, Y) | (N, N) | (N, N) | (N, N) | (N, N) |
| 7 | (N, Y) | (N, Y) | (N, N) | (N, N) | (N, N) |
| 8 | (N, Y) | (N, N) | (N, N) | (N, N) | (N, N) |

ered within five days of the case in the same hospital. The risk factors of interest are the mother's bleeding in this pregnancy ($N$ = no, $Y$ = yes) and any previous pregnancy loss by the mother ($N$ = no, $Y$ = yes). Response data (bleeding, previous loss) are given in Table 11.13.

***Example 11.11***    For each of 15 low-birth-weight babies (the cases), we have three matched controls (the number of controls per case need not be the same). Four risk factors are under investigation: weight (in pounds) of the mother at the last menstrual period, hypertension, smoking, and uterine irritability (Table 11.14); for the last three factors, a value of 1 indicates a yes and a value of 0 indicates a no. The mother's age was used as the matching variable.

### 11.7.1  Simple Regression Analysis

In this section we discuss the basic ideas of simple regression analysis when only one predictor or independent variable is available for predicting the binary response of interest. We illustrate these for the more simple designs, in which each matched set has one case and case $i$ is matched to $m_i$ controls; the number of controls $m_i$ varies from case to case.

***Likelihood Function***    In our framework, let $x_i$ be the covariate value for case $i$ and $x_{ij}$ be the covariate value the the $j$th control matched to case $i$. Then for the $i$th matched set, it was proven that the conditional probability of the outcome observed (that the subject with covariate value $x_i$ be the case) given that we have one case per matched set is

$$\frac{\exp(\beta x_i)}{\exp(\beta x_i) + \sum_j^{m_i} \exp(\beta x_{ij})}$$

If the sample consists of $N$ matched sets, the conditional likelihood function is the product of the terms above over the $N$ matched sets:

**TABLE 11.14**

| Matched Set | Case | Mother's Weight | Hypertension | Smoking | Uterine Irritability |
|---|---|---|---|---|---|
| 1 | 1 | 130 | 0 | 0 | 0 |
|   | 0 | 112 | 0 | 0 | 0 |
|   | 0 | 135 | 1 | 0 | 0 |
|   | 0 | 270 | 0 | 0 | 0 |
| 2 | 1 | 110 | 0 | 0 | 0 |
|   | 0 | 103 | 0 | 0 | 0 |
|   | 0 | 113 | 0 | 0 | 0 |
|   | 0 | 142 | 0 | 1 | 0 |
| 3 | 1 | 110 | 1 | 0 | 0 |
|   | 0 | 100 | 1 | 0 | 0 |
|   | 0 | 120 | 1 | 0 | 0 |
|   | 0 | 229 | 0 | 0 | 0 |
| 4 | 1 | 102 | 0 | 0 | 0 |
|   | 0 | 182 | 0 | 0 | 1 |
|   | 0 | 150 | 0 | 0 | 0 |
|   | 0 | 189 | 0 | 0 | 0 |
| 5 | 1 | 125 | 0 | 0 | 1 |
|   | 0 | 120 | 0 | 0 | 1 |
|   | 0 | 169 | 0 | 0 | 1 |
|   | 0 | 158 | 0 | 0 | 0 |
| 6 | 1 | 200 | 0 | 0 | 1 |
|   | 0 | 108 | 1 | 0 | 1 |
|   | 0 | 185 | 1 | 0 | 0 |
|   | 0 | 110 | 1 | 0 | 1 |
| 7 | 1 | 130 | 1 | 0 | 0 |
|   | 0 | 95 | 0 | 1 | 0 |
|   | 0 | 120 | 0 | 1 | 0 |
|   | 0 | 169 | 0 | 0 | 0 |
| 8 | 1 | 97 | 0 | 0 | 1 |
|   | 0 | 128 | 0 | 0 | 0 |
|   | 0 | 115 | 1 | 0 | 0 |
|   | 0 | 190 | 0 | 0 | 0 |
| 9 | 1 | 132 | 0 | 1 | 0 |
|   | 0 | 90 | 1 | 0 | 0 |
|   | 0 | 110 | 0 | 0 | 0 |
|   | 0 | 133 | 0 | 0 | 0 |
| 10 | 1 | 105 | 0 | 1 | 0 |
|   | 0 | 118 | 1 | 0 | 0 |
|   | 0 | 155 | 0 | 0 | 0 |
|   | 0 | 241 | 0 | 1 | 0 |
| 11 | 1 | 96 | 0 | 0 | 0 |
|   | 0 | 168 | 1 | 0 | 0 |
|   | 0 | 160 | 0 | 0 | 0 |
|   | 0 | 133 | 1 | 0 | 0 |

*(Continued)*

**TABLE 11.14** *(Continued)*

| Matched Set | Case | Mother's Weight | Hypertension | Smoking | Uterine Irritability |
|---|---|---|---|---|---|
| 12 | 1 | 120 | 1 | 0 | 1 |
|  | 0 | 120 | 1 | 0 | 0 |
|  | 0 | 167 | 0 | 0 | 0 |
|  | 0 | 250 | 1 | 0 | 0 |
| 13 | 1 | 130 | 0 | 0 | 1 |
|  | 0 | 150 | 0 | 0 | 0 |
|  | 0 | 135 | 0 | 0 | 0 |
|  | 0 | 154 | 0 | 0 | 0 |
| 14 | 1 | 142 | 1 | 0 | 0 |
|  | 0 | 153 | 0 | 0 | 0 |
|  | 0 | 110 | 0 | 0 | 0 |
|  | 0 | 112 | 0 | 0 | 0 |
| 15 | 1 | 102 | 1 | 0 | 0 |
|  | 0 | 215 | 1 | 0 | 0 |
|  | 0 | 120 | 0 | 0 | 0 |
|  | 0 | 150 | 1 | 0 | 0 |

$$L = \prod_{i=1}^{N} \frac{\exp(\beta x_i)}{\exp(\beta x_i) + \sum_{j}^{m_i} \exp(\beta x_{ij})}$$

from which we can obtain an estimate of the parameter $\beta$. The likelihood function above has the same mathematical form as the overall partial likelihood for the proportional hazards survival model with strata, one for each matched set, and one event time for each (see Sections 11.3 and 11.4). This enables us to adapt programs written for a proportional hazards model to analyze epidemiologic matched studies as seen in subsequent examples. The essential features of the adaptation are as follows:

1. Creating matched set numbers and using them as different levels of a stratification factor.
2. Assigning to each subject a number to be used in place of duration times. These numbers are chosen arbitrarily as long as the number assigned to a case is smaller than the number assigned to a control in the same matched set. This is possible because when there is only one event in each set, the numerical value for the time to event becomes irrelevant.

*Measure of Association*   Similar to the case of the logistic model of Chapter 9, $\exp(\beta)$ represents one of the following:

1. The odds ratio associated with an exposure if $X$ is binary (exposed $X = 1$ versus unexposed $X = 0$); or

**TABLE 11.15**

|  | Case | |
|---|---|---|
| Control | 1 | 0 |
| 1 | $n_{11}$ | $n_{01}$ |
| 0 | $n_{10}$ | $n_{00}$ |

2. The odds ratio due to a 1-unit increase if $X$ is continuous $(X = x + 1$ versus $X = x)$.

After $\hat{\beta}$ and its standard error have been obtained, a 95% confidence interval for the odds ratio above is given by

$$\exp[\hat{\beta} \pm 1.96 \, SE(\hat{\beta})]$$

**Special Case** Consider now the simplest case of a pair matched (i.e., 1:1 matching) with a binary covariate: exposed $X = 1$ versus unexposed $X = 0$. Let the data be summarized and presented as in Section 5.2 (Table 11.15). For example, $n_{10}$ denotes the number of pairs where the case is exposed but the matched control is unexposed. The likelihood function above is reduced to

$$L(\beta) = \left(\frac{1}{1+1}\right)^{n_{00}} \left[\frac{\exp(\beta)}{1+\exp(\beta)}\right]^{n_{10}} \left[\frac{1}{1+\exp(\beta)}\right]^{n_{01}} \left[\frac{\exp(\beta)}{\exp(\beta)+\exp(\beta)}\right]^{n_{11}}$$

$$= \frac{\exp(\beta n_{10})}{[1+\exp(\beta)]^{n_{10}+n_{01}}}$$

From this we can obtain a point estimate:

$$\hat{\beta} = \frac{n_{10}}{n_{01}}$$

which is the usual odds ratio estimate from pair-matched data.

**Tests of Association** Another aspect of statistical inference concerns the test of significance; the null hypothesis to be considered is

$$H_0: \beta = 0$$

The reason for interest in testing whether or not $\beta = 0$ is that $\beta = 0$ implies that there is no relation between the binary dependent variable and the covariate $X$ under investigation. We can simply appply a McNemar chi-square test (if the covariate is binary or categorical or a paired $t$ test or signed-rank Wilcoxon test

**TABLE 11.16**

| Variable | Coefficient | Standard Error | $z$ Statistic | $p$ Value |
|---|---|---|---|---|
| Mother's weight | −0.0211 | 0.0112 | −1.884 | 0.0593 |

(if the covariate under investigation is on a continuous scale). Of course, application of the conditional logistic model is still desirable, at least in the case of a continuous covariate, because it would provide a measure of association; the odds ratio.

**Example 11.12** Refer to the data for low-birth-weight babies in Example 11.11 (Table 11.14) and suppose that we want to investigate the relationship between the low-birth-weight problem, our outcome for the study, and the weight of the mother taken during the last menstrual period. An application of the simple conditional logistic regression analysis yields the results shown in Table 11.16. The result indicates that the effect of the mother's weight is nearly significant at the 5% level ($p = 0.0593$). The odds ratio associated with, say, a 10-lb increase in weight is

$$\exp(-0.2114) = 0.809$$

If a mother increases her weight about 10 lb, the odds on having a low-birth-weight baby are reduced by almost 20%.

*Note:* An SAS program would include these instructions:

```
INPUT SET CASE MWEIGHT;
DUMMYTIME = 2-CASE;
CARDS;
(data)
PROC PHREG DATA = LOWWEIGHT;
MODEL DUMMYTIME*CASE(0) = MWEIGHT/TIES = DISCRETE;
STRATA SET;
```

where LOWWEIGHT is the name assigned to the data set, DUMMYTIME is the name for the makeup time variable defined in the upper part of the program, CASE is the case–control status indicator (coded as 1 for a case and 0 for a control), and MWEIGHT is the variable name for the weight of the mother during the last menstrual period. The matched SET number (1 to 15 in this example) is used as the stratification factor.

### 11.7.2 Multiple Regression Analysis

The effect of some factor on a dependent or response variable may be influenced by the presence of other factors through effect modifications (i.e., interactions). Therefore, to provide a more comprehensive analysis, it is very desirable to consider a large number of factors and sort out which are most closely related to the dependent variable. In this section we discuss a multivariate method for such a risk determination. This method, multiple conditional logistic regression analysis, involves a linear combination of the explanatory or independent variables; the variables must be quantitative with particular numerical values for each patient. A covariate or independent variable such as a patient characteristic may be dichotomous, polytomous, or continuous (categorical factors will be represented by dummy variables).

Examples of dichotomous covariates are gender and presence/absence of certain comorbidity. Polytomous covariates include race and different grades of symptoms; these can be covered by the use of *dummy variables*. Continuous covariates include patient age and blood pressure. In many cases, data transformations (e.g., taking the logarithm) may be desirable to satisfy the linearity assumption. We illustrate this process for a very general design in which matched set $i$ ($1 \leq N$) has $n_i$ cases matched to $m_i$ controls; the numbers of cases $n_i$ and controls $m_i$ vary from matched set to matched set.

***Likelihood Function*** For the general case of $n_i$ cases matched to $m_i$ controls in a set, we have the conditional probability of the observed outcome (that a specific set of $n_i$ subjects are cases) given that the number of cases is $n_l$ (any $n_i$ subjects could be cases):

$$\frac{\exp[\sum_{j=1}^{n_i}(\beta^T \mathbf{x_j})]}{\sum_{R(n_i, m_i)} \exp[\sum_{j=1}^{n_i}(\beta^T \mathbf{x_j})]}$$

where the sum in the denominator ranges over the collections $R(n_i, m_i)$ of all partitions of the $n_i + m_i$ subjects into two: one of size $n_i$ and one of size $m_i$. The full conditional likelihood is the product over all matched sets, one probability for each set:

$$L = \prod_{i=1}^{N} \frac{\exp[\sum_{j=1}^{n_i}(\beta^T \mathbf{x_j})]}{\sum_{R(n_i, m_i)} \exp[\sum_{j=1}^{n_i}(\beta^T \mathbf{x_j})]}$$

which can be implemented using the same SAS program.

Similar to the univariate case, $\exp(\beta_i)$ represents one of the following:

1. The odds ratio associated with an exposure if $X_i$ is binary (exposed $X_i = 1$ versus unexposed $X_i = 0$); or
2. The odds ratio due to a 1-unit increase if $X_i$ is continuous ($X_i = x + 1$ versus $X_i = x$).

After $\hat{\beta}_i$ and its standard error have been obtained, a 95% confidence interval for the odds ratio above is given by

$$\exp[\hat{\beta}_i \pm 1.96 \, \text{SE}(\hat{\beta}_i)]$$

These results are necessary in the effort to identify important risk factors in matched designs. Of course, before such analyses are done, the problem and the data have to be examined carefully. If some of the variables are highly correlated, one or fewer of the correlated factors are likely to be as good predictors as all of them; information from similar studies also has to be incorporated so as to drop some of these correlated explanatory variables. The use of products such as $X_1 X_2$ and higher power terms such as $X_1^2$ may be necessary and can improve the goodness of fit (unfortunately, it is very difficult to tell). It is important to note that we are assuming a *linear regression model* in which, for example, the odds ratio due to a 1-unit increase in the value of a continuous $X_i$ ($X_i = x + 1$ versus $X_i = x$) is independent of $x$. Therefore, if this *linearity* seems to be violated (again, it is very difficult to tell; the only easy way is fitting a polynomial model as seen in a later example), the incorporation of powers of $X_i$ should be considered seriously. The use of products will help in the investigation of possible effect modifications. Finally, there is the messy problem of missing data; most packaged programs would delete a subject if one or more covariate values are missing.

***Testing Hypotheses in Multiple Regression***   Once we have fit a multiple conditional logistic regression model and obtained estimates for the various parameters of interest, we want to answer questions about the contributions of various factors to the prediction of the binary response variable using matched designs. There are three types of such questions:

1. *Overall test.* Taken collectively, does the entire set of explatory or independent variables contribute significantly to the prediction of response?
2. *Test for the value of a single factor.* Does the addition of one particular variable of interest add significantly to the prediction of response over and above that achieved by other independent variables?
3. *Test for contribution of a group of variables.* Does the addition of a group of variables add significantly to the prediction of response over and above that achieved by other independent variables?

*Overall Regression Test*   We now consider the first question stated above concerning an overall test for a model containg $k$ factors. The null hypothesis for this test may be stated as: "All $k$ independent variables *considered together* do not explain the variation in response any more than the size alone." In other words,

$$H_0: \beta_1 = \beta_2 = \cdots = \beta_k = 0$$

Three statistics can be used to test this *global* null hypothesis; each has a symptotic chi-square distribution with $k$ degrees of freedom under $H_0$: the likelihood ratio test, Wald's test, and the score test. All three statistics are provided by most standard computer programs such as SAS, and they are asymptotically equivalent (i.e., for very large sample sizes), yielding identical statistical decisions most of the time. However, Wald's test is used much less often than the other two.

**Example 11.13**   Refer to the data for low-birth-weight babies in Example 11.11 (Table 11.14). With all four covariates, we have the following test statistics for the global null hypothesis:

(a)  Likelihood test:

$$\chi_{LR}^2 = 9.530 \text{ with 4 df}; \ p = 0.0491$$

(b)  Wald's test:

$$\chi_W^2 = 6.001 \text{ with 4 df}; \ p = 0.1991$$

(c)  Score test:

$$\chi_S^2 = 8.491 \text{ with 4 df}; \ p = 0.0752$$

The results indicates a weak combined explanatory power; Wald's test is not even significant. Very often, this means implicitly that perhaps only one or two covariates are associated significantly with the response of interest (a weak overall correlation).

*Test for a Single Variable*   Let us assume that we now wish to test whether the addition of one particular independent variable of interest adds significantly to the prediction of the response over and above that achieved by factors already present in the model (usually after seeing a significant result for the global hypothesis above). The null hypothesis for this single-variable test may be stated as: "Factor $X_i$ does not have any value added to the prediction of the response *given that other factors are already included in the model.*" In other words,

$$H_0: \beta_i = 0$$

To test such a null hypothesis, one can use

$$z_i = \frac{\hat{\beta}_i}{SE(\hat{\beta}_i)}$$

where $\hat{\beta}_i$ is the corresponding estimated regression coefficient and $SE(\hat{\beta}_i)$ is the estimate of the standard error of $\hat{\beta}_i$, both of which are printed by standard

**TABLE 11.17**

| Variable | Coefficient | Standard Error | $z$ Statistic | $p$ Value |
|---|---|---|---|---|
| Mother's weight | −0.0191 | 0.0114 | −1.673 | 0.0942 |
| Smoking | −0.0885 | 0.8618 | −0.103 | 0.9182 |
| Hypertension | 0.6325 | 1.1979 | 0.528 | 0.5975 |
| Uterine irritability | 2.1376 | 1.1985 | 1.784 | 0.0745 |

computer-packaged programs such as SAS. In performing this test, we refer the value of the $z$ score to percentiles of the standard normal distribution; for example, we compare the absolute value of $z$ to 1.96 for a two-sided test at the 5% level.

***Example 11.14***   Refer to the data for low-birth-weight babies in Example 11.11 (Table 11.14). With all four covariates, we have the results shown in Table 11.17. Only the mother's weight ($p = 0.0942$) and uterine irritability ($p = 0.0745$) are marginally significant. In fact, these two variables are highly correlated: that is, if one is deleted from the model, the other would become more significant.

The overall tests and the tests for single variables are implemented simultaneously sing the same computer program, and here is another example.

***Example 11.15***   Refer to the data for vaginal carcinoma in Example 11.10 (Table 11.13). An application of a conditional logistic regression analysis yields the following results:

1. Likelihood test for the global hypothesis:

$$\chi^2_{LR} = 9.624 \text{ with 2 df}; \ p = 0.0081$$

2. Wald's test for the global hypothesis:

$$\chi^2_W = 6.336 \text{ with 2 df}; \ p = 0.0027$$

3. Score test for the global hypothesis:

$$\chi^2_S = 11.860 \text{ with 2 df}; \ p = 0.0421$$

For individual covariates, we have the results shown in Table 11.18.

In addition to a priori interest in the effects of individual covariates, given a continuous variable of interest, one can fit a polynomial model and use this

**TABLE 11.18**

| Variable | Coefficient | Standard Error | $z$ Statistic | $p$ Value |
|---|---|---|---|---|
| Bleeding | 1.6198 | 1.3689 | 1.183 | 0.2367 |
| Pregnancy loss | 1.7319 | 0.8934 | 1.938 | 0.0526 |

type of test to check for linearity. It can also be used to check for a single product representing an effect modification.

***Example 11.16***    Refer to the data for low-birth-weight babies in Example 11.11 (Table 11.14), but this time we investigate only one covariate, the mother's weight. After fitting the second-degree polynomial model, we obtained a result which indicates that the *curvature effect* is negligible ($p = 0.9131$).

*Contribution of a Group of Variables*    This testing procedure addresses the more general problem of assessing the additional contribution of two or more factors to the prediction of the response over and above that made by other variables already in the regression model. In other words, the null hypothesis is of the form

$$H_0: \beta_1 = \beta_2 = \cdots = \beta_m = 0$$

To test such a null hypothesis, one can perform a likelihood ratio chi-square test with $m$ df:

$$\chi^2_{LR} = 2[\ln L(\hat{\beta}; \text{ all } X\text{'s})$$

$$- \ln L(\hat{\beta}; \text{ all other } X\text{'s with } X\text{'s under investigation deleted})]$$

As with the tests above for individual covariates, this *multiple contribution procedure* is very useful for assessing the importance of potential explanatory variables. In particular, it is often used to test whether a similar group of variables, such as *demographic characteristics*, is important for prediction of the response; these variables have some trait in common. Another application would be a collection of powers *and/or* product terms (referred to as *interaction variables*). It is often of interest to assess the interaction effects collectively before trying to consider individual interaction terms in a model, as suggested previously. In fact, such use reduces the total number of tests to be performed, and this, in turn, helps to provide better control of overall type I error rates, which may be inflated due to multiple testing.

***Example 11.17***    Refer to the data for low-birth-weight babies in Example 11.11 (Table 11.14). With all four covariates, we consider collectively three interaction terms: mother's weight × smoking, mother's weight × hypertension, mother's weight × uterine irritability. The basic idea is to see if *any* of the other variables would modify the effect of the mother's weight on the response (having a low-birth-weight baby).

1. With the original four variables, we obtained ln $L = -16.030$.
2. With all seven variables, four original plus three products, we obtained ln $L = -14.199$.

Therefore, we have

$$\chi^2_{LR} = 2[\ln L(\hat{\beta};\ \text{seven variables}) - \ln L(\hat{\beta};\ \text{four original variables})]$$
$$= 3.662;\ 3\ \text{df},\ p\ \text{value} \geq 0.10$$

indicating a rather weak level of interactions.

*Stepwise Regression*    In many applications our major interest is to identify important risk factors. In other words, we wish to identify from many available factors a small subset of factors that relate significantly to the outcome (e.g., the disease under investigation). In that identification process, of course, we wish to avoid a large type I (false positive) error. In a regression analysis, a type I error corresponds to including a predictor that has no real relationship to the outcome; such an inclusion can greatly confuse interpretation of the regression results. In a standard multiple regression analysis, this goal can be achieved by using a strategy that adds to or removes from a regression model one factor at a time according to a certain order of relative importance. Therefore, the two important steps are as follows:

1. Specify a criterion or criteria for selecting a model.
2. Specify a strategy for applying the criterion or criteria chosen.

The process follows the outline of Chapter 5 for logistic regression, combining the forward selection and backward elimination in the stepwise process, with selection at each step based on the likelihood ratio chi-square test. SAS's PROC PHREG does have an automatic stepwise option to implement these features.

*Example 11.18*    Refer to the data for low-birth-weight babies in Example 11.11 (Table 11.14) with all four covariates: mother's weight, smoking, hypertension, and uterine irritability. This time we perform a stepwise regression analysis in which we specify that a variable has to be significant at the 0.10 level before it can enter into the model and that a variable in the model has to be significant at 0.15 for it to remain in the model (most standard computer programs allow users to make these selections; default values are available).

**TABLE 11.19**

| Variable | Score $\chi^2$ | $p$ Value |
|---|---|---|
| Mother's weight | 3.9754 | 0.0462 |
| Smoking | 0.0000 | 1.0000 |
| Hypertension | 0.2857 | 0.5930 |
| Uterine irritability | 5.5556 | 0.0184 |

**TABLE 11.20**

| Variable | Score $\chi^2$ | $p$ Value |
|---|---|---|
| Mother's weight | 2.9401 | 0.0864 |
| Smoking | 0.0027 | 0.9584 |
| Hypertension | 0.2857 | 0.5930 |

**TABLE 11.21**

| Factor | Coefficient | Standard Error | $z$ Statistic | $p$ Value |
|---|---|---|---|---|
| Mother's weight | −0.0192 | 0.0116 | −1.655 | 0.0978 |
| Uterine irritability | 2.1410 | 1.1983 | 1.787 | 0.0740 |

**TABLE 11.22**

| Variable | Score $\chi^2$ | $p$ Value |
|---|---|---|
| Smoking | 0.0840 | 0.7720 |
| Hypertension | 0.3596 | 0.5487 |

First, we get the individual test results for all variables (Table 11.19). These indicate that uterine irritability is the most significant variable.

- *Step 1:* Variable uterine irritability is entered. Analysis of variables not in the model yields the results shown in Table 11.20.
- *Step 2:* Variable mother's weight is entered. Analysis of variables in the model yields Table 11.21. Neither variable is removed. Analysis of variables not in the model yields Table 11.22. No (additional) variables meet the 0.1 level for entry into the model.

*Note:* An SAS program would include these instructions:

```
PROC PHREG DATA LOWWEIGHT;
MODEL DUMMYTIME*CASE(0) = MWEIGHT SMOKING HYPERT
UIRRIT/SELECTION = STEPWISE SLENTRY = .10 SLSTAY = .15;
STRATA = SET;
```

where HYPERT and UIRRIT are hypertension and uterine irritability. The default values for SLENTRY (*p* value to enter) and SLSTAY (*p* value to stay) are 0.05 and 0.10, respectively.

## EXERCISES

**11.1**  Given the small data set

$$9, \ 13, \ 13^+, \ 18, \ 23, \ 28^+, \ 31, \ 34, \ 45^+, \ 48, \ 161^+$$

calculate and graph the Kaplan–Meier curve.

**11.2**  A group of 12 hemophiliacs, all under 41 years of age at the time of HIV seroconversion, were followed from primary AIDS diagnosis until death (ideally, we should take as a starting point the time at which a person contracts AIDS rather than the time at which the patient is diagnosed, but this information is not available). Survival times (in months) from diagnosis until death of these hemophiliacs were: 2, 3, 6, 6, 7, 10, 15, 15, 16, 27, 30, and 32. Calculate and graph the Kaplan–Meier curve.

**11.3**  Suppose that we are interested in studying patients with systemic cancer who subsequently develop a brain metastasis; our ultimate goal is to prolong their lives by controlling the disease. A sample of 23 such patients, all of whom were treated with radiotherapy, were followed from the first day of their treatment until recurrence of the original tumor. Recurrence is defined as the reappearance of a metastasis in exactly the same site, or in the case of patients whose tumor never completely disappeared, enlargement of the original lesion. Times to recurrence (in weeks) for the 23 patients were: 2, 2, 2, 3, 4, 5, 5, 6, 7, 8, 9, 10, 14, 14, 18, 19, 20, 22, 22, 31, 33, 39, and 195. Calculate and graph the Kaplan–Meier curve.

**11.4**  A laboratory investigator interested in the relationship between diet and the development of tumors divided 90 rats into three groups and fed them low-fat, saturated-fat, and unsaturated-fat diets, respectively. The rats were of the same age and species and were in similar physical condition. An identical amount of tumor cells was injected into a foot pad of each rat. The tumor-free time is the time from injection of tumor cells to the time that a tumor develops; all 30 rats in the unsaturated-fat diet group developed tumors; tumor-free times (in days) were: 112, 68, 84, 109, 153, 143, 60, 70, 98, 164, 63, 63, 77, 91, 91, 66, 70, 77, 63, 66, 66, 94, 101, 105, 108, 112, 115, 126, 161, and 178. Calculate and graph the Kaplan–Meier curve.

**11.5**  Data are shown in Table 11.3 for two groups of patients who died of acute myelogenous leukemia (see Example 11.3). Patients were classified into the two groups according to the presence or absence of a morphologic characteristic of white cells. Patients termed AG positive were identified by the presence of Auer rods and/or significant granulature of the leukemic cells in the bone marrow at diagnosis. For AG-negative patients these factors were absent. Leukemia is a cancer characterized by an overproliferation of white blood cells; the higher the white blood count (WBC), the more severe the disease. Calculate and graph in the same figure the two Kaplan–Meier curves (one for AG-positive patients and one for AG-negative patients). How do they compare?

**11.6**  In Exercise 11.4 we described a diet study, and tumor-free times were given for the 30 rats fed an unsaturated-fat diet. Tumor-free times (days) for the other two groups are as follows:

- *Low-fat:* 140, 177, 50, 65, 86, 153, 181, 191, 77, 84, 87, 56, 66, 73, 119, $140^+$, and 14 rats at $200^+$
- *Saturated-fat:* 124, 58, 56, 68, 79, 89, 107, 86, 142, 110, 96, 142, 86, 75, 117, 98, 105, 126, 43, 46, 81, 133, 165, $170^+$, and 6 rats at $200^+$

($140^+$ and $170^+$ were due to accidental deaths without evidence of tumor). Calculate and graph the two Kaplan–Meier curves, one for rats fed a low-fat diet and one for rats fed a saturated-fat diet. Put these two curves and the one from Exercise 11.4 in the same figure and draw conclusions.

**11.7**  Consider the data shown in Table E11.7 (analysis date, 01/90; A, alive; D, dead). For each subject, determine the time (in months) to death (D) or to the ending date (for survivors whose status was marked as A); then calculate and graph the Kaplan–Meier curve.

**TABLE E11.7**

| Subject | Starting | Ending | Status (A/D) |
|---------|----------|--------|--------------|
| 1 | 01/80 | 01/90 | A |
| 2 | 06/80 | 07/88 | D |
| 3 | 11/80 | 10/84 | D |
| 4 | 08/81 | 02/88 | D |
| 5 | 04/82 | 01/90 | A |
| 6 | 06/83 | 11/85 | D |
| 7 | 10/85 | 01/90 | A |
| 8 | 02/86 | 06/88 | D |
| 9 | 04/86 | 12/88 | D |
| 10 | 11/86 | 07/89 | D |

**11.8**   Given the small data set:

$$Sample\ 1:\quad 24,\ 30,\ 42,\ 15^+,\ 40^+,\ 42^+$$
$$Sample\ 2:\quad 10,\ 26,\ 28,\ 30,\ 41,\ 12^+$$

compare them using both the log-rank and generalized Wilcoxon tests.

**11.9**   *Pneumocystis carinii* pneumonia (PCP) is the most common opportunistic infection in HIV-infected patients and a life-threatening disease. Many North Americans with AIDS have one or two episodes of PCP during the course of their HIV infection. PCP is a consideration factor in mortality, morbidity, and expense; and recurrences are common. In the data set given in Table E11.9 at the end of this chapter, we have:

- Treatments, coded as A and B
- Patient characteristics: baseline CD4 count, gender (1, male; 0, female), race (1, white; 2, black; 3, other), weight (lb), homosexuality (1, yes; 0, no)
- PCP recurrence indicator (1, yes; 0, no), PDATE or time to recurrence (months)
- DIE or survival indicator (1, yes; 0, no), DDATE or time to death (or to date last seen for survivors; months)

Consider each of these endpoints: relapse (treating death as censoring), death (treating relapse as censoring), and death or relapse (whichever comes first). For each endpoint:

(a) Estimate the survival function for homosexual white men.

(b) Estimate the survival functions for each treatment.

(c) Compare the two treatments; do they differ in the short and long terms?

(d) Compare men and women.

(e) Taken collectively, do the covariates contribute significantly to prediction of survival?

(f) Fit the multiple regression model to obtain estimates of individual regression coefficients and their standard errors. Draw conclusions concerning the conditional contribution of each factor.

(g) Within the context of the multiple regression model in part (b), does treatment alter the effect of CD4?

(h) Focus on treatment as the primary factor, taken collectively; was this main effect altered by any other covariates?

(i) Within the context of the multiple regression model in part (b), is the effect of CD4 linear?

(j) Do treatment and CD4, individually, fit the proportional hazards model?

(*Note*: A Web-based electronic copy is available upon request.)

**TABLE E11.9**

| TRT | CD4 | GENDER | RACE | WT | HOMO | PCP | PDATE | DIE | DDATE |
|-----|-----|--------|------|-----|------|-----|-------|-----|-------|
| B | 2 | 1 | 1 | 142 | 1 | 1 | 11.9 | 0 | 14.6 |
| B | 139 | 1 | 2 | 117 | 0 | 0 | 11.6 | 1 | 11.6 |
| A | 68 | 1 | 2 | 149 | 0 | 0 | 12.8 | 0 | 12.8 |
| A | 12 | 1 | 1 | 160 | 1 | 0 | 7.3 | 1 | 7.3 |
| B | 36 | 1 | 2 | 157 | 0 | 1 | 4.5 | 0 | 8.5 |
| B | 77 | 1 | 1 | 12 | 1 | 0 | 18.1 | 1 | 18.1 |
| A | 56 | 1 | 1 | 158 | 0 | 0 | 14.7 | 1 | 14.7 |
| B | 208 | 1 | 2 | 157 | 1 | 0 | 24.0 | 1 | 24.0 |
| A | 40 | 1 | 1 | 122 | 1 | 0 | 16.2 | 0 | 16.2 |
| A | 53 | 1 | 2 | 125 | 1 | 0 | 26.6 | 1 | 26.6 |
| A | 28 | 1 | 2 | 130 | 0 | 1 | 14.5 | 1 | 19.3 |
| A | 162 | 1 | 1 | 124 | 0 | 0 | 25.8 | 1 | 25.8 |
| B | 163 | 1 | 2 | 130 | 0 | 0 | 16.1 | 0 | 16.1 |
| A | 65 | 1 | 1 | 12 | 0 | 0 | 19.4 | 0 | 19.4 |
| A | 247 | 1 | 1 | 167 | 0 | 0 | 23.4 | 0 | 23.4 |
| B | 131 | 1 | 1 | 16 | 0 | 0 | 2.7 | 0 | 2.7 |
| A | 25 | 1 | 1 | 13 | 1 | 0 | 20.1 | 1 | 20.1 |
| A | 118 | 1 | 1 | 155 | 0 | 0 | 17.3 | 0 | 17.3 |
| B | 21 | 1 | 1 | 126 | 0 | 0 | 6.0 | 1 | 6.0 |
| B | 81 | 1 | 2 | 168 | 1 | 0 | 1.6 | 0 | 1.6 |
| A | 89 | 1 | 1 | 169 | 1 | 0 | 29.5 | 0 | 29.5 |
| B | 172 | 1 | 1 | 163 | 1 | 0 | 24.2 | 1 | 16.5 |
| B | 21 | 1 | 1 | 164 | 1 | 1 | 4.9 | 0 | 22.9 |
| A | 7 | 1 | 1 | 139 | 1 | 0 | 14.8 | 1 | 14.8 |
| B | 94 | 1 | 1 | 165 | 1 | 0 | 29.8 | 0 | 29.8 |
| B | 14 | 1 | 2 | 17 | 0 | 0 | 21.9 | 1 | 21.4 |
| A | 38 | 1 | 1 | 17 | 1 | 0 | 18.8 | 1 | 18.8 |
| A | 73 | 1 | 1 | 14 | 1 | 0 | 20.5 | 1 | 20.5 |
| B | 25 | 1 | 2 | 19 | 0 | 0 | 13.1 | 1 | 13.1 |
| A | 13 | 1 | 3 | 121 | 1 | 0 | 21.4 | 0 | 21.4 |
| B | 30 | 1 | 1 | 145 | 1 | 0 | 21.0 | 0 | 21.0 |
| A | 152 | 1 | 3 | 124 | 1 | 0 | 19.4 | 0 | 19.4 |
| B | 68 | 1 | 3 | 15 | 0 | 0 | 17.5 | 1 | 17.4 |
| A | 27 | 1 | 1 | 128 | 1 | 0 | 18.5 | 0 | 18.5 |
| A | 38 | 1 | 1 | 159 | 1 | 1 | 18.3 | 1 | 18.3 |
| B | 265 | 1 | 3 | 242 | 0 | 0 | 11.1 | 1 | 11.1 |
| B | 29 | 0 | 3 | 13 | 0 | 0 | 14.0 | 0 | 14.0 |
| A | 73 | 1 | 3 | 130 | 1 | 0 | 11.1 | 1 | 11.1 |
| B | 103 | 1 | 1 | 164 | 1 | 0 | 11.1 | 0 | 11.1 |
| B | 98 | 1 | 1 | 193 | 1 | 0 | 10.2 | 0 | 10.2 |
| A | 120 | 1 | 1 | 17 | 1 | 0 | 5.2 | 0 | 5.2 |
| B | 131 | 1 | 2 | 184 | 0 | 0 | 5.5 | 0 | 5.5 |
| A | 48 | 0 | 1 | 160 | 0 | 0 | 13.7 | 1 | 13.7 |
| B | 80 | 1 | 1 | 115 | 1 | 0 | 12.0 | 0 | 12.0 |
| A | 132 | 1 | 3 | 130 | 1 | 0 | 27.0 | 0 | 27.0 |
| A | 54 | 1 | 1 | 148 | 1 | 0 | 11.7 | 0 | 11.7 |
| B | 189 | 1 | 1 | 198 | 1 | 0 | 24.5 | 0 | 24.5 |

*(Continued)*

**TABLE E11.9** *(Continued)*

| TRT | CD4 | GENDER | RACE | WT | HOMO | PCP | PDATE | DIE | DDATE |
|-----|-----|--------|------|-----|------|-----|-------|-----|-------|
| B | 14 | 1 | 2 | 160 | 0 | 0 | 1.3 | 0 | 1.3 |
| A | 321 | 1 | 1 | 130 | 1 | 1 | 18.5 | 0 | 18.6 |
| B | 148 | 1 | 1 | 126 | 1 | 0 | 22.8 | 0 | 22.8 |
| B | 54 | 1 | 1 | 181 | 1 | 1 | 14.5 | 0 | 15.7 |
| A | 17 | 1 | 1 | 152 | 1 | 0 | 19.8 | 0 | 19.8 |
| A | 37 | 1 | 3 | 120 | 1 | 0 | 16.6 | 0 | 16.6 |
| B | 71 | 0 | 1 | 136 | 0 | 0 | 16.5 | 0 | 16.5 |
| A | 9 | 1 | 3 | 130 | 1 | 0 | 8.9 | 1 | 8.9 |
| B | 231 | 1 | 3 | 140 | 0 | 1 | 10.3 | 0 | 10.6 |
| A | 22 | 1 | 2 | 190 | 1 | 0 | 8.5 | 0 | 8.5 |
| A | 43 | 1 | 1 | 134 | 1 | 0 | 17.9 | 1 | 17.9 |
| B | 103 | 1 | 1 | 110 | 1 | 0 | 20.3 | 0 | 20.3 |
| A | 146 | 1 | 1 | 213 | 1 | 0 | 20.5 | 0 | 20.5 |
| A | 92 | 1 | 1 | 128 | 1 | 0 | 14.2 | 0 | 14.2 |
| B | 218 | 1 | 1 | 163 | 1 | 0 | 1.9 | 0 | 1.9 |
| A | 100 | 1 | 1 | 170 | 1 | 0 | 14.0 | 0 | 14.0 |
| B | 148 | 1 | 1 | 158 | 1 | 1 | 15.4 | 0 | 16.1 |
| B | 44 | 1 | 2 | 124 | 1 | 0 | 7.3 | 0 | 7.3 |
| A | 76 | 1 | 1 | 149 | 0 | 1 | 15.9 | 1 | 23.4 |
| B | 30 | 1 | 1 | 181 | 1 | 0 | 6.6 | 1 | 6.6 |
| B | 260 | 1 | 1 | 165 | 1 | 1 | 7.5 | 1 | 18.0 |
| B | 40 | 1 | 1 | 204 | 0 | 0 | 21.0 | 1 | 18.8 |
| A | 90 | 1 | 1 | 149 | 1 | 1 | 17.0 | 0 | 21.8 |
| A | 120 | 1 | 1 | 152 | 0 | 0 | 21.8 | 0 | 21.8 |
| B | 80 | 1 | 1 | 199 | 1 | 1 | 20.6 | 0 | 20.6 |
| A | 170 | 1 | 1 | 141 | 1 | 0 | 18.6 | 1 | 18.6 |
| A | 54 | 1 | 1 | 148 | 1 | 0 | 18.6 | 0 | 18.6 |
| A | 151 | 1 | 1 | 140 | 1 | 0 | 21.2 | 0 | 21.2 |
| B | 107 | 1 | 1 | 158 | 1 | 0 | 22.5 | 1 | 22.3 |
| A | 9 | 1 | 1 | 116 | 1 | 0 | 18.0 | 1 | 18.0 |
| B | 79 | 1 | 3 | 132 | 0 | 0 | 22.6 | 1 | 22.6 |
| A | 72 | 1 | 1 | 131 | 1 | 0 | 19.9 | 0 | 19.9 |
| A | 100 | 1 | 1 | 182 | 1 | 0 | 21.2 | 0 | 21.2 |
| B | 16 | 1 | 2 | 106 | 1 | 0 | 18.3 | 0 | 18.3 |
| B | 10 | 1 | 1 | 168 | 1 | 0 | 24.7 | 0 | 24.7 |
| A | 135 | 1 | 1 | 149 | 1 | 0 | 23.8 | 0 | 23.8 |
| B | 235 | 1 | 1 | 137 | 0 | 0 | 22.7 | 0 | 22.7 |
| B | 20 | 1 | 1 | 104 | 0 | 0 | 14.0 | 0 | 14.0 |
| A | 67 | 1 | 2 | 150 | 0 | 0 | 19.4 | 1 | 19.4 |
| B | 7 | 1 | 1 | 182 | 1 | 0 | 17.0 | 1 | 17.0 |
| B | 139 | 1 | 1 | 143 | 1 | 0 | 21.4 | 0 | 21.4 |
| B | 13 | 1 | 3 | 132 | 0 | 0 | 23.5 | 0 | 23.5 |
| A | 117 | 1 | 1 | 144 | 1 | 0 | 19.5 | 1 | 19.5 |
| A | 11 | 1 | 2 | 111 | 1 | 0 | 19.3 | 1 | 19.3 |
| B | 280 | 1 | 1 | 145 | 1 | 0 | 11.6 | 1 | 11.6 |
| A | 119 | 1 | 1 | 159 | 1 | 1 | 13.1 | 0 | 19.3 |
| B | 9 | 1 | 1 | 146 | 1 | 1 | 17.0 | 1 | 18.2 |

**TABLE E11.9** *(Continued)*

| TRT | CD4 | GENDER | RACE | WT | HOMO | PCP | PDATE | DIE | DDATE |
|-----|-----|--------|------|-----|------|-----|-------|-----|-------|
| A | 30 | 1 | 2 | 150 | 0 | 0 | 20.9 | 0 | 20.9 |
| B | 22 | 1 | 1 | 138 | 1 | 1 | 1.1 | 1 | 10.0 |
| B | 186 | 1 | 3 | 114 | 1 | 0 | 17.2 | 0 | 17.2 |
| A | 42 | 1 | 1 | 167 | 1 | 0 | 19.2 | 0 | 19.2 |
| B | 9 | 1 | 2 | 146 | 1 | 0 | 6.0 | 1 | 6.0 |
| B | 99 | 1 | 1 | 149 | 0 | 1 | 14.8 | 0 | 15.5 |
| A | 21 | 1 | 1 | 141 | 1 | 0 | 17.7 | 0 | 17.7 |
| A | 16 | 1 | 2 | 116 | 0 | 0 | 17.5 | 0 | 17.5 |
| B | 10 | 1 | 1 | 143 | 1 | 1 | 8.3 | 1 | 13.5 |
| B | 109 | 1 | 1 | 130 | 1 | 1 | 12.0 | 0 | 12.1 |
| B | 72 | 1 | 1 | 137 | 0 | 0 | 12.8 | 0 | 12.8 |
| B | 582 | 1 | 1 | 143 | 1 | 0 | 15.7 | 0 | 15.7 |
| A | 8 | 1 | 2 | 134 | 1 | 0 | 9.3 | 1 | 9.3 |
| A | 69 | 1 | 1 | 160 | 0 | 0 | 10.1 | 0 | 10.1 |
| A | 57 | 1 | 1 | 138 | 1 | 0 | 10.2 | 0 | 10.2 |
| A | 47 | 1 | 1 | 159 | 1 | 0 | 9.1 | 0 | 9.1 |
| A | 149 | 1 | 3 | 152 | 0 | 0 | 9.8 | 0 | 9.8 |
| B | 229 | 1 | 2 | 130 | 1 | 0 | 9.4 | 0 | 9.4 |
| A | 9 | 1 | 1 | 165 | 1 | 0 | 9.2 | 0 | 9.2 |
| A | 10 | 1 | 1 | 162 | 0 | 0 | 9.2 | 0 | 9.2 |
| A | 78 | 1 | 1 | 145 | 1 | 0 | 10.2 | 0 | 10.2 |
| B | 147 | 1 | 1 | 180 | 1 | 0 | 9.0 | 0 | 9.0 |
| B | 147 | 1 | 1 | 180 | 1 | 0 | 9.0 | 0 | 9.0 |
| B | 126 | 1 | 1 | 124 | 1 | 0 | 5.5 | 0 | 5.5 |
| A | 19 | 1 | 2 | 192 | 0 | 0 | 6.0 | 0 | 6.0 |
| A | 142 | 1 | 1 | 17 | 1 | 0 | 17.3 | 1 | 17.2 |
| B | 277 | 0 | 1 | 14 | 0 | 0 | 17.0 | 0 | 17.0 |
| B | 80 | 1 | 1 | 13 | 1 | 0 | 15.0 | 0 | 15.0 |
| A | 366 | 1 | 1 | 15 | 1 | 0 | 14.9 | 0 | 14.9 |
| A | 76 | 1 | 1 | 18 | 1 | 0 | 9.2 | 0 | 9.2 |
| A | 13 | 1 | 1 | 171 | 1 | 0 | 30.2 | 0 | 30.2 |
| B | 17 | 1 | 1 | 276 | 0 | 0 | 15.8 | 1 | 15.8 |
| B | 193 | 1 | 1 | 164 | 1 | 0 | 22.5 | 1 | 22.5 |
| A | 108 | 1 | 1 | 161 | 0 | 0 | 24.0 | 0 | 24.0 |
| B | 41 | 1 | 1 | 153 | 0 | 0 | 23.9 | 0 | 23.9 |
| A | 113 | 1 | 1 | 131 | 0 | 0 | 21.4 | 0 | 21.4 |
| B | 1 | 1 | 2 | 136 | 0 | 0 | 19.6 | 0 | 19.6 |
| A | . | 0 | 2 | 109 | 0 | 0 | 6.0 | 1 | 6.0 |
| A | 47 | 1 | 1 | 168 | 1 | 0 | 18.2 | 1 | 18.2 |
| B | 172 | 1 | 2 | 195 | 1 | 0 | 10.3 | 1 | 10.3 |
| A | 247 | 1 | 1 | 123 | 1 | 0 | 16.2 | 0 | 16.2 |
| B | 21 | 1 | 2 | 124 | 0 | 0 | 9.7 | 0 | 9.7 |
| B | 0 | 1 | 2 | 116 | 1 | 0 | 11.0 | 0 | 11.0 |
| B | 38 | 1 | 2 | 160 | 1 | 0 | 14.7 | 0 | 14.7 |
| A | 50 | 1 | 1 | 127 | 1 | 0 | 13.6 | 0 | 13.6 |
| A | 4 | 1 | 2 | 218 | 0 | 0 | 12.9 | 0 | 12.9 |
| A | 150 | 1 | 1 | 200 | 1 | 0 | 11.7 | 1 | 11.7 |

*(Continued)*

**TABLE E11.9** *(Continued)*

| TRT | CD4 | GENDER | RACE | WT | HOMO | PCP | PDATE | DIE | DDATE |
|-----|-----|--------|------|-----|------|-----|-------|-----|-------|
| B | 0 | 1 | 2 | 133 | 0 | 1 | 4.7 | 0 | 11.8 |
| A | 97 | 1 | 2 | 156 | 0 | 0 | 11.9 | 0 | 11.9 |
| B | 312 | 1 | 1 | 140 | 1 | 0 | 10.6 | 0 | 10.6 |
| B | 35 | 1 | 1 | 155 | 1 | 0 | 11.0 | 0 | 11.0 |
| A | 100 | 1 | 1 | 157 | 1 | 0 | 9.2 | 0 | 9.2 |
| A | 69 | 1 | 1 | 126 | 0 | 0 | 9.2 | 0 | 9.2 |
| B | 124 | 1 | 2 | 135 | 1 | 0 | 6.5 | 0 | 6.5 |
| B | 25 | 1 | 1 | 162 | 1 | 0 | 16.0 | 1 | 0.0 |
| A | 61 | 0 | 2 | 102 | 0 | 0 | 18.5 | 1 | 18.5 |
| B | 102 | 1 | 1 | 177 | 1 | 1 | 11.3 | 0 | 17.4 |
| A | 198 | 1 | 2 | 164 | 0 | 0 | 23.2 | 0 | 23.2 |
| B | 10 | 1 | 1 | 173 | 0 | 0 | 8.4 | 1 | 8.4 |
| A | 56 | 1 | 1 | 163 | 1 | 0 | 11.9 | 0 | 11.9 |
| A | 43 | 1 | 1 | 134 | 1 | 0 | 9.2 | 0 | 9.2 |
| B | 202 | 1 | 2 | 158 | 0 | 0 | 9.2 | 0 | 9.2 |
| A | 31 | 1 | 1 | 150 | 1 | 0 | 9.5 | 1 | 5.6 |
| B | 243 | 1 | 1 | 136 | 1 | 0 | 22.7 | 0 | 22.7 |
| B | 40 | 1 | 1 | 179 | 1 | 0 | 23.0 | 0 | 23.0 |
| A | 365 | 1 | 1 | 129 | 0 | 0 | 17.9 | 0 | 17.9 |
| A | 29 | 1 | 2 | 145 | 0 | 1 | 0.6 | 0 | 2.6 |
| A | 97 | 1 | 1 | 127 | 0 | 0 | 13.7 | 0 | 13.7 |
| B | 314 | 1 | 3 | 143 | 0 | 0 | 12.2 | 1 | 12.2 |
| B | 17 | 1 | 1 | 114 | 0 | 1 | 17.3 | 0 | 17.7 |
| A | 123 | 1 | 1 | 158 | 0 | 0 | 21.5 | 0 | 21.5 |
| B | 92 | 1 | 1 | 128 | 0 | 0 | 6.0 | 0 | 6.0 |
| A | 39 | 0 | 2 | 15 | 0 | 0 | 10.8 | 0 | 10.8 |
| A | 87 | 1 | 1 | 156 | 1 | 0 | 28.3 | 0 | 28.3 |
| A | 93 | 1 | 1 | 170 | 0 | 0 | 23.9 | 0 | 23.9 |
| A | 4 | 0 | 2 | 104 | 0 | 0 | 21.0 | 0 | 21.0 |
| A | 60 | 1 | 1 | 150 | 0 | 0 | 6.3 | 0 | 6.3 |
| B | 20 | 0 | 1 | 133 | 0 | 0 | 17.3 | 1 | 17.3 |
| A | 52 | 1 | 3 | 125 | 0 | 0 | 12.0 | 0 | 12.0 |
| B | 78 | 0 | 1 | 99 | 0 | 0 | 16.7 | 0 | 16.7 |
| B | 262 | 1 | 2 | 192 | 0 | 0 | 12.7 | 0 | 12.7 |
| A | 19 | 1 | 2 | 143 | 1 | 0 | 6.0 | 1 | 6.0 |
| A | 85 | 1 | 1 | 152 | 0 | 0 | 10.8 | 0 | 10.8 |
| B | 6 | 1 | 1 | 151 | 1 | 0 | 13.0 | 0 | 13.0 |
| B | 53 | 1 | 2 | 115 | 0 | 0 | 8.9 | 0 | 8.9 |
| A | 386 | 1 | 1 | 220 | 1 | 0 | 27.6 | 0 | 27.6 |
| A | 12 | 1 | 1 | 130 | 1 | 0 | 26.4 | 1 | 26.4 |
| B | 356 | 0 | 1 | 110 | 0 | 0 | 27.8 | 0 | 27.8 |
| A | 19 | 1 | 1 | 187 | 0 | 0 | 28.0 | 0 | 28.0 |
| A | 39 | 1 | 2 | 135 | 0 | 0 | 2.9 | 0 | 2.9 |
| B | 9 | 1 | 1 | 139 | 0 | 1 | 6.9 | 0 | 8.0 |
| B | 44 | 1 | 2 | 112 | 0 | 0 | 23.1 | 0 | 23.1 |
| B | 7 | 1 | 1 | 141 | 1 | 0 | 15.9 | 1 | 15.9 |
| A | 34 | 1 | 1 | ·110 | 1 | 1 | 0.4 | 1 | 6.1 |

**TABLE E11.9** (*Continued*)

| TRT | CD4 | GENDER | RACE | WT | HOMO | PCP | PDATE | DIE | DDATE |
|-----|-----|--------|------|-----|------|-----|-------|-----|-------|
| B | 126 | 1 | 1 | 155 | 1 | 1 | 0.2 | 0 | 6.9 |
| B | 4 | 1 | 1 | 142 | 1 | 0 | 20.3 | 0 | 20.3 |
| A | 16 | 1 | 1 | 154 | 0 | 0 | 14.7 | 1 | 14.7 |
| A | 22 | 1 | 1 | 121 | 1 | 0 | 21.4 | 0 | 21.4 |
| B | 35 | 1 | 1 | 165 | 1 | 0 | 21.2 | 0 | 21.2 |
| A | 98 | 1 | 1 | 167 | 1 | 0 | 17.5 | 0 | 17.5 |
| A | 357 | 1 | 1 | 133 | 0 | 0 | 16.6 | 0 | 16.6 |
| B | 209 | 1 | 1 | 146 | 1 | 0 | 15.6 | 0 | 15.6 |
| A | 138 | 1 | 1 | 134 | 1 | 0 | 8.8 | 0 | 8.8 |
| B | 36 | 1 | 1 | 169 | 1 | 0 | 3.4 | 0 | 3.4 |
| A | 90 | 1 | 1 | 166 | 0 | 0 | 30.0 | 0 | 30.0 |
| B | 51 | 1 | 1 | 120 | 0 | 1 | 26.0 | 0 | 26.4 |
| B | 25 | 1 | 2 | 161 | 0 | 0 | 29.0 | 0 | 29.0 |
| A | 17 | 1 | 1 | 130 | 0 | 0 | 7.3 | 1 | 7.3 |
| A | 73 | 1 | 1 | 140 | 0 | 0 | 20.9 | 0 | 20.9 |
| A | 123 | 1 | 1 | 134 | 1 | 0 | 17.4 | 0 | 17.4 |
| B | 161 | 1 | 1 | 177 | 1 | 1 | 19.3 | 1 | 19.3 |
| B | 105 | 1 | 1 | 128 | 1 | 1 | 3.7 | 0 | 23.5 |
| A | 74 | 1 | 2 | 134 | 1 | 0 | 24.8 | 0 | 24.8 |
| A | 7 | 1 | 1 | 13 | 0 | 0 | 10.3 | 1 | 10.3 |
| B | 29 | 0 | 1 | 97 | 0 | 1 | 13.1 | 0 | 23.9 |
| A | 84 | 1 | 1 | 217 | 1 | 0 | 24.8 | 0 | 24.8 |
| B | 9 | 1 | 1 | 158 | 1 | 0 | 23.5 | 0 | 23.5 |
| A | 29 | 1 | 1 | 16 | 1 | 0 | 23.5 | 0 | 23.5 |
| B | 24 | 1 | 1 | 136 | 1 | 0 | 19.4 | 0 | 19.4 |
| B | 715 | 1 | 2 | 126 | 1 | 0 | 15.7 | 0 | 15.7 |
| B | 4 | 1 | 2 | 159 | 0 | 0 | 17.1 | 0 | 17.1 |
| A | 147 | 1 | 1 | 170 | 1 | 1 | 9.8 | 0 | 16.8 |
| A | 162 | 1 | 1 | 137 | 0 | 0 | 11.0 | 0 | 11.0 |
| B | 35 | 1 | 1 | 150 | 1 | 0 | 11.9 | 0 | 11.9 |
| B | 14 | 1 | 1 | 153 | 1 | 0 | 8.2 | 1 | 8.2 |
| B | 227 | 1 | 1 | 150 | 1 | 0 | 9.5 | 0 | 9.5 |
| B | 137 | 1 | 1 | 145 | 1 | 0 | 9.0 | 0 | 9.0 |
| A | 48 | 1 | 1 | 143 | 0 | 0 | 8.3 | 0 | 8.3 |
| A | 62 | 1 | 1 | 175 | 1 | 0 | 6.7 | 0 | 6.7 |
| A | 47 | 1 | 1 | 164 | 1 | 0 | 5.5 | 0 | 5.5 |
| B | 7 | 1 | 1 | 205 | 0 | 0 | 6.9 | 0 | 6.9 |
| B | 9 | 1 | 1 | 121 | 0 | 1 | 19.4 | 0 | 23.9 |
| B | 243 | 1 | 2 | 152 | 0 | 1 | 12.0 | 1 | 0.0 |
| A | 133 | 1 | 1 | 136 | 0 | 0 | 23.1 | 0 | 23.1 |
| A | 56 | 1 | 1 | 159 | 1 | 0 | 23.2 | 0 | 23.2 |
| A | 11 | 1 | 1 | 157 | 0 | 0 | 8.7 | 1 | 8.7 |
| A | 94 | 1 | 2 | 116 | 0 | 0 | 15.1 | 1 | 15.1 |
| A | 68 | 1 | 1 | 185 | 1 | 0 | 21.3 | 0 | 21.3 |
| A | 139 | 1 | 1 | 145 | 1 | 0 | 19.1 | 0 | 19.1 |
| B | 15 | 0 | 1 | 114 | 0 | 0 | 17.4 | 0 | 17.4 |
| B | 22 | 1 | 2 | 125 | 1 | 0 | 4.4 | 1 | 4.4 |

**11.10**    It has been noted that metal workers have an increased risk for cancer of the internal nose and paranasal sinuses, perhaps as a result of exposure to cutting oils. A study was conducted to see whether this particular exposure also increases the risk for squamous cell carcinoma of the scrotum (Rousch et al., 1982). Cases included all 45 squamous cell carcinomas of the scrotum diagnosed in Connecticut residents from 1955 to 1973, as obtained from the Connecticut Tumor Registry. Matched controls were selected for each case based on the age at death (within eight years), year of death (within three years), and number of jobs as obtained from combined death certificate and directory sources. An occupational indicator of metal worker (yes/no) was evaluated as the possible risk factor in this study; results are shown in Table E11.10.

**TABLE E11.10**

|  | Controls | |
|---|---|---|
| Cases | Yes | No |
| Yes | 2 | 26 |
| No | 5 | 12 |

(a) Find a 95% confidence interval for the odds ratio measuring the strength of the relationship between the disease and the exposure.

(b) Test for the independence between the disease and the exposure.

**11.11**    Ninety-eight heterosexual couples, at least one of whom was HIV infected, were enrolled in an HIV transmission study and interviewed about sexual behavior (Padian, 1990). Table E11.11 provides a summary of condom use reported by heterosexual partners. Test to compare the reporting results between men and women.

**TABLE E11.11**

|  | Man | | |
|---|---|---|---|
| Woman | Ever | Never | Total |
| Ever | 45 | 6 | 51 |
| Never | 7 | 40 | 47 |
| Total | 52 | 46 | 98 |

**11.12**    A matched case–control study was conducted to evaluate the cumulative effects of acrylate and methacrylate vapors on olfactory function (Schwarts et al., 1989). Cases were defined as scoring at or below the 10th percentile on the UPSIT (University of Pennsylvania Smell Identification Test; Table E11.12).

**TABLE E11.12**

| | Cases | |
|---|---|---|
| Controls | Exposed | Unexposed |
| Exposed | 25 | 22 |
| Unexposed | 9 | 21 |

(a) Find a 95% confidence interval for the odds ratio measuring the strength of the relationship between the disease and the exposure.

(b) Test for the independence between the disease and the exposure.

**11.13** A study in Maryland identified 4032 white persons, enumerated in a nonofficial 1963 census, who became widowed between 1963 and 1974 (Helsing and Szklo, 1981). These people were matched, one to one, to married persons on the basis of race, gender, year of birth, and geography of residence. The matched pairs were followed in a second census in 1975.

(a) We have the overall male mortality shown in Table E11.13a. Test to compare the mortality of widowed men versus married men.

**TABLE E11.13a**

| | Married Men | |
|---|---|---|
| Widowed Men | Dead | Alive |
| Dead | 2 | 292 |
| Alive | 210 | 700 |

(b) The data for 2828 matched pairs of women are shown in Table E11.13b. Test to compare the mortality of widowed women versus married women.

**TABLE E11.13b**

| Widowed Women | Married Women | |
|---|---|---|
| | Dead | Alive |
| Dead | 1 | 264 |
| Alive | 249 | 2314 |

**11.14** Table E11.14 at the end of this chapter provides some data from a matched case–control study to investigate the association between the use of x-ray and risk of childhood acute myeloidleukemia. In each matched set or pair, the case and control(s) were matched by age, race, and county of residence. The variables are:

**TABLE E11.14**

| MatchedSet | Disease | Sex | Downs | Age | MXRay | UMXRay | LMXay | FXRay | CXRay | CNXRay |
|---|---|---|---|---|---|---|---|---|---|---|
| 1 | 1 | 2 | 1 | 0 | 1 | 0 | 0 | 1 | 1 | 1 |
|  | 2 | 2 | 1 | 0 | 1 | 0 | 0 | 1 | 1 | 1 |
| 2 | 1 | 1 | 1 | 6 | 1 | 0 | 0 | 1 | 2 | 3 |
|  | 2 | 1 | 1 | 6 | 1 | 0 | 0 | 1 | 2 | 2 |
| 3 | 1 | 2 | 1 | 8 | 1 | 0 | 0 | 1 | 1 | 1 |
|  | 2 | 2 | 1 | 8 | 1 | 0 | 0 | 1 | 1 | 1 |
| 4 | 1 | 1 | 2 | 1 | 1 | 0 | 0 | 1 | 1 | 1 |
|  | 2 | 1 | 1 | 1 | 1 | 0 | 0 | 1 | 1 | 1 |
| 5 | 1 | 1 | 1 | 4 | 2 | 0 | 1 | 1 | 2 | 2 |
|  | 2 | 1 | 1 | 4 | 1 | 0 | 0 | 1 | 1 | 1 |
| 6 | 1 | 2 | 1 | 9 | 2 | 1 | 0 | 1 | 2 | 2 |
|  | 2 | 1 | 1 | 9 | 1 | 0 | 0 | 1 | 2 | 2 |
| 7 | 1 | 2 | 1 | 17 | 1 | 0 | 0 | 1 | 1 | 1 |
|  | 2 | 2 | 1 | 17 | 1 | 0 | 0 | 1 | 1 | 1 |
| 8 | 1 | 2 | 1 | 5 | 1 | 0 | 0 | 1 | 1 | 1 |
|  | 2 | 2 | 2 | 5 | 1 | 0 | 0 | 1 | 1 | 1 |
| 9 | 1 | 1 | 2 | 0 | 1 | 0 | 0 | 2 | 1 | 1 |
|  | 2 | 2 | 1 | 0 | 2 | 1 | 0 | 1 | 1 | 1 |
| 10 | 1 | 2 | 1 | 7 | 1 | 0 | 0 | 2 | 1 | 1 |
|  | 2 | 2 | 1 | 7 | 1 | 0 | 0 | 1 | 1 | 1 |
| 11 | 1 | 1 | 1 | 15 | 1 | 0 | 0 | 1 | 1 | 1 |
|  | 2 | 1 | 1 | 15 | 1 | 0 | 0 | 1 | 1 | 1 |
| 12 | 1 | 1 | 1 | 12 | 1 | 0 | 0 | 1 | 2 | 2 |
|  | 2 | 1 | 1 | 12 | 1 | 0 | 0 | 1 | 2 | 2 |
| 13 | 1 | 1 | 1 | 4 | 1 | 0 | 0 | 1 | 1 | 1 |
|  | 2 | 2 | 1 | 4 | 1 | 0 | 0 | 1 | 1 | 1 |

(Continued)

| Row | | | | | | | | | | |
|---|---|---|---|---|---|---|---|---|---|---|
| 14 | 1 | 1 | 1 | 1 | 14 | 1 | 1 | 0 | 0 | 1 | 2 | 2 | 2 |
| | 2 | 2 | 1 | 1 | 14 | 1 | 2 | 0 | 0 | 1 | 1 | 1 | 1 |
| | 2 | 1 | 1 | 1 | 14 | 1 | 1 | 0 | 0 | 2 | 1 | 1 | 1 |
| 15 | 1 | 1 | 1 | 1 | 7 | 1 | 1 | 0 | 0 | 1 | 1 | 1 | 1 |
| | 2 | 2 | 1 | 1 | 7 | 1 | 1 | 0 | 0 | 1 | 2 | 2 | 1 |
| | 2 | 1 | 1 | 1 | 7 | 1 | 1 | 0 | 0 | 2 | 2 | 2 | 2 |
| 16 | 1 | 1 | 1 | 1 | 8 | 1 | 1 | 0 | 0 | 1 | 1 | 1 | 1 |
| | 2 | 2 | 1 | 1 | 8 | 1 | 2 | 0 | 0 | 2 | 2 | 2 | 2 |
| 17 | 1 | 1 | 1 | 1 | 6 | 1 | 1 | 0 | 0 | 1 | 1 | 1 | 1 |
| | 2 | 1 | 1 | 1 | 6 | 1 | 1 | 0 | 0 | 1 | 3 | 2 | 2 |
| 18 | 1 | 1 | 1 | 1 | 13 | 1 | 2 | 0 | 0 | 2 | 2 | 2 | 2 |
| | 2 | 2 | 1 | 1 | 13 | 1 | 2 | 0 | 0 | 2 | 1 | 1 | 1 |
| 19 | 1 | 1 | 1 | 1 | 17 | 1 | 2 | 0 | 0 | 1 | 2 | 2 | 2 |
| | 2 | 2 | 1 | 1 | 17 | 2 | 1 | 1 | 0 | 2 | 4 | 2 | 4 |
| 20 | 1 | 1 | 1 | 1 | 5 | 1 | 2 | 0 | 0 | 1 | 1 | 1 | 1 |
| | 2 | 1 | 1 | 1 | 5 | 1 | 1 | 0 | 0 | 2 | 4 | 1 | 4 |
| 21 | 1 | 1 | 1 | 1 | 13 | 1 | 1 | 0 | 0 | 1 | 2 | 2 | 2 |
| | 2 | 2 | 1 | 1 | 13 | 1 | 1 | 0 | 0 | 2 | 1 | 2 | 1 |
| 22 | 1 | 1 | 1 | 1 | 16 | 1 | 2 | 0 | 0 | 2 | 2 | 2 | 2 |
| | 2 | 2 | 1 | 1 | 16 | 1 | 2 | 0 | 0 | 2 | 1 | 1 | 1 |
| | 2 | 2 | 1 | 1 | 16 | 1 | 1 | 0 | 0 | 1 | 1 | 1 | 1 |
| 23 | 1 | 1 | 1 | 1 | 10 | 1 | 1 | 0 | 0 | 1 | 1 | 1 | 1 |
| | 2 | 2 | 1 | 1 | 10 | 1 | 2 | 0 | 0 | 1 | 1 | 1 | 1 |
| 24 | 1 | 1 | 1 | 1 | 0 | 1 | 1 | 0 | 0 | 1 | 1 | 1 | 1 |
| | 2 | 1 | 1 | 1 | 0 | 1 | 1 | 1 | 0 | 1 | 2 | 1 | 2 |
| 25 | 1 | 1 | 2 | 2 | –1 | 2 | 2 | 0 | 0 | 2 | 1 | 2 | 1 |
| | 2 | 2 | 1 | 1 | 13 | 1 | 1 | 0 | 0 | 2 | 1 | 1 | 1 |
| 26 | 1 | 1 | 2 | 2 | 13 | 1 | 2 | 1 | 1 | 1 | 1 | 1 | 1 |

**TABLE E11.14** (*Continued*)

| MatchedSet | Disease | Sex | Downs | Age | MXRay | UMXRay | LMXay | FXRay | CXRay | CNXRay |
|---|---|---|---|---|---|---|---|---|---|---|
| 27 | 1 | 1 | 1 | 11 | 1 | 0 | 0 | 1 | 1 | 1 |
|    | 2 | 1 | 1 | 11 | 1 | 0 | 0 | 1 | 2 | 2 |
| 28 | 1 | 2 | 1 | 4 | 1 | 0 | 0 | 2 | 1 | 1 |
|    | 2 | 2 | 1 | 4 | 1 | 0 | 0 | 1 | 2 | 2 |
| 29 | 1 | 2 | 1 | 1 | 1 | 0 | 0 | 2 | 1 | 1 |
|    | 2 | 1 | 1 | 1 | 1 | 0 | 0 | 1 | 1 | 1 |
| 30 | 1 | 2 | 1 | 15 | 2 | 0 | 1 | 2 | 2 | 3 |
|    | 2 | 2 | 1 | 15 | 1 | 0 | 0 | 2 | 2 | 2 |
| 31 | 1 | 1 | 1 | 9 | 1 | 0 | 0 | 1 | 1 | 1 |
|    | 2 | 2 | 1 | 9 | 1 | 0 | 0 | 1 | 1 | 1 |
| 32 | 1 | 1 | 1 | 15 | 1 | 0 | 0 | 2 | 2 | 2 |
|    | 2 | 1 | 1 | 15 | 1 | 0 | 0 | 2 | 2 | 3 |
| 33 | 1 | 2 | 1 | 5 | 1 | 0 | 0 | 1 | 2 | 3 |
|    | 2 | 2 | 1 | 5 | 2 | 1 | 0 | 1 | 1 | 1 |
| 34 | 1 | 1 | 1 | 10 | 2 | 0 | 1 | 2 | 2 | 1 |
|    | 2 | 1 | 1 | 10 | 2 | 0 | 0 | 2 | 2 | 2 |
| 35 | 1 | 2 | 1 | 8 | 1 | 0 | 0 | 1 | 2 | 2 |
|    | 2 | 2 | 1 | 8 | 1 | 0 | 0 | 1 | 1 | 1 |
| 36 | 1 | 2 | 1 | 15 | 1 | 0 | 0 | 1 | 2 | 4 |
|    | 2 | 2 | 1 | 15 | 1 | 0 | 0 | 1 | 2 | 2 |
| 37 | 1 | 2 | 2 | 1 | 1 | 1 | 0 | 1 | 1 | 1 |
|    | 2 | 2 | 1 | 1 | 2 | 0 | 0 | 2 | 1 | 1 |
| 38 | 1 | 2 | 1 | 0 | 1 | 0 | 0 | 1 | 1 | 1 |
|    | 2 | 1 | 1 | 0 | 1 | 0 | 0 | 1 | 1 | 1 |
| 39 | 1 | 1 | 1 | 6 | 1 | 0 | 0 | 2 | 2 | 2 |
|    | 2 | 2 | 1 | 6 | 1 | 0 | 0 | 1 | 1 | 1 |
| 40 | 1 | 1 | 1 | 14 | 1 | 0 | 0 | 2 | 2 | 2 |
|    | 2 | 2 | 1 | 14 | 1 | 0 | 0 | 1 | 1 | 1 |

| Case | C1 | C2 | C3 | C4 | C5 | C6 | C7 | C8 | C9 | C10 |
|------|----|----|----|----|----|----|----|----|----|-----|
| 41 | 1 | 1 | 1 | 2 | 1 | 0 | 0 | 2 | 1 | 1 |
| 42 | 1 | 2 | 2 | 1 | 2 | 0 | 0 | 1 | 1 | 1 |
| 43 | 1 | 2 | 1 | 6 | 1 | 1 | 0 | 1 | 1 | 1 |
| 44 | 1 | 1 | 1 | 16 | 1 | 0 | 0 | 1 | 1 | 1 |
| 45 | 1 | 2 | 1 | 4 | 1 | 0 | 0 | 1 | 2 | 2 |
| 46 | 1 | 1 | 1 | 1 | 1 | 1 | 0 | 1 | 1 | 1 |
| 47 | 1 | 2 | 1 | 0 | 1 | 0 | 0 | 1 | 1 | 1 |
| 48 | 1 | 1 | 1 | 0 | 1 | 0 | 0 | 2 | 2 | 4 |
| 49 | 1 | 2 | 1 | 3 | 1 | 0 | 0 | 2 | 2 | 4 |
| 50 | 1 | 1 | 1 | 5 | 1 | 0 | 0 | 1 | 1 | 1 |
| 51 | 1 | 2 | 1 | 8 | 1 | 0 | 0 | 2 | 2 | 2 |
| 52 | 1 | 1 | 1 | 9 | 1 | 0 | 0 | 2 | 2 | 2 |
| 53 | 1 | 2 | 1 | 9 | 1 | 0 | 0 | 1 | 1 | 1 |
| 54 | 1 | 2 | 2 | 2 | 2 | 1 | 0 | 1 | 1 | 1 |
| 55 | 1 | 1 | 1 | 3 | 1 | 0 | 0 | 1 | 1 | 2 |

(Continued)

**TABLE E11.14** (*Continued*)

| MatchedSet | Disease | Sex | Downs | Age | MXRay | UMXRay | LMXay | FXRay | CXRay | CNXRay |
|---|---|---|---|---|---|---|---|---|---|---|
| 56 | 1 | 1 | 1 | 17 | 1 | 0 | 0 | 1 | 1 | 1 |
|    | 2 | 1 | 1 | 17 | 2 | 1 | 0 | 1 | 2 | 2 |
|    | 2 | 1 | 1 | 17 | 1 | 0 | 0 | 1 | 1 | 1 |
| 57 | 1 | 2 | 1 | 3 | 1 | 0 | 0 | 1 | 1 | 1 |
|    | 2 | 2 | 1 | 3 | 1 | 0 | 0 | 2 | 1 | 1 |
| 58 | 1 | 1 | 1 | 10 | 1 | 0 | 0 | 1 | 2 | 4 |
|    | 2 | 1 | 1 | 10 | 2 | 0 | 1 | 1 | 1 | 1 |
| 59 | 1 | 1 | 1 | 13 | 1 | 0 | 0 | 1 | 2 | 2 |
|    | 2 | 2 | 1 | 13 | 2 | 0 | 1 | 1 | 1 | 1 |
| 60 | 1 | 1 | 1 | 0 | 1 | 0 | 0 | 1 | 1 | 1 |
|    | 2 | 1 | 1 | 0 | 1 | 0 | 0 | 1 | 1 | 1 |
| 61 | 1 | 1 | 1 | 11 | 1 | 0 | 0 | 1 | 1 | 3 |
|    | 2 | 1 | 1 | 11 | 1 | 0 | 0 | 2 | 2 | 1 |
| 62 | 1 | 1 | 1 | 14 | 1 | 0 | 0 | 1 | 2 | 2 |
|    | 2 | 1 | 1 | 14 | 1 | 0 | 0 | 1 | 2 | 4 |
| 63 | 1 | 2 | 1 | 5 | 1 | 0 | 0 | 1 | 1 | 1 |
|    | 2 | 1 | 1 | 5 | 1 | 0 | 0 | 1 | 1 | 1 |
|    | 2 | 2 | 1 | 5 | 1 | 0 | 0 | 1 | 1 | 1 |
| 64 | 1 | 1 | 1 | 12 | 1 | 0 | 0 | 2 | 2 | 2 |
|    | 2 | 2 | 1 | 12 | 1 | 0 | 0 | 1 | 1 | 1 |
| 65 | 1 | 2 | 1 | 9 | 2 | 0 | 1 | 1 | 2 | 2 |
|    | 2 | 2 | 1 | 9 | 1 | 0 | 0 | 1 | 1 | 1 |
| 66 | 1 | 2 | 1 | 5 | 1 | 0 | 0 | 1 | 2 | 4 |
|    | 2 | 1 | 1 | 5 | 1 | 0 | 0 | 2 | 1 | 1 |
| 67 | 1 | 1 | 1 | 8 | 1 | 0 | 0 | 1 | 1 | 1 |
|    | 2 | 1 | 1 | 8 | 1 | 0 | 0 | 1 | 1 | 1 |
| 68 | 1 | 1 | 1 | 15 | 1 | 0 | 0 | 1 | 2 | 4 |
|    | 2 | 1 | 1 | 15 | 1 | 0 | 0 | 2 | 1 | 1 |
|    | 2 | 2 | 1 | 15 | 2 | 0 | 1 | 1 | 2 | 2 |

(Continued)

| 69 | 70 | 71 | 72 | 73 | 74 | 75 | 76 | 77 | 78 | 79 | 80 | 81 | 82 | 83 |
|---|---|---|---|---|---|---|---|---|---|---|---|---|---|---|
| 1 | 1 | 1 | 1 | 1 | 1 | 1 | 1 | 1 | 3 2 | 1 | 1 | 2 | 1 | 1 1 | 1 1 | 1 2 |
| 1 | 1 | 1 | 1 | 1 | 1 | 1 | 1 | 2 2 | 1 | 1 | 2 | 1 | 1 1 | 1 1 | 1 2 |
| 1 | 1 | 1 | 1 | 1 | 1 2 | 1 | 2 | 1 | 2 | 1 | 2 | 1 1 | 2 | 1 1 | 1 2 |
| 0 | 0 | 0 | 0 | 0 | 0 | 0 | 0 | 0 | 0 | 0 | 0 | 0 | 0 | 0 |
| 0 | 0 | 0 | 0 | 0 | 0 | 0 | 0 | 0 | 0 | 0 | 0 | 0 | 0 | 0 |
| 1 | 1 | 1 | 1 | 1 | 1 | 1 | 1 | 1 | 1 | 1 | 1 | 1 | 1 | 1 |
| 10 10 | 3 3 | 1 1 | 1 9 | 9 1 | 8 8 | 12 12 | 1 4 | 4 11 | 11 2 | 4 4 | 1 0 | 5 5 |
| 1 | 1 | 1 | 1 | 1 | 1 | 1 | 1 | 1 | 1 | 1 | 1 | 1 | 1 | 1 |
| 2 2 | 1 2 | 1 2 | 2 1 | 2 1 | 2 2 | 1 2 | 2 2 | 2 2 | 1 2 | 2 1 | 2 1 | 1 1 |
| 1 2 | 1 2 | 1 2 | 2 1 | 2 1 | 2 1 | 2 1 | 2 1 | 2 1 | 2 1 | 2 1 | 2 1 | 2 1 | 2 |

**TABLE E11.14** (*Continued*)

| MatchedSet | Disease | Sex | Downs | Age | MXRay | UMXRay | LMXay | FXRay | CXRay | CNXRay |
|---|---|---|---|---|---|---|---|---|---|---|
| 84 | 1 | 2 | 2 | 1 | 2 | 0 | 1 | 2 | 1 | 1 |
|    | 2 | 1 | 1 | 1 | 1 | 0 | 0 | 2 | 1 | 1 |
| 85 | 1 | 1 | 1 | 12 | 1 | 0 | 0 | 2 | 2 | 2 |
|    | 2 | 1 | 1 | 12 | 1 | 0 | 0 | 1 | 1 | 1 |
|    | 2 | 1 | 1 | 12 | 1 | 0 | 0 | 2 | 1 | 1 |
| 86 | 1 | 1 | 1 | 12 | 2 | 0 | 1 | 2 | 2 | 2 |
|    | 2 | 1 | 1 | 12 | 1 | 0 | 0 | 2 | 2 | 4 |
| 87 | 1 | 1 | 1 | 1 | 1 | 0 | 0 | 1 | 1 | 1 |
|    | 2 | 1 | 1 | 1 | 1 | 0 | 0 | 1 | 1 | 1 |
| 88 | 1 | 1 | 1 | 9 | 1 | 0 | 0 | 2 | 1 | 1 |
|    | 2 | 1 | 1 | 9 | 1 | 0 | 0 | 2 | 1 | 1 |
| 89 | 1 | 2 | 1 | 2 | 1 | 0 | 0 | 2 | 1 | 1 |
|    | 2 | 2 | 1 | 2 | 1 | 0 | 0 | 1 | 1 | 1 |
| 90 | 2 | 2 | 1 | 1 | 1 | 0 | 0 | 2 | 1 | 1 |
|    | 2 | 2 | 1 | 1 | 1 | 0 | 0 | 1 | 1 | 1 |
| 91 | 1 | 2 | 1 | 2 | 1 | 0 | 0 | 2 | 2 | 4 |
|    | 2 | 2 | 1 | 2 | 1 | 0 | 0 | 1 | 1 | 1 |
| 92 | 1 | 1 | 1 | 15 | 1 | 0 | 0 | 2 | 1 | 1 |
|    | 2 | 1 | 1 | 15 | 1 | 0 | 0 | 1 | 1 | 1 |
| 93 | 2 | 1 | 1 | 13 | 1 | 0 | 0 | 2 | 1 | 1 |
|    | 2 | 1 | 1 | 13 | 1 | 0 | 0 | 1 | 1 | 1 |
| 94 | 1 | 1 | 1 | 6 | 1 | 0 | 0 | 2 | 2 | 4 |
|    | 2 | 2 | 1 | 6 | 1 | 0 | 0 | 2 | 1 | 1 |
| 95 | 1 | 2 | 2 | 1 | 1 | 0 | 0 | 1 | 1 | 1 |
|    | 2 | 2 | 1 | 1 | 1 | 0 | 0 | 1 | 1 | 1 |
| 96 | 1 | 2 | 1 | 8 | 1 | 0 | 0 | 2 | 1 | 1 |
|    | 2 | 2 | 1 | 8 | 1 | 0 | 0 | 1 | 1 | 1 |
| 97 | 1 | 2 | 1 | 4 | 1 | 0 | 0 | 2 | 2 | 4 |
|    | 2 | 2 | 1 | 4 | 1 | 0 | 0 | 2 | 1 | 1 |

| | | | | | | | | | |
|---|---|---|---|---|---|---|---|---|---|
| 98 | 1 2 | 1 | 1 | 1 | 1 | 6 6 | 1 | 2 | 2 |
| 99 | 1 2 | 1 | 1 | 2 | 1 | 1 1 | 1 | 1 | 1 |
| 100 | 1 2 | 2 | 1 | 1 | 1 | 14 14 | 1 | 1 | 1 |
| 101 | 1 2 | 1 | 1 | 2 | 1 | 1 1 | 1 | 1 | 4 |
| 102 | 1 2 | 2 2 | 2 | 1 | 2 | 1 1 | 1 | 1 | 3 |
| 103 | 1 2 | 1 | 1 | 1 | 1 | 13 13 | 1 | 1 | 1 |
| 104 | 1 2 | 2 2 | 1 | 2 | 2 | 13 13 | 1 | 2 | 1 |
| 105 | 1 2 | 2 2 | 1 | 1 | 2 | 11 11 | 1 | 2 | 1 |
| 106 | 1 2 | 2 2 | 1 | 2 | 2 | 13 13 | 1 | 2 | 4 |
| 107 | 1 2 | 2 2 | 1 | 2 | 1 | 2 2 | 1 | 1 | 3 |
| 108 | 1 2 | 2 2 | 1 | 1 | 1 | 7 7 7 | 1 | 1 | 3 |
| 109 | 1 2 | 2 2 | 1 | 2 | 1 | 16 16 | 1 | 2 | 1 |
| 110 | 1 2 | 1 | 1 | 1 | 1 | 3 3 | 1 | 1 | 1 |
| 111 | 1 2 | 2 2 | 1 | 2 | 1 | 13 13 | 1 | 2 | 4 |
| 112 | 2 | 2 | 1 | 2 | 1 | 6 6 | 1 | 1 | 1 |

- Matched set (or pair)
- Disease (1, case; 2, control)
- Some chracteristics of the child: sex (1, male; 2, female), Down's syndrome (a known risk factor for leukemia; 1, no; 2, yes), age
- Risk factors related to the use of x-ray: MXray (mother ever had x-ray during pregnancy; 1, no; 2, yes), UMXray (mother ever had upper-body x-ray during pregnancy; 0, no; 1, yes), LMXray (mother ever had lower-body x-ray during pregnancy; 0, no; 1, yes), FXray (father ever had x-ray; 1, no; 2, yes), CXray (child ever had x-ray; 1, no; 2, yes), CNXray (child's total number of x-rays; 1, none; 2, 1–2; 3, 3–4; 4, 5 or more)

**(a)** Taken collectively, do the covariates contribute significantly to the separation of cases and controls?

**(b)** Fit the multiple regression model to obtain estimates of individual regression coefficients and their standard errors. Draw conclusions concerning the conditional contribution of each factor.

**(c)** Within the context of the multiple regression model in part (b), does gender alter the effect of Down's syndrome?

**(d)** Within the context of the multiple regression model in part (b), taken collectively, does the exposure to x-ray (by the father, or mother, or child) relate significanty to the disease of the child?

**(e)** Within the context of the multiple regression model in part (b), is the effect of age linear?

**(f)** Focus on Down's syndrome as the primary factor, taken collectively; was this main effect altered by any other covariates?

# 12

# STUDY DESIGNS

Statistics is more than just a collection of long columns of numbers and sets of formulas. Statistics is a way of thinking—thinking about ways to gather and analyze data. The gathering part comes before the analyzing part; the first thing a statistician or a learner of statistics does when faced with data is to find out how the data were collected. Not only does how we should analyze data depend on how data were collected, but formulas and techniques may be misused by a well-intentioned researcher simply because data were not collected properly. In other cases, studies were inconclusive because they were poorly planned and not enough data were collected to accomplish the goals and support the hypotheses.

Study data may be collected in many different ways. When we want information, the most common approach is to conduct a survey in which subjects in a sample are asked to express opinions on a variety of issues. For example, an investigator surveyed several hundred students in grades 7 through 12 with a set of questions asking the date of their last physical checkup and how often they smoke cigarettes or drink alcohol.

The format of a survey is such that one can assume there is an identifiable, existing parent population of subjects. We act as if the sample is obtained from the parent population according to a carefully defined techinical procedure called *random sampling*. The basic steps and characteristics of a such a process were described in detail in Section 3.1.2. However, in biomedical research, a sample survey is not the common form of study; it may not be used at all. The laboratory investigator uses animals in projects, but the animals are not selected randomly from a large population of animals. The clinician, who is attempting to describe the results obtained with a particular therapy, cannot say that he or she has obtained patients as a random sample from a parent population of patients.

## 12.1   TYPES OF STUDY DESIGNS

In addition to surveys that are cross-sectional, as seen in many examples in earlier chapters, study data may be collected in many different ways. For example, investigators are faced more and more frequently with the problem of determining whether a specific factor or exposure is related to a certain aspect of health. Does air polution cause lung cancer? Do birth control pills cause thromboembolic death? There are reasons for believing that the answer to each of these and other questions is yes, but all are controversial; otherwise, no studies are needed. Generally, biomedical research data may come from different sources, the two fundamental designs being retrospective and prospective. But strategies can be divided further into four different types:

1. Retrospective studies (of past events)
2. Prospective studies (of past events)
3. Cohort studies (of ongoing or future events)
4. Clinical trials

*Retrospective studies* of past events gather past data from selected cases, persons are identified who have experienced the event in question, and controls, persons who have not experienced the event in question, to determine differences, if any, in exposure to a suspected risk factor under investigation. They are commonly referred to as *case–control studies*; each case–control study is focused on a particular disease. In a typical case–control study, cases of a specific disease are ascertained as they arise from population-based registers or lists of hospital admissions, and controls are sampled either as disease-free persons from the population at risk or as hospitalized patients having a diagnosis other than the one under study. An example is the study of thromboembolic death and birth control drugs. Thromboembolic deaths were identified from death certificates, and exposure to the pill was traced by interview with each woman's physician and a check of her various medical records. Control women were women in the same age range under the care of the same physicians.

*Prospective studies* of past events are less popular because they depend on the existence of records of high quality. In these, samples of exposed subjects and unexposed subjects are identified in the records. Then the records of the persons selected are traced to determine if they have ever experienced the event to the present time. Events in question are past events, but the method is called *prospective* because it proceeds from exposure forward to the event.

*Cohort studies* are epidemiological designs in which one enrolls a group of persons and follows them over certain periods of time; examples include occupational mortality studies and *clinical trials*. The cohort study design focuses on a particular exposure rather than a particular disease as in case–control studies. There have been several major cohort studies with significant contribution to our understanding of important public health issues, but this form of study design is not very popular because cohort studies are time and cost consuming.

In this chapter we focus on study designs. However, since in biomedical research, the sample survey is not a common form of study, and prospective studies of past events and cohort studies are not often conducted, we put more emphasis on the designs of clinical trials which are important because they are experiments on human beings, and of case–control studies, which are the most popular of all study designs.

## 12.2  CLASSIFICATION OF CLINICAL TRIALS

Clinical studies form a class of all scientific approaches to evaluating medical disease prevention, diagnostic techniques, and treatments. Among this class, trials, often called *clinical trials*, form a subset of those clinical studies that evaluate investigational drugs.

Trials, especially cancer trials, are classified into phases:

- Phase I trials focus on safety of a new investigational medicine. These are the first human trials after successful animal trials.
- Phase II trials are small trials to evaluate efficacy and focus more on a safety profile.
- Phase III trials are well-controlled trials, the most rigorous demonstration of a drug's efficacy prior to federal regulatory approval.
- Phase IV trials are often conducted after a medicine is marketed to provide additional details about the medicine's efficacy and a more complete safety profile.

Phase I trials apply to patients from standard treatment failure who are at high risk of death in the short term. As for the new medicine or drug to be tested, there is no efficacy at low doses; at high doses, there will be unavoidable toxicity, which may be severe and may even be fatal. Little is known about the dose range; animal studies may not be helpful enough. The goal in a phase I trial is to identify a maximum tolerated dose (MTD), a dose that has reasonable efficacy (i.e., is toxic enough, say, to kill cancer cells) but with tolerable toxicity (i.e., not toxic enough to kill the patient).

Phase II trials, the next step, are often the simplest: The drug, at the optimal dose (MTD) found in a phase I trial, is given to a small group of patients who meet predetermined inclusion criteria. The most common form are single-arm studies where investigators are seeking to establish the antitumor activity of a drug usually measured by a response rate. A patient responds when his or her cancer condition improves (e.g., the tumor disappears or shrinks substantially).

The response rate is the proportion or percentage of patients who respond. A phase II trial may be conducted in two stages (as will be seen in Section 12.6) when investigators are concerned about severe side effects. A second type of phase II trial consists of small comparative trials where we want to establish the efficacy of a new drug against a control or standard regimen. In these phase II trials, with or without randomization, investigators often test their validity by paying careful attention to inclusion and exclusion criteria. *Inclusion criteria* focus on the definition of patient characteristics required for entry into a clinical trial. These describe the population of patients that the drug is intended to serve. There are *exclusion criteria* as well, to keep out patients that the drug is not intended to serve.

Phase III and IV trials are designed similarly. Phase III trials are conducted before regulatory approval, and phase IV trials, which are often optional, are conducted after regulatory approval. These are larger, controlled trials, whose control is achieved by randomization. Patients enter the study sequentially and upon enrollment, each patient is randomized to receive either the investigational drug or a placebo (or standard therapy). As medication, the placebo is "blank," that is, without any active medicine. Placebo, whose size and shape are similar to those of the drug, is used to control psychological and emotional effects (i.e., possible prejudices on the part of the patient and/or investigator). *Randomization* is a technique to ensure that the two groups, the one receiving the real drug and the one receiving the placebo, are more comparable, more similar with respect to known as well as unknown factors (so that the conclusion is more valid). For example, the new patient is assigned to receive the drug or the placebo by a process similar to that of flipping a coin. Trials in phases III and IV are often conducted as a *double blind*, that is, blind to the patient (he or she does not know if a real drug is given so as to prevent psychological effects; of course, the patient's consent is required) and blind to the investigator (so as to prevent bias in measuring/evaluating outcomes). Some member of the investigation team, often designated a priori, keeps the code (the list of which patients received drug and which patients received placebo) which is broken only at the time of study completion and data analysis. The term *triple blind* may be used in some trials to indicate the blinding of regulatory officers.

A phase III or IV trial usually consists of two periods: an enrollment period, when patients enter the study and are randomized, and a follow-up period. The latter is very desirable if long-term outcome is needed. As an example, a study may consist of three years of enrollment and two years of follow-up; no patients are enrolled during the last two years. Figure 12.1 shows a description of a typical phase III or IV clinical trial.

## 12.3  DESIGNING PHASE I CANCER TRIALS

Different from other phase I clinical trials, phase I clinical trials in cancer have several main features. First, the efficacy of chemotherapy or any cancer treat-

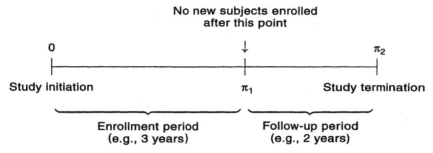

**Figure 12.1**  Phase III or IV clinical trial.

ment is, indeed, frequently associated with a nonnegligible risk of severe toxic effect, often fatal, so that ethically, the initial administration of such drugs cannot be investigated in healthy volunteers but only in cancer patients. Usually, only a small number of patients are available to be entered in phase I cancer trials. Second, these patients are at very high risk of death in the short term under all standard therapies, some of which may already have failed. At low doses, little or no efficacy is expected from the new therapy, and a slow intrapatient dose escalation is not possible. Third, there is not enough information about the drug's activity profile. In addition, clinicians often want to proceed as rapidly as possible to phase II trials with more emphasis on efficacy. The lack of information about the relationship between dose and probability of toxicity causes a fundamental dilemma inherent in phase I cancer trials: the conflict between scientific and ethical intent. We need to reconcile the risks of toxicity to patients with the potential benefit to these patients, with an efficient design that uses no more patients than necessary. Thus, a phase I cancer trial may be viewed as a problem in optimization: maximizing the dose–toxicity evaluation while minimizing the number of patients treated.

Although recently their ad hoc nature and imprecise determination of maximum tolerated dose (MTD) have been called into question, cohort-escalation trial designs, called *standard designs* have been used widely for years. In the last several years a competing design called *Fast Track* is getting more popular. These two cohort-escalation trial designs can be described as follows.

Since a slow intrapatient dose escalation is either not possible or not practical, investigators often use five to seven doses selected from "safe enough" to "effective enough." The starting dose selection of a phase I trial depends heavily on pharmacology and toxicology from preclinical studies. Although the translation from animal to human is not always a perfect correlation, toxicology studies offer an estimation range of a drug's dose–toxicity profile and the organ sites that are most likely to be affected in humans. Once the starting dose is selected, a reasonable dose escalation scheme needs to be defined. There is no single optimal or efficient escalation scheme for all drugs. Generally, dose levels are selected such that the percentage increments between successive doses diminish as the dose is increased. A modified Fibonacci sequence, with increases of 100, 67, 50, 40, and 33%, is often employed, because it follows a diminishing pattern but with modest increases.

The standard design uses three-patient cohorts and begins with one cohort at the lowest possible dose level. It observes the number of patients in the cohort who experience toxicity seriously enough to be considered a dose-limiting toxicity (DLT). The trial escalates through the sequence of doses until enough patients experience DLTs to stop the trial and declare an MTD. The dose at which the toxicity threshold is exceeded is designated the MTD. In a standard design, if no patients in a cohort experience a DLT, the trial continues with a new cohort at the next higher dose; if two or three patients experience a DLT, the trial is stopped as the toxicity threshold is exceeded and an MTD is identified; if exactly one patient experiences a DLT, a new cohort of three patients is employed at the same dose. In this second cohort, evaluated at the same dose, if no severe toxicity is observed, the dose is escalated to the next-highest level; otherwise, the trial is terminated and the dose in use at trial termination is recommended as the MTD. Note that intrapatient escalation is not used to evaluate the doses, to avoid the confounding effect of carryover from one dose to the next. We can refer to the standard design as a *three-and-three design* because at each new dose it enrolls a cohort of three patients with the option of enrolling an additional three patients evaluated at the same dose. Some slight variation of the standard design are also used in various trials.

The fast-track design is a variation of the standard design. It was created by modifying the standard design to move through low-toxic-rate doses using fewer patients. The design uses a predefined set of doses and cohorts of one or three patients, escalating through the sequence of doses using a one-patient cohort until the first DLT is observed. After that, only three-patient cohorts are used. When no DLT is observed, the trial continues at the next-highest dose with a cohort of one new patient. When a DLT is observed in a one-patient evaluation of a dose, the same dose is evaluated a second time with a cohort of three new patients, if no patient in this cohort experiences a DLT, the design moves to the next-highest dose with a new cohort of three patients, and from this point, the design progresses as a standard design. When one or more patients in a three-patient cohort experiences a DLT, the current dose is considered the MTD. If one-patient cohort is used at each dose level throughout, six patients are often tested at the very last dose. Similar to a standard design, no intrapatient escalation is allowed in a fast-track design.

There seems to be no perfect solution. The standard design is more popular and more conservative (i.e., safer); very few patients are likely to be overtreated by doses with undesirable levels of toxicity. However, in a standard design, many patients who enter early in the trial are likely to be treated suboptimally, and only a few patients may be left after an MTD is reached, especially if there were many doses below MTD. Generally, the use of a fast-track design seems very attractive because some clinicians want to proceed to a phase II trial as fast as they can, to have a first look at efficacy. The fast-track design quickly escalates through early doses that have a low expected risk of dose-limiting toxicity, thereby reducing the number of patients treated at the lowest toxicity selected in single-patient cohorts. On the other hand, the fast-track design may allow a higher percentage of patients to be treated at very high toxic doses; and the fact that it uses a single-patient cohort until the first DLT is observed seems too risky for some investigators. For more experienced investigators, the fast-track design presents an improved use of patient resources with a moderate compromise of patient safety; but safety could be a problem with inexperienced investigators who might select high doses to start with. The common problem for both designs is the lack of robustness: that the expected rate of MTD selected is strongly influenced by the doses used, and these doses may be selected arbitrarily by investigators, which makes their experience a crucial factor.

## 12.4   SAMPLE SIZE DETERMINATION FOR PHASE II TRIALS AND SURVEYS

The determination of the size of a sample is a crucial element in the design of a survey or a clinical trial. In designing any study, one of the first questions that must be answered is: How large must the sample be to accomplish the goals of the study? Depending on the study goals, the planning of sample size can be approached accordingly.

Phase II trials are the simplest. The drug, at the optimal dose (MTD) found from a previous phase I trial, is given to a small group of patients who meet predetermined inclusion criteria. The focus is often on the response rate. Because of this focus, the planning of sample size can be approached in terms of controlling the width of a desired confidence interval for the parameter of interest, the response rate.

Suppose that the goal of a study is to estimate an unknown response rate $\pi$. For the confidence interval to be useful, it must be short enough to pinpoint the value of the parameter reasonably well with a high degree of confidence. If a study is unplanned or poorly planned, there is a real possibility that the resulting confidence interval will be too long to be of any use to the researcher. In this case, we may decide to have an estimate error not exceeding $d$, an upper bound for the margin of error since the 95% confidence interval for the response rate $\pi$, a population proportion, is

$$p \pm 1.96\sqrt{\frac{p(1-p)}{n}}$$

where $p$ is the sample proportion. Therefore, our goal is expressed as

$$1.96\sqrt{\frac{p(1-p)}{n}} \le d$$

leading to the required minimum sample size:

$$n = \frac{(1.96)^2 p(1-p)}{d^2}$$

(rounded up to the next integer). This required sample size is affected by three factors:

1. The degree of confidence (i.e., 95% which yields the coefficient 1.96)
2. The maximum tolerated error or upper bound for the margin of error, $d$, determined by the investigator(s)
3. The proportion $p$ itself

This third factor is unsettling. To find $n$ so as to obtain an accurate value of the proportion, we need the proportion itself. There is no perfect, exact solution for this. Usually, we can use information from similar studies, past studies, or studies on similar populations. If no good prior knowledge about the proportion is available, we can replace $p(1-p)$ by 0.25 and use a conservative sample size estimate:

$$n_{max} = \frac{(1.96)^2(0.25)}{d^2}$$

because $n_{max} \geq n$ regardless of the value of $\pi$. Most phase II trials are small; investigators often set the maximum tolerated error or upper bound for the margin of error, $d$, at 10% (0.10) or 15%; some even set it at 20%.

***Example 12.1***   If we set the maximum tolerated error $d$ at 10%, the required minimum sample size is

$$n_{max} = \frac{(1.96)^2(0.25)}{(0.1)^2}$$

or 97 patients, which is usually too high for a small phase II trial, especially in the field of cancer research, where very few patients are available. If we set the maximum tolerated error $d$ at 15%, the required minimum sample size is

$$n_{max} = \frac{(1.96)^2(0.25)}{(0.15)^2}$$

or 43 patients.

The same method for sample size determination as above applies to surveys as well, except that for surveys we can afford to use much larger sample sizes.

We can set the maximum tolerated error at a very low level, resulting in very short confidence intervals.

***Example 12.2***   Suppose that a study is to be conducted to estimate the smoking rate among N.O.W. (National Organization for Women) members. Suppose also that we want to estimate this proportion to within 3% (i.e., $d = 0.03$) with 95% confidence.

(a) Since the current smoking rate among women in general is about 27% (0.27), we can use this figure in calculating the required sample size. This results in

$$n = \frac{(1.96)^2(0.27)(0.73)}{(0.03)^2}$$
$$= 841.3$$

or a sample of size 842 is needed.

(b) If we do not want or have the figure of 27%, we still can conservatively take

$$n_{max} = \frac{(1.96)^2(0.25)}{(0.03)^2}$$
$$= 1067.1$$

(i.e., we can sample 1068 members of N.O.W.). Note that this conservative sample size is adequate regardless of the true value $\pi$ of the unknown population proportion; values of $n$ and $n_{max}$ are closer when $\pi$ is near 0.5.

## 12.5   SAMPLE SIZES FOR OTHER PHASE II TRIALS

As pointed out in previous sections, most phase II trials are single-arm studies where we are seeking the antitumor activitity of a drug measured by response rate. But there are also a variety of other phase II trials.

Some phase II trials are randomized comparative studies. These are most likely to be cases where we have established activity for a given drug (from a previous one-arm nonrandomized trial) and wish to add another drug to that regimen. In these randomized phase II trials, the goal is to *select* the *better* treatment (the sample sizes for these phase II trials are covered in Section 12.7).

Some phase II trials deal with assessing the activity of a biologic agent where tumor response is not the main endpoint of interest. We may be attempting to determine the effect of a new agent: for example, on the prevention of a toxicity. The primary endpoint may be measured on a continuous scale. In other trials, a pharmacologic- or biologic-to-outcome correlative objective may be the target.

### 12.5.1  Continuous Endpoints

When the primary outcome of a trial is measured on a continuous scale, the focus is on the mean. Because of this focus, the planning of sample size can be approached in terms of controlling the width of a desired confidence interval for the parameter of interest, the (population) mean. The sample size determination is similar to the case when the focus is the response rate. That is, for the confidence interval to be useful, it must be short enough to pinpoint the value of the parameter, the mean, reasonably well with a high degree of confidence, say 95%. If a study is unplanned or poorly planned, there is a real possibility that the resulting confidence interval will be too long to be of any use to the researcher. In this case, we may decide to have an estimate error not exceeding $d$, an upper bound for the margin of error. With a given level of the maximum tolerated error $d$, the minimum required sample size is given by

$$n = \frac{(1.96)^2 s^2}{d^2}$$

(rounded up to the next integer). This required sample size is also affected by three factors:

1. The coefficient 1.96. As mentioned previously, a different coefficient is used for a different degree of confidence, which is set arbitrarily by the investigator; 95% is a conventional choice.
2. The maximum tolerated error $d$, which is also set arbitrarily by the investigator.
3. The variability of the population measurements, the variance. This seems like a circular problem. We want to find the size of a sample so as to estimate the mean accurately, and to do that, we need to know the variance before we have the data! Of course, the exact value of the variance is also unknown. However, we can use information from similar studies, past studies, or some reasonable upper bound. If nothing else is available, we may need to run a preliminary or pilot study. One-fourth of the range may serve as a rough estimate for the standard deviation.

### 12.5.2  Correlation Endpoints

When the parameter is a coefficient of correlation, the planning of sample size is approached differently because it is very difficult to come up with a meaningful maximum tolerated error for estimation of the coefficient of correlation. Instead of controlling the width of a desired confidence interval, the sample size determination ia approached in terms of controlling the risk of making a type II error. The decision is concerned with testing a null hypothesis,

$$H_0: \rho = 0$$

against an alternative hypothesis,

$$H_A: \rho = \rho_A$$

in which $\rho_A$ is the investigator's hypothesized value for the coefficient of correlation $\rho$ of interest. With a given level of significance $\alpha$ (usually, 0.05) and a desired statistical power $(1 - \beta$; $\beta$ is the size of type II error associated with the alternative $H_A$), the required sample size is given by

$$\sqrt{n - 3} F(\rho_A) = z_{1-\alpha} + z_{1-\beta}$$

where

$$F(\rho) = \frac{1}{2} \ln \frac{1 + \rho}{1 - \rho}$$

The transformation from $\rho$ to $F(\rho)$ is often referred to as *Fisher's transformation*, the same transformation used in forming confidence intervals in Chapter 4. Obviously, to detect a true correlation $\rho_A$ greater than 0.5, a small sample size would suffice, which is suitable in the context of phase II trials.

**Example 12.3**   Suppose that we decide to preset $\alpha = 0.05$. To design a study such that its power to detect a true correlation $\rho_A = 0.6$ is 90% (or $\beta = 0.10$), we would need only

$$F(\rho_A) = \frac{1}{2} \ln \frac{1 + \rho_A}{1 - \rho_A}$$

$$= \frac{1}{2} \ln \frac{1.6}{0.4}$$

$$= 0.693$$

$$\sqrt{n - 3} F(\rho_A) = z_{1-\alpha} + z_{1-\beta}$$

$$\sqrt{n - 3}(0.693) = 1.96 + 1.28$$

or $n = 25$ subjects.

*Example 12.4*  Perhaps it is simpler to see the sample size determination concerning a continuous endpoint in the context of a survey. Suppose that a study is to be conducted to estimate the average birth weight of babies born to mothers addicted to cocaine. Suppose also that we want to estimate this average to within $\frac{1}{2}$ lb with 95% confidence. This goal specifies two quantities:

$$d = 0.5$$

$$\text{coefficient} = 1.96$$

What value should be used for the variance? Information from normal babies may be used to estimate $s$. The rationale here is that the addiction affects every baby almost uniformly; this may result in a smaller average, but the variance is unchanged. Suppose that the estimate from normal babies is $\sigma \simeq 2.5$ lb, then the required sample size is

$$n = \frac{(1.96)^2 (2.5)^2}{(0.5)^2}$$

$$= 97 \text{ approximately}$$

## 12.6  ABOUT SIMON'S TWO-STAGE PHASE II DESIGN

Phase I trials treat only three to six patients per dose level according to the standard design. In addition, those patients may be diverse with regard to their cancer diagnosis; consequently, phase I trials provide little or no information about efficacy. A phase II trial is the first step in the study of antitumor effects of an investigational drug. The aim of a phase II trial of a new anticancer drug is to determine whether the drug has sufficient antitumor activity to warrant further development. Further development may mean combining the drug with other drugs, or initiation of a phase III trial. However, these patients are often at high risk of dying from cancers if not treated effectively. Therefore, it is desirable to use as few patients as possible in a phase II trial *if* the regimen under investigation is, in fact, low antitumor activity. When such *ethical concerns* are of high priority, investigators often choose the *Simon two-stage design*:

1. A group of $n_1$ patients are enrolled in the first stage. If $r_1$ or fewer of these $n_1$ patients respond to the drug, the drug is rejected and the trial is terminated; if more than $r_1$ responses are observed, investigators proceed to stage II and enroll $n_2$ more patients.
2. After stage II, if $r$ or fewer responses are observed, including those in stage I, the drug is rejected; If more than $r$ responses are observed, the drug is recommended for further evaluation.

Simon's design is based on testing a null hypothesis $H_0$: $\pi \leq \pi_0$, that the true response rate $\pi$ is less than some low and uninteresting level $\pi_0$, against an alternative hypothesis $H_A$: $\pi \geq \pi_A$, that the true response rate $\pi$ exceeds a certain desirable target level $\pi_A$, which, if true, would allow us to consider the drug to have sufficient antitumor activity to warrant further development. The design parameters $n_1$, $r_1$, $n_2$, and $r$ are determined so as to minimize the number of patients $n = n_1 + n_2$ if $H_0$ is true: The drug, in fact, has low antitumor activity. The option that allows early termination of the trial satisfies high-priority ethical concerns. The derivation is more advanved and there are no closed-form formulas for the design parameters $n_1$, $r_1$, $n_2$, and $r$. Beginning users can look for help to Simon's two-stage design, if appropriate.

## 12.7    PHASE II DESIGNS FOR SELECTION

Some randomized phase II trials do not fit the framework of tests of significance. In preforming statistical tests or tests of significance, we have the option to declare a trial not significant when data do not provide enough support for a treatment difference. In those cases we decide not to pursue the new treatment, and we do not choose the new treatment because it does not prove any better than the placebo effect or that of a standard therapy. In some cancer areas we may not have a standard therapy, or if we do, some subgroups of patients may have failed using standard therapies. Suppose further that we have established activity for a given drug from a previous one-arm nonrandomized trial, and the only remaining question is scheduling: for example, daily versus one-every-other-day schedules. Or we may wish to add another drug to that regimen to improve it. In these cases we do not have the option to declare the trial not significant because (1) one of the treatments or schedules has to be chosen (because patients have to be treated), and (2) it is inconsequential to choose one of the two treatments/schedules even if they equally efficacious. The aim of these randomized trials is to choose the *better* treatment.

### 12.7.1    Continuous Endpoints

When the primary outcome of a trial is measured on a continuous scale, the focus is on the mean. At the end of the study, we select the treatment or schedule with the *larger* sample mean. But first we have to define what we mean by *better treatment*. Suppose that treatment 2 is said to be *better* than treatment 1 if

$$\mu_2 - \mu_1 \geq d$$

where $d$ is the magnitude of the difference between $\mu_2$ and $\mu_1$ that is deemed to be important; the quantity $d$ is often called the *minimum clinical significant difference*. Then we want to make the *right* selection by making sure that at the end of the study, the *better* treatment will be the one with the *larger* sample mean. This goal is achieved by imposing a condition,

$$\Pr(\bar{x}_2 \geq \bar{x}_1 \mid \mu_2 - \mu_1 = d) = 1 - \alpha$$

For example, if we want to be 99% sure that the better treatment will be the one with the larger sample mean, we can preset $\alpha = 0.01$. To do that, the total sample size must be at least

$$N = 4(z_{1-\alpha})^2 \frac{\sigma^2}{d^2}$$

assuming that we conduct a balanced study with each group consisting of $n = N/2$ subjects. To calculate this minimum required total sample size, we need the variance $\sigma$. The exact value of $\sigma^2$ is unknown; we may depend on prior knowledge about one of the two arms from a previous study or use some upper bound.

### 12.7.2    Binary Endpoints

When the primary outcome of a trial is measured on a binary scale, the focus is on a proportion, the response rate. At the end of the study, we select the treatment or schedule with the *larger* sample proportion. But first we have to define what we mean by *better treatment*. Suppose that treatment 2 is said to be *better* than treatment 1 if

$$\pi_2 - \pi_1 \geq d$$

where $d$ is the magnitude of the difference between $\pi_2$ and $\pi_1$ that is deemed to be important; the quantity $d$ is often called the minimum clinical significant difference. Then we want to make the *right* selection by making sure that at the end of the study, the *better* treatment will be the one with the *larger* sample proportion. This goal is achieved by imposing a condition,

$$\Pr(p_2 \geq p_1 \mid \pi_2 - \pi_1 = d) = 1 - \alpha$$

where the $p$'s are sample proportions. For example, if we want to be 99% sure that the better treatment will be the one with the larger sample proportion, we can preset $\alpha = 0.01$. To do that, the total sample size must be at least

$$N = 4(z_{1-\alpha})^2 \frac{\bar{\pi}(1 - \bar{\pi})}{(\pi_2 - \pi_1)^2}$$

assuming that we conduct a balanced study with each group consisting of

$n = N/2$ subjects. In this formula, $\bar{\pi}$ is the average proportion:

$$\bar{\pi} = \frac{\pi_1 + \pi_2}{2}$$

It is obvious that the problem of planning sample size is more difficult, and a good solution requires a deeper knowledge of the scientific problem: a good idea of the magnitude of the proportions $\pi_1$ and $\pi_2$ themselves. In many caes, that may be impractical at this stage.

***Example 12.5*** Suppose that for a certain problem, $d = 5$ and it is estimated that $\sigma^2 = 36$. Then if we want to be 95% sure that the better treatment will be the one with the larger sample mean, we would need

$$N = 4(1.96)^2 \frac{36}{(5)^2}$$

$$\simeq 24$$

with 12 subjects in each of the two groups. It will be seen later that with similar specifications, a phase III design would require a larger sample to detect a treatment difference of $d = 5$ using a statistical test of significance.

## 12.8    TOXICITY MONITORING IN PHASE II TRIALS

In a clinical trial of a new treatment, severe side effects may be a problem, and the trial may have to be stopped if the incidence is too high. For example, bone marrow transplantation is a complex procedure that exposes patients to a high risk of a variety of complications, many of them fatal. Investigators are willing to take these risks because they are in exchange for much higher risks associated with the leukemia or other disease for which the patient is being treated; for many of these patients, standard and safer treatments have failed. Investigators often have to face this problem of severe side effects and contemplate stopping phase II trials. Phase I trials focus on safety of a new investigational medicine, and phase II trials are small trials to evaluate efficacy. However, phase I trials are conducted with a small number of patients; therefore, safety is still a major concern in a phase II trial. If either the accrual or the treatment occurs over an extended period of time, we need to anticipate the need for a decision to halt the trial if an excess of severe side effects occurs.

The following monitoring rule was derived using a more advanced statistical method called the *sequential probability ratio test*. Basically, patients are enrolled, randomized if needed, and treated; and the trial proceeds continuously until the number of patients with severe side effects meets the criterion judged as excessive and the trial is stopped. The primary parameter is the incidence rate $\pi$ of severe side effects as defined specifically for the trial: for example, toxicity grade III or IV. As with any other statistical test of significant, the decision is concerned with testing a null hypothesis

$$H_0: \pi = \pi_0$$

against an alternative hypothesis

$$H_A: \pi = \pi_A$$

in which $\pi_0$ is the investigator's hypothesized value for the incidence rate $\pi$ of severe side effects, often formulated based on knowledge from previous phase I trial results. The other figure, $\pi_A$, is the *maximum tolerated level* for the incidence rate $\pi$ of severe side effects. The trial has to be stopped if the incidence rate $\pi$ of severe side effects exceeds $\pi_A$. In addition to the null and alternative parameters, $\pi_0$ and $\pi_A$, a stopping rule also depends on the chosen level of significance $\alpha$ (usually, 0.05) and the desired statistical power $(1 - \beta)$; $\beta$ is the size of type II error associated with the alternative $H_A: \pi = \pi_A$. Power is usually preset at 80 or 90%. With these specifications, we monitor for the side effects by sequentially counting the number of events $e$ (i.e., number of patients with severe side effects) and the number of evaluable patients $n(e)$ at which the $e$th event is observed. The trial is stopped when this condition is first met:

$$n(e) = \frac{\ln(1 - \beta) - \ln \alpha + e[\ln(1 - \pi_A) - \ln(1 - \pi_0) - \ln \pi_A + \ln \pi_0]}{\ln(1 - \pi_A) - \ln(1 - \pi_0)}$$

In other words, the formula above gives us the maximum number of evaluable patients $n(e)$ at which the trial has to be stopped if $e$ events have been observed.

Some phase II trials may be randomized; however, even in these randomized trials, toxicity monitoring should be done separately for each study arm. That is, if the side effect can reasonably occur in only one of the arms of the study, probably the arm treated by the new therapy, the incidence in that group alone is considered. Otherwise, the sensitivity of the process to stop the trial would be diluted by inclusion of the other group. Sometimes the goal is to compare two treatments according to some composite hypothesis that the new treatment is equally effective but has less toxicity. In those cases, both efficacy and toxicity are endpoints, and the analysis should be planned accordingly, but the situations are not that of monitoring in order to stop the trial as intended in the rule above.

*Example 12.6* Suppose that in the planning for a phase II trial, an investigator (or clinicians in a study committee) decided that $\pi_0 = 3\%$ (0.03) based on some prior knowlege and that $\pi_A = 15\%$ (0.15) should be the upper limit that can be tolerated (as related to the risks of the disease itself). For this illustrative example, we find that $n(1) = -7$, $n(2) = 5$, $n(3) = 18$, $n(4) = 31$, and so on, when we preset the level of significance at 0.05 and statistical power at 80%. In other words, we stop the trial if there are two events among the first five evaluable patients, three events among the first 18 patients, four events among the first 31 patients, and so on. Here we use only positive solutions and the integer proportion of each solution from the equation above. The negative solution $n(1) = -7$ indicates that the first event will not result in stopping the trial (because it is judged as not excessive yet).

***Example 12.7***    The stopping rule would be more stringent if we want higher (statistical) power or if incidence rates are higher. For example:

(a) With $\pi_0 = 3\%$ and $\pi_A = 15\%$ as set previously, but if we preset the level of significance at 0.05 and statistical power at 90%, the results become $n(1) = -8$, $n(2) = 4$, $n(3) = 17$, $n(4) = 30$, and so on. That is, we would stop the trial if there are two events among the first four evaluable patients, three events among the first 17 patients, four events among the first 30 patients, and so on.

(b) On the other hand, if we keep the level of significance at 0.05 and statistical power at 80%, but if we decide on $\pi_0 = 5\%$ and $\pi_A = 20\%$, the results become $n(1) = -7$, $n(2) = 2$, $n(3) = 11$, $n(4) = 20$, and so on. That is, we would stop the trial if there are two events among the first two evaluable patients, three events among the first 11 patients, four events among the first 20 patients, and so on.

It can be seen that the rule accelerates faster with higher rates than with higher power.

## 12.9  SAMPLE SIZE DETERMINATION FOR PHASE III TRIALS

The determination of the size of a sample is a crucial element in the design of a study, whether it is a survey or a clinical trial. In designing any study, one of the first questions that must be answered is: How large must a sample be to accomplish the goals of the study? Depending on the study goals, the planning of sample size can be approached in two different ways: either in terms of controlling the width of a desired confidence interval for the parameter of interest, or in terms of controlling the risk of making type II errors. In Section 12.3, the planning of sample size was approached in terms of controlling the width of a desired confidence interval for the parameter of interest, the response rate in a phase II trial. However, phase III and IV clinical trials are conducted not for parameter estimation but for the comparison of two treatments (e.g., a new therapy versus a placebo or a standard therapy). Therefore, it is more suitable to approach the planning of sample size in terms of controlling the risk of making type II errors. Since phase I and II trials, especially phase I trials, are specific for cancer research but phase III and IV trials are applicable in any field, we will cover this part in more general terms and include examples in fields other than cancers.

Recall that in testing a null hypothesis, two types of errors are possible. We might reject $H_0$ when in fact $H_0$ is true, thus committing a type I error. However, this type of error can be controlled in the decision-making process; conventionally, the probability of making this mistake is set at $\alpha = 0.05$ or $0.01$. A type II error occurs when we fail to reject $H_0$ even though it is false. In the drug-testing example above, a type II error leads to our inability to recognize the effectiveness of the new drug being studied. The probability of committing a type II error is denoted by $\beta$, and $1 - \beta$ is called the *power* of a statistical test. Since the power is the probability that we will be able to support our research claim (i.e., the alternative hypothesis) when it is right, studies should be designed to have high power. This is achieved through the planning of sample size. The method for sample size determination is not unique; it depends on the endpoint and its measurement scale.

### 12.9.1  Comparison of Two Means

In many studies, the endpoint is on a continuous scale. For example, a researcher is studying a drug that is to be used to reduce the cholesterol level in adult males aged 30 and over. Subjects are to be randomized into two groups, one receiving the new drug (group 1), and one a look-alike placebo (group 2). The response variable considered is the change in cholesterol level before and after the intervention. The null hypothesis to be tested is

$$H_0: \mu_1 = \mu_2$$

vs.

$$H_A: \mu_2 > \mu_1 \quad \text{or} \quad \mu_2 \neq \mu_1$$

Data would be analyzed using, for example, the two-sample $t$ test of Chapter 7. However, before any data are collected, the crucial question is: How large a total sample should be used to conduct the study?

In the comparison of two population means, $\mu_1$ versus $\mu_2$, the required minimum total sample size is calculated from

$$N = 4(z_{1-\alpha} + z_{1-\beta})^2 \frac{\sigma^2}{d^2}$$

assuming that we conduct a balanced study with each group consisting of $n = N/2$ subjects. This required total sample size is affected by four factors:

1. The size $\alpha$ of the test. As mentioned previously, this is set arbitrarily by the investigator; conventionally, $\alpha = 0.05$ is often used. The quantity $z_{1-\alpha}$

in the formula above is the percentile of the standard normal distribution associated with a choice of $\alpha$; for example, $z_{1-\alpha} = 1.96$ when $\alpha = 0.05$ is chosen. In the process of sample size determination, statistical tests, such as two-sample $t$ test, are usually planned as two-sided. However, if a one-sided test is planned, this step is changed slightly; for example, we use $z = 1.65$ when $\alpha = 0.05$ is chosen.

2. The desired power $(1 - \beta)$ (or probability of committing a type II error $\beta$). This value is also selected by the investigator; a power of 80 or 90% is often used.

3. The quantity

$$d = |\mu_2 - \mu_1|$$

which is the magnitude of the difference between $\mu_1$ and $\mu_2$ that is deemed to be important. The quantity $d$ is often called the minimum clinical significant difference and its determination is a clinical decision, not a statistical decision.

4. The variance $\sigma^2$ of the population. This variance is the only quantity that is difficult to determine. The exact value of $\sigma^2$ is unknown; we may use information from similar studies or past studies or use an upper bound. Some investigators may even run a preliminary or pilot study to estimate $\sigma^2$; but estimate from a small pilot study may only be as good as any guess.

*Example 12.8*   Suppose that a researcher is studying a drug which is used to reduce the cholesterol level in adult males aged 30 or over and wants to test it against a placebo in a balanced randomized study. Suppose also that it is important that a reduction difference of 5 be detected ($d = 5$). We decide to preset $\alpha = 0.05$ and want to design a study such that its power to detect a difference between means of 5 is 95% (or $\beta = 0.05$). Also, the variance of cholesterol reduction (with placebo) is known to be about $s^2 \simeq 36$.

$$\alpha = 0.05 \rightarrow z_{1-\alpha} = 1.96 \text{ (two-sided test)}$$

$$\beta = 0.05 \rightarrow z_{1-\beta} = 1.65$$

leading to the required total sample size:

$$N = (4)(1.96 + 1.65)^2 \frac{36}{(5)^2}$$

$$\simeq 76$$

Each group will have 38 subjects.

*Example 12.9*  Suppose that in Example 12.8, the researcher wanted to design a study such that its power to detect a difference between means of 3 is 90% (or $\beta = 0.10$). In addition, the variance of cholesterol reduction (with placebo) is not known precisely, but it is reasonable to assume that it does not exceed 50. As in Example 12.8, let's set $\alpha = 0.05$, leading to

$$\alpha = 0.05 \rightarrow z_{1-\alpha} = 1.96 \text{ (two-sided test)}$$

$$\beta = 0.10 \rightarrow z_{1-\beta} = 1.28$$

Then using the upper bound for variance (i.e., 50), the required total sample size is

$$N = (4)(1.96 + 1.28)^2 \frac{50}{(3)^2}$$

$$\simeq 234$$

Each group will have 117 subjects.

Suppose, however, that the study was actually conducted with only 180 subjects, 90 randomized to each group (it is a common situation that studies are underenrolled). From the formula for sample size, we can solve and obtained

$$z_{1-\beta} = \sqrt{\frac{N}{4} \frac{d^2}{\sigma^2}} - z_{1-\alpha}$$

$$= \sqrt{\frac{180}{4} \frac{3^2}{50}} - 1.96$$

$$= 0.886$$

corresponding to a power of approximately 81%.

### 12.9.2  Comparison of Two Proportions

In many other studies, the endpoint may be on a binary scale; so let us consider a similar problem where we want to design a study to compare two proportions. For example, a new vaccine will be tested in which subjects are to be randomized into two groups of equal size: a control (nonimmunized) group (group 1), and an experimental (immunized) group (group 2). Subjects in both control and experimental groups will be challenged by a certain type of bacteria and we wish to compare the infection rates. The null hypothesis to be tested is

$$H_0: \pi_1 = \pi_2$$

versus

$$H_A: \pi_1 < \pi_2 \quad \text{or} \quad \pi_1 \neq \pi_2$$

How large a total sample should be used to conduct this vaccine study? Suppose that it is important to detect a reduction of infection rate

$$d = \pi_2 - \pi_1$$

If we decide to preset the size of the study at $\alpha = 0.05$ and want the power $(1 - \beta)$ to detect the difference $d$, the required sample size is given by the complicated formula

$$N = 4(z_{1-\alpha} + z_{1-\beta})^2 \frac{\bar{\pi}(1 - \bar{\pi})}{(\pi_2 - \pi_1)^2}$$

or

$$z_{1-\beta} = \frac{\sqrt{N}|\pi_2 - \pi_1|}{2\sqrt{\bar{\pi}(1 - \bar{\pi})}} - z_{1-\alpha}$$

In this formula the quantities $z_{1-\alpha}$ and $z_{1-\beta}$ are defined as in Section 12.9.1 and $\bar{\pi}$ is the average proportion:

$$\bar{\pi} = \frac{\pi_1 + \pi_2}{2}$$

It is obvious that the problem of planning sample size is more difficult and that a good solution requires a deeper knowledge of the scientific problem: a good idea of the magnitude of the proportions $\pi_1$ and $\pi_2$ themselves.

***Example 12.10*** Suppose that we wish to conduct a clinical trial of a new therapy where the rate of successes in the control group was known to be about 5%. Further, we would consider the new therapy to be superior—cost, risks, and other factors considered—if its rate of successes is about 15%. Suppose also that we decide to preset $\alpha = 0.05$ and want the power to be about 90% (i.e., $\beta = 0.10$). In other words, we use

$$z_{1-\alpha} = 1.96$$

$$z_{1-\beta} = 1.28$$

From this information, the total sample size required is

$$N = (4)(1.96 + 1.28)^2 \frac{(0.10)(0.90)}{(0.15 - 0.05)^2}$$

$$\simeq 378 \text{ or } 189 \text{ patients in each group}$$

*Example 12.11*  A new vaccine will be tested in which subjects are to be randomized into two groups of equal size: a control (unimmunized) group and an experimental (immunized) group. Based on prior knowledge about the vaccine through small pilot studies, the following assumptions are made:

1. The infection of the control group (when challenged by a certain type of bacteria) is expected to be about 50%:

$$\pi_2 = 0.50$$

2. About 80% of the experimental group is expected to develop adequate antibodies (i.e., at least a twofold increase). If antibodies are inadequate, the infection rate is about the same as for a control subject. But if an experimental subject has adequate antibodies, the vaccine is expected to be about 85% effective (which corresponds to a 15% infection rate against the challenged bacteria).

Putting these assumptions together, we obtain an expected value of $\pi_1$:

$$\pi_1 = (0.80)(0.15) + (0.20)(0.50)$$

$$= 0.22$$

Suppose also that we decide to preset $\alpha = 0.05$ and want the power to be about 95% (i.e., $\beta = 0.05$). In other words, we use

$$z_{1-\alpha} = 1.96$$

$$z_{1-\beta} = 1.65$$

From this information, the total sample size required is

$$N = (4)(1.96 + 1.65)^2 \frac{(0.36)(0.64)}{(0.50 - 0.22)^2}$$

$$\simeq 154$$

so that each group will have 77 subjects. In this solution we use

$$\bar{\pi} = 0.36$$

the average of 22% and 50%.

### 12.9.3  Survival Time as the Endpoint

When patients' survivorship is considered as the endpoint of a trial, the problem may look similar to that of comparing two proportions. For example, one can focus on a conventional time span, say five years, and compare the two survival rates. The comparision of the five-year survival rate from an experimental treatment versus the five-year survival rate from a standard regimen, say in the analysis of results from a trial of cancer treatments, fits the framework of a comparison of two population proportions as seen in Section 12.4.2. The problem is that for survival data, studies have staggered entry and subjects are followed for varying lengths of time; they do not have the same probability for the event to occur. Therefore, similar to the process of data analysis, the process of sample size determination should be treated differently for trials where the endpoint is binary from trials when the endpoint is survival time.

As seen in Chapter 11, the log-rank test has become commonly used in the analysis of clinical trials where event or outcome become manifest only after a prolonged time interval. The method for sample size determination, where the difference in survival experience of the two groups in a clinical trial is tested using the log-rank test, proceeds as followed. Suppose that the two treatments give rise to survival rates $P_1$ (for the experimental treatment) and $P_2$ (for the standard regimen), respectively, at some conventional time point: say, five years. If the ratio of the two hazards (or risk functions) in the two groups are assumed not changing with time in a proportional hazards model (PHM) and is $\theta{:}1$, the quantities $P_1$, $P_2$, and $\theta$ are related to each other by

$$\theta = \frac{\ln P_1}{\ln P_2}$$

1. The total number of events $d$ from both treatment arms needed to be observed in the trial is given by

$$d = (z_{1-\alpha} + z_{1-\beta})^2 \left(\frac{1+\theta}{1-\theta}\right)^2$$

   In this formula the quantities $z_{1-\alpha}$ and $z_{1-\beta}$ are defined as in Section 12.9.2, and $\theta$ is the hazards ratio.

2. Once the total number of events $d$ from both treatment arms has been estimated, the total number of patients $N$ required in the trial can be calculated from

$$N = \frac{2d}{2 - P_1 - P_2}$$

   assuming equal numbers of patients randomized into the two treatment arms, with $N/2$ in each group. In this formula the quantities $P_1$ and $P_2$ are five-year (or two- or three-year) survival rates.

*Example 12.12*  Consider the planning of a clinical trial of superficial bladder cancer. With the current method of treatment (resection of tumor at cystoscopy), the recurrence-free rate [i.e., the survival rate when the event under investigation is recurrency (the tumor comes back)] is 50% at two years. Investigators hope to increase this to at least 70% using intravesical chemotherapy (treatment by drug) immediately after surgery at the time of cystoscopy. This alternative hypothesis is equivalent to a hypothesized hazards ratio of

$$\theta = \frac{\ln(0.5)}{\ln(0.7)}$$

$$= 1.94$$

Suppose also that we decide to preset $\alpha = 0.05$ (two-sided) and want the power to be about 90% (i.e., $\beta = 0.10$). In other words, we use

$$z_{1-\alpha} = 1.96$$

$$z_{1-\beta} = 1.28$$

From this information, the required total sample size is

$$d = (z_{1-\alpha} + z_{1-\beta})^2 \left(\frac{1+\theta}{1-\theta}\right)^2$$

$$= (1.96 + 1.28)^2 \left(\frac{1 + 1.94}{1 - 1.94}\right)^2$$

$$= 103 \text{ events}$$

$$N = \frac{2d}{2 - P_1 - P_2}$$

$$= \frac{(2)(103)}{2 - 0.5 - 0.7}$$

$$= 258 \text{ patients, } 129 \text{ in each group}$$

**Example 12.13** In the example above, suppose that the study was under-enrolled and was actually conducted with only 200 patients, 100 randomized to each of the two treatment arms. Then we have

$$d = \frac{(N)(2 - P_1 - P_2)}{2}$$

$$= 80 \text{ events}$$

$$(z_{1-\alpha} + z_{1-\beta})^2 = \frac{d}{[(1+\theta)/(1-\theta)]^2}$$

$$= 8.178$$

$$z_{1-\beta} = \sqrt{8.178} - 1.96$$

$$= 0.90$$

corresponding to a power of approximately 84%.

## 12.10   SAMPLE SIZE DETERMINATION FOR CASE–CONTROL STUDIES

In a typical case–control study, cases of a specific disease are ascertained as they arise from population-based registers or lists of hospital admissions, and controls are sampled either as disease-free people from the population at risk, or as hospitalized patients having a diagnosis other than the one under study. Then in the analysis, we compare the exposure histories of the two groups. In other words, a typical case–control study fits the framework of a two-arm randomized phase III trials. However, the sample determination is somewhat more complicated, for three reasons:

1. Instead of searching for a difference of two means or proportions as in the case of a phase III trial, the alternative hypothesis of a case–control study is postulated in the form of a relative risk.
2. It must be decided whether to design a study with equal or unequal sample sizes because in epidemiologic studies, there are typically a small number of cases and a large number of potential controls to select from.
3. It must be decided whether to design a matched or an unmatched study.

For example, we may want to design a case–control study to detect a relative risk, due to a binary exposure, of 2.0, and the size of the control group is twice the number of the cases. Of course, the solution also depends on the endpoint and its measurement scale; so let's consider the two usual categories one at a time and some simple configurations.

### 12.10.1   Unmatched Designs for a Binary Exposure

As mentioned previously, the data analysis is similar to that of a phase III trial where we want to compare two proportions. However, in the design stage, the alternative hypothesis is formulated in the form of a relative risk $\theta$. Since we cannot estimate or investigate relative risk using a case–control design, we would treat the given number $\theta$ as an *odds ratio*, the ratio of the odds of being exposed by a case $\pi_1/(1 - \pi_1)$ divided by the odds of being exposed by a control $\pi_0/(1 - \pi_0)$. In other words, from the information given, consisting of the exposure rate of the control group $\pi_0$ and the approximated odds ratio due to exposure $\theta$, we can obtain the two proportions $\pi_0$ and $\pi_1$. Then the process of sample size determination can proceed similar to that of a phase III trial. For example, if we want to plan for a study with equal sample size, $N/2$ cases and $N/2$ controls, the total sample size needed should be at least

$$N = 4(z_{1-\alpha} + z_{1-\beta})^2 \frac{\bar{\pi}(1 - \bar{\pi})}{(\pi_1 - \pi_0)^2}$$

where

$$\pi_1 = \frac{\theta \pi_0}{1 + (\theta - 1)\pi_0}$$

The problem is more complicated if we plan for groups with unequal sample sizes. First, we have to specify the allocation of sizes:

$$n_1 = w_1 N$$

$$n_0 = w_0 N$$

$$w_1 + w_0 = 1.0$$

where $N$ is the total sample size needed. For example, if we want the size of the control group to be three times the number of the cases, $w_1 = 0.25$ and $w_0 = 0.75$. Then the the total sample size needed $N$ can be obtained from the formula

$$\sqrt{N}|\pi_1 - \pi_0| = z_{1-\alpha}\sqrt{\bar{\pi}(1 - \bar{\pi})(w_1^{-1} + w_0^{-1})}$$

$$+ z_{1-\beta}\sqrt{\pi_1(1 - \pi_1)w_1^{-1} + \pi_0(1 - \pi_0)w_0^{-1}}$$

where $\bar{\pi}$ is the weighted average:

$$\bar{\pi} = w_1\pi_1 + w_0\pi_0$$

***Example 12.14***   Suppose that an investigator is considering designing a case–control study of a potential association between congenital heart defect and the use of oral contraceptives. Suppose also that approximately 30% of women of childbearing age use oral contraceptives within three months of a conception, and suppose that a relative risk of $\theta = 3$ is hypothesized. We also decide to preset $\alpha = 0.05$ and want to design a study with equal sample sizes so that its power to the hypothesized relative risk of $\theta = 3$ is 90% (or $\beta = 0.10$).

First, the exposure rate for the cases, the percent of women of childbearing age who use oral contraceptives within three months of a conception, is obtained from

$$\pi_1 = \frac{\theta \pi_0}{1 + (\theta - 1)\pi_0}$$

$$= \frac{(3)(0.3)}{(3 - 1)(0.3)}$$

$$= 0.5625$$

and from

$$\bar{\pi} = \frac{0.3 + 0.5625}{2}$$

$$= 0.4313$$

$$\alpha = 0.05 \rightarrow z_{1-\alpha} = 1.96 \text{ (two-sided test)}$$

$$\beta = 0.10 \rightarrow z_{1-\beta} = 1.28$$

we obtain a required total sample size of

$$N = 4(z_{1-\alpha} + z_{1-\beta})^2 \frac{\bar{\pi}(1 - \bar{\pi})}{(\pi_1 - \pi_0)^2}$$

$$= (4)(1.96 + 1.28)^2 \frac{(0.4313)(0.5687)}{(0.2625)^2}$$

$$\simeq 146$$

or 73 cases and 73 controls are needed.

***Example 12.15***  Suppose that all specifications are the same as in Example 12.14, but we design a study in which the size of the control group is four times the number of the cases. Here we have

$$\bar{\pi} = (0.3)(0.8) + (0.5625)(0.2)$$

$$= 0.3525$$

leading to a required total sample size $N$, satisfying

$$\sqrt{N}(0.5625 - 0.3) = (1.96)\sqrt{(0.3525)(0.6475)(5 + 1.25)}$$

$$+ (1.28)\sqrt{(0.3)(0.7)(1.25) + (0.5625)(0.4375)(5)}$$

$$N \simeq 222$$

or 45 cases and 177 controls. It can be seen that the study requires a larger number of subjects, 222 as compared to 146 subjects in Example 12.14; however, it may be easier to implement because it requires fewer cases, 45 to 73.

### 12.10.2    Matched Designs for a Binary Exposure

The design of a matched case–control study of a binary exposure is also specified by the very same two parameters: the exposure rate of the control group $\pi_0$ and the relitive risk associated with the exposure $\theta$. The problem, however, becomes more complicated because the analysis of a matched case–control study of a binary exposure uses only *discordant pairs*, pairs of subjects where the exposure histories of the case and his or her matched control are different. We have to go through two steps, first to calculate the number of discordant pairs $m$, then the number of pairs $M$; the total number of subjects is $N = 2M$, $M$ cases and $M$ controls.

The exposure rate of the cases is first calculated using the same previous formula:

$$\pi_1 = \frac{\theta \pi_0}{1 + (\theta - 1)\pi_0}$$

Then, given specified levels of type I and type II errors $\alpha$ and $\beta$, the number of discordant pairs $m$ required to detect a relative risk $\theta$, treated as an approximate odds ratio, is obtained from

$$m = \frac{[(z_{1-\alpha}/2) + z_{1-\beta}\sqrt{P(1 - P)}]^2}{(P - 0.5)^2}$$

where

$$P = \frac{\theta}{1 + \theta}$$

Finally, the total number of pairs $M$ is given by

$$M = \frac{m}{\pi_0(1 - \pi_1) + \pi_1(1 - \pi_0)}$$

**Example 12.16**  Suppose that an investigator is considering designing a case–control study of a potential association between endometrial cancer and exposure to estrogen (whether ever taken). Suppose also that the exposure rate of controls is estimated to be about 40% and that a relative risk of $\theta = 4$ is hypothesized. We also decide to preset $\alpha = 0.05$ and to design a study large enough so that its power regarding the hypothesized relative risk above is 90% (or $\beta = 0.10$). We also plan a 1:1 matched design; matching criteria are age, race, and county of residence.

First, we obtain the exposure rate of the cases and the $z$ values using specified levels of type I and type II errors:

$$\pi_1 = \frac{\theta \pi_0}{1 + (\theta - 1)\pi_0}$$

$$= \frac{(4)(0.4)}{1 + (3)(0.4)}$$

$$= 0.7373$$

$$\alpha = 0.05 \rightarrow z_{1-\alpha} = 1.96 \text{ (two-sided test)}$$

$$\beta = 0.10 \rightarrow z_{1-\beta} = 1.28$$

The number of discordant pairs is given by

$$P = \frac{\theta}{1 + \theta}$$

$$= 0.80$$

$$m = \frac{[(z_{1-\alpha}/2) + z_{1-\beta}\sqrt{P(1 - P)}]^2}{(P - 0.5)^2}$$

$$= \frac{[(1.96/2) + 1.28\sqrt{(0.80)(0.20)}]^2}{0.30^2}$$

$$\simeq 25$$

$$M = \frac{m}{\pi_0(1 - \pi_1) + \pi_1(1 - \pi_0)}$$

$$= \frac{38}{(0.4)(0.2627) + (0.7373)(0.6)}$$

$$\simeq 46$$

that is pairs, 46 cases and 46 matching controls.

### 12.10.3   Unmatched Designs for a Continuous Exposure

When the risk factor under investigation in a case–control study is measured on a continuous scale, the problem is similar to that of a phase III trial where we want to compare two population means as seen in Section 12.7.1. Recall that in a comparison of two population means, $\mu_1$ versus $\mu_2$, the required minimum total sample size is calculated from

$$N = 4(z_{1-\alpha} + z_{1-\beta})^2 \frac{\sigma^2}{d^2}$$

assuming that we conduct a balanced study with each group consisting of $n = N/2$ subjects. Besides the level of significance $\alpha$ and the desired power $(1 - \beta)$, this required total sample size is affected by the variance $\sigma^2$ of the population and the quantity

$$d = |\mu_1 - \mu_0|$$

which is the magnitude of the difference between $\mu_1$, the mean for the cases, and $\mu_0$, the mean for the controls, that is deemed to be important. To put it in a different way, besides the level of significance $\alpha$ and the desired power $(1 - \beta)$, the required total sample size depends on the ratio $d/\sigma$. You will see this similarity in the design of a case–control study with a continuous risk factor. When the risk factor under investigation in a case–control study is measured on a continuous scale, the data are analyzed using the method of logistic regression (Chapter 9). However, as pointed out in Section 9.3, when the cases and controls are assumed to have the same variance $\sigma^2$, the logistic regression model can be written as

$$\text{logit} = \ln \frac{p_x}{1 - p_x}$$

$$= \text{constant} + \frac{\mu_1 - \mu_0}{\sigma^2} x$$

Under this model, the log of the odds ratio associated with a 1-unit increase in the value of the risk factor is

$$\beta_1 = \frac{\mu_1 - \mu_0}{\sigma^2}$$

Therefore, the log of the odds ratio associated with a 1-standard deviation increase in the value of the risk factor is

$$\beta_1 = \frac{\mu_1 - \mu_0}{\sigma}$$

which is the same as the ratio $d/\sigma$ above.

In the design of case–control studies with a continuous risk factor, the key parameter is the log of the odds ratio $\theta$ associated with a 1-standard deviation change in the value of the covariate. Consider a level of significance $\alpha$ and statistical power $1 - \beta$.

1. If we plan a balanced study with each group consisting of $n = N/2$ subjects, the total sample size $N$ is given by

$$N = \frac{4}{(\ln \theta)^2} (z_{1-\alpha} + z_{1-\beta})^2$$

2. If we allocate different sizes to the cases and the controls,

$$n_1 = w_1 N$$

$$n_0 = w_0 N$$

$$w_1 + w_0 = 1.0$$

the total sample size $N$ is given by

$$N = \frac{1}{(\ln \theta)^2} \left( \frac{1}{w_1} + \frac{1}{w_0} \right) (z_{1-\alpha} + z_{1-\beta})^2$$

For example, if we want the size of the control group to be three times the number of the cases, $w_1 = 0.25$ and $w_0 = 0.75$.

***Example 12.17*** Suppose that an investigator is considering designing a case–control study of a potential association between coronary heart disease and serum cholesterol level. Suppose further that it is desirable to detect an odds ratio of $\theta = 2.0$ for a person with a cholesterol level 1 standard deviation above the mean for his or her age group using a two-sided test with a significance level of 5% and a power of 90%. From

$$\alpha = 0.05 \rightarrow z_{1-\alpha} = 1.96 \text{ (two-sided test)}$$
$$\beta = 0.10 \rightarrow z_{1-\beta} = 1.28$$

the required total sample size is:

$$N = \frac{4}{\theta^2} (z_{1-\alpha} + z_{1-\beta})^2$$
$$= \frac{4}{(\ln 2)^2} (1.96 + 1.28)^2$$
$$\simeq 62$$

if we plan a balanced study with each group consisting of 31 subjects.

***Example 12.18*** Suppose that all specifications are the same as in Example 12.17 but we design a study in which the size of the control group is three times the number of cases. Here we have

$$N = \frac{1}{[\ln(\theta)]^2} \left( \frac{1}{w_1} + \frac{1}{w_0} \right) (z_{1-\alpha} + z_{1-\beta})^2$$
$$= \frac{1}{[\ln(2)]^2} \left( \frac{1}{0.25} + \frac{1}{0.75} \right) (1.96 + 1.28)^2$$
$$\simeq 84$$

or 21 cases and 63 controls.

## EXERCISES

**12.1**   Some opponents of the randomized, double-blinded clinical trial, especially in the field of psychiatry, have argued that a necessary or at least important component in the efficacy of a psychoactive drug, for example a tranquilizer, is the confidence that the physician and the patient have in this efficacy. This factor is lost if one of the drugs in the trial is a placebo and the active drug and placebo cannot be identified. Hence active drug that would be efficacious if identified appears no better than placebo, and therefore is lost to medical practice. Do you agree with this position? Explain.

**12.2**   Suppose that we consider conducting a phase I trial using the standard design with only three prespecified doses. Suppose further that the toxicity rates of these three doses are 10%, 20%, and 30%, respectively. What is the probability that the middle dose will be selected as the MTD? (*Hint:* Use the binomial probability calculations of Chapter 3.)

**12.3**   We can refer to the standard design as a three-and-three design because at each new dose, it enrolls a cohort of three patients with the option of enrolling an additional three patients evaluated at the same dose. Describe the dose-escalation plan for a three-and-two design and describe its possible effects on the toxicity rate of the resulting maximum tolerated dose (MTD).

**12.4**   Repeat Exercise 12.2 but assuming that the toxicity rates of the three doses used are 40%, 50%, and 60%, respectively.

**12.5**   Refer to Example 12.9, where we found that 234 subjects are needed for a predetermined power of 90% and that the power would be 81% if the study enrolls only 180 subjects. What would be the power if the study was substantially underenrolled and conducted with only 120 patients?

**12.6**   Refer to Example 12.10, where we found that 378 subjects are needed for a predetermined power of 90%. What would be the power if the study was actually conducted with only 300 patients, 150 in each group?

**12.7**   Refer to Example 12.14, where we found that 73 cases and 73 controls are needed for a predetermined power of 90%. What would be the power if the study was actually conducted with only 100 subjects, 50 in each group?

**12.8**   Refer to Example 12.18, where we found that 21 cases and 63 controls are needed for a predetermined power of 90%. What would be the power if the study was actually conducted with only 15 cases but 105 controls?

**12.9**  Refer to Example 12.12. Find the total sample size needed if we retain all the specifications except that the hypothesized hazards ratio is increased to 3.0.

**12.10**  The status of the axillary lymph node basin is the most powerful predictor of long-term survival in patients with breast cancer. The pathologic analysis of the axillary nodes also provides essential information used to determine the administration of adjuvant therapies. Until recently, an axillary lymph node dissection (ALND) was the standard surgical procedure to identify nodal metastases. However, ALND is associated with numerous side effects, including arm numbness and pain, infection, and lymphedema. A new procedure, sentinal lymp node (SLN) biopsy, has been proposed as a substitute and it has been reported to have a successful identification rate of about 90%. Suppose that we want to conduct a study to estimate and confirm this rate to identify nodal metastases among breast cancer patients because previous estimates were all based on rather small samples. How many patients are needed to confirm this 90% success rate with a margin of error of $\pm 5\%$? Does the answer change if we do not trust the 90% figure and calculate a conservative sample size estimate?

**12.11**  Metastasic melanoma and renal cell carcinoma are incurable malignancies with a median survival time of less than a year. Although these malignancies are refractory to most chemotherapy drugs, their growth may be regulated by immune mechanisms and there are various strategies for development and administration of tumor vaccines. An investigator considers conducting a phase II trial for such a vaccine for patients with stage IV melanoma. How many patients are needed to to estimate the response rate with a margin of error of $\pm 10\%$?

**12.12**  Suppose that we consider conducting a study to evaluate the efficacy of prolonged infusional paclitaxel (96-hour continuous infusion) in patients with recurrent or metastatic squamous carcinoma of the head and neck. How many patients are needed to to estimate the response rate with a margin of error of $\pm 15\%$?

**12.13**  Normal red blood cells in humans are shaped like biconcave disks. Occasionally, hemoglobin, a protein that readily combines with oxygen, is formed imperfectly in the cell. One type of imperfect hemoglobin causes the cells to have a caved-in, or sicklelike appearance. These sickle cells are less efficient carriers of oxygen than normal cells and result in an oxygen deficiency called *sickle cell anemia*. This condition has a significant prevalence among blacks. Suppose that a study is to be conducted to estimate the prevalence among blacks in a certain large city.

(a)  How large a sample should be chosen to estimate this proportion to within 1 percentage point with 99% confidence? with 95% confidence? (Use a conservative estimate because no prior estimate of the prevalence in this city is assumed available.)

**(b)** A similar study was recently conducted in another state. Of the 13,573 blacks sampled, 1085 were found to have sickle cell anemia. Using this information, resolve part (a).

**12.14**  A researcher wants to estimate the average weight loss obtained by patients at a residential weight-loss clinic during the first week of a controlled diet and exercise regimen. How large a sample is needed to estimate this mean to within 0.5 lb with 95% confidence? Assume that past data indicate a standard deviation of about 1 lb.

**12.15**  Suppose that a study is designed to select the better of two treatments. The endpoint is measured on a continuous scale and a treatment is said to be better if the true mean is 10 units larger than the mean associated with the other treatment. Suppose also that the two groups have the same variance, which is estimated at about $\sigma^2 = 400$. Find the total sample size needed if we want 99% certainty of making the right selection.

**12.16**  Suppose that in planning for a phase II trial, an investigator believes that the incidence of severe side effects is about 5% and the trial has to be stopped if the incidence of severe side effects exceeds 20%. Preset the level of significance at 0.05 and design a stopping rule that has a power of 90%.

**12.17**  A study will be conducted to determine if some literature on smoking will improve patient comprehension. All subjects will be administered a pretest, then randomized into two groups: without or with a booklet. After a week, all subjects will be administered a second test. The data (differences between prescore and postscore) for a pilot study without a booklet yielded (score are on a scale from 0 to 5 points)

$$n = 44; \quad \bar{x} = 0.25, \quad s = 2.28$$

How large should the total sample size be if we decide to preset $\alpha = 0.05$ and that it is important to detect a mean difference of 1.0 with a power of 90%?

**12.18**  A study will be conducted to investigate a claim that oat bran will reduce serum cholesterol in men with high cholesterol levels. Subjects will be randomized to diets that include either oat bran or cornflakes cereals. After two weeks, LDL cholesterol level (in mmol/L) will be measured and the two groups will be compared via a two-sample $t$ test. A pilot study with cornflakes yields

$$n = 14; \quad \bar{x} = 4.44, \quad s = 0.97$$

How large should a total sample size be if we decide to preset $\alpha = 0.01$ and that it is important to detect an LDL cholesterol level reduction of 1.0 mmol/L with a power of 95%?

**12.19**  Depression is one of the most commonly diagnosed conditions among hospitalized patients in mental institutions. The primary measure of depression is the CES-D scale developed by the Center for Epidemiologic Studies, in which each person is scored on a scale of 0 to 60. The following results were found for a group of randomly selected women: $n = 200$, $\bar{x} = 10.4$, and $s = 10.3$. A study is now considered to investigate the effect of a new drug aimed at lowering anxiety among hospitalized patients in similar mental institutions. Subjects would be randomized to receive either the new drug or placebo, then averages of CES-D scores will be compared using the two-sided two-sample $t$ test at the 5% level. How large should the total sample size be if it is important to detect a CES-D score reduction of 3.0 with a power of 90%?

**12.20**  A study will be conducted to compare the proportions of unplanned pregnancy between condom users and pill users. Preliminary data show that these proportions are approximately 10% and 5%, respectively. How large should the total sample size be so that it would be able to detect such a difference of 5% with a power of 90% using a statistical test at the two-sided level of significance of 0.01?

**12.21**  Suppose that we want to compare the use of medical care by black and white teenagers. The aim is to compare the proportions of teenagers without physical checkups within the last two years. Some recent survey shows that these rates for blacks and whites are 17% and 7%, respectively. How large should the total sample be so that it would be able to detect such a 10% difference with a power of 90% using a statistical test at the two-sided level of significance of 0.01?

**12.22**  Among ovarian cancer patients treated with cisplatin, it is anticipated that 20% will experience either partial or complete response. If adding paclitaxel to this regimen can increase the response by 15% without undue toxicity, that would be considered as clinically significant. Calculate the total sample size needed for a randomized trial that would have a 80% chance of detecting this magnitude of treatment difference while the probability of type I error for a two-sided test is preset at 0.05.

**12.23**  Metastatic breast cancer is a leading cause of cancer-related mortality and there has been no major change in the mortality rate over the past few decades. Therapeutic options are available with active drugs such as paclitaxel. However, the promising response rate is also accompanied by a high incidence of toxicities, especially neurotoxicity. An investigator considers testing a new agent that may provide significant prevention, reduction, or mitigation of drug-related toxicity. This new agent is to be tested against a placebo in a double-blind randomized trial among patients with mestastatic breast cancer who receive weekly paclitaxel. The rate of neurotoxicity over the period of the trial is estimated to be about 40% in the placebo group, and the hypothesis is that this new agent lowers the toxicity rate by one-half, to 20%. Find the total sample size needed using a two-sided level of significance of 0.05 and the assumption that the hypothesis would be detectable with a power of 80%.

12.24   In the study on metastatic breast cancer in Exercise 12.23, the investigator also focuses on tumor response rate, hoping to show that this rate is comparable in the two treatment groups. The hypothesis is that addition of the new agent to a weekly paclitaxel regimen would reduce the incidence of neurotoxicity without a compromise in its efficacy. At the present time, it is estimated that the tumor response rate for the placebo group, without the new agent added, is about 70%. Assuming the same response rate for the treated patients, find the margin of error of its estimate using the sample size obtained from Exercise 12.23.

12.25   The primary objective of a phase III trial is to compare disease-free survival among women with high-risk operable breast cancer following surgical resection of all known disease and randomized to receive as adjuvant therapy either CEF or a new therapy. CEF has been established successfully as a standard adjuvant regimen in Canada; the aim here is to determine whether the addition of a taxane (to form the new regimen) can improve survival outcome over CEF alone. The five-year disease-free survival rate for women receiving CEF alone as adjuvant therapy was estimated at about 60%, and it is hypothesized that the newly formed regimen would improve that rate from 60% to 70%. Find the total sample size needed using a two-sided test at the 0.05 level of significance and a statistical power of 80%.

12.26   Ovarian cancer is the fourth most common cause of cancer deaths in women. Approximately 75% of patients present with an advanced stage, and because of this, only a minority of patients will have surgically curable localized disease, and systematic chemotherapy has become the primary treatment modality. A randomized phase III trial is considered to compare paclitaxel–carboplatin versus docetaxel–carboplatin as first-line chemotherapy in stage IV epithelial ovarian cancer. Suppose that we plan to use a statistical test at the two-sided 5% level of significance and the study is designed to have 80% power to detect the alternative hypothesis that the two-year survival rates in the paclitaxel and docetaxel arms are 40% and 50%, respectively. Find the total sample size required.

**12.27**  A phase III double-blind randomized trial is planned to compare a new drug versus placebo as adjuvant therapy for the treatment of women with metastatic ovarian cancer who have had complete clinical response to their primary treatment protocol, consisting of surgical debulking and platium-based chemotherapy. Find the total sample size needed using a two-sided test at the 0.05 level of significance and a statistical power of 80% to detect the alternative hypothesis that the two-year relapse rates in the placebo and the new drug arms are 60% and 40%, respectively.

**12.28**  Suppose that an investigator considers conducting a case–control study to evaluate the relationship between invasive epithelial ovarian cancer and the history of infertility (yes/no). It is estimated that the proportion of controls with a history of infertility is about 10% and the investigator wishes to detect a relative risk of 2.0 with a power of 80% using a two-sided level of significance of 0.05.

  **(a)** Find the total sample size needed if the two groups, cases and controls, are designed to have the same size.

  **(b)** Find the total sample size needed if the investigator likes to have the number of controls four times the number of cases.

  **(c)** Find the number of cases needed if the design is 1:1 matched; matching criteria are various menstrual characteristics, exogenous estrogen use, and prior pelvic surgeries.

**12.29**  Suppose that an investigator considers conducting a case–control study to evaluate the relationship between cesarean section delivery (C-section) and the use of electronic fetal monitoring (EFM, also called ultrasound) during labor. It is estimated that the proportion of controls (vaginal deliveries) who were exposed to EFM is about 40% and the investigator wishes to detect a relative risk of 2.0 with a power of 90% using a two-sided level of significance of 0.05.

  **(a)** Find the total sample size needed if the two groups, cases and controls, are designed to have the same size.

  **(b)** Find the total sample size needed if the investigator likes to have the number of controls three times the number of cases.

  **(c)** Find the number of cases needed if the design is 1:1 matched; matching criteria are age, race, socioeconomic condition, education, and type of health insurance.

**12.30**  When a patient is diagnosed as having cancer of the prostate, an important question in deciding on treatment strategy for the patient is whether or not the cancer has spread to the neighboring lymph nodes. The question is so critical in prognosis and treatment that it is customary to operate on the patient (i.e., perform a laparotomy) for the sole purpose of examining the nodes and removing tissue samples to examine under the microscope for evidence of cancer. However, certain variables that can be measured without surgery may be predictive of the nodal involvement; one of which is level of serum acid phosphatase. Suppose that an investigator considers conducting a case–control study to evaluate this possible relationship between nodal involvement (cases) and level of serum acid phosphatase. Suppose further that it is desirable to detect an odds ratio of $\theta = 1.5$ for a person with a serum acid phosphatase level 1 standard deviation above the mean for his age group using a two-sided test with a significance level of 5% and a power of 80%. Find the total sample size needed for using a two-sided test at the 0.05 level of significance.

# BIBLIOGRAPHY

Agresti, A. (1990). *Categorical Data Analysis*. New York: Wiley.

Ahlquist, D. A., D. B. McGill, S. Schwartz, and W. F. Taylor (1985). Fecal blood levels in health and disease: A study using HemoQuant. *New England Journal of Medicine* 314: 1422.

Anderson, J. W. et al. (1990). Oat bran cereal lowers serum cholesterol total and LDL cholesterol in hypercholesterolemic men. *American Journal of Clinical Nutrition* 52: 495–499.

Armitage, P. (1977). *Statistical Methods in Medical Research*. New York: Wiley.

Arsenault, P. S. (1980). Maternal and antenatal factors in the risk of sudden infant death syndrome. *American Journal of Epidemiology* 111: 279–284.

Bamber, D. (1975). The area above the ordinal dominance graph and the area below the receiver operating graph. *Journal of Mathematical Psychology* 12: 387–415.

Begg, C. B. and B. McNeil (1988). Assessment of radiologic tests: Control of bias and other design considerations. *Radiology* 167: 565–569.

Berkowitz, G. S. (1981). An epidemiologic study of pre-term delivery. *American Journal of Epidemiology* 113: 81–92.

Berry, G. (1983). The analysis of mortality by the subject–years method. *Biometrics* 39: 173–184.

Biracree, T. (1984). Your intelligence quotient, in *How You Rate*. New York: Dell Publishing.

Blot, W. J. et al. (1978). Lung cancer after employment in shipyards during World War II. *New England Journal of Medicine* 299: 620–624.

Breslow, N. (1970). A generalized Kruskal–Wallis test for comparing $K$ samples subject to unequal patterns of censorship. *Biometrika* 57: 579–594.

Breslow, N. (1982). Covariance adjustment of relative-risk estimates in matched studies. *Biometrics* 38: 661–672.

Breslow, N. E. and N. E. Day (1980). *Statistical Methods in Cancer Research*, Vol. I: *The Analysis of Case–Control Studies*. Lyons, France: International Agency for Research on Cancer.

Brown, B. W. (1980). Prediction analyses for binary data. In *Biostatistics Casebook*, edited by R. G. Miller, B. Efron, B. W. Brown, and L. E. Moses. pp. 3–18. New York: Wiley.

Centers for Disease Control (1990). *Summary of Notifiable Diseases: United States 1989 Morbidity and Mortality Weekly Report* 38.

Chin, T., W. Marine, E. Hall, C. Gravelle, and J. Speers (1961). The influence of Salk vaccination on the epidemic pattern and the spread of the virus in the community. *American Journal of Hygiene* 73: 67–94.

Cohen, J. (1960). A coefficient of agreement for nominal scale. *Educational and Psychological Measuments* 20: 37–46.

Coren, S. (1989). Left-handedness and accident-related injury risk. *American Journal of Public Health* 79: 1040–1041.

Cox, D. R. (1972). Regression models and life tables. *Journal of the Royal Statistical Society, Series B* 34: 187–220.

Cox, D. R. and D. Oakes (1984). *Analysis of Survival Data*. New York: Chapman & Hall.

Cox, D. R. and E. J. Snell (1989). *The Analysis of Binary Data*, 2nd ed. London: Chapman & Hall.

D'Angelo, L. J., J. C. Hierholzer, R. C. Holman, and J. D. Smith (1981). Epidemic keratoconjunctivitis caused by adenovirus type 8: Epidemiologic and laboratory aspects of a large outbreak. *American Journal of Epidemiology* 113: 44–49.

Daniel, W. W. (1987). *Biostatistics: A Foundation for Analysis in the Health Sciences*. New York: Wiley.

Dienstag, J. L. and D. M. Ryan (1982). Occupational exposure to hepatitis B virus in hospital personnel: Infection or immunization. *American Journal of Epidemiology* 15: 26–39.

Douglas, G. (1990). Drug therapy. *New England Journal of Medicine* 322: 443–449.

Einarsson, K. et al. (1985). Influence of age on secretion of cholesterol and synthesis of bile acids by the liver. *New England Journal of Medicine* 313: 277–282.

Engs, R. C. and D. J. Hanson (1988). University students' drinking patterns and problems: Examining the effects of raising the purchase age. *Public Health Reports* 103: 667–673.

Fiskens, E. J. M. and D. Kronshout (1989). Cardiovascular risk factors and the 25 year incidence of diabetes mellitus in middle-aged men. *American Journal of Epidemiology* 130: 1101–1108.

Fowkes, F. G. R. et al. (1992). Smoking, lipids, glucose intolerance, and blood pressure as risk factors for peripheral atherosclerosis compared with ischemic heart disease in the Edinburgh Artery Study. *American Journal of Epidemiology* 135: 331–340.

Fox, A. J. and P. F. Collier (1976). Low mortality rates in industrial cohort studies due to selection for work and survival in the industry. *British Journal of Preventive and Social Medicine* 30: 225–230.

Freeman, D. H. (1980). *Applied Categorical Data Analysis*. New York: Marcel Dekker.

Freireich, E. J. et al. (1963). The effect of 6-mercaptopurine on the duration of steroid-induced remissions in acute leukemia: A model for evaluation of other potentially useful therapy. *Blood* 21: 699–716.

Frerichs, R. R. et al. (1981). Prevalence of depression in Los Angeles County. *American Journal of Epidemiology* 113: 691–699.

Frome, E. L. (1983). The analysis of rates using Poisson regression models. *Biometrics* 39: 665–674.

Frome, E. L. and H. Checkoway (1985). Use of Poisson regression models in estimating rates and ratios. *American Journal of Epidemiology* 121: 309–323.

Fulwood, R. et al. (1986). Total serum cholesterol levels of adults 20–74 years of age: United States, 1976–1980. *Vital and Health Statistics, Series M* 236.

Gehan, E. A. (1965a). A generalized Wilcoxon test for comparing arbitrarily singly-censored samples. *Biometrika* 52: 203–223.

Gehan, E. A. (1965b). A generalized two-sample Wilcoxon test for doubly censored data. *Biometrika* 52: 650–653.

Grady, W. R. et al. (1986). Contraceptives failure in the United States: Estimates from the 1982 National Survey of Family Growth. *Family Planning Perspectives* 18: 200–209.

Graham, S. et al. (1988). Dietary epidemiology of cancer of the colon in western New York. *American Journal of Epidemiology* 128: 490–503.

Greenwood, M. (1926). The natural duration of cancer. *Reports on Public Health and Medical Subjects, Her Majesty's Stationary Office* 33: 1–26.

Gwirtsman, H. E. et al. (1989). Decreased caloric intake in normal weight patients with bulimia: Comparison with female volunteers. *American Journal of Clinical Nutrition* 49: 86–92.

Hanley, J. A. and B. J. McNeil (1982). The meaning and use of the area under a receiver operating characteristic (ROC) curve. *Radiology* 143: 29–36.

Hanley, J. A. and B. J. McNeil (1983). Method for comparing the area under the ROC curves derived from the same cases. *Radiology* 148: 839.

Helsing, K. J. and M. Szklo (1981). Mortality after bereavement. *American Journal of Epidemiology* 114: 41–52.

Herbst, A. L., H. Ulfelder, and D. C. Poskanzer. (1971). Adenocarcinoma of the vagina. *New England Journal of Medicine* 284: 878–881.

Hiller, R. and A. H. Kahn (1976). Blindness from glaucoma. *British Journal of Ophthalmology* 80: 62–69.

Hlatky, M. A., D. B. Pryor, F. E. Harrell, R. M. Califf, D. B. Mark, and R. A. Rosati (1984). Factors affecting sensitivity and specificity of exercise electrocardiography: Multivariate analysis. *American Journal of Medicine* 77: 64–71.

Hollows, F. C. and P. A. Graham (1966). Intraocular pressure, glaucoma, and glaucoma suspects in a defined population. *British Journal of Ophthalmology* 50: 570–586.

Hosmer, D. W., Jr. and S. Lemeshow (1989). *Applied Logistic Regression.* New York: Wiley.

Hsieh, F. Y. (1989). Sample size tables for logistic regression. *Statistics in Medicine* 8: 795–802.

Jackson, R. et al. (1992). Does recent alcohol consumption reduce the risk of acute myocardial infarction and coronary death in regular drinkers? *American Journal of Epidemiology* 136: 819–824.

Kaplan, E. L. and P. Meier (1958). Nonparametric estimation from incomplete observations. *Journal of the American Statistical Association* 53: 457–481.

Kay, R. and S. Little (1987). Transformations of the explanatory variables in the logistic regression model for binary data. *Biometrika* 74: 495–501.

Kelsey, J. L., V. A. Livolsi, T. R. Holford, D. B. Fischer, E. D. Mostow, P. E. Schartz, T. O'Connor, and C. White (1982). A case–control study of cancer of the endometrium. *American Journal of Epidemiology* 116: 333–342.

Khabbaz, R. et al. (1990). Epidemiologic assessment of screening tests for antibody to human T lymphotropic virus type I. *American Journal of Public Health* 80: 190–192.

Kleinbaum, D. G., L. L. Kupper, and K. E. Muller (1988). *Applied Regression Analysis and Other Multivariate Methods.* Boston: PWS-Kent.

Kleinman, J. C. and A. Kopstein (1981). Who is being screened for cervical cancer? *American Journal of Public Health* 71: 73–76.

Klinhamer, P. J. J. M. et al. (1989). Intraobserver and interobserver variability in the quality assessment of cervical smears. *Acta Cytologica* 33: 215–218.

Knowler, W. C. et al. (1981). Diabetes incidence in Pima Indians: Contributions of obesity and parental diabetes. *American Journal of Epidemiology* 113: 144–156.

Koenig, J. Q. et al. (1990). Prior exposure to ozone potentiates subsequent response to sulfur dioxide in adolescent asthmatic subjects. *American Review of Respiratory Disease* 141: 377–380.

Kono, S. et al. (1992). Prevalence of gallstone disease in relation to smoking, alcohol use, obesity, and glucose tolerance: A study of self-defense officials in Japan. *American Journal of Epidemiology* 136: 787–805.

Kushi, L. H. et al. (1988). The association of dietary fat with serum cholesterol in vegetarians: The effects of dietary assessment on the correlation coefficient. *American Journal of Epidemiology* 128: 1054–1064.

Lachin, J. M. (1981). Intoduction to sample size determination and power analysis for clinical trials. *Controlled Clinical Trials* 2: 93–113.

Le, C. T. (1997a). *Applied Survival Analysis.* New York: Wiley.

Le, C. T. (1997b). Evaluation of confounding effects in ROC studies. *Biometrics* 53: 998–1007.

Le, C. T. and B. R. Lindgren (1988). Computational implementation of the conditional logistic regression model in the analysis of epidemiologic matched studies. *Computers and Biomedical Research* 21: 48–52.

Lee, M. (1989). Improving patient comprehension of literature on smoking. *American Journal of Public Health* 79: 1411–1412.

Li, D. K. et al. (1990). Prior condom use and the risk of tubal pregnancy. *American Journal of Public Health* 80: 964–966.

Mack, T. M. et al. (1976). Estrogens and endometrial cancer in a retirement community. *New England Journal of Medicine* 294: 1262–1267.

Makuc, D. et al. (1989). National trends in the use of preventive health care by women. *American Journal of Public Health* 79: 21–26.

Mantel, N. and W. Haenszel (1959). Statistical aspects of the analysis of data from retrospective studies of disease. *Journal of the National Cancer Institute* 22: 719–748.

Matinez, F. D. et al. (1992). Maternal age as a risk factor for wheezing lower respiratory illness in the first year of life. *American Journal of Epidemiology* 136: 1258–1268.

May, D. (1974). Error rates in cervical cytological screening tests. *British Journal of Cancer* 29: 106–113.

McCusker, J. et al. (1988). Association of electronic fetal monitoring during labor with caesarean section rate with neonatal morbidity and mortality. *American Journal of Public Health* 78: 1170–1174.

Negri, E. et al. (1988). Risk factors for breast cancer: Pooled results from three Italian case–control studies. *American Journal of Epidemiology* 128: 1207–1215.

Nischan, P. et al. (1988). Smoking and invasive cervical cancer risk: Results from a case–control study. *American Journal of Epidemiology* 128: 74–77.

Nurminen, M. et al. (1982). Quantitated effects of carbon disulfide exposure, elevated blood pressure and aging on coronary mortality. *American Journal of Epidemiology* 115: 107–118.

Ockene, J. (1990). The relationship of smoking cessation to coronary heart disease and lung cancer in the Multiple Risk Factor Intervention Trial. *American Journal of Public Health* 80: 954–958.

Padian, N. S. (1990). Sexual histories of heterosexual couples with one HIV-infected partner. *American Journal of Public Health* 80: 990–991.

Palta, M. et al. (1982). Comparison of self-reported and measured height and weight. *American Journal of Epidemiology* 115: 223–230.

Pappas, G. et al. (1990). Hypertension prevalence and the status of awareness, treatment, and control in the Hispanic health and nutrition examination survey (HHANES). *American Journal of Public Health* 80: 1431–1436.

Peto, R. (1972). Contribution to discussion of paper by D. R. Cox. *Journal of the Royal Statistical Society, Series B* 34: 205–207.

Renaud, L. and S. Suissa (1989). Evaluation of the efficacy of simulation games in traffic safety education of kindergarten children. *American Journal of Public Health* 79: 307–309.

Renes, R. et al. (1981). Transmission of multiple drug-resistant tuberculosis: Report of school and community outbreaks. *American Journal of Epidemiology* 113: 423–435.

Rosenberg, L. et al. (1981). Case–control studies on the acute effects of coffee upon the risk of myocardial infarction: Problems in the selection of a hospital control series. *American Journal of Epidemiology* 113: 646–652.

Rossignol, A. M. (1989). Tea and premenstrual syndrome in the People's Republic of China. *American Journal of Public Health* 79: 67–68.

Rousch, G. C. et al. (1982). Scrotal carcinoma in Connecticut metal workers: Sequel to a study of sinonasal cancer. *American Journal of Epidemiology* 116: 76–85.

Salem-Schatz, S. et al. (1990). Influence of clinical knowledge, organization context and practice style on transfusion decision making. *Journal of the American Medical Association* 25: 476–483.

Sandek, C. D. et al. (1989). A preliminary trial of the programmable implantable medication system for insulin delivery. *New England Journal of Medicine* 321: 574–579.

Schwarts, B. et al. (1989). Olfactory function in chemical workers exposed to acrylate and methacrylate vapors. *American Journal of Public Health* 79: 613–618.

Selby, J. V. et al. (1989). Precursors of essential hypertension: The role of body fat distribution. *American Journal of Epidemiology* 129: 43–53.

Shapiro, S. et al. (1979). Oral contraceptive use in relation to myocardial infarction. *Lancet* 1: 743–746.

Strader, C. H. et al. (1988). Vasectomy and the incidence of testicular cancer. *American Journal of Epidemiology* 128: 56–63.

Strogatz, D. (1990). Use of medical care for chest pain differences between blacks and whites. *American Journal of Public Health* 80: 290–293.

Tarone, R. E. and J. Ware (1977). On distribution-free tests for equality of survival distributions. *Biometrika* 64: 156–160.

Taylor, H. (1981). Racial variations in vision. *American Journal of Epidemiology* 113: 62–80.

Taylor, P. R. et al. (1989). The relationship of polychlorinated biphenyls to birth weight and gestational age in the offspring of occupationally exposed mothers. *American Journal of Epidemiology* 129: 395–406.

Thompson, R. S. et al. (1989). A case–control study of the effectiveness of bicycle safety helmets. *New England Journal of Medicine* 320: 1361–1367.

Tosteson, A. A. and C. B. Begg (1988). A general regression methodology for ROC-curve estimation. *Medical Decision Making* 8: 204–215.

True, W. R. et al. (1988). Stress symptomology among Vietnam veterans. *American Journal of Epidemiology* 128: 85–92.

Tuyns, A. J. et al. (1977). Esophageal cancer in Ille-et-Vilaine in relation to alcohol and tobacco consumption: Multiplicative risks. *Bulletin of Cancer* 64: 45–60.

Umen, A. J. and C. T. Le (1986). Prognostic factors, models, and related statistical problems in the survival of end-stage renal disease patients on hemodialysis. *Statistics in Medicine* 5: 637–652.

Weinberg, G. B. et al. (1982). The relationship between the geographic distribution of lung cancer incidence and cigarette smoking in Allegheny County, Pennsylvania. *American Journal of Epidemiology* 115: 40–58.

Whittemore, A. S. (1981). Sample size for logistic regression with small response probability. *Journal of the American Statistical Association* 76: 27–32.

Whittemore, A. S. et al. (1988). Personal and environmental characteristics related to epithelial ovarian cancer. *American Journal of Epidemiology* 128: 1228–1240.

Whittemore, A. S. et al. (1992). Characteristics relating to ovarian cancer risk: Collaborative analysis of 12 U.S. case–control studies. *American Journal of Epidemiology* 136: 1184–1203.

Yassi, A. et al. (1991). An analysis of occupational blood level trends in Manitoba: 1979 through 1987. *American Journal of Public Health* 81: 736–740.

Yen, S., C. Hsieh, and B. MacMahon (1982). Consumption of alcohol and tobacco and other risk factors for pancreatitis. *American Journal of Epidemiology* 116: 407–414.

# APPENDICES

## APPENDIX A: TABLE OF RANDOM NUMBERS

| | | | | | | | | | |
|---|---|---|---|---|---|---|---|---|---|
| 63271 | 59986 | 71744 | 51102 | 15141 | 80714 | 58683 | 93108 | 13554 | 79945 |
| 88547 | 09896 | 95436 | 79115 | 08303 | 01041 | 20030 | 63754 | 08459 | 28364 |
| 55957 | 57243 | 83865 | 09911 | 19761 | 66535 | 40102 | 26646 | 60147 | 15704 |
| 46276 | 87453 | 44790 | 67122 | 45573 | 84358 | 21625 | 16999 | 13385 | 22782 |
| 55363 | 07449 | 34835 | 15290 | 76616 | 67191 | 12777 | 21861 | 68689 | 03263 |
| 69393 | 92785 | 49902 | 58447 | 42048 | 30378 | 87618 | 26933 | 40640 | 16281 |
| 13186 | 29431 | 88190 | 04588 | 38733 | 81290 | 89541 | 70290 | 40113 | 08243 |
| 17726 | 28652 | 56836 | 78351 | 47327 | 18518 | 92222 | 55201 | 27340 | 10493 |
| 36520 | 64465 | 05550 | 30157 | 82242 | 29520 | 69753 | 72602 | 23756 | 54935 |
| 81628 | 36100 | 39254 | 56835 | 37636 | 02421 | 98063 | 89641 | 64953 | 99337 |
| 84649 | 48968 | 75215 | 75498 | 49539 | 74240 | 03466 | 49292 | 36401 | 45525 |
| 63291 | 11618 | 12613 | 75055 | 43915 | 26488 | 41116 | 64531 | 56827 | 30825 |
| 70502 | 53225 | 03655 | 05915 | 37140 | 57051 | 48393 | 91322 | 25653 | 06543 |
| 06426 | 24771 | 59935 | 49801 | 11082 | 66762 | 94477 | 02494 | 88215 | 27191 |
| 20711 | 55609 | 29430 | 70165 | 45406 | 78484 | 31639 | 52009 | 18873 | 96927 |
| 41990 | 70538 | 77191 | 25860 | 55204 | 73417 | 83920 | 69468 | 74972 | 38712 |
| 72452 | 36618 | 76298 | 26678 | 89334 | 33938 | 95567 | 29380 | 75906 | 91807 |
| 37042 | 40318 | 57099 | 10528 | 09925 | 89773 | 41335 | 96244 | 29002 | 46453 |
| 53766 | 52875 | 15987 | 46962 | 67342 | 77592 | 57651 | 95508 | 80033 | 69828 |
| 90585 | 58955 | 53122 | 16025 | 84299 | 53310 | 67380 | 84249 | 25348 | 04332 |
| 32001 | 96293 | 37203 | 64516 | 51530 | 37069 | 40261 | 61374 | 05815 | 06714 |
| 62606 | 64324 | 46354 | 72157 | 67248 | 20135 | 49804 | 09226 | 64419 | 29457 |
| 10078 | 28073 | 85389 | 50324 | 14500 | 15562 | 64165 | 06125 | 71353 | 77669 |
| 91561 | 46145 | 24177 | 15294 | 10061 | 98124 | 75732 | 00815 | 83452 | 97355 |
| 13091 | 98112 | 53959 | 79607 | 52244 | 63303 | 10413 | 63839 | 74762 | 50289 |
| 73864 | 83014 | 72457 | 22682 | 03033 | 61714 | 88173 | 90835 | 00634 | 85169 |
| 66668 | 25467 | 48894 | 51043 | 02365 | 91726 | 09365 | 63167 | 95264 | 45643 |
| 84745 | 41042 | 29493 | 01836 | 09044 | 51926 | 43630 | 63470 | 76508 | 14194 |
| 48068 | 26805 | 94595 | 47907 | 13357 | 38412 | 33318 | 26098 | 82782 | 42851 |
| 54310 | 96175 | 97594 | 88616 | 42035 | 38093 | 36745 | 56702 | 40644 | 83514 |
| 14877 | 33095 | 10924 | 58013 | 61439 | 21882 | 42059 | 24177 | 58739 | 60170 |
| 78295 | 23179 | 02771 | 43464 | 59061 | 71411 | 05697 | 67194 | 30495 | 21157 |
| 67524 | 02865 | 39593 | 54278 | 04237 | 92441 | 26602 | 63835 | 38032 | 94770 |
| 58268 | 57219 | 68124 | 73455 | 83236 | 08710 | 04284 | 55005 | 84171 | 42596 |
| 97158 | 28672 | 50685 | 01181 | 24262 | 19427 | 52106 | 34308 | 73685 | 74246 |
| 04230 | 16831 | 69085 | 30802 | 65559 | 09205 | 71829 | 06489 | 85650 | 38707 |
| 94879 | 56606 | 30401 | 02602 | 57658 | 70091 | 54986 | 41394 | 60437 | 03195 |
| 71446 | 15232 | 66715 | 26385 | 91518 | 70566 | 02888 | 79941 | 39684 | 54315 |
| 32886 | 05644 | 79316 | 09819 | 00813 | 88407 | 17461 | 73925 | 53037 | 91904 |
| 62048 | 33711 | 25290 | 21526 | 02223 | 75947 | 66466 | 06232 | 10913 | 75336 |
| 84534 | 42351 | 21628 | 53669 | 81352 | 95152 | 08107 | 98814 | 72743 | 12849 |
| 84707 | 15885 | 84710 | 35866 | 06446 | 86311 | 32648 | 88141 | 73902 | 69981 |
| 19409 | 40868 | 64220 | 80861 | 13860 | 68493 | 52908 | 26374 | 63297 | 45052 |

## APPENDIX B: AREA UNDER THE STANDARD NORMAL CURVE

Entries in the table give the area under the curve between the mean and $z$ standard deviations above the mean. For example, for $z = 1.25$, the area under the curve between the mean and $z$ is 0.3944.

| $z$ | .00 | .01 | .02 | .03 | .04 | .05 | .06 | .07 | .08 | .09 |
|-----|-----|-----|-----|-----|-----|-----|-----|-----|-----|-----|
| .0 | .0000 | .0040 | .0080 | .0120 | .0160 | .0199 | .0239 | .0279 | .0319 | .0359 |
| .1 | .0398 | .0438 | .0478 | .0517 | .0557 | .0596 | .0636 | .0675 | .0714 | .0753 |
| .2 | .0793 | .0832 | .0871 | .0910 | .0948 | .0987 | .1026 | .1064 | .1103 | .1141 |
| .3 | .1179 | .1217 | .1255 | .1293 | .1331 | .1368 | .1406 | .1443 | .1480 | .1517 |
| .4 | .1554 | .1591 | .1628 | .1664 | .1700 | .1736 | .1772 | .1808 | .1844 | .1879 |
| .5 | .1915 | .1950 | .1985 | .2019 | .2054 | .2088 | .2123 | .2157 | .2190 | .2224 |
| .6 | .2257 | .2291 | .2324 | .2357 | .2389 | .2422 | .2454 | .2486 | .2518 | .2549 |
| .7 | .2580 | .2612 | .2642 | .2673 | .2704 | .2734 | .2764 | .2794 | .2823 | .2852 |
| .8 | .2881 | .2910 | .2939 | .2967 | .2995 | .3023 | .3051 | .3078 | .3106 | .3133 |
| .9 | .3159 | .3186 | .3212 | .3238 | .3264 | .3289 | .3315 | .3340 | .3365 | .3389 |
| 1.0 | .3413 | .3438 | .3461 | .3485 | .3508 | .3531 | .3665 | .3577 | .3599 | .3621 |
| 1.1 | .3643 | .3554 | .3686 | .3708 | .3729 | .3749 | .3770 | .3790 | .3810 | .3830 |
| 1.2 | .3849 | .3869 | .3888 | .3907 | .3925 | .3944 | .3962 | .3980 | .3997 | .4015 |
| 1.3 | .4032 | .4049 | .4066 | .4082 | .4099 | .4115 | .4131 | .4147 | .4162 | .4177 |
| 1.4 | .4192 | .4207 | .4222 | .4236 | .4251 | .4265 | .4279 | .4292 | .4306 | .4319 |
| 1.5 | .4332 | .4345 | .4357 | .4370 | .4382 | .4394 | .4406 | .4418 | .4429 | .4441 |
| 1.6 | .4452 | .4463 | .4474 | .4484 | .4495 | .4505 | .4515 | .4525 | .4535 | .4545 |
| 1.7 | .4554 | .4564 | .4573 | .4582 | .4591 | .4599 | .4608 | .4616 | .4625 | .4633 |
| 1.8 | .4641 | .4649 | .4656 | .4664 | .4671 | .4678 | .4686 | .4693 | .4699 | .4706 |
| 1.9 | .4713 | .4719 | .4726 | .4732 | .4738 | .4744 | .4750 | .4756 | .4761 | .4767 |
| 2.0 | .4772 | .4778 | .4783 | .4788 | .4793 | .4798 | .4803 | .4808 | .4812 | .4817 |
| 2.1 | .4821 | .4826 | .4830 | .4834 | .4838 | .4842 | .4846 | .4850 | .4854 | .4857 |
| 2.2 | .4861 | .4864 | .4868 | .4871 | .4875 | .4878 | .4881 | .4884 | .4887 | .4890 |
| 2.3 | .4893 | .4896 | .4898 | .4901 | .4904 | .4906 | .4909 | .4911 | .4913 | .4916 |
| 2.4 | .4918 | .4920 | .4922 | .4925 | .4927 | .4929 | .4931 | .4932 | .4934 | .4936 |
| 2.5 | .4938 | .4940 | .4941 | .4943 | .4945 | .4946 | .4948 | .4949 | .4951 | .4942 |
| 2.6 | .4953 | .4955 | .4956 | .4957 | .4959 | .4960 | .4961 | .4962 | .4963 | .4964 |
| 2.7 | .4965 | .4966 | .4967 | .4968 | .4969 | .4970 | .4971 | .4972 | .4973 | .4974 |
| 2.8 | .4974 | .4975 | .4976 | .4977 | .4977 | .4978 | .4979 | .4979 | .4980 | .4981 |
| 2.9 | .4981 | .4982 | .4982 | .4983 | .4984 | .4984 | .4985 | .4985 | .4986 | .4986 |
| 3.0 | .4986 | .4987 | .4987 | .4988 | .4988 | .4989 | .4989 | .4989 | .4990 | .4990 |

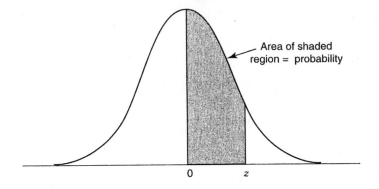

Area of shaded
region = probability

## APPENDIX C: PERCENTILES OF THE $t$ DISTRIBUTION

Entries in the table give $t_\alpha$ values, where $\alpha$ is the area or probability in the upper tail of the $t$ distribution. For example, with 10 degrees of freedom and a 0.025 area in the upper tail, $t_{.025} = 2.228$.

| Degrees of Freedom | Area in Upper Tail | | | | |
|---|---|---|---|---|---|
| | 0.10 | 0.05 | 0.025 | 0.01 | 0.005 |
| 1 | 3.078 | 6.314 | 12.706 | 31.821 | 63.657 |
| 2 | 1.886 | 2.920 | 4.303 | 6.965 | 9.925 |
| 3 | 1.638 | 2.353 | 3.182 | 4.541 | 5.841 |
| 4 | 1.533 | 2.132 | 2.776 | 3.747 | 4.604 |
| 5 | 1.476 | 2.015 | 2.571 | 3.365 | 4.032 |
| 6 | 1.440 | 1.943 | 2.447 | 3.143 | 3.707 |
| 7 | 1.415 | 1.895 | 2.365 | 2.998 | 3.499 |
| 8 | 1.397 | 1.860 | 2.306 | 2.896 | 3.355 |
| 9 | 1.383 | 1.833 | 2.262 | 2.821 | 3.250 |
| 10 | 1.372 | 1.812 | 2.228 | 2.764 | 3.169 |
| 11 | 1.363 | 1.796 | 2.201 | 2.718 | 3.106 |
| 12 | 1.356 | 1.782 | 2.179 | 2.681 | 3.055 |
| 13 | 1.350 | 1.771 | 2.160 | 2.650 | 3.012 |
| 14 | 1.345 | 1.761 | 2.145 | 2.624 | 2.977 |
| 15 | 1.341 | 1.753 | 2.131 | 2.602 | 2.947 |
| 16 | 1.337 | 1.746 | 2.120 | 2.583 | 2.921 |
| 17 | 1.333 | 1.740 | 2.110 | 2.567 | 2.898 |
| 18 | 1.330 | 1.734 | 2.101 | 2.552 | 2.878 |
| 19 | 1.328 | 1.729 | 2.093 | 2.539 | 2.861 |
| 20 | 1.325 | 1.725 | 2.086 | 2.528 | 2.845 |
| 21 | 1.323 | 1.721 | 2.080 | 2.518 | 2.831 |
| 22 | 1.321 | 1.717 | 2.074 | 2.508 | 2.819 |
| 23 | 1.319 | 1.714 | 2.069 | 2.500 | 2.807 |
| 24 | 1.318 | 1.711 | 2.064 | 2.492 | 2.797 |
| 25 | 1.316 | 1.708 | 2.060 | 2.485 | 2.787 |
| 26 | 1.315 | 1.706 | 2.056 | 2.479 | 2.779 |
| 27 | 1.314 | 1.703 | 2.052 | 2.473 | 2.771 |
| 28 | 1.313 | 1.701 | 2.048 | 2.467 | 2.763 |
| 29 | 1.311 | 1.699 | 2.045 | 2.462 | 2.756 |
| 30 | 1.310 | 1.697 | 2.042 | 2.457 | 2.750 |
| 40 | 1.303 | 1.684 | 2.021 | 2.423 | 2.704 |
| 60 | 1.296 | 1.671 | 2.000 | 2.390 | 2.660 |
| 120 | 1.289 | 1.658 | 1.980 | 2.358 | 2.617 |
| $\infty$ | 1.282 | 1.645 | 1.960 | 2.326 | 2.576 |

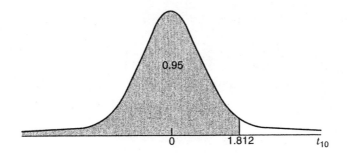

## APPENDIX D: PERCENTILES OF CHI-SQUARE DISTRIBUTIONS

Entries in the table give $\chi_\alpha^2$ values, where $\alpha$ is the area or probability in the upper tail of the chi-square distribution. For example, with 20 degrees of freedom and a 0.05 area in the upper tail, $\chi_{0.05}^2 = 31.410$.

| Degrees of Freedom | Area in Upper Tail | |
|---|---|---|
| | 0.05 | 0.01 |
| 1 | 3.841 | 6.635 |
| 2 | 5.991 | 9.210 |
| 3 | 7.815 | 11.350 |
| 4 | 9.488 | 13.277 |
| 5 | 11.071 | 15.086 |
| 6 | 12.592 | 16.812 |
| 7 | 14.067 | 18.475 |
| 8 | 15.507 | 20.090 |
| 9 | 16.919 | 21.666 |
| 10 | 18.307 | 23.209 |
| 11 | 19.675 | 24.725 |
| 12 | 21.026 | 26.217 |
| 13 | 22.362 | 27.688 |
| 14 | 23.685 | 29.141 |
| 15 | 24.996 | 30.578 |
| 16 | 26.296 | 32.000 |
| 17 | 27.587 | 33.408 |
| 18 | 28.869 | 34.805 |
| 19 | 30.144 | 36.191 |
| 20 | 31.410 | 37.566 |
| 21 | 32.671 | 38.932 |
| 22 | 33.924 | 40.289 |
| 23 | 35.173 | 41.638 |
| 24 | 36.415 | 42.980 |
| 25 | 37.653 | 44.314 |
| 26 | 38.885 | 45.642 |
| 27 | 40.113 | 46.963 |
| 28 | 41.337 | 48.278 |
| 29 | 42.557 | 49.588 |
| 30 | 43.773 | 50.892 |
| 40 | 55.759 | 63.691 |
| 50 | 67.505 | 76.154 |
| 60 | 79.082 | 88.380 |
| 70 | 90.531 | 100.425 |
| 80 | 101.879 | 112.329 |
| 90 | 113.145 | 124.116 |
| 100 | 124.342 | 135.807 |

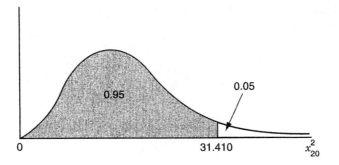

## APPENDIX E: PERCENTILES OF THE $F$ DISTRIBUTION

Entries in the table give $F_\alpha$ values, where $\alpha$ is the area or probability in the upper tail of the $F$ distribution. For example, with 3 numerator degrees of freedom, 20 denominator degrees of freedom, and a 0.01 area in the upper tail, $F_{0.01} = 4.94$. at df $= (3, 20)$.

| Denominator Degrees of Freedom | (Numerator Degrees of Freedom, Area in Upper Tail) | | | | | | | |
|---|---|---|---|---|---|---|---|---|
| | (2, .05) | (2, .01) | (3, .05) | (3, .01) | (4, .05) | (4, .01) | (5, .05) | (5, .01) |
| 5 | 5.79 | 13.27 | 5.41 | 12.06 | 5.19 | 11.39 | 4.82 | 10.97 |
| 6 | 5.14 | 10.92 | 4.76 | 9.78 | 4.53 | 9.15 | 4.39 | 8.75 |
| 7 | 4.74 | 9.55 | 4.35 | 8.45 | 4.12 | 7.85 | 3.97 | 7.46 |
| 8 | 4.46 | 8.65 | 4.07 | 7.59 | 3.84 | 7.01 | 3.69 | 6.63 |
| 9 | 4.26 | 8.02 | 3.86 | 6.99 | 3.63 | 6.42 | 3.48 | 6.06 |
| 10 | 4.10 | 7.56 | 3.71 | 6.55 | 3.48 | 5.99 | 3.33 | 5.64 |
| 11 | 3.98 | 7.21 | 3.59 | 6.22 | 3.36 | 5.67 | 3.20 | 5.32 |
| 12 | 3.89 | 6.93 | 3.49 | 5.95 | 3.26 | 5.41 | 3.11 | 5.06 |
| 13 | 3.81 | 6.70 | 3.41 | 5.74 | 3.18 | 5.21 | 3.03 | 4.86 |
| 14 | 3.74 | 6.51 | 3.34 | 5.56 | 3.11 | 5.24 | 2.96 | 4.69 |
| 15 | 3.68 | 6.36 | 3.29 | 5.42 | 3.06 | 4.89 | 2.90 | 4.56 |
| 16 | 3.63 | 6.23 | 3.24 | 5.29 | 3.01 | 4.77 | 2.85 | 4.44 |
| 17 | 3.59 | 6.11 | 3.20 | 5.18 | 2.96 | 4.67 | 2.81 | 4.34 |
| 18 | 3.55 | 6.01 | 3.16 | 5.09 | 2.93 | 4.58 | 2.77 | 4.25 |
| 19 | 3.52 | 5.93 | 3.13 | 5.01 | 2.90 | 4.50 | 2.74 | 4.17 |
| 20 | 3.49 | 5.85 | 3.10 | 4.94 | 2.87 | 4.43 | 2.71 | 4.10 |
| 21 | 3.47 | 5.78 | 3.07 | 4.87 | 2.84 | 4.37 | 2.68 | 4.04 |
| 22 | 3.44 | 5.72 | 3.05 | 4.82 | 2.82 | 4.31 | 2.66 | 3.99 |
| 23 | 3.42 | 5.66 | 3.03 | 4.76 | 2.80 | 4.26 | 2.64 | 3.94 |
| 24 | 3.40 | 5.61 | 3.01 | 4.72 | 2.78 | 4.22 | 2.62 | 3.90 |
| 25 | 3.39 | 5.57 | 2.99 | 4.68 | 2.76 | 4.18 | 2.60 | 3.85 |
| 26 | 3.37 | 5.53 | 2.98 | 4.64 | 2.74 | 4.14 | 2.59 | 3.82 |
| 27 | 3.35 | 5.49 | 2.96 | 4.60 | 2.73 | 4.11 | 2.57 | 3.78 |
| 28 | 3.34 | 5.45 | 2.95 | 4.57 | 2.71 | 4.07 | 2.56 | 3.75 |
| 29 | 3.33 | 5.42 | 2.93 | 4.54 | 2.70 | 4.04 | 2.55 | 3.73 |
| 30 | 3.32 | 5.39 | 2.92 | 4.51 | 2.69 | 4.02 | 2.53 | 3.70 |
| 35 | 3.27 | 5.27 | 2.87 | 4.40 | 2.64 | 3.91 | 2.49 | 3.59 |
| 40 | 3.23 | 5.18 | 2.84 | 4.31 | 2.61 | 3.83 | 2.45 | 3.51 |
| 50 | 3.18 | 5.06 | 2.79 | 4.20 | 2.56 | 3.72 | 2.40 | 3.41 |
| 60 | 3.15 | 4.98 | 2.76 | 4.13 | 2.53 | 3.65 | 2.37 | 3.34 |
| 80 | 3.11 | 4.88 | 2.72 | 4.04 | 2.49 | 3.56 | 2.33 | 3.26 |
| 100 | 3.09 | 4.82 | 2.70 | 3.98 | 2.46 | 3.51 | 2.31 | 3.21 |
| 120 | 3.07 | 4.79 | 2.68 | 3.95 | 2.45 | 3.48 | 2.29 | 3.17 |
| $\infty$ | 3.00 | 4.61 | 2.60 | 3.78 | 2.37 | 3.32 | 2.21 | 3.02 |

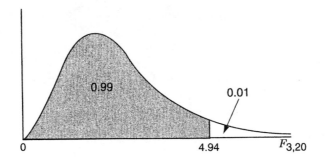

# ANSWERS TO SELECTED EXERCISES

*Chapter 1*

**1.1** For left-handed: $p = 0.517$; for right-handed: $p = 0.361$.

**1.2** For factory workers: $x = 49$; for nursing students: $x = 73$.

**1.3** For cases: $p = 0.775$; for controls: $p = 0.724$.

**1.4** For nursing home A: $p = 0.250$; for nursing home B: $p = 0.025$. The proportion in nursing home A where the "index" nurse worked is 10 times higher than the proportion in nursing home B.

**1.5** **(a)** For high exposure level: $p = 0.488$; for low exposure level: $p = 0.111$. Students in the high-exposure group have a much higher proportion of cases, more than four times higher.

   **(b)** OR $= 7.66$; it supports the conclusion in part (a) showing that high exposure is associated with higher odds, and so a higher proportion, of positive cases.

**1.6** **(a)** With EFM: $p = 0.126$; without EFM: $p = 0.077$. The EFM-exposed group has a higher proportion of cesarean deliveries.

   **(b)** OR $= 1.72$; it supports the conclusion in part (a) showing that EFM exposure is associated with higher odds, and so a higher proportion, of cesarean deliveries.

**1.7** **(a)** With helmet: $p = 0.116$; without helmet: $p = 0.338$. The group without helmet protection has a higher proportion of head injuries.

**(b)** OR = 0.26; it supports the conclusion in part (a) showing that helmet protection is associated with reduced odds, and so a lower proportion, of head injuries.

**1.8** **(a)** Men: OR = 0.94, indicating a slightly lower risk of myocardial infarction. Women: OR = 0.55, indicating a lower risk of myocardial infarction.

**(b)** The effect of drinking is stronger in women, reducing the risk more: 45% versus 6%.

**(c)** Men: OR = 0.57 indicating a lower risk of coronary death; women: OR = 0.39 indicating a lower risk of coronary death.

**(d)** The effect of drinking is stronger in women, reducing the risk more: 61% versus 43%.

**1.9** **(a)** Zero or one partner: OR = 2.70; two or more partners: OR = 1.10.

**(b)** Both odds ratios indicate an elevated risk associated with smoking, but the effect on those with zero or one partner is clearer.

**(c)** Combined estimate of odds ratio: $OR_{MH} = 1.26$.

**1.15** Sensitivity = 0.733; specificity = 0.972.

**1.16** For Dupont's EIA: sensitivity = 0.938; specificity = 0.988; for Cellular Product's EIA: sensitivity = 1.0; specificity = 0.952.

**1.17** **(a)** Heart disease, 30.2%; cancer, 24.1%; coronary disease, 8.2%; accidents, 4.0%; others, 33.4%; **(b)** population size: 3,525,297; **(c)** rates per 100,000: cancer, 235.4; coronary disease, 80.3; accidents, 39.2; others, 325.5.

**1.19** OR = 1.43.

**1.20** **(a)** For nonsmokers: OR = 1.28; **(b)** for smokers: OR = 1.61; **(c)** the risk seems to be higher for smokers; **(d)** combined estimate: $OR_{MH} = 1.53$.

**1.21** **(a)** Odds ratios associated with race (black vs. white, nonpoor): **(i)** 25–44: 1.07; **(ii)** 45–64: 2.00; **(iii)** 65+: 1.54. Different ratios indicate possible effect modifications by age.

**(b)** Odds ratios associated with income (poor vs. nonpoor, black): **(i)** 25–44: 2.46; **(ii)** 45–64: 1.65; **(iii)** 65+: 1.18. Different ratios indicate a possible effect modification by age.

**(c)** Odds ratios associated with race (black vs. white) **(i)** for 65+ years + poor: 1.18; **(ii)** for 65+ years + nonpoor: 1.54. The difference (1.18 ≠ 1.52) indicates a possible effect modification by income.

**1.22** **(a)**

| Age Group | Odds Ratio |
| --- | --- |
| 25–44 | 7.71 |
| 45–64 | 6.03 |
| 65+ | 3.91 |

(b) The odds ratios decrease with increasing age; (c) $OR_{MH} = 5.40$.

**1.23** (a)

| Weight Group | Odds Ratio |
|---|---|
| <57 | 5.00 |
| 57–75 | 2.75 |
| >75 | 1.30 |

(b) The odds ratios decrease with increasing weight; (c) $OR_{MH} = 2.78$.

**1.24**   For age at first live birth:

| Age | Odds Ratio |
|---|---|
| 22–24 | 1.26 |
| 25–27 | 1.49 |
| 28+ | 1.95 |

General odds ratio is 1.43.
For age at menopause:

| Age | Odds Ratio |
|---|---|
| 45–49 | 1.10 |
| 50+ | 1.40 |

General odds ratio is 1.27.

**1.25**   Results for parts (b) and (c):
For smoking:

| Status | Odds Ratio |
|---|---|
| Past | 1.24 |
| Current | 1.40 |

General odds ratio is 1.23.
For alcohol:

| Status | Odds Ratio |
|---|---|
| Past | 1.08 |
| Current | 0.87 |

General odds ratio is 0.88.

For body mass index:

| BMI Level | Odds Ratio |
| --- | --- |
| 22.5–24.9 | 1.41 |
| 25+ | 1.39 |

General odds ratio is 1.22.

**1.26**    For duration of unprotected intercourse:

| Years | Odds Ratio |
| --- | --- |
| 2–9 | 0.94 |
| 10–14 | 1.04 |
| 15+ | 1.54 |

General odds ratio is 1.21, showing an upward trend of risks.
For history of infertility:

| History | Odds Ratio |
| --- | --- |
| Yes, no drug use | 1.13 |
| Yes, with drug use | 3.34 |

General odds ratio is 1.33, showing an upward trend of risks.

**1.27**    For boys:

| Maternal Age | Odds Ratio |
| --- | --- |
| 21–25 | 0.97 |
| 26–30 | 0.67 |
| 30+ | 0.43 |

General odds ratio is 0.62, showing an downward trend of risks, decreasing with increasing maternal age.
For girls:

| Maternal Age | Odds Ratio |
| --- | --- |
| 21–25 | 0.80 |
| 26–30 | 0.81 |
| 30+ | 0.62 |

General odds ratio is 0.84, showing an downward trend of risks, decreasing with increasing maternal age, but the trend is weaker than that for boys.

**1.28**    Sensitivity = 0.650; specificity = 0.917.

**1.29** (a) For 1987: 24,027; for 1986: 15,017; (b) number of cases of AIDS transmitted from mothers to newborns in 1988: 468.

**1.30** Follow-up death rates:

| Age (yr) | Deaths/1000 months |
|----------|--------------------|
| 21–30 | 3.95 |
| 31–40 | 5.05 |
| 41–50 | 11.72 |
| 51–60 | 9.80 |
| 61–70 | 10.19 |
| 70+ | 20.09 |

RR(70+ years vs. 51–60 years) = 2.05.

**1.31** Death rates for Georgia (per 100,000):
  (a) Crude rate for Georgia:                                             908.3
  (b) Adjusted rate (United States as standard), Georgia:   1060.9
      vs. adjusted rate for Alaska:                                       788.6
      vs. adjusted rate for Floria:                                        770.6
  (c) Adjusted rate (Alaska as standard):                          560.5
      vs. crude rate for Alaska:                                           396.8

**1.32** With Georgia as standard: Alaska, 668.4; Florida, 658.5 vs. crude rate for Georgia of 908.3.

**1.33** Standardized mortality ratios:

| Years Since Entering the Industry | 1–4 | 5–9 | 10–14 | 15+ |
|-----------------------------------|-----|-----|-------|-----|
| SMR | 0.215 | 0.702 | 0.846 | 0.907 |

RR(15+ years vs. 1–4 years): 4.22.

**1.35** (c) General odds ratio is 1.31, showing an upward trend with coffee consumption: given two person who were admitted for different conditions, the odds that the one with the acute condition will consume more coffee is 1.31.

**1.36** (a)

| Group | Proportion of HBV-Positive Workers |
|-------|-------------------------------------|
| Physicians | |
|   Frequent | 0.210 |
|   Infrequent | 0.079 |
| Nurses | |
|   Frequent | 0.212 |
|   Infrequent | 0.087 |

**(b)** Odds ratios associated with frequent contacts: for physicians, 3.11; for nurses, 2.80; **(c)** odds ratios are similar but larger for physicians; **(d)** $OR_{MH} = 2.93$.

**1.38** **(a)** For age:

| Group | Odds Ratio |
|-------|-----------|
| 14–17 | 2.09 |
| 18–19 | 1.96 |
| 20–24 | 1.69 |
| 25–29 | 1.02 |

Yes, the younger the mother, the higher the risk.

**(b)** For socioeconomic level:

| Group | Odds Ratio |
|-------|-----------|
| Upper | 0.30 |
| Upper middle | 0.34 |
| Middle | 0.56 |
| Lower middle | 0.71 |

Yes, the poorer the mother, the higher the risk.

**(c)** General odds ratio is 0.57: given two mothers with different economic levels, the odds that the richer one having premature delivery is 0.57. Yes, it supports the results in part (b).

**1.39** **(a)**

| Group | SMR |
|-------|-----|
| Male | 1.22 |
| Female | 0.78 |
| Black | 2.09 |
| White | 0.85 |

**(b)** Relative risks: associated with gender, 1.57; associated with race, 2.47.

**1.40**

| Group | Odds Ratio |
|-------|-----------|
| Protestants | 0.50 |
| Catholics | 4.69 |
| Others | 0.79 |

Yes, there is clear evidence of an effect modification $(4.69 \neq 0.50, 0.79)$.

**1.41**  For age at first live birth, with "28 or older" group as baseline:

| Age | Odds Ratio |
|-----|-----------|
| <22 | 0.51 |
| 22–24 | 0.65 |
| 25–27 | 0.76 |

For age at menopause, with "50 or older" group as baseline:

| Age | Odds Ratio |
|-----|-----------|
| <45 | 0.72 |
| 45–49 | 0.79 |

**1.42**

| Symptom | Odds Ratio |
|---------|-----------|
| Nightmares | 3.72 |
| Sleep problems | 1.54 |
| Troubled memories | 3.46 |
| Depression | 1.46 |
| Temper control problems | 1.78 |
| Life goal association | 1.50 |
| Omit feelings | 1.39 |
| Confusion | 1.57 |

**1.43**

| Group | Odds Ratio |
|-------|-----------|
| Males | |
| Low fat, low fiber | 1.15 |
| High fat, high fiber | 1.80 |
| High fat, low fiber | 1.81 |
| Females | |
| Low fat, low fiber | 1.72 |
| High fat, high fiber | 1.85 |
| High fat, low fiber | 2.20 |

Yes, there are evidences of effect midifications. For example, for males: high vs. low fat, the odds ratio is 1.8 with high fiber and 1.57 (=1.81/1.15) with low fiber; for females: high vs. low fat, the odds ratio is 1.85 with high fiber and 1.28 (=2.20/1.72) with low fiber.

**1.45**  (a) Choosing "never" as baseline, here are the odds ratios associated with being a resident (vs. attending physician):

| Action Level | Odds Ratio |
|---|---|
| Rarely | 1.93 |
| Occasionally | 24.3 |
| Frequently | 33.8 |
| Very frequently | 5.2 |

**(b)** The general odds ratio is 6.15; given two physician of different types, the odds that the resident committing more unnecessary transfusion is 6.15; Yes, this agrees with the results in part **(a)**.

**1.46**

| Factor | Odds Ratio |
|---|---|
| X-ray | 8.86 |
| Stage | 5.25 |
| Grade | 3.26 |

## Chapter 2

**2.1** Median = 193.

**2.2** Median for 1979 is 47.6, for 1987 is 27.8.

**2.8** Median from graph is 83, exact is 86.

**2.9**

| | $\bar{x}$ | $s^2$ | $s$ | $s/\bar{x}$ |
|---|---|---|---|---|
| Men | 84.71 | 573.68 | 23.95 | 28.3% |
| Women | 88.51 | 760.90 | 27.58 | 31.2% |

**2.11** $\bar{x} = 168.75$, $s^2 = 1372.75$, $s = 37.05$.

**2.12** $\bar{x} = 3.05$, $s^2 = 0.37$, $s = 0.61$.

**2.13** $\bar{x} = 112.78$, $s^2 = 208.07$, $s = 14.42$.

**2.14** $\bar{x} = 100.58$, $s^2 = 196.08$, $s = 14.00$.

**2.15** $\bar{x} = 0.718$, $s^2 = 0.261$, $s = 0.511$.

**2.16**

| | $\bar{x}$ | $s^2$ | $s$ |
|---|---|---|---|
| Age | 65.6 | 243.11 | 15.19 |
| SBP | 146.2 | 379.46 | 19.48 |

**2.17**

| | $\bar{x}$ | $s^2$ | $s$ |
|---|---|---|---|
| Females | 107.6 | 4373.5 | 66.1 |
| Males | 97.8 | 1635.9 | 40.4 |

**2.18** $\bar{x} = 0.22$, $s^2 = 0.44$, $s = 0.66$.

**2.19**

|  | $\bar{x}$ | $s^2$ | $s$ |
|---|---|---|---|
| Treatment | 651.9 | 31,394.3 | 177.2 |
| Control | 656.1 | 505.1 | 22.5 |

**2.20**  (a) $\bar{x} = 169.0$, mean is lower than LA's; $s^2 = 81.5$, $s = 9.0$; (b) CV = 5%.

**2.21**  $\bar{x} = 2.5$, $s^2 = 9.5$, $s = 3.1$.

**2.22**  Mean = 14.1, geometric mean = 10.3, median = 12.5.

**2.23**  Mean = 21.3, geometric mean = 10.6, median = 10.0.

**2.24**  Mean = 98.5, geometric mean = 93.4, median = 92.5.

**2.25**

|  | $\bar{x}$ | $s^2$ |
|---|---|---|
| Bulimic | 22.1 | 21.0 |
| Healthy | 29.7 | 42.1 |

Bulimic group has smaller mean and smaller variance.

**2.26**

|  | $\bar{x}$ | $s^2$ | $s$ | Median |
|---|---|---|---|---|
| Drug A | 133.9 | 503.8 | 22.4 | 130.0 |
| Drug R | 267.4 | 4449.0 | 66.7 | 253.0 |

**2.27**  (a) For survival time:

|  | $\bar{x}$ | $s^2$ | $s$ |
|---|---|---|---|
| AG positive | 62.5 | 2954.3 | 54.4 |
| AG negative | 17.9 | 412.2 | 20.3 |

(b) For WBC:

|  | $\bar{x}$ | Geometric Mean | Median |
|---|---|---|---|
| AG positive | 29,073.5 | 12,471.9 | 10,000 |
| AG negative | 29,262.5 | 13,771.9 | 20,000 |

**2.28**

| Measure | Value |
|---|---|
| Pearson's | 0.931 |
| Kendall's | 0.875 |
| Spearman | 0.963 |

**2.29**

| Measure | Men | Women |
|---|---|---|
| Pearson's | 0.514 | 0.718 |
| Kendall's | 0.282 | 0.714 |
| Spearman | 0.377 | 0.849 |

**2.30**

| Measure | Value |
|---|---|
| Pearson's | −0.786 |
| Kendall's | −0.619 |
| Spearman | −0.767 |

**2.31**

| Measure | Standard | Test |
|---|---|---|
| Pearson's | 0.940 | 0.956 |
| Kendall's | 0.896 | 0.955 |
| Spearman | 0.960 | 0.988 |

**2.32**  20 patients with nodes, 33 patients without nodes.

| Factor | $\bar{x}$ | $s^2$ | $s$ |
|---|---|---|---|
| Age | | | |
|   Without node | 60.1 | 31.4 | 5.6 |
|   With node | 58.3 | 49.1 | 7.0 |
| Acid | | | |
|   Without node | 64.5 | 744.2 | 27.3 |
|   With node | 77.5 | 515.3 | 22.7 |

**2.33**  All patients: 0.054; with nodes: 0.273; without nodes: −0.016. Nodal involvement seems to change the strength of the relationship.

**2.34**  **(a)** 12 females, 32 males; 20 with residency, 24 without.

| Factor | $\bar{x}$ | $s$ |
|---|---|---|
| Gender | | |
|   Females | 0.00116 | 0.00083 |
|   Males | 0.00139 | 0.00091 |
| Residency | | |
|   Without | 0.00147 | 0.00099 |
|   With | 0.00116 | 0.00073 |

**(b)** Between complaints and revenue: 0.031; between complaints and work load: 0.279.

**2.35**  **(a)** Between y and $x_1$ is 0.756; **(b)** between y and $x_2$ is 0.831.

### Chapter 3

**3.1**  Odds   ratio $= 0.62$;   $\Pr(\text{Pap} = \text{yes}) = 0.82$,   $\Pr(\text{Pap} = \text{yes} \mid \text{black}) = 0.75 \neq 0.82$.

**3.2**  Odds   ratio $= 5.99$;   $\Pr(\text{second} = \text{present}) = 0.65$,   $\Pr(\text{second} = \text{present} \mid \text{first} = \text{present}) = 0.78 \neq 0.65$.

**3.3**  (a) 0.202; (b) 0.217; (c) 0.376; (d) 0.268.

**3.4**  Positive predictive value for population A: 0.991, for population B: 0.900.

**3.5**  (a) Sensitivity = 0.733, specificity = 0.972; (b) 0.016; (c) 0.301.

**3.6**

| Prevalence | Positive Predictive Value |
|---|---|
| 0.2 | 0.867 |
| 0.4 | 0.946 |
| 0.6 | 0.975 |
| 0.7 | 0.984 |
| 0.8 | 0.991 |
| 0.9 | 0.996 |

Yes:

$$\text{prevalence} = \frac{(\text{PPV})(1 - \text{specificity})}{(\text{PPV})(1 - \text{specificity}) + (1 - \text{PPV})(\text{sensitivity})}$$

$$= 0.133 \quad \text{if PPV} = 0.8$$

**3.7**  (a) 0.1056; (b) 0.7995.

**3.8**  (a) 0.9500; (b) 0.0790; (c) 0.6992.

**3.9**  (a) 0.9573; (b) 0.1056.

**3.10**  (a) 1.645; (b) 1.96; (c) 0.84.

**3.11**  $102.8 = 118.4 - (1.28)(12.17)$.

**3.12**  20–24 years:  101.23   106.31   141.49   146.57
25–29 years:  104.34   109.00   141.20   145.86
For example, $123.9 - (1.645)(13.74) = 101.3$.

**3.13**  (a) 200+ days: 0.1587; 365+ days: $\simeq 0$; (b) 0.0228.

**3.14**  (a) 0.0808; (b) 82.4.

**3.15**  (a) 17.2; (b) 19.2%; (c) 0.0409.

**3.16**  (a) 0.5934; (b) 0.0475; (c) 0.0475.

**3.17**  (a) $\simeq 0$ ($z = 3.79$); (b) $\simeq 0$ ($z = -3.10$).

**3.18**  (a) 0.2266; (b) 0.0045; (c) 0.0014.

**3.19**  (a) 0.0985; (b) 0.0019.

**3.20**  Rate = 13.89 per 1000 live births; $z = 13.52$.

**3.21**  (a) Left of 2.086: 0.975; left of 2.845: 0.995; (b) right of 1.725: 0.05; right of 2.528: 0.01; (c) beyond $\pm 2.086$: 0.05; beyond $\pm 2.845$: 0.01.

**3.22** **(a)** Right of 5.991: 0.05; right of 9.210: 0.01; **(b)** right of 6.348: between 0.01 and 0.05; **(c)** between 5.991 and 9.210: 0.04.

**3.23** **(a)** Right of 3.32: 0.05; right of 5.39: 0.01; **(b)** right of 2.61: $<0.01$; **(c)** between 3.32 and 5.39: 0.04.

**3.24** $\kappa = 0.399$, marginally good agreement.

**3.25** $\kappa = 0.734$, good agreement—almost excellent.

## Chapter 4

**4.1** $\mu = 0.5$; 6 possible samples with $\mu_{\bar{x}} = 0.5 \; (=\mu)$.

**4.2** $\Pr(\mu - 1 \le \bar{x} \le \mu + 1) = \Pr(-2.33 \le z \le 2.33)$
$$= 0.9802$$

**4.3**

| Group | 95% Confidence Interval |
|---|---|
| Left-handed | $(0.444, 0.590)$ |
| Right-handed | $(0.338, 0.384)$ |

**4.4**

| Group | 95% Confidence Interval |
|---|---|
| Students | $(0.320, 0.460)$ |
| Workers | $(0.667, 0.873)$ |

**4.5**

| Year | 95% Confidence Interval |
|---|---|
| 1983 | $(0.455, 0.493)$ |
| 1987 | $(0.355, 0.391)$ |

**4.6**

| Level | 95% Confidence Interval |
|---|---|
| High | $(0.402, 0.574)$ |
| Low | $(0.077, 0.145)$ |

**4.7** For whites: $25.3 \pm (1.96)(0.9) = (23.54\%, 27.06\%)$
For blacks: $38.6 \pm (1.96)(1.8) = (35.07\%, 42.13\%)$

No, sample sizes were already incorporated into standard errors.

**4.8**

| Parameter | 95% Confidence Interval |
|---|---|
| Sensitivity | $(0.575, 0.891)$ |
| Specificity | $(0.964, 0.980)$ |

**4.9**

| Assay | Parameter | 95% Confidence Interval |
|---|---|---|
| Dupont's | Sensitivity | $(0.820, 1.0)$ |
| | Specificity | $(0.971, 1.0)$ |
| Cellular Product's | Sensitivity | not available |
| | Specificity | $(0.935, 0.990)$ |

*Note:* Sample sizes are very small here.

**4.10**

| Personnel | Exposure | 95% Confidence Interval |
|---|---|---|
| Physicians | Frequent | $(0.121, 0.299)$ |
| | Infrequent | $(0.023, 0.135)$ |
| Nurses | Frequent | $(0.133, 0.291)$ |
| | Infrequent | $(0.038, 0.136)$ |

**4.11** **(a)** Proportion: $(0.067, 0.087)$; odds ratio: $(1.44, 2.05)$.

**4.12** **(a)** Proportion: $(0.302, 0.374)$; odds ratio: $(0.15, 0.44)$.

**4.13**

| Event | Gender | 95% Confidence Interval for OR |
|---|---|---|
| Myocardial infarction | Men | $(0.69, 1.28)$ |
| | Women | $(.30, .99)$ |
| Coronary death | Men | $(.39, .83)$ |
| | Women | $(.12, 1.25)$ |

No clear evidence of effect modification, intervals are overlapped.

**4.14**

| Religion Group | 95% Confidence Interval for OR |
|---|---|
| Protestant | $(0.31, 0.84)$ |
| Catholic | $(1.58, 13.94)$ |
| Others | $(0.41, 1.51)$ |

The odds ratio for the Catholics is much higher.

**4.15** **(a)** Proportion: $(0.141, 0.209)$; odds ratio: $(1.06, 1.93)$.

**4.16** **(a)** Odds ratio $= \dfrac{(177)(31)}{(11)(249)}$

$$= 2.0$$

$$\exp\left(\ln 2.0 \pm 1.96\sqrt{\frac{1}{177} + \frac{1}{31} + \frac{1}{11} + \frac{1}{249}}\right) = (0.98, 4.09)$$

**(b)** Odds ratio $= \dfrac{(177)(26)}{(11)(233)}$

$$= 1.78$$

$$\exp\left( \ln 1.78 \pm 1.96\sqrt{\frac{1}{177} + \frac{1}{26} + \frac{1}{11} + \frac{1}{233}} \right) = (0.86, 3.73)$$

**4.17**

| Maternal Age (years) | OR (95% Confidence Interval) | |
|---|---|---|
| | Boys | Girls |
| <21 | 2.32 (0.90, 5.99) | 1.62 (0.62, 4.25) |
| 21–25 | 2.24 (1.23, 4.09) | 1.31 (0.74, 2.30) |
| 26–30 | 1.55 (0.87, 2.77) | 1.31 (0.76, 2.29) |

**4.18** **(a)**

| Duration (years) | OR (95% Confidence Interval) |
|---|---|
| 2–9 | 0.94 (0.74, 1.20) |
| 10–14 | 1.04 (0.71, 1.53) |
| ≥15 | 1.54 (1.17, 2.02) |

**(b)**

| Group | OR (95% Confidence Interval |
|---|---|
| No drug use | 1.13 (0.83, 1.53) |
| Drug use | 3.34 (1.59, 7.02) |

**4.19**

$$\bar{x} = 3.05$$

$$s = 0.61$$

$$SE(\bar{x}) = 0.19$$

$$3.05 \pm (2.262)(0.19) = (2.61, 3.49)$$

**4.20**

$$\bar{x} = 112.78$$

$$s = 14.42$$

$$SE(\bar{x}) = 3.40$$

$$112.78 \pm (2.110)(3.40) = (105.61, 119.95)$$

**4.21**

$$\bar{x} = 100.58$$

$$s = 14.00$$

$$SE(\bar{x}) = 4.04$$

$$100.58 \pm (2.201)(4.04) = (91.68, 129.48)$$

**4.22**

| Group | 95% Confidence Interval |
|---|---|
| Females | $(63.2, 152.0)$ |
| Males | $(76.3, 119.3)$ |

**4.23**

$$\bar{x} = 0.22$$

$$s = 0.66$$

$$SE(\bar{x}) = 0.18$$

$$0.22 \pm (2.160)(0.18) = (-0.16, 0.60)$$

**4.24**

| Group | 95% Confidence Interval |
|---|---|
| Treatment | $(653.3, 676.9)$ |
| Control | $(681.1, 722.7)$ |

**4.25** **(a)** Mean: $(0.63, 4.57)$; **(b)** correlation coefficient: $(0.764, 0.981)$.

**4.26** On log scale: $(1.88, 2.84)$; in weeks: $(6.53, 17.19)$.

**4.27** Control:          $7.9 \pm (1.96)(3.7)/\sqrt{30} = (6.58, 9.22)$
Simulation game:   $10.1 \pm (1.96)(2.3)/\sqrt{33} = (9.32, 10.88)$

**4.28** **(a)** $25.0 \pm (1.96)(2.7)/\sqrt{58} = (24.31, 25.69)$; **(b)** on the average, people with large body mass index are more likely to develop diabetes mellitus.

**4.29**

| DBP Level | Exposure | 95% Confidence Interval |
|---|---|---|
| <95 | Yes | $(213.2, 226.8)$ |
|  | No | $(216.0, 226.0)$ |
| 95–100 | Yes | $(215.3, 238.6)$ |
|  | No | $(223.6, 248.4)$ |
| $\geq 100$ | Yes | $(215.0, 251.0)$ |
|  | No | $(186.2, 245.8)$ |

**4.30** $(-0.220, 0.320)$.

**4.31** $(0.545, 0.907)$.

**4.32** Standard: $(0.760, 0.986)$, test: $(0.820, 0.990)$.

**4.33** With smoke: $(0.398, 0.914)$, with sulfur dioxide: $(0.555, 0.942)$.

**4.34** Men: $(-0.171, 0.864)$, women: $(0.161, 0.928)$.

**4.35** AG positive: $(0.341, 0.886)$, AG negative: $(-0.276, 0.666)$.

### Chapter 5

**5.1** (a) $H_0$: $\mu = 30$; $H_A$: $\mu > 30$; (b) $H_0$: $\mu = 11.5$; $H_A$: $\mu \neq 11.5$; (c) hypotheses are for population parameters; (d) $H_0$: $\mu = 31.5$; $H_A$: $\mu < 31.5$; (e) $H_0$: $\mu = 16$; $H_A$: $\mu \neq 16$; (f) Same as part (c).

**5.2** $H_0$: $\mu = 74.5$.

**5.3** $H_0$: $\mu = 7250$; $H_A$: $\mu < 7250$.

**5.4** $H_0$: $\pi = 38$; $H_A$: $\pi > 38$.

**5.5** $H_0$: $\pi = 0.007$; $H_A$: $\mu > 0.007$.

$$p \text{ value} = \Pr(\geq 20 \text{ cases out of } 1000 \mid H_0)$$

$$H_0\text{: mean} = (1000)(0.007) = 7$$

$$\text{variance} = (1000)(0.007)(0.993) = (2.64)^2$$

$$z = \frac{20 - 7}{2.64}$$

$$= 4.92; \quad p \text{ value} \simeq 0$$

**5.6** The mean difference; $H_0$: $\mu_d = 0$, $H_A$: $\mu_d \neq 0$.

**5.7** $H_0$: $\pi = \dfrac{20.5}{5000}$

$H_A$: $\pi > \dfrac{20.5}{5000}$

**5.8** $H_0$: $\sigma = 20$, $H_A$: $\sigma \neq 20$.

**5.9** One-tailed.

**5.10** Under $H_0$: $\pi = 0.25$; variance $= \pi(1 - \pi)/100 = (0.043)^2$

$$z = \frac{0.18 - 0.25}{0.043}$$

$$= -1.63; \quad \alpha = 0.052$$

Under $H_A$: $\pi = 0.15$; variance $= \pi(1 - \pi)/100 = (0.036)^2$

$$z = \frac{0.18 - 0.15}{0.036}$$

$$= 0.83; \quad \beta = 0.2033$$

The change makes $\alpha$ smaller and $\beta$ larger.

**5.11**   Under $H_0$: $z = \dfrac{0.22 - 0.25}{0.043}$

$$= -0.70; \quad \alpha = 0.242$$

Under $H_A$: $z = \dfrac{0.22 - 0.15}{0.036}$

$$= 1.94; \quad \beta = 0.026$$

The new change makes $\alpha$ larger and $\beta$ smaller.

**5.12**   $p$ value $= \Pr(p < 0.18 \text{ or } p > 0.32)$

$$= 2 \Pr\left(z \geq \frac{0.32 - 0.25}{0.043} = 1.63\right)$$

$$= 0.1032$$

**5.13**
$$p = 0.18$$
$$\mathrm{SE}(p) = 0.0385$$

$0.18 \pm (1.96)(0.0385) = (0.105, 0.255)$, which includes 0.25

**5.14**   $\mathrm{SE}(\bar{x}) = 0.054$; cut point for $\alpha = 0.05$: $4.86 \pm (1.96)(0.054) = 4.75$ and 4.96.

$$\mu = 4.86 + 0.1$$
$$= 4.96$$
$$z = \frac{4.96 - 4.96}{0.054}$$
$$= 0; \quad \text{power} = 0.5$$

**5.15**   $\mathrm{SE}(\bar{x}) = 1.11$; cut points for $\alpha = 0.05$: $128.6 \pm (1.96)(1.11) = 126.4$ and 130.8.

$$\mu = 135$$
$$\text{power} = 1 - \Pr(126.4 \leq \bar{x} \leq 130.8 \mid \mu = 135)$$
$$\simeq 1.0$$

## Chapter 6

**6.1**   Proportion is $p = 0.227$; $z = -1.67$; $p$ value $= (2)(0.0475) = 0.095$.

**6.2**   $\chi^2 = 75.03$; $p$ value $\cong 0$.

**6.3**    $z = (264 - 249)/\sqrt{264 + 249}$

$= 0.66;$    $p$ value $= 0.5092$

or $\chi^2 = 0.44.$

**6.4**    $z = 3.77$ or $\chi^2 = 14.23;$ $p$ value $< 0.01.$

**6.5**    $z = 4.60;$ $p$ value $\simeq 0.$

**6.6**    $H_0$: consistent report for a couple, i.e., man and woman agree.

$z = (6 - 7)/\sqrt{6 + 7}$

$= -0.28;$    $p$ value $= 0.7794$

**6.7**    $H_0$: no effects of acrylate and methacrylate vapors on olfactory function.

$z = (22 - 9)/\sqrt{22 + 9}$

$= 2.33;$    $p$ value $= 0.0198$

**6.8**    $z = 4.11$ or $\chi^2 = 16.89;$ $p$ value $< 0.01.$

**6.9**    $z = 5.22$ or $\chi^2 = 27.22;$ $p$ value $< 0.01.$

**6.10**    $z = 7.44$ or $\chi^2 = 55.36;$ $p$ value $< 0.01.$

**6.11**    $\chi^2 = 44.49;$ $p$ value $< 0.01.$

**6.12**    $\chi^2 = 37.73;$ $p$ value $< 0.01.$

**6.13**    $\chi^2 = 37.95;$ $p$ value $< 0.01.$

**6.14**    $\chi^2 = 28.26;$ $p$ value $< 0.01.$

**6.15**    $\chi^2 = 5.58;$ $p$ value $< 0.05.$

**6.16**    $p$ value $= 0.002.$

**6.17**    **(a)** Pearson's: $\chi^2 = 11.35;$ $p$ value $= 0.001;$ **(b)** Pearson's with Yates' correction: $\chi_c^2 = 7.11;$ $p$ value $= 0.008;$ **(c)** Fisher's exact: $p$ value $= 0.013.$

**6.18**    $p$ value $= 0.018.$

**6.19**    $\chi^2 = 14.196;$ df $= 3;$ $p$ value $< 0.05.$

**6.20**    For males: $\chi^2 = 6.321;$ df $= 3;$ $p$ value $= 0.097;$ for females: $\chi^2 = 5.476;$ df $= 3;$ $p$ value $= 0.140.$

**6.21**    For men:
Myocardial infarction: $\chi^2 = 0.16;$ $p$ value $> 0.05.$
Coronary death: $\chi^2 = 8.47;$ $p$ value $< 0.01.$

For Women:
Myocardial infarction: $\chi^2 = 4.09$; $p$ value $< 0.05$.
Coronary death: $\chi^2 = 2.62$; $p$ value $> 0.05$.

**6.22**  Myocardial infarction:

For men: $a = 197$

$$\frac{r_1 c_1}{n} = 199.59$$

$$\frac{r_1 r_2 c_1 c_2}{n^2(n-1)} = 40.98$$

For women: $a = 144$

$$\frac{r_1 c_1}{n} = 150.95$$

$$\frac{r_1 r_2 c_1 c_2}{n^2(n-1)} = 12.05$$

$$z = \frac{(197 - 199.59) + (144 - 150.95)}{\sqrt{40.98 + 12.05}}$$

$$= -1.31; \quad p \text{ value} = 0.0951$$

Coronary death:

For men: $a = 135$

$$\frac{r_1 c_1}{n} = 150.15$$

$$\frac{r_1 r_2 c_1 c_2}{n^2(n-1)} = 27.17$$

For women: $a = 89$

$$\frac{r_1 c_1}{n} = 92.07$$

$$\frac{r_1 r_2 c_1 c_2}{n^2(n-1)} = 3.62$$

$$z = \frac{(135 - 150.15) + (89 - 92.07)}{\sqrt{27.17 + 3.62}}$$

$$= -3.28; \quad p \text{ value} < 0.001$$

**6.23**    For 25–44 years: $a = 5$

$$\frac{r_1 c_1}{n} = 1.27$$

$$\frac{r_1 r_2 c_1 c_2}{n^2(n-1)} = 1.08$$

For 45–64 years: $a = 67$

$$\frac{r_1 c_1}{n} = 32.98$$

$$\frac{r_1 r_2 c_1 c_2}{n^2(n-1)} = 17.65$$

For 65+ years: $a = 24$

$$\frac{r_1 c_1}{n} = 13.28$$

$$\frac{r_1 r_2 c_1 c_2}{n^2(n-1)} = 7.34$$

$$z = \frac{(5 - 1.27) + (67 - 32.98) + (24 - 13.28)}{\sqrt{1.08 + 17.65 + 7.34}}$$

$$= 9.49; \quad p \text{ value} \simeq 0$$

**6.24**    <57 kg: $a = 20$

$$\frac{r_1 c_1}{n} = 9.39$$

$$\frac{r_1 r_2 c_1 c_2}{n^2(n-1)} = 5.86$$

57–75 kg: $a = 37$

$$\frac{r_1 c_1}{n} = 21.46$$

$$\frac{r_1 r_2 c_1 c_2}{n^2(n-1)} = 13.60$$

>75 kg: $a = 9$

$$\frac{r_1 c_1}{n} = 7.63$$

$$\frac{r_1 r_2 c_1 c_2}{n^2(n-1)} = 4.96$$

$$z = \frac{(20 - 9.39) + (37 - 21.46) + (9 - 7.63)}{\sqrt{5.86 + 13.60 + 4.96}}$$

$$= 5.57; \quad p \text{ value} \simeq 0$$

**6.25** For smoking: $\chi^2 = 0.93$; $p$ value $> 0.05$; for alcohol: $\chi^2 = 0.26$; $p$ value $> 0.05$; for body mass index: $\chi^2 = 0.87$; $p$ value $> 0.05$.

**6.26** $C = 24{,}876$

$D = 14{,}159$

$S = 10{,}717$

$\sigma_S = 2341.13$

$z = 4.58; \quad p \text{ value} \simeq 0$

(after eliminating category "unknown").

**6.27** For boys: $C = 19{,}478$

$D = 11{,}565$

$S = 7913$

$\sigma_S = 2558.85$

$z = 3.09; \quad p \text{ value} = 0.001$

For girls: $C = 17{,}120$

$D = 14{,}336$

$S = 2784$

$\sigma_S = 2766.60$

$z = 1.01; \quad p \text{ value} = 0.1587$

**6.28** For duration (years):

$C = 237{,}635$

$D = 240{,}865$

$S = 16{,}770$

$\sigma_S = 17{,}812.60$

$z = 0.94; \quad p \text{ value} = 0.1736$

For history of infertility:

$$C = 95{,}216$$

$$D = 71{,}846$$

$$S = 23{,}270$$

$$\sigma_S = 11{,}721.57$$

$$z = 1.99; \quad p \text{ value} = 0.0233$$

**6.29**    $C = 2{,}442{,}198$

$$D = 1{,}496{,}110$$

$$S = 946{,}088$$

$$\sigma_S = 95{,}706$$

$$z = 9.89; \quad p \text{ value} \simeq 0$$

**6.30**    $C = 364$

$$D = 2238$$

$$S = -1874$$

$$\sigma_S = 372$$

$$z = -5.03; \quad p \text{ value} \simeq 0$$

## Chapter 7

**7.1**    $\text{SE}(\bar{x}) = 0.5$

$$t = (7.84 - 7)/(0.5)$$

$$= 1.68, 15 \text{ df}; \quad 0.05 < p \text{ value} < 0.10$$

**7.2**    $\bar{d} = 200$

$$s_d = 397.2$$

$$\text{SE}(\bar{d}) = 150.1$$

$$t = (200 - 0)/(150 - 1)$$

$$= 1.33, 6 \text{ df}; \quad p \text{ value} > 0.20$$

**7.3**    $\bar{d} = 39.4$

$$s_d = 31.39$$

$$\text{SE}(\bar{d}) = 11.86$$

$$t = 3.32, 6 \text{ df}; \quad p \text{ value} = 0.016$$

**7.4**  $SE(\bar{d}) = 1.13$

$t = 3.95, 59$ df;    $p$ value $= 0.0002$

**7.5**    $\bar{d} = -1.10$

$s_d = 7.9$

$SE(\bar{d}) = 1.72$

$t = -0.64, 20$ df;    $p$ value $> 0.20$

**7.6**    $\bar{d} = 0.36$

$s_d = 0.41$

$SE(\bar{d}) = 0.11$

$t = 3.29, 13$ df;    $p$ value $= 0.006$

**7.7**  **(a)**  $t = 11.00$, $p$ value $< 0.001$.
    **(b)**  $t = 7.75$, $p$ value $< 0.001$.
    **(c)**  $t = 2.21$, $p$ value $= 0.031$.

**(d)**  $s_p = 1.47$

$t = 2.70$;    $p < 0.01$

**(e)**  $s_p = 1.52$

$t = 1.19$;    $p > 0.20$

**(f)**  $\leq$High school:  $s_p = 1.53$

$t = 6.53$;    $p \simeq 0$

$\geq$College:  $s_p = 1.35$

$t = 3.20$;    $p \simeq 0$

**7.8**  SBP: $t = 12.11$, $p \cong 0$; DBP: $t = 10.95$, $p \cong 0$; BMI: $t = 6.71$, $p \cong 0$.

**7.9**  $s_p = 0.175$

$t = 3.40$;    $p \simeq 0$

**7.10**  $t = 0.54, 25$ df, $p$ value $> 0.2$.

**7.11**  $t = 2.86, 61$ df, $p$ value $= 0.004$.

**7.12**  $s_p = 7.9$

$t = 4.40$;    $p \simeq 0$

**7.13** Treatment: $\bar{x}_1 = 701.9$, $s_1 = 32.8$; control: $\bar{x}_2 = 656.1$, $s_2 = 22.5$.

$s_p = 29.6$

$t = 3.45, 17$ df; $\quad p < 0.01$

**7.14** $s_p = 9.3$

$t = 4.71$; $\quad p \simeq 0$

**7.15** Weight gain: $s_p = 14.4$

$t = 2.30$; $\quad p < 0.05$

Birth weight: $s_p = 471.1$

$t = 2.08$; $\quad p < 0.05$

Gestational age: $s_p = 15.3$

$t \simeq 0$; $\quad p \simeq 0.5$

**7.16** Sum of ranks: bulimic adolescents, 337.5; healthy adolescents, 403.5.

$$\mu_H = \frac{(15)(15 + 23 + 1)}{2}$$

$$= 292.5$$

$$\sigma_H = \sqrt{\frac{(15)(23)(15 + 23 + 1)}{12}}$$

$$= 33.5$$

$$z = \frac{403.5 - 292.5}{33.5}$$

$$= 3.31; \quad p \simeq 0$$

**7.17** Sum of ranks: experimental group, 151; control group, 39.

$$\mu_E = \frac{(12)(12 + 7 + 1)}{2}$$

$$= 120$$

$$\sigma_E = \sqrt{\frac{(12)(7)(12 + 7 + 1)}{12}}$$

$$= 11.8$$

$$z = \frac{151 - 120}{11.8}$$

$$= 262; \quad p = 0.0088$$

**7.18**

| Source of Variation | SS | df | MS | F Statistic | p Value |
|---|---|---|---|---|---|
| Between samples | 55.44 | 2 | 27.72 | 2.62 | 0.1059 |
| Within samples | 158.83 | 15 | 10.59 | | |
| Total | 214.28 | 17 | | | |

**7.19**

| Source of Variation | SS | df | MS | F Statistic | p Value |
|---|---|---|---|---|---|
| Between samples | 74.803 | 3 | 24.934 | 76.23 | 0.0001 |
| Within samples | 4.252 | 13 | 0.327 | | |
| Total | 79.055 | 16 | | | |

**7.20**

| Source of Variation | SS | df | MS | F Statistic | p Value |
|---|---|---|---|---|---|
| Between samples | 40.526 | 2 | 20.263 | 3.509 | 0.0329 |
| Within samples | 527.526 | 126 | 5.774 | | |
| Total | 568.052 | 128 | | | |

**7.21**

| Source of Variation | SS | df | MS | F Statistic | p Value |
|---|---|---|---|---|---|
| Between samples | 57.184 | 3 | 19.061 | 37.026 | <0.0001 |
| Within samples | 769.113 | 1494 | 0.515 | | |
| Total | 826.297 | 1497 | | | |

**7.22** For exposed group:

| Source of Variation | SS | df | MS | F Statistic | p Value |
|---|---|---|---|---|---|
| Between samples | 5,310.032 | 2 | 2,655.016 | 0.987 | 0.3738 |
| Within samples | 845,028 | 314 | 2,691.172 | | |
| Total | 850,338.032 | 316 | | | |

For nonexposed group:

| Source of Variation | SS | df | MS | F Statistic | p Value |
|---|---|---|---|---|---|
| Between samples | 10,513.249 | 2 | 5,256.624 | 2.867 | 0.0583 |
| Within samples | 607,048 | 331 | 1,833.982 | | |
| Total | 617,561.249 | 333 | | | |

**7.23** Factor: age
(a) $t = 1.037$, $p$ value $= 0.3337$; (b) $z = -0.864$, $p$ value $= 0.3875$.
Factor: acid
(a) $t = -1.785$, $p$ value $= 0.080$; (b) $z = 2.718$, $p$ value $= 0.0066$.

## Chapter 8

**8.1** **(b)** $\hat{\beta}_0 = 0.0301$, $\hat{\beta}_1 = 1.1386$, $\hat{y} = 0.596$; **(c)** $t = 5.622$, $p = 0.0049$; **(d)** $r^2 = 0.888$.

**8.2** **(b)** $\hat{\beta}_0 = 6.08$, $\hat{\beta}_1 = 0.35$, $\hat{y} = 27.08$; **(c)** $t = 2.173$, $p = 0.082$; **(d)** $r^2 = 0.115$.

**8.3** **(b)** $\hat{\beta}_0 = 311.45$, $\hat{\beta}_1 = -0.08$, $\hat{y} = 79.45$; **(c)** $t = -5.648$, $p = 0.0001$; **(d)** $r^2 = 0.618$.

**8.4** **(a)** $F = 16.2$, df $= (2.19)$, $p = 0.0001$; **(b)** $r^2 = 0.631$.

**(c)**

| Term | $\hat{\beta}$ | SE($\hat{\beta}$) | $t$ Statistic | $p$ Value |
|------|------|------|------|------|
| $x$ (food) | −0.007 | 0.091 | −0.082 | 0.936 |
| $x^2$ | −0.00001 | 0.00001 | −0.811 | 0.427 |

**8.5** For men: **(b)** $\hat{\beta}_0 = -22.057$, $\hat{\beta}_1 = 0.538$, $\hat{y} = 64.09$; **(c)** $t = 1.697$, $p = 0.1282$; **(d)** $r^2 = 0.265$.
For women: **(b)** $\hat{\beta}_0 = -61.624$, $\hat{\beta}_1 = 0.715$, $\hat{y} = 52.78$; **(c)** $t = 2.917$, $p = 0.0194$; **(d)** $r^2 = 0.516$; **(e)** relationship is stronger and statistically significant for women but not so for men.

**8.6** **(a)**

| Term | $\hat{\beta}$ | SE($\hat{\beta}$) | $t$ Statistic | $p$ Value |
|------|------|------|------|------|
| Sex | −39.573 | 81.462 | −0.486 | 0.634 |
| Height | 0.361 | 0.650 | 0.555 | 0.587 |
| Sex-by-height | 0.177 | 0.492 | 0.360 | 0.724 |

**(b)** No $(p = 0.724)$; **(c)** $F = 22.39$, df $= (3, 16)$, $p = 0.0001$; **(d)** $r^2 = 0.808$.

**8.7** For standard: **(b)** $\hat{\beta}_0 = 0.292$, $\hat{\beta}_1 = 0.362$, $\hat{y} = 1.959$; **(c)** $t = 7.809$, $p = 0.0001$; **(d)** $r^2 = 0.884$.
For test: **(b)** $\hat{\beta}_0 = 0.283$, $\hat{\beta}_1 = 0.280$, $\hat{y} = 1.919$; **(c)** $t = 9.152$, $p = 0.0001$; **(d)** $r^2 = 0.913$; **(e)** no indication of effect modification.

**8.8** **(a)**

| Term | $\hat{\beta}$ | SE($\hat{\beta}$) | $t$ Statistic | $p$ Value |
|------|------|------|------|------|
| Preparation | −0.009 | 0.005 | −1.680 | 0.112 |
| Log(dose) | 0.443 | 0.088 | 5.051 | <0.001 |
| Prep-by-log(dose) | −0.081 | 0.056 | −1.458 | 0.164 |

**(b)** No, weak indication $(p = 0.164)$; **(c)** $F = 46.23$, df $= (3, 16)$, $p = 0.0001$; **(d)** $R^2 = 0.897$.

**8.9** For AG positive: **(b)** $\hat{\beta}_0 = 4.810$, $\hat{\beta}_1 = -0.818$, $\hat{y} = 19.58$; **(c)** $t = -3.821$, $p = 0.002$; **(d)** $r^2 = 0.493$.

For AG negative: (b) $\hat{\beta}_0 = 1.963$, $\hat{\beta}_1 = -0.234$, $\hat{y} = 9.05$; (c) $t = -0.987$, $p = 0.340$; (d) $r^2 = 0.065$; (e) relationship is stronger and statistically significant for AG positives but not so for AG negatives.

**8.10** (a)

| Term | $\hat{\beta}$ | SE($\hat{\beta}$) | $t$ Statistic | $p$ Value |
|---|---|---|---|---|
| AG | 2.847 | 1.343 | 2.119 | 0.043 |
| Log(WBC) | -0.234 | 0.243 | -0.961 | 0.345 |
| AG-by-log(WBC) | -0.583 | 0.321 | -1.817 | 0.080 |

(b) Rather strong indication ($p = 0.080$); (c) $F = 7.69$, df = $(3, 29)$, $p = 0.0006$; (d) $r^2 = 0.443$.

**8.11** For revenue: (b) $\hat{\beta}_0 = (1.113)(10^{-3})$, $\hat{\beta}_1 = (0.829)(10^{-6})$, $\hat{y} = (1.329)(10^{-3})$; (c) $t = 0.198$, $p = 0.844$; (d) $r^2 = 0.001$.
For work load hours: (b) $\hat{\beta}_0 = (0.260)(10^{-3})$, $\hat{\beta}_1 = (0.754)(10^{-6})$, $\hat{y} = (1.252)(10^{-3})$; (c) $t = 1.882$, $p = 0.067$; (d) $r^2 = 0.078$.

**8.12** (a)

| Term | $\hat{\beta}$ | SE($\hat{\beta}$) | $t$ Statistic | $p$ Value |
|---|---|---|---|---|
| Residency | $(3.176)(10^{-3})$ | $(1.393)(10^{-3})$ | 2.279 | 0.028 |
| Gender | $(0.348)(10^{-3})$ | $(0.303)(10^{-3})$ | 1.149 | 0.258 |
| Revenue | $(1.449)(10^{-6})$ | $(4.340)(10^{-6})$ | 0.334 | 0.741 |
| Hours | $(2.206)(10^{-6})$ | $(0.760)(10^{-6})$ | 2.889 | 0.006 |
| Residency by hours | $(-2.264)(10^{-6})$ | $(0.930)(10^{-6})$ | -2.436 | 0.020 |

(b) Yes, rather strong indication ($p = 0.020$); (c) $F = 2.14$, df = $(5, 38)$, $p = 0.083$; (d) $r^2 = 0.219$.

*Chapter 11*

**11.8** Log-rank test: $p = 0.0896$; generalized Wilcoxon test: $p = 0.1590$.

**11.10** 95% confidence interval for odds ratio: $(1.997, 13.542)$; McNemar's chi-square: $X^2 = 14.226$; $p$ value $= 0.00016$.

**11.11** McNemar's chi-square: $X^2 = 0.077$; $p$ value $= 0.78140$.

**11.12** 95% confidence interval for odds ratio: $(1.126, 5.309)$; McNemar's chi-square: $X^2 = 5.452$; $p$ value $= 0.02122$.

**11.13** For men: McNemar's chi-square, $X^2 = 13.394$; $p$ value $= 0.00025$; for women: McNemar's chi-square, $X^2 = 0.439$; $p$ value $= 0.50761$.

*Chapter 12*

**12.2** $[0.9^3 + 3(0.1)(0.9)^2(0.9)^3]\{0.2^3 + 3(0.2)^2(0.8) + 3(0.2)(0.8)^2[1 - (0.8)^3]\}$
$= 0.264$.

**12.3** At each new dose level, enroll three patients; if no patient has DLT, the trial continues with a new cohort at the next higher dose; if two or three experience DLT, the trial is stopped. If one experiences DLT, a new cohort of two patients is enrolled at the same dose, escalating to next-higher dose only if no DLT is observed. The new design helps to escalate a little easier; the resulting MTD would have a little higher expected toxicity rate.

**12.4**  $[0.6^3 + 3(0.4)(0.6)^2(0.6)^3]\{0.5^3 + 3(0.5)^3 + 3(0.5)^3[1 - (0.5)^3]\} = 0.256.$

**12.5**  $z_{1-\beta} = 0.364$, corresponding to a power of 64%.

**12.6**  $z_{1-\beta} = 0.927$, corresponding to a power of 82%.

**12.7**  $z_{1-\beta} = 0.690$, corresponding to a power of 75%.

**12.8**  $z_{1-\beta} = 0.551$, corresponding to a power of 71%.

**12.9**  $d = 42$ events and we need

$$N = \frac{(2)(42)}{2 - 0.5 - 0.794} = 120 \text{ subjects}$$

or 60 subjects in each group.

**12.10**  $n = \dfrac{(1.96)^2(0.9)(0.1)}{(0.05)^2} = 139 \text{ subjects}$

If we do not use the 90% figure, we would need

$$n_{\max} = \frac{(1.96)^2(0.25)}{(0.05)^2} = 385 \text{ subjects}$$

**12.11**  $n_{\max} = \dfrac{(1.96)^2(0.25)}{(0.1)^2} = 99 \text{ subjects}$

**12.12**  $n_{\max} = \dfrac{(1.96)^2(0.25)}{(0.15)^2} = 43 \text{ subjects}$

**12.13**  **(a)** With 95% confidence, we need

$$n_{\max} = \frac{(1.96)^2(0.25)}{(0.01)^2} = 9604 \text{ subjects}$$

With 99% confidence, we need

$$n_{\max} = \frac{(2.58)^2(0.25)}{(0.01)^2} = 16,641 \text{ subjects}$$

(b) With 95% confidence, we need

$$n_{max} = \frac{(1.96)^2(0.08)(0.92)}{(0.01)^2} = 2827 \text{ subjects}$$

With 99% confidence, we need

$$n_{max} = \frac{(2.58)^2(0.08)(0.92)}{(0.01)^2} = 4900 \text{ subjects}$$

**12.14**  $n = \frac{(1.96)^2(1)}{(0.5)^2} \approx 16 \text{ subjects}$

**12.15**  $N = (4)(1.96)^2 \frac{400}{(10)^2} = 62 \text{ or } 31 \text{ per group}$

**12.16**  $n(e) = \frac{\ln(0.9) - \ln(0.05) + e(\ln(0.8) - \ln(0.95) - \ln(0.2) + \ln(0.05)}{\ln(0.8) - \ln(0.95)}$

$n(1) = -8, \quad n(2) = 2, \quad n(3) = 11, \quad n(4) = 20, \quad n(5) = 29, \quad \text{and so on}$

**12.17**  $N = (4)(1.96 + 1.28)^2 \frac{(2.28)^2}{(1)^2} = 220 \text{ or } 110 \text{ per group}$

**12.18**  $N = (4)(1.96 + 1.65)^2 \frac{(0.97)^2}{(1)^2} = 50 \text{ or } 25 \text{ per group}$

**12.19**  $N = (4)(1.96 + 1.28)^2 \frac{(10.3)^2}{(3)^2} = 496 \text{ or } 248 \text{ per group}$

**12.20**  $N = (4)(2.58 + 1.28)^2 \frac{(0.075)(0.925)}{(0.05)^2} = 1654 \text{ or } 827 \text{ per group}$

**12.21**  $N = (4)(2.58 + 1.28)^2 \frac{(0.12)(0.88)}{(0.1)^2} = 630 \text{ or } 315 \text{ per group}$

**12.22**  $N = (4)(1.96 + 0.84)^2 \frac{(0.275)(0.725)}{(0.15)^2} = 70 \text{ or } 35 \text{ per group}$

**12.23**  $N = (4)(1.96 + 0.84)^2 \frac{(0.3)(0.7)}{(0.2)^2} = 42 \text{ or } 21 \text{ per group}$

**12.24**   $d = 0.196$, almost 20%.

**12.25**   $\theta = \dfrac{\ln(0.6)}{\ln(0.7)}$

$= 1.432$

$d = (1.96 + 0.84)^2 \left(\dfrac{1 + 1.432}{1 - 1.432}\right)^2$

$= 249$ events

$N = \dfrac{(2)(249)}{2 - 0.6 - 0.7}$

$= 710$ subjects or 355 per group

**12.26**   $\theta = \dfrac{\ln(0.4)}{\ln(0.5)}$

$= 1.322$

$d = (1.96 + 0.84)^2 \left(\dfrac{1 + 1.322}{1 - 1.322}\right)^2$

$= 408$ events

$N = \dfrac{(2)(408)}{2 - 0.4 - 0.5}$

$= 742$ subjects or 371 per group

**12.27**   $\theta = \dfrac{\ln(0.4)}{\ln(0.6)}$

$= 1.794$

$d = (1.96 + 0.84)^2 \left(\dfrac{1 + 1.794}{1 - 1.794}\right)^2$

$= 98$ events

$N = \dfrac{(2)(98)}{2 - 0.4 - 0.6}$

$= 98$ subjects or 49 per group

**12.28**   $\pi_1 = 0.18$: **(a)** $N = 590$, 245 cases and 245 controls; **(b)** $N = 960$, 192 cases and 768 controls; **(c)** $m = 66$ discordant pairs and $M = 271$ case–control pairs.

**12.29**  $\pi_1 = 0.57$: **(a)** $N = 364$, 182 cases and 182 controls; **(b)** $N = 480$, 120 cases and 350 controls; **(c)** $m = 81$ discordant pairs and $M = 158$ case–control pairs.

**12.30**  $$N = \frac{4}{[\ln(1.5)]^2}(1.96 + .84)^2$$

$= 192;\ 96$ cases and 96 controls

# INDEX